Published by:
Scientific Publishers (India)
5-A, New Pali Road, P.O. Box 91,
Jodhpur – 342 001, India

E-mail: info@scientificpub.com
Web: www.scientificpub.com

Print : 2021

ISBN: 978-81-7233-674-5

Citation:
Nath, Prem, & Gaddagimath, P.B. (Eds.) 2010, Horticulture and Livelihood Security, Scientific Publishers, Jodhpur, Rajasthan.

Lasertype set : Rajesh Ojha

Binding at: Rajasthan Law Book Binding Works, Jodhpur
Printed in India

Dedicated to
One Billion People
Suffering From
Hunger And Malnutrition
Across The Globe

FOREWORD

I find myself fortunate to have the opportunity to live and experience the working life of the farmers and the farming in Kansas (USA), Nigeria (Africa), Saudi Arabia and Yemen (Middle East), Italy (Europe) and India, Myanmar, and Thailand (Asia). It was further strengthened by visits to farmers in another about 100 countries. Day after day and month after month and even years, it made me to realize that community culture, language and farming in each state, country and continent are different from each other but some things are common and that they are all doing farming, they are growing food and they all doing for the livelihood of their families, neighbors and communities near and far.

Further, my assignment with the global FAO's Special Programme on Food Security for Low-Income Food Deficit Countries launched during 1994 gave deeper insight into the important dimension of livelihood i.e. food security. It was not only the question of making adequate food available but make it available all the time. But at that time more than 800 million people were suffering from hunger and malnutrition, evidently they had no access to food and now the number of those hungry people has increased to one billion, in spite of efforts to mitigate it. The question is how to access the food to those who have no capacity to have it.

The farmers have worked to their capacity within the resources available to produce food, and now it is the turn of the governance of countries to enhance the capacity of the people to access the food for their livelihood. The answer is, increased employment and income generation.

Thus, livelihood has different dimensions and facets, but agriculture and food buying capacity are major components because food is the basic. In the present time with modern life and living, food has emerged with important components like adequacy, nutrition and health requirements which have been designated as food security, nutrition security and health security. Hence, horticulture is a serious stake holder as described in the present publication.

With its mandate on food and nutrition security, the Dr. Prem Nath Agricultural Science Foundation (PNASF), Bangalore has endeavored to invite the national and international well-known authors to contribute to this book on this subject. Grateful thanks are due to authors of each paper presented in this book for their voluntary contribution. It is expected that the present publication will enlighten the readers and disseminate the message keeping in line with the title and theme of the book and subject.

The editors of this book have done tremendous job in preparing the blueprint of this book, in organizing the papers and in editing them, Ms.

Vanita N. Shetty of the PNASF, Bangalore spent long hours to prepare the whole manuscript in an excellent manner.

Acknowledgement is due to the Scientific Publishers (India), Jodhpur for their best efforts in quality printing of this book in a timely manner.

Lastly, many thanks to those who have contributed in some form or other in the preparation of this book, but their name does not appear in this publication.

September 25, 2010

Prem Nath
Chairman, PNASF
Former Asst. Director General, FAO
Chairperson, Executive Council, VEGINET

RAJ BHAVAN
BANGALORE

No. GS 57 MSG 2010
June 9, 2010

MESSAGE

I am happy to learn that the Dr. Prem Nath Agricultural Science Foundation (PNASF), Bangalore is preparing and publishing a book entitled "Horticulture and Livelihood Security" which is first of its kind on this aspect of the people particularly of the developing world.

Horticultural crops are emerging as important source of food next to cereals and are providing food, nutrition, health with a potential for income generation. In this millennium, about one billion people across the globe are suffering from hunger and malnutrition, out of which about 230 million live in India. It is an appropriate time for our nation to rise to the occasion and provide adequate food and nutrition to all people.

I heartily congratulate Dr. Prem Nath, for taking the lead and to the well-known authors for contributing valuable articles for such a publication which is so timely and useful in the interest of horticulture and agriculture sector of our country and abroad.

(H.R. Bhardwaj)

M.V. RAJASEKHARAN
FORMER UNION MINISTER OF STATE
FOR PLANNING
GOVERNMENT OF INDIA

"Shri Mallikarjuna Nilaya",
No.20/1, Kanakapura Road,
Basavanagudi,
BANGALORE - 560 004 INDIA
Phone :- Res - 080-26576763
Fax : Res - 91 - 080 - 26578366
Fax : Off - 91 - 080 - 26690860
E-mail :asiancarebgl@vsnl.net

June 10, 2010

MESSAGE

I am indeed very happy to know that Dr. P.N. Agricultural. Science Foundation (PNASF), Bangalore, has endeavored to publish a new book entitled "Horticulture and Livelihood Security" for the benefit of Policymakers, Universities, Development Officers and farmers.

The PNSAF has always been on the forefront in demonstrating the most advanced technologies both in the field of Agriculture and Horticulture, for the benefit of not only to the Policy-makers, but also the Researchers, and more than that to, the farming communities who, are the real beneficiaries of such innovative technologies. I am sure, the proposed new Publication would be most useful, not only for the Policymakers, but also to the Scientific Community and the farming community, not only in India, but also all over the world. In recent years, with the growth of population, Horticulture has assumed greater importance in providing food, nutrition and livelihood security to the vast millions of people throughout the world. I am sure, this new publication would be most valuable document for everyone.

The PNSF has been doing yeomen service under the inspiring leadership of it's Chairman Dr. Prem Nath. I wish the Foundation all success in the years to come.

(M.V. RAJASEKHARAN)

New Delhi Address: Chairman, Vishwa Yuvak Kendra (International Youth Centre), Circular Road, Chanakyapuri, New Delhi-110021. Tel.: 011-23013631 Fax: 011-23016604 E-mail: vyk@vykonline.org

منظمة الأغذية والزراعة للأمم المتحدة

联合国粮食及农业组织

Food and Agriculture Organization of the United Nations

Organisation des Nations Unies pour l'alimentation et l'agriculture

Продовольственная и сельскохозяйственная организация Объединенных Наций

Organización de las Naciones Unidas para la Agricultura y la Alimentación

Regional Office for Asia and the Pacific
Maliwan Mansion, 39 Phra Atit Rd
Bangkok 10200, Thailand

June 28, 2010

MESSAGE

While the world is projected to need a major increase in crop production to feed a population of around 9 billion people in 2050, it must do so against a challenging backdrop, in particular the decreasing availability of and competition for land and water as well as climate change. Changing dietary and nutritional needs and requirements as a result of urbanization and niche markets also present a challenge.

Previous attempts at managed intensification – such as the Green Revolution of the 1960s and 1970s – have been a qualified success. In some cases it is now recognized that the yield increases achieved through increased use of fertilizers, high yielding varieties, irrigation, pesticides and intensive tillage were made at the expense of the environment or in ways which were otherwise unsustainable. Also some smaller-scale farmers were unable to participate or reap the rewards of scale.

FAO is working to develop strategies and programmes to assist member countries to take sustainable horticultural production intensification forward while recognized that suitable adaptation over time will be needed to take account of the uncertainties related to agriculture, food production, climate change, environmental health, population demographics, changing consumption patterns, the sustainability of value chains and global economies.

For sustainable horticulture production intensification to become a reality, concerted action at different levels is required, from the farm level, to science and technology, and up to the economic and policy levels. FAO is connected to all of these levels and can support the required changes at each. FAO has also, over time, developed tools and methods in participation, rural development and institution-building which are highly relevant to the adoption of new approaches.

Contd...

In this context, promoting horticultural crops as a vital source of income and livelihood security for farmers and rural people is a priority issue, and in the Asia-Pacific region FAO has for many decades advocated for as well as cooperated in the promotion of a wide range of horticultural crops such as fruits, vegetables, spices, flowers, medicinal and aromatic crops.

Despite considerable success in this field, the challenges confronting us are still many. It thus pleases me to see that scientists from across the globe participated and presented papers in the workshop and discussed a wide range of issues covering almost all aspects of horticultural crops. I believe that the compilation of this valuable information along with its recommendations will undoubtedly have enormous impact on furthering the link between horticultural development and livelihood security in this region.

A wealth of information and recommendations are contained in this timely initiative by PNASF. Furthermore, *Horticulture and Livelihood Security* provides valuable directions for use by concerned stakeholders towards increasing production and productivity of horticultural crops in the region in our joint endeavor to achieve the right to food and a dignified life for all.

Hiroyuki Konuma
Assistant Director-General and FAO Regional
Representative for Asia and the Pacific
Food and Agriculture Organization of the United Nations (FAO)

ايكاردا المركز الدولي للبحوث الزراعية في المناطق الجافة
ICARDA International Center for Agricultural Research in the Dry Areas

June 29, 2010

MESSAGE

Horticulture (including vegetables, fruits, nuts and medicinal, herbal and aromatic plants) is playing an increasingly important role in food security, improving livelihoods and in economic growth.

Horticultural crops generally have a higher product value per unit of land or water than field crops, are relatively labor intensive and, therefore, can generate productive employment and incomes from a relatively small area, particularly when produced under intensive, plastic house, conditions. In dry areas, production of horticultural crops under protected agriculture can make the most productive use of scarce water resources. ICARDA has demonstrated that greenhouse production of high value horticultural crops is a feasible option for substantially increasing household incomes from small holdings in marginal areas. Similarly, it has been demonstrated that including locally adapted tree crops in farming systems is an important risk mitigation strategy.

Increasing the consumption of fruit and vegetables is also critical to improving human nutrition. Lack of nutrients and vitamins is the "hidden hunger" and contributes to a range of diet-related chronic diseases. Diversification into horticulture provides rural households with greater nutritional security.

In recent years, average annual production growth in horticultural crops has been larger than in cereal crops and there have been substantial increases in total volumes of fruit and vegetables traded worldwide. Population growth, higher incomes, globalization and demands for more diverse and nutritious diets are major drivers of change. There are significant opportunities to link horticultural producers with domestic and international markets and the growing demand for horticultural products.

Not only do horticultural crops potentially provide greater profitability, but there is great potential for adding further value through post-harvest processing. Increasing horticultural production contributes to the commercialization of the rural economy by generating associated employment in input supply, processing, packaging, and marketing. There are opportunities to take advantage of international market niches and premium prices for organically produced horticultural products, specialty crops and high quality products. Globally, there is increasing demand for processed, ready-to-eat convenience products, including cut fruit and salad mixes. Developing countries are taking advantage of this trend, and over the past decade the increase of their processed food exports has exceeded that from the developed regions.

This book is a comprehensive introduction to the contribution of horticulture to food security, livelihoods and economic growth and is a valuable source of international experience and information on improving horticultural systems.

Mahmoud Solh
Director General, ICARDA

**DEPARTMENT OF HORTICULTURE
AND LANDSCAPE ARCHITECTURE**

June 21, 2010

MESSAGE

Horticulture has and an important role to play in both the developed and developing worlds. It provides nourishing and delicious food, beautiful and endearing ornamentals and landscapes, employment for millions of people, and a source of delight for gardeners, urban dwellers, young and old. Horticulture truly provides food for body and soul.

Jules Janick
James Troop Distinguished
Professor in Horticulture

Horticulture Building * 625 Agriculture Mall Drive * West Lafayette, IN 47907-2010
(765) 494-1300 * Fax: (765) 494-0391 * http://www.hort.purdue.edu

ABOUT THE EDITORS

Dr. Prem Nath is the founder of the PNAS Foundation established in 2000. Dr. Nath obtained his Ph.D. degree in Horticulture from the Kansas State University in U.S.A. during 1962. The same year, he returned to India and served at the Indian Agricultural Research Institute (New Delhi), University of Udaipur (Rajasthan), Indian Institute of Horticultural Research (Bangalore) for about 15 years before joining the Food and Agriculture Organization of the United Nations (FAO) during 1976 and served in different capacities in a number of countries, like Indonesia, Nigeria, Saudi Arabia, Yemen Arab Republic, Italy, Myanmar, and Thailand for about 25 years, with the last posting as FAO Assistant Director-General and Regional Representative for Asia and the Pacific and headed FAO Regional Office in Bangkok, Thailand from where he retired in 2000. Even after his retirement with the FAO, Dr. Nath continues his activities on issues of food and agriculture/ horticulture through the Dr. Prem Nath Agricultural Science Foundation (PNASF) to which he has donated his life-savings.

The vegetable varieties developed by him and his colleagues are still popular in India and abroad. Dr. Nath made significant achievements in developing a number of vegetable varieties viz Arka Suryamakhi in pumpkin, Arka Sheel and Arka Kusumkar in brinjal, Arka Manik and Arka Jyoti in watermelon, Arka Harit in bitter gourd, Arka Rajhans and Arka Jeet in musk melon, and Arka Tinda in roundmelon in India. He guided researches on cucurbits for a number of M.Sc and PhD students.

In 1965, Dr. Nath authored the book entitled 'Cucurbitaceous Crops in Northern India' published by the University of Udaipur. Among several publications, the book on 'Vegetable Crops for the Tropical Region' published by the Indian Council for Agricultural Research (ICAR), New Delhi, in 1976 and 1987 and reprinted in 1994 and 2004 are still popular among the Agricultural Universities in India. The book entitled "Food Security and Vegetables – A Global Perspective" co-edited by him and published by the PNASF in 2004 and written by the well-known national and international author's acclaimed high recognition nationally and world-wide. He is also a co-author of the book –"State of the Indian Farmer: A Millennium Study" published in 2004 by the Ministry of Agriculture and Cooperation, Government of India. He is the senior author of the book entitled "Vegetable Crops-Improvement and Production" published by PNASF, Bangalore during 2008; co-author of the book entitled "Foods and Nutrition" published by PNASF, Bangalore and Studium Press, New Delhi during 2009. Dr. Nath has hundreds of publications and technical reports published in India and abroad.

Dr. Prem Nath has the credit of establishing horticultural/agricultural research institutes in India, Nigeria, Saudi Arabia and Yemen and of guiding horticulture research programmes in Africa, near East, Asia and East European countries. In 1994, he was responsible for launching successfully the FAO's Special Programme on Food Security for Low-Income Food Deficit Countries (LIFDC's) across the globe.

Presently, Dr. Prem Nath is the Chairman of the Dr. Prem Nath Agricultural Science Foundation (PNASF), Bangalore – 560 094. Also, he is the Chairperson of the Vegetable Science International Network (VEGINET), Bangalore. As a Chairman of the Organizing Council of the International Conference on Vegetables (ICV-2002), he successfully organized the Conference held during November 2002 at Bangalore. He organized FAO/PNASF International Workshop on Urban and Peri-Urban Agriculture for Asian Countries during June 29 to July 2, 2005 in Bangalore, India; and International Conference on Horticulture (ICH-2009) held during November 9-12, 2009, Bangalore, India; where about 750 delegates from about 40 countries participated. He served as the Chairman of the WTO Cell, Government of Karnataka and as the member of the various committees in food and agriculture at the state, national and international levels. He is the Fellow of a number of national and international professional societies/ academy in Horticulture and Agriculture.

Pulakeshi Basayya Gaddagimath is a post graduate in Organic Chemistry from Karnataka University, Dharwad in 1978 and served as a lecturer in Kittle Science College, Dharwad till 1982.His interest in Mass Communication made him to leave the teaching job and join post graduate course in Mass Communication and Journalism in 1982 as a regular student and graduated from Karnataka University Dharwad in 1984.He had been a scholarly student through out his career.

During his post graduation in Mass communication and Journalism, he served many national news papers until he joined as Jr.Scientific Officer, at Directorate of Oilseed Research, Hyderabad in 1985.He joined the Indian Institute of Horticultural Research, Bengaluru in 1989 and is continuing his services till to date as Information and Publication Officer.

A man who has extensively traveled and well read in various fields, is bestowed with excellent, both written and spoken communication skills. As a media person he has anchored /conducted more than 40 T.V.programmes and live programmes on Doordarshan on various aspects of agriculture. He was also a consultant to Dept. of Horticulture, and Water Resources Dept., Govt. of Karnataka for developing scripts for documentaries and interactive CDs and has scripted more than 10 documentaries on rural water supply and sanitation. Apart from a good number of official publications at IIHR, Bengaluru and DOR, Hyderabad, he has authored two books, edited four books (two international), published three books and written many technical and popular papers.

CONTENTS

III
CONTRIBUTION OF HORTICULTURE TECHNOLOGY

V
GOVERNMENT POLICY AND DEVELOPMENT PROGRAMMES

VI
NATIONAL AND INTERNATIONAL TECHNICAL COOPERATION

I
IMPACT OF HORTICULTURE ON LIVELIHOOD SECURITY

1. ROLE OF HORTICULTURE IN LIVELIHOOD SECURITY

Prem Nath[1]

Former Asst. Director General, Food and Agriculture Organization (FAO), Chairman, Dr. P.N. Agricultural Science Foundation (PNASF), Bangalore - 560 094, INDIA

1. AGRICULTURE AND LIVELIHOOD SECURITY

Food remains the prime focus for the survival of human beings and its production is propelled by efficient farming for which the agricultural/horticultural science is the engine. At this time, among 6 billion people on earth, about one billion are still suffering from hunger and malnutrition across the globe and of which 230 millions are homed in India. Thus, addressing issues of food and nutrition security attains its priority among the major issues of food, clothing and shelter for human kind. Moreover, dealing with health issues becomes equally important in the wake of prevailing diseases and health disorders.

Among all sectors of development, agriculture is essential and major contributor to livelihood security of the people, more so because more than 70 percent of the population in developing world is dependent on agriculture and the food security of world population is directly linked with the successful agriculture. The World Bank highlighted the critical role agriculture and rural development can play in helping people move out of poverty and that the sector has the powerful influence on economic growth, poverty reduction and environmental sustainability (Anonymous, 2008). It is further stated that about one billion people live in rural areas in South Asia alone and the regions economy depends on agriculture (Anonymous, 2008).

Attempts have been made by authors to define Livelihood Security but accepted one seems to be as stated by Chambers and Conway (1992) '*a livelihood comprises the capabilities, assets (stores, resources, claims and access) and activities required for a means of living: a livelihood is sustainable which can cope with and recover from stress and shocks, maintain or enhance its capabilities and assets, and provide sustainable livelihood*

---

1. Assistance provided by Dr. M. Sudha, IIHR, Bangalore, is gratefully acknowledged.

opportunities for the next generation; and which contributes net benefits to other livelihoods at the local and global levels and in the long and short term.'

Development organizations world over, involved in poverty alleviation, food and livelihood security issues, have adopted this concept and have moved from a concern for regional and national food security to a consideration of household and individual food security issues. At the household level the concern shifted from "food first" or food production to a wider focus on the ability of households to secure the food that they required. This then led to a widening of the scope and recognition that food was just one of the ranges of factors that determined poor people's decisions. Thus the evolution of the concepts and issues related to household food and nutritional security led to the development of the concept of household livelihood security and then, more broadly, to livelihoods.

In the present context, the main elements of livelihood security considered are food security, nutrition security, health security and employment and income generation where agriculture is the lead contributor. It is also recognized that the food security and the livelihood security are inseparable, as also the income generation and the livelihood security. It need not be over-emphasized that nutrition and health have now assumed more importance than before in the life of the people.

The recent paper of Bisalaiah (2009) has focused on the strong linkage of agriculture and horticulture to livelihood security which needs consideration and attention.

According to World Bank much needs to be done to alleviate global crises. An estimated 2.5 billion people are trying to survive on $ 2 or less a day. Skyrocketing food prices are a harsh reality, resulting in even greater hunger and malnutrition worldwide (Anonymous, 2008). Most recent global poverty projections anticipate that the proportion of people living in extreme poverty-on less than $ 1 a day will fall from 29 percent in 1990 to 10 percent in 2015. (Anonymous, 2008)

2. HORTICULTURE AND LIVELIHOOD SECURITY

There is an observable shift in the food habits of people in the last two decades that is reflected as changing cropping pattern, away from cereal base to other commercial crops. During the same period, economic growth in horticulture has far exceeded than that in most agricultural commodities. Annual growth rates for vegetable supplies have surpassed cereals by 200 percent to 800 percent since the 70s, while this has been accelerated in the 1990s. This trend is attributed largely to the changing consumer preferences, powered by increased income and general standard of living led awareness towards diversified and balanced nutrition.

Growth potential for the production of horticultural commodities is strong in the developing countries and emerging economies. It is reported that China recorded the highest increase in area under horticultural crops with 20 percent of the arable land under this set of crops. The total value of horticultural crops traded at present is more than double that of cereal crops, reaching nearly 21% of total value exported from developing countries.

Experience suggests that development of horticultural crop production promises these countries the ability not only to meet their domestic food needs, but offer diversified income sources to their rural economies. Further, horticultural crop based products are proven supplements of human health, besides being sources of farmer household economic and social advancement. Thus, horticultural commodities on the whole seem to be ideally suited to accomplish the objective of providing 'sustainable livelihood security'.

According to IFPRI (1999), about 52 percent of the developing world's population will be living in urban areas by 2020, up from 38 percent in 1995 due to tremendous growth of mega cities. The rapid urbanization of the developing world and associated changes in life-style will have significant bearing on food preference and food demand. Hence, the peri-urban and urban agriculture will play an important role and horticultural crops will dominate because of unparallel advantage of cultivation and production coupled with its nutrition values. The FAO/PNASF International Workshop on Urban and Peri-Urban Agriculture (UPA) for Asian Countries held during June 29 to July 02, 200 in Bangalore highlighted the importance and significance of UPA and suggested policies and programmes for its implementation (Anonymous, 2005).

During last two decades, horticultural crops have emerged as the second line of defense in the food domain whereas cereals and legumes continue to be in the front line (Nath, 2008). The growing importance of horticulture lies in its significant support to food security; as major contributor to nutrition security; as promoter of health security and as an important source of employment and income generation (Nath, 2009). The International Conference on Vegetables (ICV-2002) held during November 2002 in Bangalore, has made an earnest effort in demonstrating the growing significance of vegetables towards food and nutrition security (Anonymous, 2003).

The International Conference on Horticulture (ICH-2009) with the theme *"Horticulture for Livelihood Security and Economic Growth"* held in Bangalore during November 9-12, 2009, has highlighted its importance and salient features projecting the future steps to be undertaken by all stakeholders. The conference recognized horticulture as the second line of defence with cereals and legumes serving on the front line of the food domain. It was stated that horticulture (a) supports significantly to food security, (b) is the major contributor to nutrition security (c) promotes good health, and (d) is source of employment generation- all leading to livelihood security and economic growth. Some of the salient recommendations were (Anonymous, 2010);

- Awareness and education programme on food, nutrition and health security be augmented;
- The university syllabus be upgraded; the quality of research should be farmer/demand driven and priority and quality of research improved;

- More investment and planning in Transfer of Technology (TOT) were recommended;
- Appropriate policy and programmes formulated to augment food, nutrition and health security and employment and income generation;
- There is need to establish strong linkages among countries and even among states of large countries by Networking of Institutes and Organization for exchange of information, seed and plant materials and expertise in horticulture. Awareness and education programme on International Cooperation among Developing Countries (TCDC) be augmented.

3. FOOD, NUTRITION AND HEALTH SECURITY

- **Food Security**

Food security exists when all people, at all times have access to sufficient, safe and nutritious food to meet their dietary needs and food preferences for an active and healthy life, as stated by the FAO and accepted by the World Food Summit 1996 (FAO, 2009a). This is in consistent with the right to adequate food and the fundamental right of everyone to be free from hunger.

The main components of food security as recognized are (a) adequate availability of food, (b) stability of food supply, (c) accessibility to food (Fresco and Baudoin, 2004). Adequate and consistent supply of food demand consistent production as well as quality and safe handling of food. Among other components, access to food remained a dominating force in the frame of food security, where economic accessibility played an important role (Nath, 2004).

Within last decade or so various countries and international organizations have taken important steps to augment food security based on their commitments made during the World Food Summit (1996). The progress made has been reflected in the proceeding section on Millennium Development Goals and Progress.

Towards achieving the goal, the FAO sponsored Special Programme for Food Security (SPFS) initiated during 1994 needs mention when it was launched in 14 Low Income Food Deficit Countries in different regions of the world which was extended to 71 countries by 2002 (Kobakiwal, 2004). Among the commodities utilized fruits, vegetables and tuber crops figured prominently in increasing the income of the beneficiaries.

There has been significant increase in production of fruits and vegetables world wide. The world production of fruits and vegetables was about 1,158 metric tons in 2000, which had further increased during 1990-2009, export of fresh and processed vegetables had arose from 19.7 million ton to 32 million tons (Fresco and Baudoin, 2004).

In recent report of FAO (2010), the food prices will remain above the pre-crisis level of early 2008 in developing countries, negatively affecting access to food of vulnerable populations.

- **Nutrition**

Proper nutrition means the intake of many different kinds of food with a proper balance and in optional amounts. The term 'undernourished' is used to designate those people lacking or having inadequate calorie intake. The term 'malnourished' is used to designate people lacking minimum daily intake of proteins including essential amino-acids, vitamins, and essential minerals and micro-nutrients. Undernourished people are likely to be also malnourished (Fresco and Baudoin, 2004).

Fruits and vegetables play an important role in providing vitamins, minerals, micro-nutrients, as well as proteins to obtain a balanced diet. Starchy vegetables and more specifically the root and tuber crops play an important role as a source of calories to meet the energy requirements.

Undernourishment usually relates to insufficient levels of energy intake. Malnutrition is used to refer to a number of diseases, each with a specific cause and related to one or more nutrients (protein, iodine or calcium) and each characterized by cellular imbalance between the supply of nutrients and energy on the one hand, and body's demand for them to ensure growth, maintenance and special functions, on the other. According to WHO (1996) an estimated 334 million children in developing countries were malnourished while in the same region 6.6 million out of 12.2 million deaths of children under-five were associated with malnutrition. Further, in 2020, one out of every four children in developing countries will still be malnou-rished (IFPRI, 1999).

It is reported that the malnutrition results in protein energy deficiency, low birth weight, major micronutrient deficiencies such as iron deficiency anemia, iodine deficiency and others (Nandi and Bhattacharya, 2004).

According to World Bank, child undernutrition in India is responsible for 22 percent of the country burden of diseases. More than 75 percent of pre-school children suffer from iron-deficiency anemia, 57 percent have sub clinical vitamin' A deficiency and iodine deficiency is endemic in 85 percent of districts. Micro-nutrient deficiencies alone may be costing India $2.5 billion a year. (Gragnolati *et al.,* 2006).

When farmers have better yields by growing cereals, they are no longer poor and hungry. When they shift to horticulture, by growing fruits and vegetables, they can get more income and this will reflect on a better diet, education and livelihood of the family. In Kenya, the income of $ 1 a day was raised to $ 3 and 4 a day for those farmers who have been successful in diversifying their agriculture partly or entirely towards horticulture (Fresco and Baudoin, 2004).

Household gardening is regarded as a way to contribute to food and nutrition security for a part of the population in a number of developing countries as a subsistence production. The annual yield of Indonesian village garden had a cash value equal to 60% of the annual rice production; in China, private family gardens provided 21% of the country vegetables, poultry and swine; in Peru, in one lowland tropical region, gardens and intercropping provided most of the vegetables consumed on regular basis; home gardening and subsistence agriculture was estimated to provide almost 40% of Russia's agriculture output (Fresco and Baudoin, 2004).

- **Health**

It is reported that non-communicable diseases especially cardiovascular diseases (CVDs), cancer, obesity and type 2 diabetes mellitus, currently kill more people every year than any other cause of death. Four factors in the epidemiology of these diseases-poor diet, physical inactivity, tobacco and alcohol use are of overwhelming importance to public health (FAO/WHO, 2004).

According to FAO/WHO (2004), fruits and vegetables are an important components of a healthy diet and if consumed daily in sufficient amounts, could help prevent major diseases such as CVDs and certain cancers. According to the World Health Report 2002, low fruit and vegetables intake is estimated to cause about 31% of ischaemic heart disease and 11% of stroke worldwide. It is estimated that up to 2.7 millions lives could potentially be saved each year if fruit and vegetable consumption was sufficiently increased.

Recommendations in this direction tend to complement and reinforce other valid messages based on this long-known health benefits of consuming vegetables and fruits as dietary sources of fiber, vegetable proteins and protective micronutrients. The Joint FAO/WHO Expert Consultation on diet, nutrition and the prevention of chronic diseases, recommended the intake of a minimum of 400g of fruit and vegetables per day (excluding potatoes and other starchy tubers) for the prevention of chronic diseases such as heart disease, cancer, diabetes and obesity as well as for the prevention and alleviation of several micro-nutrient deficiencies, especially in the less developed countries. The recommendation thus adds to the already strong case for the health benefits to be gained from the consumption of fruit and vegetables and paves the way for concrete action advocating increased consumption of these commodities (FAO/WHO, 2004).

In developing countries, contaminated food plays a major role in the epidemiology of cholera and other forms of epidemic diarrhea, substantially contributing to malnutrition. On a global level, an estimated 70 percent of the approx. 1.5 billion annual cases of diarrhea in the world are caused by contamination in food (Fresco and Baudoin, 2004). The "Green Food" concept launched in China during 1990 has become a success which is basically characterized by non-contamination, safety, good quality and nutritional value. Among countries, there are big differences in food safety standards and with the liberalization of international trade, the food safety standards are getting stricter.

4. EMPLOYMENT AND INCOME GENERATION

In developing countries, household sustainable livelihood is directly correlated to the household income generation capacity, which has direct relevance to the employment generation. Labor intensive horticultural crops provide up to five times higher per capita farm incomes world over for the farmers involved in horticultural crop production in comparison to cereal crops (Chand *et al.,* 2008).

In Kenya, mid 80s, more than 16,000 small holders producing for the fresh fruit and vegetable export market, accounted for 40-65 per cent of the suppliers of French beans, Asian vegetables, mango, avocado and passion fruit for export. The advantage of lower labour and land costs, combined with the rising need for suppliers, who can provide produce throughout the year resulted in a shift towards sourcing French beans and other vegetables in North Africa and sub-Saharan Africa. This growth was the result of continuous experimentation by farmers and traders with alternative institutional arrangements, which included supply of inputs on credit (Diop *et al.,* 2005).

Cross border supply chains involving horticultural products have proved to be instruments that stimulate development of local agro-industry, employment generation, improve access to technology if the connection between the producers and the different actors across the chain is well managed (Rockel *et al.,* 2002). Such value added chains have also become instruments for women empowerment along with horticultural development across continents. A partnership between groups of poor women farmers, a supply company and a marketing company has managed to produce process, pack, distribute and export high quality fresh vegetables for the Asian market, popular as "Bali Fresh" (Campell *et al.,* 2009).

In India, several multi national companies have been taking up hybrid seed production of tomato and other vegetables on farmers' fields in Haveri district in Karnataka. By allocating a small proportion land holding, as low as 0.2 ha, farmers are assured of a net return upto Rs. 45,000. Evidence show that one hectare under horticultural crops such as fruits can generate an annual income upto Rs. 20,000, while flowers like jasmine and tuberose fetch up to Rs. 25,000/ ha, in contrast to hardly Rs. 4000 in case of *ragi* or Rs. 10,000 in case of rice (Varadarajan and Ravindran, 1976, Thimmaiah, 2001). Also, experience of farmers have shown that a well maintained papaya orchard can fetch up to Rs. 82,298 /ha as against a banana orchard at Rs. 52,082, compared to paddy fetching only Rs. 2,248/ha and groundnut at Rs. 2,289/ ha. (Nageswara Rao, 1997).

Studies have shown that growing vegetables like tomato and leafy vegetables in combination or other wise with cereals increased the cropping intensity and income. While crop diversification to horticulture has been a rational behavior from the producer's point of view, this trend has also offered other benefits such as nutritional security and improved health as well, besides enhanced employment opportunities. Crop diversification in favour of horticulture and allied activities such as hybrid seed production, nursery management, protected cultivation have proved to be the most successful models for assured employment as well as higher profits across regions in India (Sudha *et al.,* 2006).

5. GDP AND HORTICULTURE

By and large agriculture growth around the globe appears to be heading towards a plateau and the growth rate in agricultural production has stagnated around 1.5% in the last decade and the agriculture share in the GDP is in the declining trend and has dipped to 15-17%. In this situation, horticultural crops besides being the cheapest sources of essential nutrients, vitamins and minerals, have emerged the key sector contributing to more than 25-30% of the gross value of agricultural output. (MOA, GOI, 10th Five Year Plan) In many developed and developing countries, horticulture and their value added products are the main contributors for the chunk of agricultural export earning. In India horticulture sector contributes about 28% to the agricultural exports, while the same is true with other emerging economies like China and Brazil (MOA, GOI, 10th Five Year Plan). Having recognized the enhanced role of horticulture towards economic growth, Government and policy support for this sector also has been increasing in a number of countries such as India, Thailand, China, Brazil, Japan, Vietnam, South Korea and others.

6. MILLENNIUM DEVELOPMENT GOALS AND PROGRESS

Eradication of extreme poverty and hunger, which are the major components of livelihood security, occupy the first place in the list of eight Millennium Development Goals. The World Bank reports that from 1990 through 2004, the proportion of people living in extreme poverty fell from almost a third to less than a fifth. Although results vary widely within regions and countries, the trend indicates that the world as a whole can meet the goal of halving the percentage of people living in poverty (Anonymous, 2008).

Since 1996, the FAO monitored the number of undernourished people on the basis of statistical estimates of the proportion of a country's population that is consuming less than the minimum number of calories required for an active and healthy life. Of the seventy-nine countries, thirty one exhibited a trend decline in the number of undernourished between 1990-92 and 2004-06, of these, eight have already halved both the proportion and the number of undernourished, thus achieving both the Millennium Development Goal and World Food Summit Targets for 2015 (FAO, 2009).

In order to locate factors responsible for the above success in selected countries for the period 1991-2005, the FAO undertook studies which revealed that the factors which attributed were (a) creating an enabling environment for economic growth and human well being, (b) reaching out to the most vulnerable and investing in the poor, (c) protecting gains and (d) planning for a sustainable future. Further, studies on selected countries on track to achieve 2015 food security targets confirmed the overall pattern but their path to attain the target were different. In Armenia, creating an enabling environment for change from traditional small to present commercial agriculture and upgrading water delivery system proved successful. The improved policies on food distribution, food supplies, food and nutrition education, information and product labeling, food quality and safety for the entire population brought success to Brazil. Nigerian farmers tripled production and now earning a good living by focusing on growing two to three crops instead of one in Vietnam, the nationwide security system, investment in rural infra-structure and targeted programmes of direct benefit to vulnerable mountain people cut the poverty gap ratio by more than half (FAO, 2009).

Besides concerns for domestic food security, many countries with good agriculture potential are promoting investment in production of commodities and processed agriculture products to meet the growing regional and global demand (FAO, 2009).

It is estimated that 85 percent of farm worldwide (or 450 million farms) measure less than two hectors. These farmers have the capacity to produce both for home consumption and for generating supplementary income but their capacity to do so is constrained by insecure access to land, lack of access to training, credit, extension and agribusiness services, managerial skills and non-affordability of skills. According to FAO, small farmers and their families represent some 2 billion people, about one-third of the global population, and are central to any solution to the long term problems of hunger and poverty (FAO, 2009).

7. CONCLUSION

While the effort is on to refine the definition of livelihood, the livelihood of the people continues since time immemorial and its growing components have multiplier effect. The food is basic to livelihood, and agriculture is basic to food but access to food overshadows food security which is plagued by lack of food buying capacity. It is essential that food buying capacity of the people is enhanced where the governance of a country is the prime mover.

Along with agriculture, horticulture is serious stakeholder in enhancing the income generation capacity of the farmers as well as in providing food and nutrition to all. Also, horticulture is important contributor to health security.

A serious attempt by governments in articulating the policies and programmes in horticulture, food, nutrition, health security and employment and income generation will go longway in overcoming insecurities. Growers be encouraged and facilitated to augment production with efficient farming and consumers be educated on food, nutrition and health benefits of horticulture products.

REFERENCES

Anonymous, 2003. Proceedings of the International Conference on Vegetables (ICV-2002), Bangalore; ICV-2002 Secretariat, P.N. Agriculture Science Foundation, Bangalore, India. E-mail: drpremnath@vsnl.net.

Anonymous, 2005. Proc. FAO/PNASF, International Workshop on Urban and Peri-Urban Agriculture for Asian Countries, June 29-July 02, 2005, Bangalore, India, p. 257.

Anonymous, 2008. Perspective on Development Fall 2008, The World Bank, Washington DC. 20433

Anonymous, 2010. Proceedings and Recommendations of the International Conference on Horticulture (ICH-2009), Bangalore; ICV-2002 Secretariat, P.N. Agriculture Science Foundation, Bangalore, India. E-mail: drpremnath@vsnl.net

Campbell, A.A., A Thorne-Lyman, K Sun, S de Pee, K Kramer, R Moench-Pfanner, M Sari, N Akhter, MW Bloem and RD Semba. 2009. Indonesian women of childbearing age are at greater risk of clinical vitamin A deficiency in families that spend more on rice and less on fruits/vegetables and animal-based foods, *Nutrition Research,* 29: 75-81.

Chambers, R., and Conway, G. 1992. Sustainable rural livelihood; Practical concepts for the 21st Century: IDS Discussion Paper No: 296, Brighton, IDS. p. 78.

Chambers, R., Conway, G. 1992. Sustainable rural livelihoods: practical concepts for the 21st century. IDS Discussion Paper No. 296. Brighton, IDS, p. 7 -8)

Chand, R., Raju, S.S. and Pandey, L.M. 2008. Progress and potential of horticulture in India, *Indian J. of Agri. Econ.,* **63**(3): 237-245.

Diop, N., Jaffee, S.M., 2005. Fruits and Vegetables: Global Trade and Competition in Fresh and Processed Product Markets, In: (eds), Aksoy, M.A, Beghin, J.C., Global

FAO, 2009a. Pathways to Success-Success Stories in Agricultural Production and Food Security, FAO, Rome, p. 1-34.

FAO, 2009b. More people than ever are victims of hunger, Background note for press conference 19 June 2009. (Accessed 27 June 2009, ww.fao.org/fileadmin/user_upload/newsroom/docs/Press%20release%20june-en.pdf).

FAO/WHO, 2004. Fruit and Vegetables for Health-Report of a Joint FAO/WHO Workshop, 1-3 September 2004, Kobe, Japan.

Fresco, Louisce O. and Baudoin, Wilfried O., 2004. Food and Nutrition Security towards Human Security, Food Security and Vegetables-A Global Perspective, P.N. Agricultural Science Foundation, Bangalore, India, p.7-42.

Gragnolati, M., Bredenkamp, C. Shekar, M., Das Gupta, M. and Lee, Y. 2006. India's Undernourished Children-A Call for and Population series, The World Bank, Washington, DC.

IFPRI, 1999. World Food Prospects: Critical Issues for the Early Twenty First Century; 2020 Vision-Food Policy Report, IFPRI, Washington, USA.

Kobakiwal, A.Q. 2004, Special Programme for Food Security (SPFS), Food Security and Vegetables-A Global Perspective, P.N. Agricultural Science Foundation, Bangalore, p. 85-100.

MOA, Government of India, 2009. Working Paper on Agriculture of 10th Five Year Plan

Nageswar Rao, A. 1997. Higher Employment and Income Potential of Horticultural Crops, *IJAE*, **52** (3)

Nandi, B.K. and Bhattacharya, L. 2004. Emerging Issues in Nutrition Security, In: Food Security and Vegetables-A Global Perspective, P.N. Agri. Sci. Foundation, Bangalore, p. 55-84

Nath, Prem, 2006. Food and Nutrition Security and Horticulture, ASHS Annual Conference, New Orleans, Louisiana, USA. Hort. Sci. vol. 41, (4), July 2006. p. 1003.

Nath, P., 2008. Vegetables for Food and Nutrition Security, Proc. ISHS 4th Balkan Symposium on Vegetables and Potatoes Vol. 1, Acta Horticulturae 830, June 2009, p. 27-32.

Nath, Prem, 2008. PNASF News Vol. III, Issue X, P.N. Agri. Sci. Foundation, Bangalore, p. 1-11.

Nath, Prem, 2009. PNASF News Vol. III, Issue XI and XII, P.N. Agri. Sci. Foundation, Bangalore, p. 1-13.

Nath, Prem, and Dutta, O.P., 2004. Food and Nutrition Security and Vegetables, In: Food Security and Vegetables-A Global Perspective, P.N. Agricultural Science Foundation, Bangalore, India, p.43-54.

Roekel, J. van, Willems, S. and Boselie, D.M., 2002. Agri Supply Chain Management - to Stimulate Cross-Border Trade in Developing Countries and Emerging Economies, World Bank Paper. WWW.IFAMA.ORG/CHAINLETTER/LETTERS/V212.PDF

Sudha, M., Gajanana, T.M., Sreenivasa, D., Murthy, 2006. Economic impact of commercial hybrid seed production in vegetables on farm income, employment and welfare – A case of tomato and okra in Karnataka, *Agri. Econ. Res. Rev.*, **19** (2): 251-268

Thimmaiah, P., 2001. Evaluation of Soft Loan Schemes for the Development of Post-Harvest Infrastructure for Horticultural Crops in Karnataka, ADRTI, ISEC, Bangalore.

Vardarajan, S. and Ravindran, N., 1976. Economics of Flower Marketing in Coimbatore City, Dept. Agric. Economics, TNAU, Coimbatore (Mimeo).

2. GROWTH OF AGRICULTURE - HORTICULTURE SECTORS AND LIVELIHOOD SECURITY

S. Bisalaiah

Former-Vice Chancellor, University of Agricultural Sciences, Bangalore, Karnataka, India

ABSTRACT

As per Global Hunger Map and Hunger Index, the overwhelming majority of the world's poor are in South Asia and Sub-Saharan Africa, mainly in rural areas. The worst affected of poverty and hunger are women and children, with far-reaching implications for quality of human resource. The experiences and evidences on poverty reduction and livelihood security (much more than food security) are suggestive of some tested development policies, programmes, strategies and approaches. First, poverty responds more to inclusive rural growth than urban economic growth. Agri-Horti and allied sectors could be leading sectors for rural economic growth as well as rural development. Second, development strategies for augmenting productivity of land, labour and capital in rural areas are crucial for livelihood security of the rural poor. Third, supportive rural infrastructure would form yet another catalyst for rural economic growth. Fourth, development alliance between farmers and corporate sector as partners would address the problems of technology, credit, input supplies and market for farm produce, with well defined legal safeguards and with the formation of farmer groups/ associations.

In this context, there are some development issues that need to be addressed. First, how is the nexus between general economic growth and poverty reduction and between agri-horti growth and poverty reduction? Second, what do the empirical evidences suggest to validate the hypothesis that horti-led growth would contribute for higher growth rate in agriculture sector? Third, what are the essential and sufficient conditions for making agri-horti a leading sector for rural development? Fourth, what are the possible macro-economic policy interventions to accelerate the role of agri-horti-led growth as drivers of poverty reduction and sustainable livelihood security? This paper is an attempt to address some of these issues.

1. THE THEME

The number of people without enough food to eat on a regular basis remains stubbornly high, touching one billion at the global level, yet another estimate (FAO, 2003) has indicated that world wide around over 850 million people are chronically hungry due to extreme poverty, while upto 2 billion people lack food security intermittently due to varying degrees of poverty. In fact, IFRI's Global Hunger Index (2006) - which captures essentially three dimensions of hunger, viz, availability of food, nutritional status of children and child mortality – has become a challenge to global civilized society. The recent UNICEF report (2008) "our Climate, our Children and our Responsibility", has brought out implications of climate change to global hunger and poverty, indicating more damaging effect on children.

Poverty, the inability to attain a "minimum" level of well being is the most fundamental economic and social problem facing humanity. The U.N. endorsement of Millennium Development Goal to have the number of chronically underunnourished people on the earth by 2015 is doubtful of achievement in many countries, including India. In the mean while, FAO has started deliberations on how to make the world food secure by 2050.

There have been paradigm shifts from food security to nutrition security, and further to sustainable livelihood security. The deliberations and research have centered around some major issues. First, how is the spatial and social dimension of hunger and poverty? Second, what are the operational counter parts of livelihood security concept for policy and programme designs? Third, how is the nexus between GDP growth, employment and poverty? Fourth, if hunger and poverty are dominantly rural phenomena, are not rural economic growth and rural development the right impact points to address the problems of hunger and poverty? Of agriculture and allied sectors (horticulture, livestock and fishery), how is the potential of horticulture for strengthening the nexus between rural growth, poverty and hunger? Fifth, what are the appropriate technological, institutional and policy interventions needed to drive the rural growth to high trajectory?

This paper is an attempt to evaluate some of these issues, with available empirical evidences and development experiences.

2. SPATIAL AND SOCIAL DISTRIBUTION OF POVERTY

Poverty is not a single phenomenon with a simple foundation, invariant across geographic location and social condition (Hughes *et al.,* 2008). In fact, measurement of poverty itself is still debated, even though income or consumption based measures are widely used. Many researchers would prefer broader "capabilities" - based approach of Sen (1999).

2.1. *Spatial Distribution*

With respect to spatial distribution of poverty at the global level, South Asia and Sub-Saharan Africa are the hunger hot spots. One of the assessments of FAO (2003) has inferred that since the 1990s, the number of food insecure people has been increasing in both Sub-Saharan Africa and South Asia, where the majority of food insecure people live, and poverty in this region is mainly a rural phenomenon. Yet other startling results (ESCAP, 2008) are that over 600 million of the world's poor still live in Asia, 97 million children remain undernourished, and four million children die before reaching the age of five. Even in USA, between 2003-05, about 17% of households in New Mexico and Mississippi and 16% in Texas were food insecure. If one goes by the GHI of IFRI (2006), Sub-Saharan Africa made least progress from 1990 to 2004, with its GHI declining by 2.2 from 26.8 to 24.6. Even though South Asia made most progress with its GHI declining by 7.9 points during the same period, the hunger situation there remains almost as bad as in Sub-Saharan Africa with an index of 24.5 in 2004.

The estimates of poverty on India are quite confronting in the face of high GDP growth rates. About 50% of those living in extreme poverty in Asia and Pacific are in India (ESCAP, 2008). A study of recent NSS results of 61st round indicates that 28% of the populations of India are below poverty line if only food component of poverty is considered. But if both food and non-food components (health and education) are considered, 36% of India population is below poverty line (Dev.*et al.,* 2008). With only food component, 47% of people of Orissa and 43% of Bihar are below poverty line in India. 2009 UNDP Human Development Report (with a composite index of life expectancy, educational attainment and adjusted real income) reinforces this estimate by placing India in 134th Rank, even below Thailand, Sri Lanka and Egypt. The recent GHI by IFRI (2008) ranks India as 66th among 88 countries surveyed, placing it below Nigeria and Cameroon and slightly above Bangladesh. The recent "Global Economic Prospects" report projects that 20% of the India's population would be below extreme poverty line even by 2015.

One could expect demographic disasters such as stunted growth, premature failure of vital organs, and defects in cognitive development in case of food insecure people of these regions.

2.2. Social Distribution

It is recalled from the discussion preceded that poverty is dominantly a rural phenomenon in food insecured regions; it is estimated that 70% of the people of the food insecure regions are in rural areas, of which half of hungry people are small holders and two tenths are landless labourers (Joachin, *et al.,* 2008).

Further, household food security is more than access and entitlement, because household food security also depends on allocative process in the family to different members of the family. In fact, women represent the distressing manifestations of poverty and gender inequality facing excess mortality and are called "missing women". The caste system in India gives rise to another manifestation of social nature of poverty, and the children and aged as well.

An analysis of incidence of poverty in India for the year 2004-2005 (Dev, 2008) indicates that 41% people belonging to schedule caste / schedule tribe category were below poverty line and among muslims the incidence of poverty was 33%, as against the national average of 28% for the same year.

The National Family Health Survey for the year 2005-2006 has clearly highlighted children and women as food insecure people. About 46% of children are malnourished, 56% of married women anemic, and 58% of pregnant women are anemic. In fact in rural areas, the intersection of gender and caste of lower social hierarchy can make women very vulnerable (Hughes *et al.*, 2008)

3. PARADIGM SHIFTS: FOOD-NUTRITION-LIVELIHOOD SECURITY CONCEPTS

One could identify four phases in the evolution of the concept of livelihood security during the last three decades or so (Frankenberger *et al.*, 1998).

3.1. In 1970s, inadequate food supplies was considered as major cause of food insecurity (Davies, *et al.*, 1991)

3.2. In 1980s, it was brought to the fore that physical availability (supply side) was only one side of food insecurity, but the reason for food insecurity at the household and individual level was lack of poor people's entitlement (accessibility) to food that is demand side. (Sen, 1981 and Borton, *et al.*, 1991). Sens theory on food entitlement (1981) had a considerable influence on this paradigm shift from supply side alone to demand side.

3.3. In 1990's thanks to the research results (UNICEF, 1990), the realization was that nutritional component should be factored into food security deliberations. The study of causes of malnutrition has revealed that food is only one factor in the malnutrition equation, and in addition to dietary intake and diversity, health and disease and maternal and child care are also important determinants. Hence, household food security is a necessary condition for food insecure people, but not a sufficient condition for nutritional security.

3.4. The evolution of the concepts of food and nutrition security has led to broaden the concept as livelihood security, treating food and nutritional security as only subsets of livelihood security. Livelihood security implies adequate and sustainable access to income and resources to meet basic needs – adequate access to food, portable water, health facilities, educational opportunities, housing, time for community participation and social integration. Thus, livelihood is made up of a range of on-farm and off-farm activities; each household can have several possible sources of entitlement which would constitute its livelihood. When one can say that the livelihood security of the household is assured? Turning to Chambers (1989), livelihood are secure when households have security of ownership of, access to resources and income earning activities, including reserves and assets, to offset risks, ease shocks and meet contingencies. When one can say that livelihood is sustainable? A livelihood is sustainable when it can cope with and recover from the stress and shocks, maintain its capability and assets, and provide sustainable livelihood opportunities for the next generation (Chambers, 1992).

3.5. The research on sustainable livelihood security has led to policy change regarding pro-poor growth strategies, programmes and approaches (Ellis, 2000 and Salesbury 2003); but the debate is still on relating to livelihood programmes, sustainable livelihood, livelihood assets, livelihood outcomes, livelihood support interventions and so on. What needs to be done to induct development orientation to livelihood promotion activities, livelihood protection, and livelihood provisioning to chronically vulnerable like aged, disabled, women and children? What strategies and approaches under growth mediated livelihood security and social security-led (support-led) livelihood security streams? All these issues would warrant further research on the concept of livelihood security and its operational counterparts for policy as well as programme design.

4. GROWTH-EMPLOYMENT-POVERTY NEXUS

It needs to be emphasized that GDP growth per se will not trickle down benefits to the poor. The benefits of growth will have to reach the poor via employment. Hence it is the nature of employment which mediates between economic growth and poverty reduction. Even though there is no invariant relationship between growth and poverty reduction (Rizwanul, 2004), following figure illustrates the conceptual relationship among employment growth with rising productivity, accrual of higher income to poor, and higher expenditure on health, education and skill development which is postulated to reduce poverty.

Two development experiences during the last two decades or so would provide better perspectives into the nexus between growth – employment – poverty. First, response of poverty to growth varies across countries. There has been a weak response in India, but in East Asian countries the response has been stronger (FAO, 1998). Despite acceleration of overall growth rate in post-reform period in India, there is no evidence of higher rate of decline in poverty, when compared to pre-reform period. Further, higher growth rate has accentuated income and consumption inequalities in post-reform period (Dev, 2008). Second, the relationship between growth, employment and poverty depends on pattern of growth, that is the origin of higher rate of GDP growth. In other words, if the higher growth rate is service sector-led as it has been in India, if pattern of employment growth has decelerated in the sector (rural) where over whelming majority of the poor are concentrated, if structural transformation has not occurred enough to shift "surplus" labour in rural sector to urban sector, the nexus between growth, employment and poverty is bound to be weak as in India. Further, service-sector-led growth is likely to give rise to less multiplier effect on employment prospects for the poor (Saxena *et al.,* 2003). As against the experience of India, the experience of China and Vietnam has indicated the positive nexus between growth, employment and poverty due to adequate emphasis placed on rural sector, especially Vietnam which has effected transition from low value to high value crops through crop diversification with higher employment and income opportunities.

The cross country study by Rizwanul (2004) on Growth-Employment-Poverty nexus, using data from Bangladesh, Bolavia, Ethiopia, India and Vietnam has derived five important inferences. First, concentration of workers in agriculture causes poverty. Second, a shift of workers from agricultural sector to manufacturing reduces poverty. Third, education of the population contributes to poverty reduction. Fourth higher dependency ratio (lower labour force participation) causes poverty. Fifth, an increase in inequality of income would lead to an increased incidence of poverty. All these five inferences could be of highly useful in policy and programme design, in view of empirical evidence of different combinations of growth and poverty reduction; while on the one hand high rates of growth can be accompanied by moderate or slow rates of poverty reduction, it is also possible to have rapid rates of poverty reduction with moderate rates of growth-if the pattern of growth is sufficiently employment intensive and the poor can readily integrate into the growth process and benefit from the income earning opportunities.

5. RURAL GROWTH, SUNRISE SECTORS AND LIVELIHOOD SECURITY

5.1. Development Experiences and Premises

The issue of poverty and livelihood security could well be placed in the perspective of some development experiences and some empirically proved premises. First, the knowledge of spatial and social distribution of poverty is crucial for policy and programme design. Spatially, rural areas of sub-Saharan Africa and South Asia (especially India) are the hot-spots of poverty. Socially, women and children are the most deprived in addition to some caste groups of lower social hierarchy in India. Second, high GDP growth rate is a necessary condition, but not a sufficient condition of livelihood security of the poor people. It is the pattern of growth, particularly the growth of agriculture and allied sectors which could make a dent on poverty. In fact, reduction in poverty needs the growth of rural farm sector and non-farm sector in rural areas. Third, service-led GDP growth in India and growth deceleration in rural sector have slowed down the rate of poverty reduction in India. One of the causes of rural poverty is declining share (about 18%) of agriculture to GDP as well as higher percentage of people depending on rural sector for their livelihood. The empirical experience that a shift away from agriculture to higher productivity sectors is associated with a reduction in poverty, does validate the importance of structural transformation. Added to this structural shift, crop and non-crop diversification towards sunrise sectors such as horticulture, livestock, fishery and other high value enterprises could also contribute for poverty reduction and promotion of livelihood security. Fourth, yet another development experience is that the countries that have succeeded in reducing poverty are characterized by more rapid economic growth, more importantly rapid growth of rural sectors, lower population growth and higher ranking in the Human Development Index.

5.2. Rapid growth of sunrise sectors such as horticulture, livestock and fishery would appear to have more potential for employment, income and livelihood opportunities for the poor. Since the focus of the present paper is on horticulture, there is a need for examining the argument in favour of horticulture sector. The arguments for mobilizing "horticulture remedy" for the poor could be articulated in terms of nutrition contents, change in the food basket of consumers, proper use of diverse agro-climatic resources of the country, higher employment and net income, foreign exchange earnings, and move towards more inclusive growth.

- First, fruits and vegetables are an important component of healthy diet and if consumed daily in sufficient amount, could also prevent major diseases. The long run health benefits of consuming fruits and vegetables as dietary sources of fiber, vegetable proteins and protective micronutrients are well documented (FAO, 2004 and Prem Nath *et al.,* 2004) For example, FAO Special Programme on Food Security, focusing on marginalized micro regions of Low income Food Deficit Countries is a case in point (chapter 5 of Prem Nath *et al.,* 2004).

- Second, crop diversification towards high value horticulture crops known for their short cycle, and high value providing source of income leading to improved livelihood security would be a key strategy for increasing agricultural productivity and reducing incidence of poverty. The development of horticulture sector is essential to achieve 4% targets of agricultural growth (Details later).

- Third, higher returns on horticultural crops vis-à-vis other crop in crop production sector is well documented (Chand, R. 2008). At micro level, per hectar net returns are found to be more in horticultural crops like banana, papaya, tomato, brinjal and flowers vis-à-vis paddy, sugarcane and groundnut. Added to this, labour intensity is found to be more in horticulture crops (Nageswara Rao, 1997 and Sudha, *et al.,* 2006).

- Fourth, the natural resource base argument is that the varied agro climatic conditions of India are conducive for cultivation of fruits and vegetables round the year (Gajanana *et al.,* 2008). In fact, India ranks first in production of mango, banana, papaya, lime and lemons and the second largest producer of brinjal, cauliflower, onion and cabbage, even thought the productivity performance of Indian horticulture crops is low. The natural advantage for growing horticulture crops could be mobilized much more to make horticulture a leading sector in rural areas with liberalized international market.

- Fifth, horticulture could be one of the key sectors to make a move towards inclusive growth to benefit small and marginal farmers and landless labour. Small and marginal farmers who form over 82% of farm households in India could be beneficiaries of horticulture crops. Among horticulture crops, vegetable growing is preferred more by small and marginal farmers due to quick returns, less capital intensive and more labour intensive which would match their resource endowments (Birthal *et al.*, 2008).
- Sixth, in addition to growing international market for horticulture products (which formed about 18% of India's agriculture exports during T.E 2005-06), within the country, there is a change in food basket of consumers from staple food grains to high value food products. The share of fruits and vegetables in urban food expenditure increased from 12% in 1983 to about 16% in 2004. Rural consumers spent about 14.5% of their total food expenditure on fruits and vegetables during 2004, compared to about 9% in 1983. Since 1995-96, demand for imported fresh fruits like apples has been on the increase, indicating large demand potential in the domestic market.

6. GROWTH PERFORMANCE OF CROP PRODUCTION SECTOR: AGRICULTURE VS HORTICULTURE

The preceding section has analyzed the arguments in favour of "horticulture remedy" for poverty and livelihood security. The focus of this section is to analyze the production and productivity performance of crop groups in agriculture and horticulture.

6.1 *Sources of Output Growth in Crop Sector*

The annual growth rate of total value of crop output in agriculture and horticulture, and the shifts in the relative importance different sources of crop output growth in these sectors are analyzed (Table 1).

Considered ten commodities / commodity groups forming 91% of total cropped area and 85% of gross value of output of crop sector.

Four major results could be extracted from this table;

- There was a deceleration in annual growth rate of real value of crop output from 3.8% during 1981/82-1995/96 to 2.1% during 1995/96-2004/05.
- Contribution of productivity, a proxy for contribution of technology, declined from 62 to 45 percent, even though this source of output growth continued to be important.

- The contribution of crop diversification (diversification mainly towards high value horticulture crops) increased from 20% during first period to 42%, implying the increased significant contribution of horticulture to crop output growth in agriculture (More on this later).
- The importance of price factor as a source of crop output growth had declined during second period, implying a greater role of non-price factors in output growth.

Table 1. Major sources of output growth in crop sector*

Growth source	% Share in crop growth sector	
	1981/82 – 1995/96	1995/96 – 2004/05
Annual growth rate (%) of total value of crop output	3.8	2.1
Productivity	62	45
Crop diversification	20	42
Price	12	07
Others	06	06
Total	100.00	100.00

*SOURCE: Adapted from Birthal *et al.* (2008).

6.2. *Area-Value Share of Horticulture Crops in Crop Sector*

Following the analysis under section 6.1, it is also relevant to analyze the area and value share of horticulture crops in crop sector (Table 2). Three results are relevant in the context of the issue whether growth of crop sector was mainly horticulture - led growth or not.

Table 2. Area share, value share and share of horticulture crops in of crop sector

Crop groups	Area share (%)	Value share* (%)	Annual growth rates (%)		Share in over all growth (%)	
			P.I	P.II	P.I	P.II
T.E 2004/05						
Condiments	1.2	3.7	9.6	2.6	7.7	5.5
Fruits & Vegetables	5.8	25.6	2.9	5.6	20.7	63.7

Source: Adapted from Birthal *et al.* (2008)

*Computed at 1993-94 prices.

NOTES:

1. Other crops in crop sector contributed for the remaining percentage share
2. Area share in gross cropped area
3. Value share in value of crop sector output
4. P.I : 1981/82 – 1995/96
5. P. II: 1995/96 – 2004/05

- With an area share of about 7%, horticulture contributed about 29% of value of crop output. Obviously, the contribution of fruits and vegetables was quite dominant.
- The annual compound growth rates of condiments had declined considerably during second period, where as the growth rate of fruits and vegetables had almost doubled during the same period.
- The share of horticulture in the total value of crop output had increased from about 28% in the first period to about 69% in the second period. It is recalled from Table 1 that the annual output growth rate of crop sector decelerated from 3.8% during the first period to 2.1% during the second period. Agriculture growth would have decelerated further, but for the robust growth in fruits and vegetables. This robust growth in horticulture sector was mainly demand-led due to change in the composition of consumer basket of goods within the country, and growing demand in the international market.

6.3. *Growth Rate Analysis of Crop Groups in Agriculture and Allied Sectors*

Table 3. Annual growth rate (%) analysis of crop groups in agriculture and allied sectors

Crop groups	P.I	P.II	P.III
Cereals	3.52	2.36	1.39
Pulses and oilseeds	5.41	2.92	1.43
Fruits and Vegetables	2.84	6.07	3.54
All crops	2.97	3.09	2.36
Live stock	4.78	4.00	3.61
Fishery	5.74	7.05	2.93

SOURCE: Adapted from Singh *et al* . (2008)

NOTES:

1. Growth rates are computed with gross value of output at 1999 - 2000 prices
2. P.I : 1981/82 – 1990/91
   P.II: 1991/92 – 1996/97
   P.III: 1997/98 – 2006/07

It could be observed from Table 3 that

- The annual growth rates of cereals declined from 3.52% in first period (i.e. 1980s) to 1.39% in third period (1997/98-2006/07). In case of pulses and oilseeds also, there was deceleration in growth rates from 5.41% to 1.43%.
- In case of fruits and vegetables, the growth rate increased from 2.84% in first period to 6.07% during period two, and declined to 3.54% during third period. This is a disturbing trend, which needs to

be arrested. Details on the needed interventions will be discussed in the next section.

- In case of other two sunrise sectors viz. livestock and fishery, there was a trend towards growth deceleration during third period.

6.4. *Disaggregative Analysis of Growth Performance of Horticulture Crops*

Table 4 provides insights into the production and productivity performance of horticulture crop groups for two periods of post-reform regime:

Table 4. Growth rates (%) of output and productivity of major horticulture crop groups during two periods

Crop Groups	Production		Productivity	
	P.I	P.II	P.I	P.II
Fruits	7.27	2.89	3.49	-1.96
Vegetables	4.71	3.40	4.34	0.89
Plantation	5.36	1.81	0.83	0.01
Spices	6.14	7.70	4.33	2.69
Total	5.66	3.27	3.34	-0.45

SOURCE: Singh *et al.* (2008)

NOTES: P.I: 1991/93 – 1996/97

P.II: 1997/98 – 2005/06

- In aggregate, the growth rate of horticulture crops had declined from 5.66% during 1991/93-1996/97 to 3.27% during 1997/98-2005/06. This decelerated growth is a disturbing trend, as discussed earlier.
- But with disaggregate analysis, the output growth rates decelerated more in case of fruits and plantation crops than in the case of vegetables.
- With respect to negative productivity of horticulture crops in general, the trend is observed to be much more in case of fruits; and in case of vegetables and spices, productivity even though positive, had decelerated considerably. What intervention strategies needed for reversing this trend towards negative growth rates and decelerated growth rates would form the theme for the next section.

7. AGRICULTURE-HORTICULTURE SECTORS FOR LIVELIHOOD SECURITY: NEED INTERVENTIONS

Recapitulation of some results/ inferences from the discussion preceeded would form the basis for indicating needed interventions to mobilize agriculture and horticulture sectors for livelihood security of the poor.

- First, special distribution of poverty in terms of rural areas, and social distribution of poverty in terms of small and marginal farmers, landless labourers and women and children.

- Second, sustainable livelihood security, instead of only food and nutrition security, would be another crucial thing.
- Third, growth- employment- poverty nexus is not uniform across countries. Thus growth per say is a necessary condition; but the major sufficient condition is rural growth which could address the poverty and livelihood security of majority of the people "parked" in rural areas.
- Fourth, "horticulture remedy" for addressing the problem of poverty and livelihood security is very well recognized.
- Fifth, growth performance of crop production sector (both agricultural and horticultural crops) has led to the inference that horticulture- led growth of crop production sector is quite essential, even to prevent growth deceleration of agriculture; further negative productivity growth rates of horticulture crops in general and particularly incase of fruits need to be kept in view.

All these five major results would be of help in identifying the needed major interventions in the areas of technology, institutions and policies.

7.1. *Technological Interventions*

- First, growth- mediated orientation to livelihood security would need improvement in productivity of agriculture and horticulture, and one of the disquieting features of Indian horticulture sector is the low and declining productivity of fruits and vegetables. The average productivity is much lower in India than in other fruit producing countries in the world. As regards vegetables, average productivity is the lowest in the world (Gajanana *et al.,* 2008). With this low and decelerated productivity in horticulture sector, the Indian horticulture sector cannot perform well in International market.
- Second, due to recent decline in growth rates of fruits and vegetables, the "ability" of horticulture sector as a remedy for poverty and hunger is reduced (Chand, 2008).
- Third, post harvest losses of fruits and vegetables is estimated to be 17 to 35 percent. In fact, the reduction in post harvest losses is a complementary means for increasing productivity (Srinivasamurthy *et al.,* 2002).
- Fourth, depletion and degradation of natural resource base in crop production sector would very much require technologies to maintain long run productivity of natural resource base.
- Fifth, there are several unsolved chronic disorders like bacterial wilt and blight in pomegranate, irregular bearing in mango, wilt in guava and so on (Gajanana *et al.*, 2008).

Hence, research and development efforts should be priority item of agenda in promoting and sustaining the growth and productivity of horticulture sector. Agriculture research and development has the potential to boost productivity of agriculture and allied sectors, promote sustainable management of natural resources, and support livelihood security of poor people through the development of new technologies (Swaminathan, 2008). The present allocation of about 0.5% of GDP originating in agriculture should be increased at least to one percent, though not to 2% spent on agriculture on research and development as in developed countries. While enhancing allocation of budget, a critical review of technology subsystems viz., technology generation (research), technology transfer (extension) and technology utilization (farm level) would be indispensable to set the resource and development priorities, streamline the extension, and to release the constraints facing farmers in technology adoption.

7.2. *Institutional Interventions*

There are three major areas in which institutional interventions are needed to support the rural growth:

- First, the horticulture product market structure is exploitative, giving rise to low share of producers in consumers' rupee. Studies conducted at IIHR, Bangalore have brought out the benefits of institutions such as cooperative marketing, growers associations, and integration of production and marketing through processing and contract farming (Subramanya *et al.*, 2000, Birthal, 2007 and Sudha, 2003).
- Second, formation of Farmers' Group as a development unit could be an important institutional structure for supporting small and marginal farmers in input and output marketing (including export of horticulture commodities), in dealing with corporate sector under contract farming, and in rebuilding and managing natural resource base in a sustainable manner.
- Third, capacity building of poor people is quite essential for managing their livelihood assets like human assets (ex: knowledge, skill and health), social assets (status in society to lead livelihood secured life with dignity.), natural assets (ex: land and water resources), physical/ biological assets (like agricultural equipment and machinery and livestock), and financial assets like income, access to credit and investment. Capacity building of farmers is required for adoption of technologies as well (Prem Nath *et al.,* 2004). The capacity building of farmers in these directions could take place through Government institutions, initiatives of corporate sector and through NGO's.

7.3. *Policy- Programme Interventions*

Policy- programme interventions are needed for effecting transition from safety- net (support- led) strategy for livelihood security to growth- mediated livelihood security. It is postulated that agri- horti- led growth strategy in rural sector is the major area of support to livelihood insecure poor and hungry. A few areas of interventions could be elaborated as below:

7.3.1. *Food self- reliance and removal of agriculture outside the purview of WTO*

It is rightly argued that food- self sufficiency at the national level is desirable, and reliance on trade can be kept within limits (Rao *et al.,* 2000). This is needed in the face of falling productivity of agriculture and horticulture, and rising prices of food articles in the International markets so as to meet the challenge towards food secure India (Dev *et al.,* 2003). Related to this is the need for monitoring the control of multinationals over food resources, the impact of emergence of new cartels like organization of Rice Exporting Countries for price setting, and WTO prescribed reforms in input subsidy without realizing the positive impact of selective subsidies on productivity and production in rural sectors. Added to these is the argument that food is not to be treated as an internationally tradable commodity like other commodities. To focus on livelihood security of the poor and hungry as well as to ensure food secure India, it is desirable to remove agriculture from the purview of WTO, if the developed nations continue to protect the interest of their farmers through "unreasonable" subsidies and use non-tariff barriers for entry of products from other countries.

7.3.2. *Diversification of livelihood options*

The subject of "Livelihood Diversification" (Daniel *et al.,* 2004) is becoming more important to provide multiple and diverse livelihoods for rural people. This is a part of structural transformation relating to crop diversification towards high value crops, development of other allied sectors like livestock and fisheries and development of non-farm sector in rural areas in a partnership mode so as to diversify the livelihood options for livelihood insecure households and individuals. This kind of diversification could accelerate the tempo of rural growth and faster reduction in density and intensity of livelihood insecure people.

7.3.3. *Interventions for gender inequality*

It is recalled that household food security, a subset of livelihood security, is more than availability, access and entitlement; because household food security also depends on allocative process in the family to different members of the family. Further, it is recalled that among the poor and hungry, women and children are the most deprived ones. This gender inequality draws the attention to ensure that women get equitable access to land and other natural resources, credit and other public- supplied inputs, and government supported output marketing agencies. In this context, it is realized that land is a proxy to absolute poverty, and is one of the most important assets for the rural people. It provides a basis for livelihoods and socio-economic power relations in rural areas. Hence the importance of equitable land tenure policy.

7.3.4. *Watershed plus strategy*

The potential of watershed project for improvement of natural resource base for agriculture and allied sectors and enhancement of livelihoods (Cathryn, 2000) is well recognized. There is a need for broadening the scope of watershed development activities so as to improve the productivity of land and water, increase the combination of land based and non-land based activities, and enhancement of capability of landless, women, small and marginal farmers for entitlement negotiation in a participatory mode. This kind of watershed plus strategy is likely to contribute substantially for sustainable livelihood for the poor and the by passed groups.

7.3.5. *Public spending for inclusive growth*

It is high time to treat livelihood as a part of inclusive growth, and rural growth is more important than urban growth to address the problem of poor and hungry. It is argued that inclusive growth, emphasizing livelihood security of the poor needs effective targeted fiscal incentives, safety nets, and public investment for less advantaged areas and social groups (Krishan *et al.*, 2009). Falling public investment for capital formation should be reversed. Inclusive growth requires broad- based public investment in agriculture, rural infrastructure, in public goods, and in human capabilities. With respect to increasing capital formation on farmers' field, farm assets like farm ponds on the land of small and marginal farmers, and community assets like restoration of waterbodies, roads, school and hospital buildings could be given priority under National Rural Employment Guarantee Scheme. Regarding development of human capabilities, investment in basic schooling, health and nutrition to enable the poor to get access would not only directly enhance their well-being, but also see improvement in income distribution and higher average incomes over the long run (Dreze *et al.,* 1989). This kind of investment, among others, would support the development premise that growth-mediated strategy for the poor should replace relief type of support- led strategy in the form of social safety nets.

7.3.6. *Horticulture development strategies for differentiated target groups and in supply-chain mode*

In addition to technological and institutional interventions for increasing productivity and production of horticulture sector and for value addition through processing, variety of strategies have to be implemented for differentiated groups engaged in horticulture production and marketing (FAO 2004 and FAO- PNASF 2004). First, development support to the rural poor, through the promotion of homestead garden and small scale commercial horticulture. Second, promotion of urban and peri-urban horticulture for home consumption and neighbourhood marketing. Third, commercial farmers through market-oriented fruit and vegetable production responding to national and international market opportunities.

Further, production and consumption of fruits and vegetables should be seen in holistic supply chain mode, i.e., from seed to table. This kind of supply chain mode would require technical guidance, credit, input supplies, well designed marketing system to reduce transport costs and post harvest losses and to food processing so as to add value. This could be done by promoting development alliance between farmers and corporate sector with defined rules of the "game".

Added to all these interventions are needed for production and price risk management through crop insurance with wider coverage and forward trading, skill-upgradation of rural labour to make them "fit" for non-farm employment, improved governance for good support-delivery system, and more emphasis on non-price factors to promote the development of agriculture and allied sectors to enable the poor to move out of livelihood insecurity.

8. RECAPITULATIONS AND CONCLUSIONS

Recent campaigns for the right to food, education, employment and information have brought the issues of deprivation and livelihood to the central stage as never before. These rights to food, health, education and employment are inter-dependent and incomplete without one another for the poor to lead a livelihood secured life with dignity.

Keeping the above backdrop in view, this paper has been an attempt to analyze spatial and social distribution of poverty, transition from food-nutrition security to livelihood security, growth- employment- poverty nexus, the role of rural growth and sunrise sectors like horticulture in promoting and sustaining livelihood security, growth performance of crop production sector, and needed technological, institutional and policy interventions for mobilizing "horticulture remedy" for the problem of livelihood insecurity.

The major conclusions and inferences are:

- South Asia (especially India) and sub-Saharan Africa are the hunger-hot-spots, and majority of the hungry and poor are in rural areas.
- The small and marginal farmers, landless labour, women and children, and some social groups in a stratified Indian society are the most deprived and livelihood insecured social groups.
- Quite a distance has been travelled in evolving the concept and components of livelihoods security. Food and nutrition security are only the subsets of livelihood security.
- Growth- employment- poverty nexus analysis has led to the inference that growth of GDP per say is a necessary condition for addressing the problems of poverty and hunger, but the most important sufficient condition is pattern of growth. For example, despite high growth rate of GDP during post-reform period in India, there is no evidence of higher rate of decline in poverty; because

higher GDP growth rate has been mainly service- led, and in fact there has been a decelerated growth in crop production sector. Hence, rural growth is more important than urban growth for livelihood security of the poor.

- There are potentials in sunrise sectors especially horticulture sector to promote rural growth and to enhance the scope for poor people to experience livelihood security. Even though growth performance of horticulture has been responsible for preventing high rate of productivity and production growth deceleration in crop production sector, production and productivity rates of horticulture sector have started decelerating since 1997/98. This is a concern that needs to be addressed.
- Major technological, institutional and policy interventions have been indicated to reverse the trend of decelerated production and productivity growth in horticulture sector, and to put it on a high growth trajectory so as to support the poor people especially in rural areas to experience a quick transition to livelihood secured status.

The following chart illustrates the major components of household livelihood security:

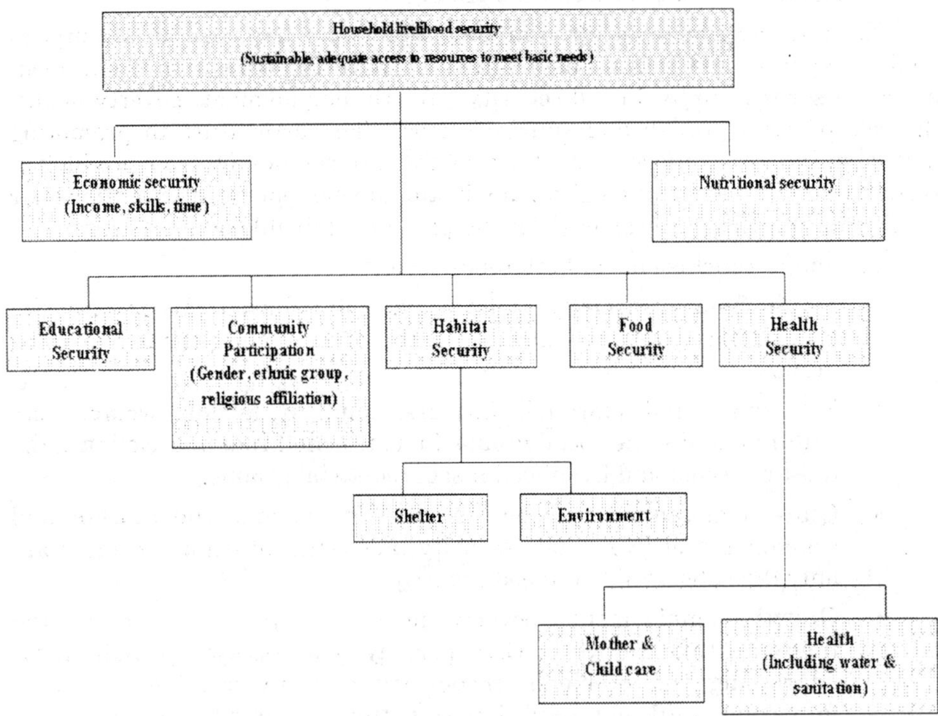

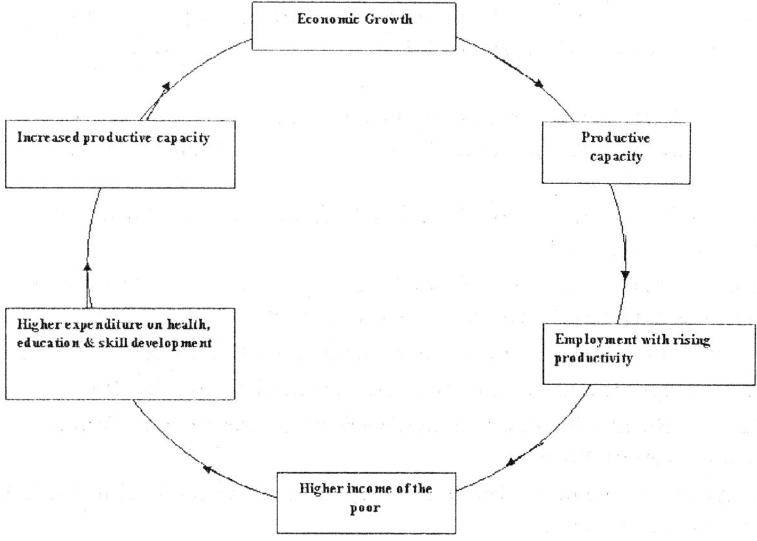

Figure 1: Virtuous circle of links between growth, employment and poverty reduction

The author is thankful to Dr. Prem Nath for his intellectual "tips" on evolution of the concept and components of livelihood security, and to Dr. Gajanana for making some useful literature available on the theme of this paper. The author is solely responsible for errors and omissions in the paper.

REFERENCES

Basudeb G H *et al.* (2007): Food Security – Indicators, Measurement and the Impact of Trade Openness. Oxford University Press.

Birthal P S *et al.* (2007): Linking Farmers to Markets for High Value Agricultural Commodities. *Agricultural Economics Research Review,* Conference Number.

Birthal P S *et al.* (2007): Institutional Innovations for Improving Smallholder Participation in High-Value Agriculture: A Case of Fruit and Vegetable Grower's Association, *Q J of International Agriculture,* **46**, 1.

Birthal P S *et al.* (2008): Can Horticulture Revitalise Agricultural Growth? *Indian Journal of Agricultural Economics,* **63:** 3, July – September.

Borton *et al.* (1991): Mapping Vulnerability to Food Insecurity. Relief and Development Institute, London (Mimeo).

Chambers R (1989): Editorial Introduction: Vulnerability, Coping and Policy. IDS Bulletin, Vol. 2, No. 2.

Chambers *et al.* (1992): Sustainable Rural Livelihoods: Practical Concepts for the 21st Century. IDS Discussion Paper No.296, Brighton, U.K, Institute of Development Studies.

Chand Ramesh *et al.* (2008): Progress and Potential of Horticulture in India. *Indian Journal of Agricultural Economics,* **63,** 3, July – September.

Cathryn T (2000): Sustainable Livelihood and Project Design in India. Working Paper 127, Overseas Development Institute, London.

Daniel S *et al.* (2004): Livelihood Options? The Political Economy of Access, Opportunity and Diversification. Working Paper 233, Overseas Development Institute, London, April.

Davies S *et al.* (1991): Early Warning in the Sahel and the Horn of Africa: A Review of the Literature. IDS Research Reports No. 20, Brighton, U.K, Institute of Development Studies.

Dev Mahendra S *et al.* (Ed) 2003: Towards Food Secure India. Institute for Human Development, New Delhi.

Dev Mahendra S (2008): Inclusive Growth in India: Agriculture, Poverty and Human Development, Oxford University Press, New Delhi.

Dev M S *et al.* (2008): Revisiting Poverty Estimates, EPM, March.

Dreze J *et al.* (1989): Hunger and Public Action, Oxford, Clarendon Press.

Ellis F (2000): Rural Livelihoods and Diversity in Developing Countries, Oxford, Oxford University Press.

E. S. C. A. (2008): Economic and Social Survey of Asia and the Pacific 2008. Academic Foundation, New Delhi.

F.A.O. (1998): Poverty Alleviation and Food Security in Asia, Bangkok.

F.A.O. (1999): Report of the Ministerial Roundtable Beyond the Asian Crisis: Sustainable Agricultural Development and Poverty Alleviation in the Next Millennium. June, 14-15, Bangkok.

F.A.O. (2003): The State of Food Insecurity in the World, Rome.

F.A.O. (2004): Fruit and Vegetables for Health, Rome.

F.A.O. (2005): International Workshop on Urban and Peri-Urban Agriculture for Asian Countries, Bangalore, June – July.

F.A.O. (2006): State of Food Insecurity in the World, Rome.

Frankenberger T R *et al.* (1998): The Household Livelihood Security Concept, CARE, U.S.A.

Gajanana T M *et al.* (2002): Marketing Practices and Post-Harvest Loss Assessment of Banana Va Poovan in Tamil Nadu. *Agricultural Economics Re. Review,* **15,** 1.

Gajanana *et al.* (2008): Horticulture Sector in India – Problems and Prospects. Country Paper on Horticulture for NABARD Training Programme, Bangalore.

Hughes B B *et al.* (2008): Reducing Global Poverty: Patterns of Potential Human Progress, Vol. 1, Oxford University Press, New Delhi.

IFRI (2006): Global Hunger Index, Washington, D.C.

Joachin V B *et al.* (2008): Globalisation Food and Agriculture and the Poor, Oxford University Press, New Delhi.

Joshi P K *et al.* (2004): Agricultural Diversification in South Asia: Patterns, Determinants and Implications. EPW **39,** 24, June, 12-18.

Kar A *et al.* (2003): Linkage Between Crop Diversification and Poverty in India. Agricultural Situation in India, **60,** 8, November.

Krishna K L *et al.* (2009): Readings in Indian Agriculture and Industry. Academic Foundation, New Delhi.

Nageswara Rao A (1997): Higher Employment and Income Potential of Horticultural Crops, *Indian Journal of Agricultural Economics*, **52,** 3, July–September.

Prem Nath *et al.* (2004): Food Security and Vegetables – A Global Perspective. Dr. Prem Nath Agricultural Sciences Foundation, Bangalore.

Rao V M *et al.* (2000): Modernising Indian Agriculture: Priority Tasks and Critical Policies Study No. 21. Department of Economic Analysis and Policy. Reserve Bank of India, Mumbai, August 29.

Ravallion M *et al.* (1996): How Important to India's Poor is the Sectoral Composition of Economic Growth? World Bank Economic Review, 11 (May).

Rizwanul I (2004): The Nexus of Economic Growth, Employment and Poverty Reduction. Discussion Paper 14, ILO, Geneva.

Saxena N C *et al.* (2003): Trends and Prospects for Poverty Reduction in Rural India: Context and Options. Working Paper 198, ODI, London, May.

Sen Amarthya (1981): Poverty and Famines: An Essay on Entitlements and Deprivation. Oxford, Carendon Press.

Sen A (1999): Development as Freedom, Knopf, New York.

Singh M *et al.* (2008): Structural Changes in Horticulture Sector in India. *Indian Journal of Agricultural Economics,* **63:** 3, July – September.

Salesbury W (2003): Sustainable Livelihoods, Working Paper 217, ODI, London, June.

Sreenivasa Murthy D *et al.* (2002): Marketing Practices and Post-Harvest Estimation in Mango Va Baganapalli at Different Stages of Marketing – A Methodological Perspective. *Agricultural Economics Re. Review,* **15:** 2.

Sreenivasa Murthy D *et al.* (2009): Marketing and Post-Harvest Losses in Fruits: Its Implications on Availability and Economy. *Indian Journal of Agricultural Economics,* **64,** 2, April – June.

Subrahmanyam K V *et al.* (2000): Cooperative Marketing of Fruits and Vegetables in India. Concept Publishing Company, New Delhi.

Sudha M *et al.* (2003): Institutional Arrangements for Linking Horticultural Production, Marketing and Processing. In Pal Suresh *et al.* (2003), Institutional Change in Indian Agriculture, NCAP, New Delhi.

Sudha M *et al.* (2006): Economic Impact of Commercial Hybrid Seed Production in Vegetables on Farm Income Employment and Farm Welfare – A Case of Tomato and Okva in Karnataka. *Agricultural Economics Res. Review,* **19,** July – December.

Sudha Narayan *et al.* (2008): Globalisation and Small Holders: A Review of Issues, Approaches and Tentative Conclusions in Joachin *et al.*

Swaminathan M S (2008): Marking Globalization Work for the Poor: Technology and Trade.

UNICEF (1990): Strategy for Improved Nutrition of Children and Women in Developing Countries.

UNICEF (2008): "Our Climate, Our Children and Our Responsibility", News Report, April 29.

3. IMPROVING NUTRITION SECURITY AND HEALTH - THE KEY ROLE OF HORTICULTURE

Janice Albert

Nutrition Officer, Nutrition and Consumer Protection Division, Food and Agriculture Organization of the United Nations (FAO), Viale delle Terme di Caracalla, Rome – 00100, Italy

ABSTRACT

Since the field of horticulture is enormous, in this presentation, we will focus on the latest information pertaining to the impact of fruits and vegetables on human health. However, this should not be interpreted as lack of recognition of the importance of roots and tubers, nuts, whole grains and other plant foods. All of these foods are needed to achieve healthy diets for all people.

The severity of the problems of micronutrient deficiencies is a compelling reason to call for greater production and consumption of these vegetable and fruits. It is well- documented that these foods can contribute to the prevention of deficiencies which can be debilitating and can lead to premature deaths. Although we do not have all the evidence we would like regarding the health benefits of consuming vegetables and fruits in relation to cancer and other chronic diseases, it is certainly clear that these foods are associated with healthier populations. Research on the impact of these foods on chronic disease is ongoing and with time, we will have a better understanding of the mechanisms that link specific components of foods to health.

FAO and WHO are collaborating in the Fruit and Vegetables for Health Initiative and through this we are encouraging countries to promote the production and consumption of these horticulture products. Nutritionists and health promoters are actively engaged in raising awareness about the benefits of consuming fruits and vegetables through dietary guidelines and nutrition education campaigns. We have reason to believe that interventions to promote gardening and educate young people will lead to increased awareness and demands for these foods in the future (Laurie and Faber, 2008 and Parmer *et al.*, 2009). As consumer awareness rises, the horticulture sector will be challenged to meet the growth in demand.

INTRODUCTION

On behalf of the Nutrition and Consumer Protection Division of FAO, we would like to thank the editors for inviting us to give an update about the emerging role of horticulture in nutrition security and health improvement. The field of horticulture is enormous, in this presentation, the focus is primarily on the latest information pertaining to the impact of fruits and vegetables on human health. However, this should not be interpreted as lack of recognition of the importance of roots and tubers, nuts, whole grains and other plant foods. All of these foods are needed to achieve healthy diets for all people.

GLOBAL FOOD AND NUTRITION SECURITY SITUATION

To fully grasp the importance of horticulture in addressing global nutrition problems, we would like to begin by drawing attention to the issue of food and nutrition security. In spite of all the advances in science and technology and economic and social progress in many countries, the number of hungry people is growing worldwide. In June 2009, FAO announced that the number of hungry people has reached 1020 million people (FAO 2009). Nearly one-sixth of all humanity does not have sufficient food. Two of the main reasons that progress in eliminating hunger are slackening and high food prices and the financial crisis, which substantially limit access to food. Figure 1 illustrates the global pattern of hunger and the fact that the situation is worsening since last year.

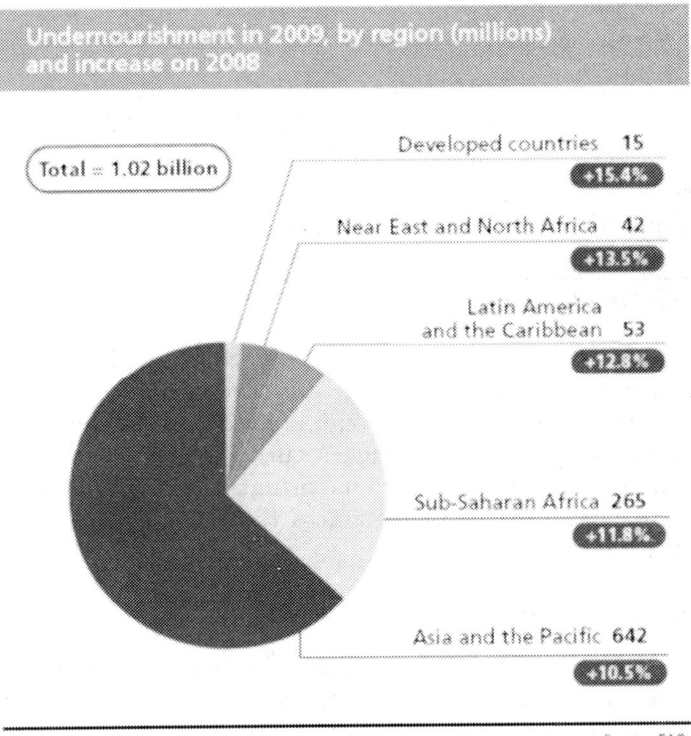

These numbers indicate the millions of people who do not have access to sufficient foods that meet their dietary energy needs. Most of our dietary energy comes from the staple foods, which are usually grains. When food prices rise and households must spend more money on their staples, it is common for them to forego consumption of animal-based foods and fruits and vegetables (Campbell *et al.,* 2009). This reduction in consumption of a range of foods can lead to vitamin and mineral deficiencies that seriously affect health.

Vitamin A Deficiency

One vitamin deficiency that is a major public health concern in lower income countries is vitamin A deficiency (VAD). VAD is the leading cause of preventable childhood blindness and it leads to night blindness in adults, particularly pregnant women. The weakened resistance to infection caused by this deficiency makes children more vulnerable to infectious diseases and death. Table 2 shows the prevalence of this deficiency worldwide.

Table 2. Vitamin A deficiency prevalence (serum retinol under 0.70 μmol/l) in 156 countries with incomes under $15000/year

WHO Regions	Pre-school age children		Pregnant women	
	Prevalence (%)	# affected (millions)	Prevalence (%)	# affected (millions)
Africa	44.4	56.4	13.5	4.18
Americas	15.6	8.68	2.0	0.23
South-East Asia	49.9	91.5	17.3	6.69
Europe	19.7	5.81	11.6	0.72
Eastern Mediterranean	20.4	13.2	16.1	2.42
Western Pacific	12.9	14.3	21.5	4.90
Global	33.3	190	15.3	19.1

Source: World Health Organization 2009

Consumption of vegetables and fruits that are rich in carotenoid precursors (mainly beta-carotene) such as green leaves, carrots, ripe mangoes, and orange-yellow vegetables and fruits can help to prevent this deficiency. This is just one example of how nutrition security is related to access to staple foods and access to foods which are important sources of particular essential nutrients. All people need to have access to a diversity of foods in order to have healthy diets.

Health Improvement

The contributions of fruits and vegetables to prevention of deficiencies is well-understood by scientists, the obstacles to resolving these problems are usually related to the availability, affordability and consumer awareness. Now, we would like to turn to the role of these foods in preventing several diet-related chronic diseases such as cardiovascular diseases, type 2 diabetes, and cancers.

Aging Populations

Chronic diseases such as cardiovascular diseases, type 2 diabetes and cancer generally affect the middle aged and elderly segments of the population. Globally, the number of persons aged 60 years or over is expected nearly to triple, increasing from 673 million in 2005 to 2 billion by 2050 (United Nations, 2007). The vast majority of these people will be in developing countries. To ensure that these people enjoy health and that their medical needs do not become a devastating drain on economies, it is critically important to make efforts today to prevent chronic diseases.

Thus, the world faces two sets of challenges, the undernutrition which usually afflicts the very young and vulnerable groups, and diet-related chronic diseases that affect adults in all social strata. Horticulture must play an important role in both arenas.

Preventive Action is Needed

Although the scientific evidence is still emerging and caution is needed in making sweeping statements about the potential of foods to prevent or treat chronic illnesses such as heart disease, type 2 diabetes and cancer, we do know that there are constituents of plant foods that play a role in protecting health. Health experts are convinced of the multiple benefits of consuming fruits and vegetables and the urgency of the need to take preventive actions and they are encouraging increased consumption of these foods. Public health agencies have stressed that at least 400 grams of fruits and vegetables should be eaten per day (WHO, 2003). Some recommend that the amount should be raised to 600 grams per day (World Cancer Research Fund, 2009).

Research on Fruits and Vegetables

Fruits and vegetables are very complex and this presentation focuses on a few components where substantial research has been carried out recently. The particular components are beta carotene (precursor for vitamin A), lycopene, alpha- tocopherol (vitamin E), folate and flavonoids. Current research suggests that each of these compounds may have great public health significance in terms of preventing disease, disability and premature mortality. This research includes epidemiological studies, animal experiments and clinical trials with humans.

Cardiovascular Diseases

Although cardiovascular diseases (CVD) usually affect middle age and elderly adults, the diets consumed early in life can influence the risk of stroke and coronary heart disease. Children's consumption of a heart healthy diet can reduce the prevalence of obesity, high cholesterol and high blood pressure in adults. When fruits and vegetables replace foods that are high in animal fat, salt and sugar this can lower health risks.

Consumption of these foods can help people to manage their weight and salt and sugar content; however, this is not the main reason that these foods are believed to be protective. In adolescents and adults, high consumption of fruits and vegetables has been associated with lower risk of cardiovascular diseases. Flavonoids, a subclass of polyphenols, are found in parsley, spinach, peppers, cauliflower, brussel sprouts, onions, apples, berries and grapes, among other foods (Holt *et al.,* 2009). They are present in the skins and peels of fruits and vegetables and give these foods their deep colours. Some flavonoids have antioxidant properties.

Vitamins C (ascorbic acid), E (alpha-tocopherol) and folate have favourable effects on vascular system and risk factors for heart disease (George *et al.,* 2009). Several studies have found that lycopene, a carotenoid which is found in tomato, has a preventive effect on cardiovascular disease and risk of myocardial infarction. Foods containing beta-carotene and vitamin A derivatives also reduce the risk of CVD.

Type 2 Diabetes

It is estimated that there are 180 million cases of diabetes mellitus (DM) worldwide (over 90% are type 2 DM); without strong preventive actions, this number may double by 2030 (Barnard *et al.,* 2009 and Wylie-Rosett and Vinicor, 2007). This increase is largely the result of changing diets and lifestyles, often associated with urban life. Observational and clinical studies have shown that diets that are comprised of an appropriate diversity of plant foods (whole grains, legumes, vegetables and fruits) can improve the management of type 2 diabetes. Some of the benefits of plant based diets include: reduced obesity, reduced fat intake and increased intake of dietary fibre (Barnard *et al.,* 2009).

Cancers

The International Agency for Research on Cancer, the World Cancer Research Fund and American Institute for Cancer Research believe that it is probable that high fruit and vegetable intakes may reduce a number of cancers (namely, oesophageal, colo-rectal, oral, pharynx, stomach, larynx, lung, ovary and kidney cancers). However, the evidence over the past two decades is conflicting and not yet considered to be convincing. More research is needed to identify the specific types of fruits and vegetables, and their particular components, that have an impact on disease.

Based on current knowledge, it appears that dark green leafy vegetables, yellow and red vegetables, cruciferous vegetables, as well as red and purple fruits and vegetables may be protective against cancer. It is plausible that the vitamins C, E and folate have an effect on cancer. The phytochemicals such as carotenoids, phenolics, isoflavonoids, isothiocyannates (ITC) and indoles also have anticarcinogenic effects (Kim and Park, 2009).

Cruciferous vegetables, such as broccoli, cauliflower, cabbage, kale and mustard are common worldwide and they are rich in carotenoids, vitamin C, folate and soluble fiber. There is evidence from animal studies that these foods contain phytochemicals that have anticarcinogenic effects (Kim and Park 2009). The latest research suggests that these vegetables, especially Brassicas, may lower the risk of gastric and lung cancers.

CONCLUSION

It is also important to note that genetics, age, gender and environmental exposures affect the influence of these foods on disease. Food preparation and storage can also affect the health benefits of vegetables and fruits.

REFERENCES

Barnard, ND, H I Katcher, D JA Jenkins, J Cohen and G Turner-McGrievy 2009. Vegetarian and vegan diets in type 2 diabetes management, *Nutrition Reviews,* **67** (5): 255-263.

Campbell, AA, A Thorne-Lyman, K Sun, S de Pee, K Kramer, R Moench-Pfanner, M Sari, N Akhter, MW Bloem and RD Semba. 2009. Indonesian women of childbearing age are at greater risk of clinical vitamin A deficiency in families that spend more on rice and less on fruits/vegetables and animal-based foods, *Nutrition Research,* **29**: 75-81.

Food and Agriculture Organization. 2009. "More people than ever are victims of hunger" Background note for press conference 19 June 2009. (Accessed 27 June 2009, www.fao.org/fileadmin/user_upload/ newsroom/docs/Press%20release%20 june-en.pdf)

George, TW, C Niwat, S Waroonphan, MH Gordon, JA Lovegrove. 2009. "Postgraduate Symposium Effects of chronic and acute consumption of fruit-and vegetable-puree-based drinks on vasodilation, risk factors for CVD and the response as a result of the *e*NOS G298T polymorphism" Conference on 'Multidisciplnary approaches to nutritional problems" *Proceedings of the Nutrition Society,* 68: 148-161.

Holt, EM, LM Steffen, A Moran, S Basu, J Steinberger, J Ross, C-P Hong and A Sinaiko. 2009. Fruit and vegetable consumption and its relation to markers of inflammation and oxidative stress in adolescents, *Journal of the American Dietetic Association,* **109** (3), March: 414-421.

Kim, MK and JH Yoon Park 2009. "Cruciferous vegetable intake and the risk of human cancer: epidemiological evidence" Conference on 'Multidisciplinary approaches to nutritional problems' Symposium on 'Nutrition and Health' *Proceedings of the Nutrition Society.* 68: 103-110.

Laurie, SM and M Faber. 2008. Integrated community-based growth monitoring and vegetable gardens focusing on crops rich in Beta- carotene: project evaluation in a rural community in the Eastern Cape, South Africa, *Journal of the Science of Food and Agriculture,* **88**: 2093-2101.

Parmer, SM, J Salisbury-Glennon, D Shannon, B Struempler. 2009. School gardens: An experiential learning approach for a nutrition education program to increase fruit and vegetable knowledge, preference and consumption among second-grade students, *Journal of Nutrition Education and Behavior,* **41** (3): 212-217.

United Nations, Department of Economic and Social Affairs, Population Division 2007. *World Population Prospects: The 2006 Revision, Highlights,* Working Paper No. ESA/P/WP.202.

World Cancer Research Fund/American Institute for Cancer Research. 2007. Food, Nutrition, Physical Activity, and the Prevention of Cancer: A Global Perspective Washington DC AICR.

World Cancer Research Fund/American Institute for Cancer Research. 2009. Policy and Action for Cancer Prevention, Food, Nutrition and Physical Activity: a Global Perspective. Washington DC AICR.

World Health Organization. 2003. Diet, Nutrition and the Prevention of Chronic Diseases Report of a Joint WHO/FAO Expert Consultation WHO Technical Report Series 916, Geneva.

4. HORTICULTURE AND HEALTH: ANCIENT AND MEDIEVAL VIEWS

Jules Janick[1], Marie-Christine Daunay[2], and Harry S. Paris[3]

[1] *Department of Horticulture and Landscape Architecture, Purdue University, West Lafayette IN, USA.*
[2] *INRA, UR1052 Unite de Génétique et d'Amélioration des Fruits et Légumes, F=84140, Montfavet, France.*
[3] *Department of Vegetable Crops & Plant Genetics, Agricultural Research Organization, Newe Ya'ar Research Center, PO Box 1021, Ramat Yishay 30-095, Israel.*

ABSTRACT

The present day emphasis of horticulture and health is an extension of ancient and medieval concerns in the East and West. An 11th-century Arabic manuscript known as the *Taqwim al-Sihha bi al-Ashab al-Sitta* (*Rectifying Health by Six Causes*) was a guide for healthy living based on the Hippocratic-Galenic system, written by the Nestorian Christian physician and philosopher Ibn Butlan (d. 1063). Ibn Butlan was born and educated in Baghdad but traveled extensively into localities that are today in Iraq, Syria, Egypt, Israel, and Turkey. The *Taqwim* was translated into Latin in the 12th century and became known as the *Tacuinum Sanitatis*. In the last decades of the 14th century, lavishly illustrated versions of the *Tacuinum Sanitatis* were commissioned by northern Italian nobility. These expensive tomes portray a utopian feudal society in which nobles are engaged in play and romance whilst feudal laborers work the estate. Rich in horticultural imagery, they include vivid scenes of the harvest of vegetables, fruits, fl owers, and culinary and medicinal herbs. Each scene is accompanied by a brief summary of the health aspects of the subject. These documents connect vegetables and fruits with human health and well-being, similar to modern medicine and nutrition. The presentday focus on the connection between horticulture and health in the West and in the East can be seen as an extension of ancient and medieval regimens for a healthy lifestyle.

INTRODUCTION

Medieval medicine in the Western world was based on the ancient philosophical concepts of Greek science, particularly Hippocrates and Galen. Ancient Greek medicine may have been influenced by Ayurvedic medicine, the tradition of holistic medicine that originated in India, and both systems are based on related humoral traditions. Both of these medical systems developed a strong nexus with nutrition, much of it based on vegetables, fruits and herbs, and developed a rationale for diet as a basis for human health and well-being. The relationships between plants and health have been and continue to be of great concern for humankind based on both diet and medicinal uses. In this paper, we analyze 12 images and associated text of an illustrated manuscript from the late Middle Ages known as *Tacuinum Sanitatis* and located in the Austrian National Library (Paris *et al.*, 2009; Daunay *et al.*, 2009). This historical document provides information on the interrelationship of horticulture and health, as understood at that time. We further compare modern and medieval feelings about the relation between horticultural plants and health.

ANCIENT MEDICINE AND HEALTH

The prehistoric discovery that certain plants cause harm and others have curative powers is the origin of the healing professions and its practitioners, as well as the profession, science, and art devoted to plants. The earliest medico- botanical treatises of the West date to antiquity (Janick, 2003). In the Middle East a Sumerian tablet from about 2100 BCE (before current era) contains a dozen prescriptions and proscribes plant sources. *The Ebers Papyrus*, a medical treatise from ancient Egypt dates to 1550 BCE but contains earlier material.

In Greece, the great botanical treatise *Enquiry into Plants* of Theophrastus, written in the 5th century BCE devotes book IX to the medicinal value of herbs. The Greek philosopher Hippocrates (460–377 BCE), now considered the Father of Medicine, proposed that health was based on a balance of humours — three observable entities, phlegm, blood, and yellow bile, and a theoretic entity, black bile. The concept of the humours was based on observational experience related to observation of symptoms (Bates, 1995). For example, phlegm was a general term for bodily fluids such as cerebrospinal fluid, saliva, nasal mucus, gastric juices, semen, and menstrual blood. The concept of humoralism was that their disequilibrium resulted in pain or disease. The Greeks associated significance to the number four. Thus each of the four humours was associated with four temperaments or *complexion*: the phlegmatic, the sanguine, the choleric, and the melancholic. The humours also correlated to four qualities (hot, dry, cold, wet) each varying in four degrees of intensity, as well as to the four seasons, and the four ages of man— infancy, youth, adulthood and old age (Fig. 1). The medical concepts of Hippocrates were further developed by Galen (129–200 CE) who proposed six causes (called "non-naturals" in the Latin translations) that influence health: food and drink, ambient air, movement and rest, sleep and wakefulness, elimination and retention, and psychological states. These concepts, adopted, elaborated upon, and further developed by Byzantine and Arab physicians dominated medicine until the 18th century. For a sick patient, the physician's role was to diagnose the elements responsible for the loss of balance and design a treatment with elements having opposite characteristics to those of the defecting humour, according to the compensation principle *contraria contrariis* (Ullmann, 1978).

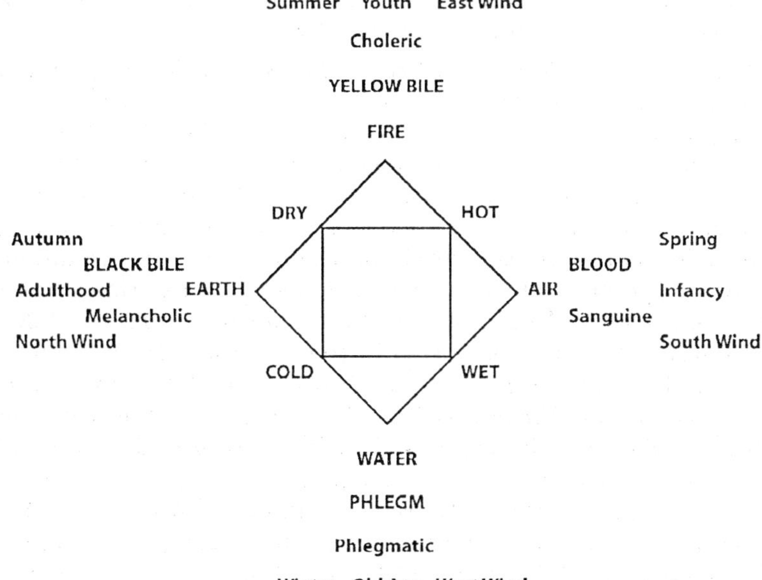

Fig. 1. Cosmology of ancient medicine based on Hippocrates and Galen (Source: Bovey, 2005).

Hippocrates had indicated that diet influenced health and recommended moderation. Food was incorporated into the humoral tradition and each food item was associated with one of the four humours as well as with the four qualities (hot, cold, wet, and dry). Remnants of this classification persist as we consider hot peppers, cool cucumbers, and dry wines. Foods associated with one given humour were assumed to balance the defecting humour responsible for sickness.

In the East there are two main medical traditions, Indian medicine known as Ayurveda, "the science of life" and Chinese medicine. Ayurveda is based on Sanskrit texts dated between 200 BCE and 200 CE which link prevention and healing to spirituality and interpret health as the result of external influences such as lifestyle and diet, and internal influences such as the bodily humours (Trawick, 1995). Disease is seen as disequilibrium and the use of herbs contributes to the restoration of the patient's balance.

Intriguingly, ayurvedic medicine of India shares the concept of elements and humours with Greek medicine. However, ayurveda considered five rather than four basic elements [*aakash* (ether or vacuum), *vayu* (air or wind), *aapa* (water), *pita* (fire), and *prithvi* (earth)] and three humors (*doshas*): *vata* (corresponding to breath), *pitta* (corresponding to bile) and *kapha* (correspon-ding to phlegm). Clearly, the concepts of Greek medicine were somehow transplanted into Indian medicine or vice versa.

In China, the traditional medicine includes also herbal and dietary therapy. The Pen T'Sao Ching, assumed to be authored by the legendary

Emperor Shen Nung in 2700 BCE, but probably written in the first century, contains about 100 herbal remedies. Humoralism was not a part of the third great medical cosmology, Chinese medicine, which developed independently of both Greek and Indian influences (Bates, 1995).

The Greek herbal *Peri Ylis Iatrikis* (*De Materia Medica* in Latin; *Of Medical Matters* in English) written by Pedanios Dioscorides of Anazarba, a Roman army physician, in the year 65, listed health giving properties of over 500 plants, many of which were to become horticultural crops (Beck, 2005). This medical treatise, the most famous ever written, was slavishly referred to, copied, and commented on for 1500 years. Compendia focused on plants, their properties and virtues, based on the Dioscoridean tradition, were referred to as herbals in Medieval and Renaissance times, and were invaluable resources for the physician and apothecary (Janick, 2003). The great epoch of printed herbals started in the late 15th century, and some of the most notable are: *Das Buch zu Distillieren*, 1500, by Hieronymus Brunschwig; *Herbarum Vivae Eicones*, 1530, 1532, 1536, by Otto Brunfels; *Kreüter Buch*, 1546, by Hieronymus Bock; *De Historia Stirpium*, 1542, of Leonhart Fuchs; *New Herbal*, 1551, 1562, 1568, by William Turner; *Commentarii*, 1544, by Pier Andrea Matthioli; *Crôÿdeboeck*, 1554, by Rembert Dodoens; and the *Herball*, 1597, by John Gerard. It culminated in the great compendia of Bauhin (1651), and Chabrey (1666). Botany and medicine were essentially in step until the 18tth century when both arts turned scientific and, from this juncture, botanical works would essentially ignore medicinal uses while medical works were devoid of plant lore. Yet, the medicinal use of plants continues as an alternate form of medicine and remains popular to the present day despite the questionable efficacy of many popular herbs and the reliance of many herbal recommendations on superstition and astrology. The fact that most drugs were originally plant-based has encouraged a new look at the medicinal properties of plants from traditional medicine.

MEDIEVAL MEDICINE: THE TACUIMUM SANITATIS

Historical

The course of medieval medicine as it relates to horticulture can be followed in an influential Arabic treatise entitled the *Taqwim al-Sihha bi al-Ashab al-Sitta* (*Rectifying Health by Six Causes*) authored by a Baghdad-born Christian physician and philosopher Ibn Butlan who died in 1063 (Fig. 2). The *Taqwim* was a synthesis of the long tradition of ancient medical science and essentially a guide for healthy living. Widely popular in the Arabic world, it stressed preventive medicine. A French translation (Elkhadem, 1990) exists of the original Arabic text. The *Taqwim* summarized information concerning food, and especially vegetables, fruits and condimentary herbs, in a tabular form that was easy to understand and use. It emphasized that a balanced life style was the prerequisite to prevent disorders and diseases. Ibn Butlan coordinated the doctrine of the six "non-naturals" by detailing 280 health-related items, their name, their nature according to the four elements, the degree, the best variety, the usefulness, the noxious effects and contra-indications, the remedies to the noxiousness, the humours produced, the proper temperaments, the season, the age, the country, the authorities cited and additional notes.

Fig. 2. Ibn Butlan from the Vienna *Tacuinum Sanitatis* folio 4r.

A Latin translation of the *Taqwim* commissioned by the Court of Sicily was completed in 1266, and became known as the *Tacuinum Sanitatis*. One hundred years later, the first illuminated copies of the *Tacuinum Sanitatis* were commissioned by Giangaleazzo Visconti, the Count of Milan in the Po Valley of northern Italy (Hoeniger, 2006), as expensive gift volumes. They portray an utopian feudal society in which nobles are engaged in play and romance whilst feudal laborers work the estate (Cogliati Arano, 1976; Witthoft, 1978; Opsomer-Halleux, 1991; Segre Rutz, 2002; Bertiz, 2003; Bovey, 2005; Hoeniger, 2006; Mane, 2006). Rich in horticultural imagery, they include vivid scenes of the harvest of vegetables, fruits, flowers, and culinary and medicinal herbs. Each scene is accompanied by a brief summary of the health aspects of the subject that derives from the *Taiwan* of Ibn Butlan. These medieval handbooks on health and wellness connect horticultural plants as well as the garden itself with human health (Bertiz, 2003) and were intended for the elite and intelligentsia of their epoch.

A dozen *Tacuinum Sanitatis* manuscripts have survived, of which six have full-page illustrations and are considered archetypical (Paris *et al.,* 2009). They are located in libraries in Austria, Belgium, France, Italy, and Spain. Four of them can be viewed on-line (mandragore.bnf.fr/html/accueil.html, bibliotheque.rouen.fr and www. casanatense.it) and a number of facsimile editions copies provide a rich source of information on agriculture and horticulture of the late medieval period, containing vivid images of plants being harvested in fields and gardens.

The images of these Tacuinum Sanitatis recessions include plants of the Taqwim supplemented with those grown in northern Italy during the late 14th century. The plant images depict harvest, emphasizing when the horticultural product has reached the proper stage for culinary use. The manuscript (Cod. Ser. N. 2644) of the Osterreichische Nationalbibliothek, Vienna contains the most accurate depictions and is the source of all plant images presented here. It includes about 26 vegetables, 33 fruits, 3 flowers, 21 culinary and medicinal herbs, and 1 mushroom (truffles) in addition to 9 cereals. Although botanical inaccuracies and lack of detail obscure precise identification in some cases, and although the images in many cases are idealized, they, overall, are of far better quality than those of most medieval manuscripts. In this paper we present illustrations of four vegetables, four fruits, and four culinary herbs along with the associated health information provided in the Latin text of the Vienna manuscript.

Vegetable Images

Cucumber: Native to India and as part of ayurvedic medicine, the first cucumber reached Europe in the medieval period (Janick et al., 2007; Paris *et al.,* 2009). In Fig. 3A, a well-dressed couple examines yellow cylindrical fruit from a viney plant. Identification of the plant is based on the tuberculate (warty) surface of the fruit, a diagnostic feature of cucumber. The bipartite label *Cucumeres et citruli* can be traced back to Table 10, line 66 of the *Taqwim*, the Arabic being *al-qitha* (chate melons) *wa* (and) *al-khiyar* (cucumbers) indicating similarity of culinary usage of these two cucurbits. Although the illustration shows mature fruits that have turned consume. The text indicates that these fruits reduce burning fevers as a result of their cold moist nature, are diuretic, and produce watery blood, but cause stomach aches.

Eggplant: Domesticated in the Indo-Burma, eggplant has been an important crop in India as early as the 3rd century BCE (Daunay and Janick, 2007) and has also a place in ayurvedic medicine. It was domesticated in China by the 1st century BCE (Wang *et al.,* 2008) and reached Europe in the medieval period (Daunay and Janick, 2007). In Fig. 3B, a dramatic scene labelled *Melongiana* displays a fondling, amorous couple in front of a row of eggplants while being severely admonished by an elegantly gowned lady. The illustration clearly suggests that eggplant had aphrodisiac properties. The plants bear a prolific crop of globular, purple fruits, that appear similar to the present-day 'Black Beauty' market type. The undulate leaf laminae are depicted accurately, but the attractive purple flowers are absent. Ibn Butlan wrote that the young, non-bitter, medium-sized Persian fruits are best, and are beneficial to blood vessels and for stomach weakness (Elkhadem, 1990).

Fig. 3. Vegetables from the Vienna *Tacuinum Sanitatis*: (A) cucumber (*Cucumis sativus*) folio 23v; (B) eggplant/aubergine (*Solanum melongena*) folio 31v; (C) onion (*Allium cepa*) folio 25v; (D) chickpea (*Cicer arietinum*) folio 49r.

Onion: Native to western Asia, onions were known in Egypt as early as 3200 BCE and were considered medicinal plants in India in the sixth century BCE (Simmonds, 1976). In Fig. 3C, a labourer standing in a field of onions with one hand outstretched expecting payment offers a bunch of white bulbs to a middle aged lady whose head is covered suggesting she is married and who already holds a bunch of purple bulbs. The illustration is labelled *Cepe.* The onions are neatly planted in rows in a garden which also contains two trees, one of which is a cherry. The text informs that onions were considered to be a diuretic, sharpen eyesight, promote coitus, generate milk in nursing mothers and semen in men, but cause headaches.

Chickpea: Native to Western Asia, progenitors of the present day chickpea have been known in India from 2000 BCE (Simmonds, 1976). Chickpea is now a traditional dryland crop in the Mideast. In Fig. 3D a well dressed woman in a red gown protected by an apron is sampling pods labelled *Cicera* in preparation for harvesting to a basket in the foreground. The crop appears spaced planted. The text indicates that chickpeas are warm and humid to the first degree and the best are large seeded. Chickpeas were supposed to increase sperm but to be noxious for the kidney and bladder.

Fruit Images

Grape: Old World species of *Vitis* are indigenous to the south Caspian belt, Turkey and the Balkans and were widely distributed to the Mediterranean basin (Janick, 2005) and were very early used to make wine. Grapes have long been a component of ayurvedic medicine as *Darakchasava* where a major ingredient is made from raisins (Paul *et al.,* 1999). Grape is illustrated three times in this *Tacuinum* manuscript, either listed as *Uve*, or for the production of verjuice (juice of immature grapes used in ancient and medieval cuisine), or as a symbol of autumn, the time of wine making. Depending on the images, the vines grow on trees or are trellised, and bear black, purple, light red or gold grapes. In Fig. 4A, a man in red tights is getting ready to cut a cluster of purple grapes with a knife, while a young woman in aproned blue dress has just handed a basket of harvested grapes to a third men in a large barrel who is stomping them to extract their juice being collected in a wooden container below that will be used to prepare wine. Grapes were considered to cause thirst and cleanse the intestine.

Apple: The large, sweet-smelling domestic apple originates in Kazakhstan, migrated to Europe in Antiquity and was widely planted in the Roman era (Janick, 2005). In Fig 4B labeled *mala acetosa* (sour apple), a courtier brings down large apples, with a stick while an elegant lady carrying a basket watches in amazement. The best sour apples are very juicy, alleviate fainting and hepatitis but adversely affect joint articulation. Sweet apple (*mala dulcis*), shown in another illustration, strengthens the heart.

Cherry: The origin of sweet and tart cherries is central Europe and areas surrounding the Black Sea, with sweet cherries occurring as far east as central Russia (Janick, 2005). In Fig. 4C, tart cherries, labelled *cerosa acetosa*, are being harvested by a child who has climbed into the tree. A well-dressed lady opens her gown to catch fallen fruits and another elegant lady carries off two full baskets balanced on a pole. The very sour ones were considered best. The fruits were considered to cure bilious attack and dry out and settle upset stomachs. *Cerosa dulcia*, sweet cherry, which softens the stomach, is displayed in a separate image.

Fig: The common fig is a classic fruit of Mediterranean climates and signs of cultivation are found in Neolithic sites (Zohary and Spiegel-Roy, 1975). Native to Western Asia, figs were early introduced to India where they have become a major fruit. In Fig. 4D, a couple harvest figs (*Fichus*); the young man in the tree throws down fruit to a lady companion in a red gown who catches the fruit in her skirt as she daringly shows her white stockinged legs. The text indicates that figs are warm and humid in the second degree and that white peeled kinds are optimum, figs cleanse the kidneys but inflates and fattens.

Fig. 4. Its from the Vienna *Tacuinum Sanitatis*: (A) grape (*Vitis vinifera*) folio 54v; (B) apple (*Malus* ×*domestica*) folio 9r; (C) tart cherry (*Prunus cerasus*) folio 12r; (D) Fig (*Ficus carica*) folio 4v.

Herb Images

Sage: This *aromatic* spice is found originating in southern Europe and Asia Minor (Simon *et al.*, 1984). In Fig. 5A, two well dressed ladies are shown picking a plant labelled *Salvia* growing in an enormous basket. Domestic sage was considered best and was good for paralysis and for the nerves, although slow to be digested.

Fig. 5. Culinary herbs from the Vienna *Tacuinum Sanitatis*: (A) sage (*Salvia officinalis*) folio 37v; (B) marjoram (*Origanum marjorana*) folio 33v; (C) dill (*Anethum graveolens*) folio 32r; (D) saffron (*Crocus sativus*) folio 40v.

Marjoram: This is a spice originating in the North Africa and Southwest Asia (Simon *et al.*, 1984). Two elegantly gowned women tend marjoram (*Maiorana*) growing in a vase resting on a bench (Fig. 5B).Very small, aromatic marjoram was considered optimum, good for cold and moist stomachs, and to purify the blood. No noxious effects are described.

Dill: Native to southwest Asia and India (Brickell and Cathey, 1996), this herb is well- known in ayurvedic medicine. The image in Fig. 5C, labelled *Aneti,* shows a woman and a child harvesting dill, in a weedy setting. The herb was considered best when green, fresh, and tender and was recommended for cold and flatulent stomachs but was harmful to the kidneys and nausea-provoking.

Saffron: The saffron crocus, an ancient perennial herb, is known only in cultivation (Simon *et al.*, 1984). Originating in southwest Asia and perhaps native to Crete, the saffron crocus has long been cultivated in India. It is one of the most expensive spices on a weight basis; it takes 210 thousand stigmas from 70 thousand flowers to make one pound of true saffron. The dried red three branched stigmas are used in cooking as a colorant and seasoning. In Fig. 5D a blue gowned woman carrying a basket harvests flowers in a basket from saffron crocus (*Crochus*) neatly planted in rows. The text advises that saffron is hot in the second degree and dry in the first degree. The best is freshly cut, strong in color, and with a delicate scent. Saffron helps hearing but may induce vomiting.

DISCUSSION AND CONCLUSION

The horticultural crops illustrated in the *Tacuinum* manuscripts are familiar to us and have changed hardly at all in their appearance over the centuries. The garden scenes are not basically different from traditional horticulture practiced today in many parts of the world. Vegetables are shown harvested mostly in gardens and fields from dense plantings while fruit trees are pictured alone, in order to show as much detail as possible. Aromatics such as sage and marjoram are generally depicted as growing in tended conditions (beds or pots), indicating that these plants were prized. Peasants and laborers are frequently pictured with the root and leafy vegetables, which were their basic food (Mane, 2006). Conversely, courtiers are more frequently found next to fruit vegetables such as cucumber and eggplant, of relatively recent introduction in Europe (Daunay *et al.*, 2008; Paris *et al.,* 2009) which were probably considered fit for the elite. The upper class is also pictured next to fruit trees, produce reserved first for the privileged (Mane, 2006). The medieval idea that the garden was a place for healing, relaxing, and physical and mental well-being was a premonition of modern horticultural therapy.

Interestingly, all of the horticultural crops presented in the illustrated *Tacuinum* manuscripts were allocated physiological effects on the body and were fully part of the Western medieval pharmacoepia. Clearly, then as now, the population was concerned with general health and bodily functions, including flatulence, bladder and kidney stones, ill-health, and sexuality.

Remarkably, the very short text that accompanies each illustration in the *Tacuinum* provides information for balancing the beneficial and noxious effects of each plant.

All crops but two (sage and marjoram) discussed here, are also found in the Ayurvedic materia medica as described in the *Madanapala's Nighantu* written in 1374 (Dash, 2008), a work contemporary to the Vienna *Tacuinum Sanitatis*. Their properties are described in terms which differ somewhat from those found in the *Tacuinum Sanitatis* and include taste, attribute, potency, specific action, and therapeutic usage. In most cases, the medicinal properties described in the Indian pharmacopoeia differ from those described in the *Tacuinum*.

Although the basis of modern medicine has completely changed from the ancient European and Asian philosophical concepts that date to antiquity, the results are amazingly the same. All cultures, ancient, medieval, and modern have come to the same conclusion: horticulture is basic to a healthy life. Although modern knowledge of medicine and nutrition has changed our understanding of food through advances in biology, chemistry, and genetics, and as we now view foods in terms of calories, vitamins, antioxidants, polyphenols, anthocyanins, polysaccharides, proteins, and lipids (Goldman, 2003), it does appear as if all the analyses, ancient and modern, come to the same result: the horticultural crops we eat are sustaining, nutritious, healthy, and delicious. Although we no longer evaluate them on the basis of hot, cold, wet, or dry, or the effect on the "humours," yet some of the ancient feelings still persist. We still say colloquially "cool as a cucumber," we refer to pungent chili peppers as "hot," we associate spicy foods with passionate temperaments, and consider aroma and fragrance to affect our well-being. The classic cures of bloodletting and emetics (vomiting) have been discarded but purging is still being promoted by some under the guise of "clean colon." At times it may seem, as we investigate the healthful attributes of our food through chemistry, that we merely seek accreditation for our preferences. While we smile indulgently at some extravagant claims of the ancients, it appears that in a number of cases their experience was prescient. For example, it turns out that horseradish does have a factor that affects urinary health (Shehata *et al.,* 2008), and we have confirmed the health giving properties of the alliums (Desjardins, 2008) and crucifers (Monteiro and Rosa, 2008). We have also found that horticultural crops need to be considered carefully for detrimental effects as in the case of allergies (5% of the population is allergic to apples), excessive soluble fibers (such as persimmon), and toxic substances such as solanine in potato.

It is clear that the present-day emphasis on the connection between horticulture and health is an extension of ancient and medieval concerns. We suggest that our present understanding of nutrition will probably be considered naïve in the coming centuries and in the future the approach for attributing health benefits to horticultural crops will likely be very different from the present one. We predict however that the conclusions will be similar: horticultural crops are important for good health!

REFERENCES

Bates, D. (ed.). 1995. *Knowledge and the scholarly medical traditions.* Cambridge Univ. Press, Cambridge.

Bauhin, J. 1651. Historia plantarum universalis. Yverdon.

Beck, L.Y. (transl.) 2005. *Pedanius Dioscorides of Anazarbus: De Materia Media.* Olms- Weidmann, Hildesheim, Germany.

Bertiz, A.A. 2003. Picturing health: The garden and courtiers at play in the late fourteenth-century illuminated Tacuinum Sanitatis. Ph.D. diss., University of Southern California, UMI 3116666.

Bovey, A. 2005. Tacuinum Sanitatis: An early renaissance guide to health. Paul Holberton, London.

Brickell, C., and Cathey, H.M. 2008. *A–Z encyclopedia of garden plants.* DK Publishing, New York.

Chabrey, D. 1666. Stirpium sciagraphia et icons. Génève.

Cogliati Arano, L. 1976. *The medieval health handbook Tacuinum Sanitatis.* G. Braziller, New York.

Dash, V.B. 2008. *Materia medica of Ayurveda, based on Madanapala's Nighantu.* Reprint ed. B. Jain Publ. (P) Ltd., New Delhi.

Daunay, M.C. and J. Janick. 2007. History and iconography of eggplant. *Chronica Hort.* **47**(3):16–22.

Daunay, M.C., Janick, J. and Paris, H.S. 2009. *Tacuinum Sanitatis*: Horticulture and Health in the Late Middle Ages. *Chronica Hort.,* **49**(3):22–29.

Daunay, M.C., Laterrot, H. and Janick, J. 2008. Iconography and history of Solanaceae: Antiquity to the XVIIth century. *Hort. Rev.,* **34**:1–111 + 8 plates.

Desjardins, Y. 2008. Onion as a nutraceutical and functional food. *Chronica Hort.,* **48**(2):8–14.

Elkhadem, H. 1990. Le Taqwim al Sihha (Tacuini Sanitatis) d'Ibn Butlan: un traité médical du XIe siècle. Peeters, Louvain.

Goldman, I.L. 2003. Recognition of fruits and vegetables as healthful: vitamins, minerals, fiber, and phytonutrients. *HortTechnology,* **13**:252-258

Hoeniger, C. 2006. The illuminated Tacuinum Sanitatis manuscripts from northern Italy ca. 1380–1400: Sources, patrons, and the creation of a new pictorial genre. p. 51–81. In: J.A. Givens, K.M. Reeds and A. Touwaide (eds.), *Visualizing medieval medicine and natural history, 1200–1550.* Aldershot, Ashgate, UK.

Janick, J. 2003. Herbals: The connection between horticulture and medicine. *HortTechnology,* **13**:229–238.

Janick, J. 2005. The origins of fruits, fruit growing, and fruit breeding. *Plant Breed. Rev.,* **25**:255–320.

Janick, J., Paris, H.S. and Parrish, D.C. 2007. The cucurbits of Mediterranean antiquity: identification of taxa from ancient images and descriptions. *Ann. Bot.,* **100**:1441– 1457.

King, L.S. 1978. *The philosophy of medicine: The early eighteenth century.* Harvard Univ. Press, Cambridge, MA. p. 189–183.

Mane, P. 2006. *Le travail à la campagne au Moyen Âge*. Étude iconographique. Paris: Picard.

Monteiro, A.A. and E. Rosa 2008. Caldo verde: A story of Portuguese brassicas. *Chronica Hort.*, **48**(1):9–11.

Opsomer-Halleux, C. 1991. *L'art de vivre en santé, images et recettes du Moyen Âge: le Tacuinum sanitatis*, manuscript 1041 de la Bibliothèque de L'Université de Liège. Liège: Ed. Du Perron.

Paris, H.S., Daunay, M.C. and Janick, J. 2009. The Cucurbitaceae and Solanaceae illustrated in medieval manuscripts known as the *Tacuinum Sanitatis*. *Ann. Bot.*, **103**: 1187–2005.

Paul B., Masih, I. Deopujari, J. and Charpentier, C. 1999. Occurrence of resveratrol and pterostilbene in age-old darakchasava, an ayurvedic medicine from India. *J. Ethnopharmacology*, **68**(1–3):71–76.

Segre Rutz, V. 2002. Historia plantarum, erbe, oro e medicina nei codici medievali, volume di commento. Modena: Franco Cosimo Panini. p. 123–170.

Shehata, A., R.M.S. Mujlwa, M. Babadoost, M. Uchanski, Norton, M.A. Skirvin, R. and Walters, S.A. 2008. Horseradish, botany, horticulture, breeding. *Hort. Rev.*, **35**:221–261.

Simmonds, N.W. (ed.) 1976. *Evolution of crop plants*. Longman, New York.

Simon, J.E, Chadwick, A.F, and Craker, L.E. 1984. *Herbs: An indexed bibliography*. Archon Books, Hamden, Connecticut.

Trawick, M. 1995. Writing the body and ruling the land: Western reflections on Chinese and Indian medicine. p. 279–296. In: D. Bates (ed.), *Knowledge and the scholarly medical traditions*. Cambridge Univ. Press, Cambridge.

Ullmann, M. 1978. *Islamic medicine (Islamic Surveys II)*. Edinburgh University Press, Edinburgh.

Wang, J.X., Gao, T.G. and Knapp, S. 2008. Ancient Chinese literature reveals pathways of eggplant domestication. *Ann. Bot.*, **102**: 891–897.

Witthoft, B. 1978. The Tacuinum Sanitatis: A Lombard panorama. *Gesta*, **17**:49–60.

Zohary, D. and Spiegel-Roy, P. 1975. Beginning of fruit growing in the Old World. *Science*, **187**: 319–327.

II
Studies on Food, Nutrition and Health Security and Income Generation

5. URBAN AGRICULTURE FOR FOOD SECURITY

Abdul Kobakiwal

Chief, Special Programmes for Food Security (SPFS), Food and Agriculture Organization of the United Nations (FAO), Viale delle Terme di Caracalla, Rome – 00100, Italy

ABSTRACT

The paper provides an insight of FAO's Special Programme for Food Security (SPFS), its scope and strategy in support of National Food Security Initiatives (NPFS). The Special Programme for Food Security helps Governm-ents replicate successful food security practices on a national scale. It encou-rages investment in rural infrastructure, off-farm income generation, urban agriculture and safety nets. Since it inception in 1994 US$770 million from donors and national governments have been invested in FAO-designed food security programmes. Over 100 countries have engaged in the SPFS and nearly half had begun moving to National Programmes, and others are preparing to do so.

The paper illustrates practical achievements in selected countries and analyses how horticulture based activities have been instrumental to making a difference in the poor peoples' food and nutrition security and livelihood. An attempt is made to assess lessons learnt and draw conclusions for the way forward in support of horticulture based interventions to alleviate poverty and improve food and nutrition security as a contribution to the Millennium Development Goals (MDGs) and more specifically MDG 1 and 7 related to poverty reduction, food security, and environmental sustainability.

INTRODUCTION AND STATUS OF THE FAO'S SPFS

In 1994, FAO launched the Special Programme for Food Security (SPFS) pilot phase to assist countries in promoting and disseminating simple, low cost technologies that improved the yields and income of poor farming households. The pilot phase focused attention on four components: small scale water control; intensification of production; diversification of production; and analysis of the socio-economic constraints to production and productivity. Urban and peri-urban agriculture (UPA) was one of the important elements of the pilot activities in many countries. UPA intervention produced good results in terms of providing supply and enhancing income of the most vulnerable and food insecure people living in city peripheries. School and hospital gardens were another area of intervention with this component of the SPFS pilot activities. This SPFS pilot phase lasted longer than anticipated – 14 rather than 3 or 4 years – but by the time it terminated in 2008 it had reached 106 countries and had achieved its objective in a number of them.

In 2001, governments that benefited from early SPFS support began transiting from pilot projects to large-scale, nationally-owned programmes for food security (NPFS). These programmes are formulated and implemented locally, with FAO playing mainly a facilitative role. FAO also offers facilitative support to regional economic integration organizations for the formulation and implementation of Regional Programmes for Food Security (RPFS) aimed at complementing and reinforcing national efforts to achieve food security by increasing smallholder productivity and improving nutrition. Globally, 17 NPFS and 4 RPFS are under implementation, 6 NPFS have been approved and are awaiting funding, 18 NPFS and 8 RPFS have been formulated and are under review, and 14 NPFS and 2 RPFS are under formulation. All NPFS share the common objective of eradicating hunger by improving smallholder productivity and rural livelihoods, but each is designed to respond to local needs and priorities. UPA has become an important activity within most of the NPFS.

The South-South Cooperation initiative, launched by FAO in 2006, has been the main source of outside technical expertise to the SPFS and continues to provide cost-effective expertise to the NPFS and RPFS. The initiative has provided recipient countries with the opportunity to have access to specialist experts from other developing countries in areas such as water control, crop production, livestock, aquaculture, UPA and agro-processing, with the possibility of posting them directly to beneficiary communities where they demonstrate improved technologies in farmers' fields. Strategic alliances with SSC providers, notably China, Indonesia and Argentina, offer the possibility of fielding large numbers of experts and technicians at a relatively low cost. By mid-2009, a total of 1 500 SSC experts and technicians had been fielded.

THE URBANIZATION OF POVERTY

[...] "urban poverty tends to be fuelled by people migrating towards the cities in an attempt to escape the deprivations associated with rural livelihoods. Partly due to the rural decline, the world is urbanizing at a fast pace and it will not be long before a greater part of developing country populations is living in large cities. Therefore, urban food security and its related problems should also be placed high on the agenda in the years to come".

——— *Jacques Diouf, Director General, FAO.*
The State of Food Insecurity in the World 2006, FAO

The world urban population is expected to double in the next 30 years, leading to a growing number of urban poor. According to Habitat, urban population expansion will be more pronounced in developing countries as a result of high birth rates and immigration from rural areas, as people flock to cities in search of food, employment and security. The information available from Habitat suggests that the increase in population will materialize as an increase in urban slum areas, with high levels of unemployment, food insecurity and malnutrition. It is expected that by 2020, 85% of the poor in Latin America, and about 40–45% of the poor in Africa and Asia will be concentrated in towns and cities. This situation is of great concern to central and local government authorities, not only in terms of food requirements but also for the social and security implications.

Already, in 1999, the FAO Committee on Agriculture (COAG) drew the attention of the international community to the growing food requirements in urban areas. Since then, FAO has multiplied its interventions in support of urban and peri-urban agriculture as an integral part of agricultural production systems, and has acknowledged its role in feeding the cities, creating employment and generating income for the urban poor.

Ten years have past during which a better understanding has been gained of the different assets of UPA, its constraints and opportunities and how it can and does effectively contribute to improving the livelihood of the resource-poor segment of the urban population.

In 1960, two thirds of the world population was rural and one third was urban. In 2008, half of the world population is rural and half is urban. In 2050, one third of the world population will be rural and two thirds will be urban. FAO data indicate that, as a result of the food crisis, world hunger is projected to reach a historic high in 2009, with 1 020 billion people going hungry every day (...). The global recession of 2008/09 and soaring prices have severely affected the urban poor.

Unfortunately, the food dimension of poverty in urban areas has not yet been translated into sufficient policy action in many countries.

FAO recommends that the opportunities for local food production within the city boundaries and its immediate surroundings within a broader approach of rural-urban linkages be given due attention as a component of the countrywide food security strategy.

"By 2020, the developing countries from Africa, Asia and Latin America will be home to 75 percent of all urban dwellers... renewed emphasis will have to be given to urban and peri-urban agriculture practices within and around the cities, which compete for land, water, energy and labour with the objective of contributing to the requirements of the urban population through horticulture, livestock, fodder and milk production, aquaculture and forestry".

——— *Dr. Jacques Diouf,*
Director-General FAO, World Food Day
16 October 2009

A key challenge for central and local governments is to integrate UPA into policy and strategies and have it formally adopted in national programmes within a broader framework of urban development planning, land and water-use management, urban waste recycling and product safety assurance.

Urban consumers generally rely on purchased foods, mainly coming from rural areas or being imported into the country. However, without financial resources, many urban poor have no alternative than to turn to urban and peri-urban agriculture activities as the only survival and livelihood option. Support in terms of access to land and quality water is the essential prerequisite for integrating UPA in the city development plan. In addition, technical guidance and supervision, training and consumer education are equally essential to ensure productivity and best practices, including risk management linked to the unconsidered use of untreated waste water for irrigation and agrochemicals for pest and disease control.

OPPORTUNITIES, CONSTRAINTS AND SUSTAINABILITY

Urban and peri-urban agriculture generally refers to food production systems that occur within or in the surroundings of cities and effectively contribute to the food access and supply while creating job and income opportunities for the poorer section of the population. It entails an important environmental protection dimension with adequate watershed management to preserve productive capacity.

Constraints relate to land tenure issues and competition for water, but also to the potential risk for consumers and the environment through overuse and misuse of pesticides and unconsidered use of treated and non-treated wastewater for irrigation. However these risk factors are largely offset by opportunities at the different levels of the value chain from field to table.

A survey carried out in 25 countries in Africa suggests that horticulture-based activities are predominant in urban agriculture, while short-cycle livestock, dairy production, aquaculture and forestry are also practised according to the opportunities. Horticulture comprises the cultivation of high value crops, such as vegetables, fruits, roots and tuber crops, as well as herbs and spices, flowers, ornamental plants and mushrooms. The comparative advantage of growing and marketing horticulture crops against other crops or other agriculture activities is based on their efficiency in terms of land and water use, job creation and income generation.

Intensified horticulture units can offer from 10 to 40 remunerative jobs per hectare for smallholder producers. Considering also other derived jobs upstream and downstream related to the production-supply chain, it is generally admitted that urban horticulture and related services generate jobs at a ratio of 1 for every 50 to 100 citizens, depending on the level of consumption per caput and the productivity of the production systems. Various typologies for horticulture production can be implemented, that range from "allotment garden schemes" with organized growers' associations, to individual homestead and kitchen gardens and micro-gardens in densely populated areas where no more agricultural land is available for crop production.

Depending on the local context, consumer preferences and sanitary regulations, different types of livestock are commonly part of the city: goat, sheep, guinea pig, pigs, poultry and cattle (usually for dairy products). Other options relate to bee-keeping, floriculture, mushroom production and aquaculture. Urban forestry schemes have their place in UPA to serve as wood lots or as "lung space" for the city. General recommendations suggest that at least 30 percent of the urban space be kept "green" in order to contribute to the absorption of CO_2, the release of O_2, the cooling of the air temperature, for enhancing the rainwater infiltration rate and for erosion control. Further opportunities exist for building synergies with city urban waste management. Organic waste can be usefully converted into high quality organic fertilizer by mechanical composting and lombriculture. Composted organic urban waste contributes to solving the waste disposal problem and to improving sustainable soil fertility management. Similarly, treated wastewater can be recycled to irrigate fruit orchards, wood lots and ornamental plants, as well as fodder crops. Through these recycling processes, UPA will contribute to keeping the city clean and have the benefit of a renewable source of water and organic fertilizers thereby reducing the reliance on external inputs.

SPFS INTERVENTIONS AND SAMPLE SUCCESS STORIES

Globally, FAO through its SPFS is working with decision makers in member countries with the objective of strengthening national capacity for policy and strategy development related to urban and peri-urban agriculture and its integration in National Food Security Programmes.

In several countries NFSP have been adopted as the overall framework through which agriculture sector development projects and strategies are being implemented eventually with support from FAO's Technical Cooperation Programme (TCP), or in the context of Government Cooperation Programmes (GCP), Decentralised Cooperation (GDCP) or within Unilateral Trust Funds agreements (UTF), projects funded through emergency operations (OSRO) or with the support of investment projects funded by International Development Banks.

The SPFS launched a series of UPA initiatives for integration into NPFS;

- to urban and peri-urban horticulture production in several cities over the world: viz. Kinshasa, Lubumbashi, Libreville ; Buenos Aires, Cairo, Alexandria, Medellin, Caracas, La Paz, Windhoek, Rundu, Oshakati, Teresina, Kigali, Bujumbura; Banjul, Bissau, Abidjan, Conakry, Port au Prince, Kigali, Nouakchott and Dakar.
- to small urban and peri-urban livestock producers in DR Congo and Djibouti
- to access and use of quality irrigation water for UPA (Dakar, Luanda, Guatemala, Managua, Bissau, Sanaa)
- to urban and peri-urban forestry (Brazzaville, N'Djamena, Bangui, Nouakchott)

- In addition, support has been provided at regional or global level for project identification and formulation and to create awareness about UPA and provide guidelines for the organizations of small holder Urban and Peri-Urban agriculture producers.

List of FAO field projects with qualifiers: UPA and Food Security [1/]

Funding source	Number of projects	Amount in USD
SPFS	6	12,062,815.00
GCP and GDCP	16	27,899,007.00
TCP	18	5,195,614.00
UTF	20	182,327.00
OSRO	22	20,131,698.00
OTHER	5	13,639,863.00
TOTAL	87	79,111,324

1/ Source: FAO Field Programme Monitoring and Information System (FPMIS)

These projects have allowed demonstrating adapted technologies, production systems that have successively been adapted by the beneficiaries.

In Dakar, Senegal on the west coast of Africa, "Micro-garden technologies for improved food and nutrition security of the urban population" is being implemented. The program originated from a pilot SPFS project which successively benefited from donor support from Italy. It is an example of good practices in urban and peri-urban agriculture. Its objective aims at improving food security and livelihood of the urban poor in Dakar through the introduction of micro-garden technologies supported by training of beneficiaries in horticultural practices. The project has achieved remarkable results in sustaining urban production of vegetables and initiating market practices for the poor families of the city and its surroundings. Over 4000 households were involved in the first phase. Almost 65 per cent of them have established individual micro-gardens while 35 per cent are exploiting 150 Community Production Centers (CPC) in different areas of Dakar. Micro-gardens are small-scale gardens that can be accommodated in any little space available around a dwelling or on a balcony, patio or roof of a house or building. The standard micro-garden unit is a table of 1 m^2 assembled with wooden laths obtained from pallets or other second-hand material. Alternative containers, such as used truck and car tyres, bags, halved drums, baskets and old buckets, can also be usefully recycled for vegetable growing. The technology is specifically adapted for high-density city areas and slums where agricultural land is no longer accessible. With micro-gardens, the urban poor can harvest at home every day fresh, safe and highly nutritious vegetables to improve their livelihood. Micro-gardens have also proven to be excellent training tools in educational school gardens for creating awareness of the health benefits of consuming fruit and vegetables. Recently the programme was selected among 500 other candidates as the winner of the "Dubai Award for Best Practice to Improve the Living Environment (DIABP)" by UN Habitat in December 2008.

A prize of USD 30 000 has been assigned to support activities for the expansion and strengthening of the programme. In fact, further expansion will be sought in the second phase to focus on organizational sustainability of the micro-garden programme under the operational leadership of the Mayor of Dakar, and marketing of micro gardens products. Micro-garden technologies will be extended to other African countries (Gambia and Niger). It is also intended to present this experience as a flagship to the Universal Expo 2015 in Milan.

In "El Alto" a suburb of La Paz in Bolivia at 4 000 m, the altiplano of the Andes region of Latin America, the project "Micro Huertas Populares" is being implemented to improve the livelihood of the indigent families living in the municipality. El Alto is located at a high altitude (between 3 850 and 4 150 m) in the flat altiplano and is characterized by low temperatures, drying winds, strong solar radiation and frequent frost, limiting the range of open field crops to short season crops. Its current population is estimated at 1 million, of which 73% live in poverty and around 60 000 families have no adequate access to basic services such as education and health. Some 40 percent of children under five are malnourished, 20% severely, and there is an infant mortality rate of around 5 per cent.

Against this daunting background, the project started its activities in August 2003 with the aim of introducing micro-garden technologies that could be managed by individual poor families. Micro-gardens are incorporated as a component of the food security strategy through the increase of food availability in quantity and quality produced locally. In Bolivia, micro-gardens are cultivated in small-scale greenhouse units of 40 m^2 each by individual families, providing fresh vegetables all year round for family consumption or for neighbourhood marketing.

The initiative has been taken up by the municipality of El Alto and the Bolivian Government with assistance from FAO and financial support provided by the Belgian Government (GCP/BOL/035/BEL). Project interventions consist of organizing a programme of small-scale production of fresh vegetables, fruits, herbs and condiments to improve food diversity and generate new incomes through the marketing of production surplus. Families are trained to apply organic cultivation as well as innovative growing techniques such as soilless culture to make the best use of limited areas where no good soil is available. Input distribution units and joint marketing initiatives have been successfully organized and are operated by the farming families themselves. More then 500 families have to date benefited directly from the project and an additional 400 families with funding raised form WFP, EU, UNESCO and INTERVIDA, a Spanish NGO.

Some countries, such as the Democratic Republic of the Congo, in the tropical zone of Africa, have duly acknowledged the importance of urban horticulture; with the help of FAO project GCP/DRC/028/BEL, a nationwide UPA programme has been implemented to cover several cities.

A city like Kinshasa, the capital of the Democratic Republic of the Congo (DRC) with a population of eight million, needs daily some 400 tons of fruit and vegetables for a low consumption level of only 50 g per caput per day (12.5 percent of the WHO 400 g recommended daily intake). It is estimated that the potential employment ratio is 1 to 100 inhabitants, representing an occupation with income for 80 000 people. From a small plot of 100 to 250 m^2, people get a net income of up to USD 200 per month, which exceeds the wage of a public employee. However, this urban and peri-urban horticulture is at risk if no adequate planning and conservation measures are put in place. Land and water are key resources that are in fierce competition from other uses. Uncontrolled application of pesticides and organic waste may result in soil and water table pollution. Following a request from the DRC Government, FAO is providing assistance in order to preserve and structure UPA for long-term sustainability. With the help of the project, 900 ha have so far been legally protected and documented for UPA activities. Infrastructure and equipment have been put in place to secure safe irrigation and drainage for 1 500 ha allowing for 45 000 direct and indirect jobs to the benefit of 225 000 people. A microcredit scheme is operational and has to date empowered 15 500 farmers and entrepreneurs.

A total of 255 farmers' organizations have benefited from training and technical advice. The institutional context has been strengthened with the creation of a "National Urban Horticulture Service" (SENAHUP) with decentralized technical offices at the level of the different cities and municipalities. UPA is a key component of the NPFS and is part of the new agriculture development plan. Market collection and sales points have been constructed in 12 neighborhoods. Homestead gardening has been introduced in 25 schools for educational scope. Training has been provided for 9 500 students and 80 school teachers. The farmers' field school has been adopted as a training and extension methodology to ensure the disseminating and adoption of Good Agriculture Practices, product safety and environment preservation

LESSONS LEARNED AND THE WAY FORWARD FOR FUTURE ACTION

UPA is still an informal sector in many cities of the world although it is making a significant contribution to urban food and nutrition security. Its role is still largely undervalued. UPA definitely has a place in the context of expanding urbanization and growing poverty.

FAO has been instrumental in formulating UPA Strategies (UPAS) and has guided their implementation. The successful formulation and implementation of UPAS is based on a participatory consultation and monitoring process to ensure multipartner involvement and cooperation from various ministries in charge of agriculture, education, health, environment, land tenure and social development, as well as private sector and financial partners, international organizations and NGOs.

Lessons learned from FAO field projects addressing UPA for sustainable development have led to proposing the following standard framework, which could serve as a guideline for national UPAs.

The framework is based on the following five pillars:

(i) Securing the two primary resources: land and water

(ii) Securing the quality and safety of produce and environment protection

(iii) Securing sustainable appropriation by actors and partners

(iv) Securing the market outlet

(v) Securing political and institutional anchorage and monitoring

Pillar 1: Securing land and water resources

Securing space for UPA activities requires a series of steps which are to be taken as soon as possible ... It's never too early and it's never too late...

It is essential to take "conservative measures" to safeguard available open spaces for UPA activities and to "reserve and protect" them against urbanization through appropriate legal documents.

Access to good quality safe water for irrigation is increasingly difficult and expensive within and around cities. While water might be available from rivers and the underground water table, there is considerable potential to increase rainwater harvesting through surface catchments or roofwater collection. The other source of water in cities is treated wastewater. About two million ha are already being irrigated with wastewater of varying quality. At the same time measures have to improve water-use efficiency in the field through localized irrigation systems, whenever feasible, and also to avoid losses in the water distribution systems from source to field, which in several countries may be as high as 50 percent.

Pillar 2: Securing product quality and environment protection

UPA products need to conform to international and national quality and food safety standards. Product quality and safety starts in the field by applying Good Agricultural Practices (GAPs). GAPs are based on a series of crop and product management and handling principles all along the producer-consumer chain.

The way forward consists of putting in place the required capacity-building mechanism to allow farmers and stakeholders in the production-consumption chain to apply GAPs.

Pillar 3: Securing the sustainable appropriation of UPA by its actors along the food chain

The successful and sustainable implementation of the UPA strategy depends on the empowerment of its actors all along the food chain.

Interventions in support of pillar No 3 include (a) securing access to inputs in line with the recommended production protocols including quality seed and planting materials of the adapted cultivars; (b) providing access to microcredit; (c) organizing farmers' organization; (d) capitalizing experiences and ensuring their dissemination.

Pillar 4: Securing market outlet for UPA

While an important proportion of UPA production may be consumed by the family (up to 60 percent of the micro-garden harvest in Senegal), marketing for income remains a major driving force for UPA. Lessons learned indicate that successful marketing of UPA products may require the "grouping" of products to meet the demand for sustainable supply in terms of quantity and quality. Successful marketing of UPA produce could be according to different modalities, namely (a) doorstep marketing, (b) door-to-door distribution and (c) community sales points.

Securing market outlet also requires proactive interventions to stimulate consumer demand and confidence, based on the traceability, distinctness and quality of the produce, but also by advocating the health benefits that can derive from a diversified diet including the daily consumption of fruits and vegetables.

Pillar 5: Securing the political and institutional anchorage of UPA

UPA is a reality in many cities in developing countries. It should not remain an informal food production sector but should become duly acknowledged for the many benefits and opportunities it offers in support of the livelihood of the poor living in and the around cities. It is recommended that the necessary policy decisions be made to ensure the orderly development and performance of UPA. The aim is to acknowledge and integrate UPA as a component of the national agriculture development plan and National Food Security Strategy.

FAO has been instrumental in convening and facilitating expert meetings and creating public awareness for policy-makers which have resulted in various declarations, for example, the Dakar Declaration for Africa and the Quito Declaration for Latin America and the Caribbean.

Securing the political and institutional anchorage of UPA can be pursued through the formulation and implementation of an UPA Strategy as part of the City Master Plan.

CONCLUSION

FAO encourages member countries, municipal authorities and relevant institutions to:

1. be aware of the need to protect and improve urban and peri-urban agriculture systems as a component of food and nutrition security strategies, specially for the benefit of poor households;
2. make available technical guidance and provide capacity building to improve the effectiveness and sustainability of UPA production and post-production systems with special attention to improving livelihoods and increasing food availability, safety and accessibility;
3. make policy decisions and provide guidance at municipal and other levels to improve the efficiency of UPA as an integral component of

the overall food supply and distribution system based on rural-urban linkages;

4. promote the protection and improvement of the urban and peri-urban environment including the watershed and surrounding agro-forestry.

6. CHANGING PATTERNS OF ECONOMIC GROWTH, CONSUMPTION AND FOOD SECURITY: EXAMINING INDIA'S EXPERIENCE IN A GLOBAL CONTEXT

Keith Wiebe*, Daniel Gustafson, David Dawe and Kostas Stamoulis

**Deputy Director, Agricultural Development Economics Division, Department of Economic and Social Department, Food and Agriculture Organization of the United Nations, Viale delle Terme di Caracalla, Rome – 00100, Italy*

ABSTRACT

High food prices in 2007 and 2008 and the current financial and economic crisis have had significant impacts on food security, particularly among poor consumers in developing countries. FAO estimates that more than one billion people are chronically undernourished in 2009, representing nearly one sixth of the world's population. In the coming decades, growth in population, income and urbanization will further increase demand for food and dietary diversity, while climate change and rising demand for energy (including biofuels) will add to the challenge of feeding the world's people. This paper explores changing consumption patterns in India as well as globally, the potential to meet future demands, and the steps needed to ensure that all people have secure access to sufficient food for active and healthy lives.

INTRODUCTION

FAO estimates that more than one billion people are chronically undernourished in 2009, representing around one sixth of the world's population. This number is estimated to have increased by over 150 million in the past several years as a result of two crises that were quite different in their nature and origins, but which have had similar impacts on food security. The first was the "food price crisis" that peaked in mid-2008, and the second was the global financial and economic crisis that began immediately thereafter. Both sharply reduced access to food, particularly for the poorest. As a result, earlier progress in reducing hunger — towards both the 1996 World Food Summit goal (reducing the number of hungry people in the world) and

Millennium Development Goal #1 (reducing the proportion of hungry people in the world)—has been reversed.

While the recent crises and their impacts have focused public and private attention on food and agriculture after decades of neglect, attention must look beyond these crises if hunger is to be reduced and eventually eliminated.

Looking to the future, a variety of factors affecting both demand and supply of food and other agricultural commodities will increase the challenge of feeding the world. Growth in population, income and urbanization will increase the total demand for food and change the composition of diets. Agriculture will be linked ever more closely with the energy sector, both as a consumer and—increasingly—as a producer of fuel. Most experts agree that the world has sufficient potential to meet these demands, if appropriate investments are made and policies introduced. But climate change will complicate the task of increasing agricultural production—improving growing conditions in some areas, making them worse in many others, but changing them virtually everywhere in ways that are still uncertain.

As anticipated by Amartya Sen nearly three decades ago, perhaps the biggest challenge to reducing hunger lies not in increasing production to meet the demands of those with sufficient incomes to buy food, whether today or in 2050, but in meeting the needs of those who lack the income to buy enough food or the resources to produce it. This challenge is illustrated by two key facts. First, the recent increase in the number of hungry people occurred even as farmers responded to higher food prices with record global cereal production in 2008. And second, even before the recent food price and economic crises, over 800 million people were undernourished—a number that has remained stubbornly and unacceptably high for the past half century, and will remain unacceptably high over the next half century unless the poor are enabled to participate in broad-based economic growth.

THE "FOOD PRICE CRISIS"

Food prices rose rapidly from 2005 to 2008 after several decades of relative stability (in nominal terms) and decline (in real terms). The FAO food price index rose by 7 percent in 2006 and 27 per cent in 2007, and that increase persisted and accelerated in the first half of 2008 (FAO, *The State of Agricultural Commodity Markets*, 2009). In June 2008, the prices of basic foods on international markets reached their highest levels in 30 years, raising widespread concerns about hunger and political instability at local, national and global scales.

There has been considerable debate about the causes of the recent rise in food prices. Growth in population and income and the changes in diet associated with urbanization, particularly in large developing countries like China and India, have certainly contributed to increased demand for food over the past several decades, but these long-term trends did not change significantly in the past few years so were not a key factor in the recent price spike. Declining investment in agriculture is also a longer-term phenomenon, and in fact is not surprising given the long-term decline in real prices of food and other agricultural commodities, but like growth in food demand did not change significantly in recent years. Likewise global cereal stocks have been declining over the past decade—but not in themselves changing sufficiently in the past several years to cause the spike in prices. Nevertheless these factors helped to set the stage for greater volatility in food prices when harvests were reduced in several key countries in 2005 and 2006, and the impact of these supply shortfalls was compounded by rising energy prices, which had the dual effect of increasing the cost of producing agricultural commodities and simultaneously making the production of biofuels more attractive—and thus increasing the demand for feedstocks from agricultural commodities.

The role of biofuels in the rise in food prices has been the subject of considerable discussion and debate, but its effect is difficult to disentangle from that of the various other factors that contributed. It is not surprising that estimates vary, given differences in methods, commodities, and time periods covered. At one extreme, a working paper by the World Bank suggested that 75% of the price increase was attributable to the increase in biofuel production, while at the other extreme, the White House Council of Economic Advisors estimated that biofuels were responsible for 3% of the increase in a global food index. But most studies have found that biofuels accounted for about a quarter to a third of the increase in the prices of maize and other staple commodities in 2007-2008. For poor consumers in developing countries, for whom basic staple foods represent a large share of their consumption and expenditures, the rapid expansion in the production of biofuels seems to have had a significant effect.

Finally, exchange rate movements and policy responses to high prices in some countries—particularly restrictions on exports in an attempt to keep domestic prices low—helped trigger a contagion effect and exacerbated the increase in prices of food and other agricultural commodities. A survey of policy responses to high food prices by governments in around 100 developing and transition economies indicates a mix of measures to support consumers (*e.g.*, food subsidies), producers (*e.g.*, subsidies on inputs), and trade-related measures (*e.g.*, relaxation of import requirements and restrictions on exports) (FAO, *Crop Prospects and Food Situation*, December 2008). In sum, multiple factors interacted to contribute to the rapid increase in prices, and it is difficult to isolate the impact of any one factor.

WHAT DID HIGH FOOD PRICES MEAN FOR FOOD SECURITY?

Most of the world's poor live in rural areas, and most of them depend at least in part on agriculture for their livelihoods, so it is not unreasonable to expect that they might be insulated from the effects of rising prices because they produce their own food, or perhaps that they might even benefit from higher prices if they produce a surplus of food for sale.

In fact, however, most of the world's poor, like most of the world's population in general, are net buyers of staple foods. This is not surprising in the case of the urban poor, but even in rural areas, the majority of poor households are net buyers of staple foods in most developing countries—whether because they lack sufficient land to produce enough food for their own consumption, or because they produce other crops for sale and purchase food.

This means that most of the world's poor rely on other sources of access to food besides direct food production, including non-agricultural income (which depends on the levels of both employment and wages) and remittances. This also means that rising food prices reduce the purchasing power of such households, and thus affect them adversely.

Reliance on income from on-farm agricultural production is highest among rural households in sub-Saharan Africa, but even there agricultural wages, off-farm employment and other sources of income account for a quarter to a third of rural household income on average (FAO, Rural Income Generating Activities database). The share of rural household income from sources other than agricultural production is typically higher (half to two-thirds) in other regions of the world. In India, the 59th Round of the National Sample Survey Organisation (NSSO) on farmers' incomes clearly shows the importance of the non-farm share within total farm household earnings. For all India and all farm sizes in 2003, the share of farm family income from agriculture was 50.1 per cent (cultivation and livestock) and for non-farm activities 49.9 per cent (wages 38.7 and non-farm businesses 11.2 percent). With the changes underway in rural India's economic landscape, the importance of linking agriculture growth to non-farm employment represents an outstanding opportunity to balance pro-poor growth and poverty reduction.

While diversification of farm household income helps sustain real incomes when agricultural prices are low, however, it may increase vulnerability when food prices are high (or when those alternative sources of income are adversely affected by economic conditions, as we will see below). When food prices rise, poor households, who may spend as much as 70-80% of their total expenditures on food, are forced to adjust in various ways (FAO, *The State of Food Insecurity in the World*, 2008). They may reduce consumption of preferred staples and higher-value foods such as fruits, vegetables, meat and dairy products (and may even eat greater quantities of cheaper staples), and they may also increase time spent working (or seeking work) and reduce expenditures on health care and other non-food items. Even if these responses succeed in maintaining caloric intake, they may still result in reduced nutritional status due to reduced consumption of protein and micronutrients, as well as diminished health status. As labour is often the most important asset held by the poor, this has implications for incomes and access to food in the future as well as the current period.

Households also have access to food through other means, for example by drawing down savings or by borrowing, or by depleting or (in extreme cases) even selling assets such as land. These options may also come at the high price of compromising future ability to generate income and access to food, but for the poorest households these may be the only options that remain.

Reduced access to food as a result of high food prices, among other factors, is estimated by FAO to have increased the number of undernourished people in the world to 915 million in 2008, up from 848 million in 2003-05 (FAO, SOFI 2008). Most of the world's undernourished are in Asia (indeed 43% in India and China alone), while the proportion of undernourished is highest in sub-Saharan Africa (where one person in three suffers from chronic hunger).

While the number of undernourished people in the world is estimated to have increased, high food prices also gave farmers incentives to increase production. In fact, global cereal harvests in 2008 reached record levels, up 7% over 2007 (FAO, *Food Outlook*, June 2009). But the supply response was uneven; cereal production rose 13% in developed countries but only 3% in developing countries (FAO, *Crop Prospects and Food Situation*, July 2009). Part of the difference may be explained by shifts between crops in 2008, but it also reflects weaker incentives for farmers in developing countries—both due to policy measures taken by governments in some developing countries to limit domestic price increases, and due to poor infrastructure that hinders farmers' access to markets for agricultural inputs as well as outputs.

Food prices fell sharply in the second half of 2008, in part due to farmer responses and a strong global harvest in 2008. Given that most of the world's poor are net buyers of staple food commodities, this suggests that their access to food would improve. However, while world food prices have retreated from their mid-2008 peak levels, they are still high by historical standards. And prices have been slower to fall locally in many developing countries than they have on international markets. In sub-Saharan Africa, in more than 80 per cent of all the local price series analyzed by FAO, latest prices are more than 25 percent higher than 24 months earlier, while in Asia and Latin America and the Caribbean, from 40 to 80 per cent of prices analyzed remain more than 25 percent higher than in the pre-food crisis period (FAO, *Crop Prospects and Food Situation*, July 2009)—and prices are projected to remain higher over the next decade than they were in the past decade (OECD-FAO 2009).

THE CURRENT FINANCIAL AND ECONOMIC CRISIS

Importantly, prices fell in 2008 not only as a result of increased supply but also as a result of declining demand due to the financial crisis that began in the developed countries and the sharp global economic downturn that followed. Falling wages, rising unemployment and declining remittances have reduced incomes and offset the effects of lower food prices, causing a sharp deterioration in access to food for many. Foreign investment in developing countries (including in the agricultural sector) has fallen sharply after increasing rapidly in recent years, and official development assistance remains uncertain as public budgets in developed countries face unprecedented demands for stimulus funds in those economies.

The current crisis is historically unprecedented in several ways. First, because it follows so closely on the heels of the sharp increases in staple food prices during 2006-2008, the mechanisms that are normally used by poor countries and households to cope with economic shocks are already stretched thin. Second, the crisis is affecting large parts of the world simultaneously. When economic crises are confined to individual countries, or several countries in a particular region, governments can make recourse to instruments such as currency devaluation, borrowing or increased use of official assistance to face the effects of the crisis. The scope of such instruments is more limited in the case of a global crisis. And third, with developing countries more financially and commercially integrated into the world economy, they are far more exposed to changes in international markets. A drop in global demand or supply or restrictions in credit availability as a result of the crisis has immediate repercussions on developing countries.

The global economic slowdown has hit developing countries through both financial and commercial channels, with important implications for their food security and overall economic growth. The main transmission channels are as follows.

Foreign Direct Investment (FDI): The International Monetary Fund (IMF) estimates that foreign investments going to developing countries will decline by 32 per cent in 2009. Although most FDI is in mining, industry and services, the agricultural sector will also be affected. Reduced employment will have economy-wide ripple effects and might force some unsuccessful urban job-seekers to return to rural areas.

Remittances: The World Bank projects a decline of around 5 to 8 per cent in migrant's transfers to developing countries in 2009. Remittances had previously grown by as much as 20 percent per year, becoming an important source of capital for developing countries. In low-income countries, for example, they represent six per cent of total economic output. What is more, remittances have usually been resistant to shocks and often even increased during economic crises in recipient countries. The countercyclical effect of these transfers is unlikely to happen this time due to the global dimension of the current crisis.

Official Development Assistance (ODA): Foreign aid is the principal source of capital inflows for many of the poorest countries, but the global economic slowdown has affected the budgets of donor countries along with the needs of recipients. After a substantial increase at the global level in 2008, the IMF expects ODA to drop for the poorest 71 countries by about 25 per cent in 2009.

Financial Markets: As the crisis evolves, developing counties face higher costs for external credit, from both private and public sources. The risk premium for loaning money to developing countries has increased, and in many cases, credit might not be available at all as banks ration their financial reserves and lend only to those perceived as the most reliable borrowers.

International Trade: Trade volume levels are expected to fall between five percent (according to the IMF) and nine percent (according to the World Trade Organization) in response to the current economic crisis. Prices of products exported by developing countries are also projected to decline in 2009. The IMF expects the fall in the value of exports to be greater for developing countries, and to be particularly damaging for those economies that rely on exports as their main source of foreign exchange.

Reduced international capital flows, tighter credit conditions, lower remittance transfers and shrinking export opportunities cut into investment and lead to reduced growth prospects in developing countries. Unless concrete steps are taken to increase investment, this can have long-term adverse effects on poor countries, even after the current global recession has passed.

WHAT DOES THE CURRENT FINANCIAL AND ECONOMIC CRISIS MEAN FOR FOOD SECURITY?

Following on the heels of the food price crisis, these threats to food security come at a time when the traditional response strategies of poor households have already been strained, meaning that few options to protect current consumption levels remain and longer-term prospects have already been compromised.

Faced with increased unemployment, declining wages and reduced demand for their work, households try to maintain income by migrating, selling assets such as livestock, borrowing money, or participating in new types of economic activity. International evidence indicates that women and even children may enter the labour force in greater numbers as a response to economic downturns.

Furthermore, households alter spending patterns and reduce expenditures on durable goods relative to food and other essential items. Food expenditure also tends to shift towards cheaper, caloric-rich, energy-dense foods such as grains, and away from more expensive protein- and nutrient-rich foods such as meat, dairy products, or fruits and vegetables.

Coping mechanisms thus involve undesirable but often unavoidable compromises for the poor: poor working mothers, for example, are less likely to seek health care for themselves or their children; migration might weaken community cohesion; withdrawing children from schools destroys long-term human capital; the sale of assets reduces the stock of physical or financial resources and is not easily reversible; and shifting from more nutritious foods towards less nutritious items or simply eating less causes malnutrition, affects labour productivity and reduces children's cognitive potential.

Poorer households are naturally hit harder by rising prices and falling incomes. The rural landless and urban poor are the most affected by the crisis because they typically spend the greatest share of their incomes on food (and the greatest share of their food expenditures on basic staples), and because they lack the resources to produce food themselves. Other vulnerable groups include female headed households whose family responsibilities (*e.g.*, caring for children and the elderly, collecting water and fuel, and cooking) limit their ability to devote a greater share of their time for paid employment.

As a result of these factors, FAO estimates that 1.02 billion people are undernourished in 2009—the highest figure in recent decades and nearly one sixth of the world's total population (FAO 2009). It is important to note that the increase in undernourishment is *not* a result of limited international food supplies. As noted previously, 2008 saw record global cereal production, partly in response to high food prices, and the latest projections indicate that global cereal production will decline only modestly in 2009, down 3.4 per cent from last year's record high but still the second largest crop ever gathered (FAO, *Crop Prospects and Food Situation*, July 2009).

It is also critical to remember that more than 800 million people were undernourished already before the recent food price and economic crises began, and will remain undernourished even after those crises have past. While some regions (*e.g.*, East and Southeast Asia) have made notable progress, in many other regions and at the global scale relatively little progress has been made towards achieving the 1996 World Food Summit goal of reducing the number of hungry people in the world by half by 2015 (Table 1).

Table 1. Number of hungry people in the world

Unless substantial and sustained remedial actions are taken immediately, the World Food Summit (WFS) target of reducing the number of hungry people by half—to no more than 420 million by 2015—will not be reached (Figure 1).

The latest hunger estimates of the Food and Agriculture Organization of the United Nations (FAO) show a significant deterioration in the already disappointing trend witnessed over the past ten years. The spike in food insecurity in 2009 underlines the urgency of tackling the root causes of hunger swiftly and effectively.

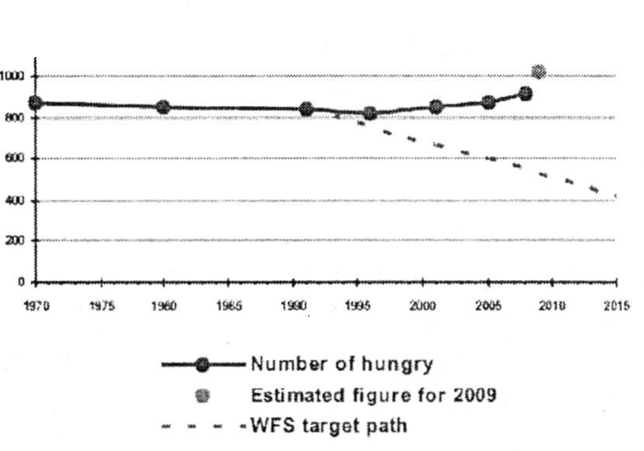

Number of hungry people in the world (in millions)

Source: FAO 2009

Somewhat better progress has been made towards achieving the Millennium Development Goal of reducing the proportion of hungry people by half by 2015, but here too, earlier progress at the global scale has been reversed in recent years, and many regions are lagging.

LONGER-TERM PROSPECTS

Looking beyond the recent and current crises, a variety of factors affecting both demand and supply of food and other agricultural commodities will increase the challenge of feeding the world in the future. Growth in population is slowing, but will still add 2.3 billion people to the world's total by mid-century. Virtually all of this growth will occur in the developing countries, and most of it will be in urban areas. Combined with increased income, which will also grow most rapidly in the developing countries, this is expected to increase global food demand by about 70% by 2050.

India exemplifies many of these trends. Between 1950 and 1980, India's real net national product per-capita increased by about half, but in the subsequent three decades it has more than tripled. Economic growth was accompanied, as expected, by a decline in the share of agricultural value added in GDP from 37% in 1981 to 18% in 2005. World Bank (2007) projections for economic growth in South Asia indicate a growth rate of 4.7% per year for real GDP at market exchange rates between 2008 and 2030. As India's long-run growth rate from 1991-92 to 2005-06 has been over 6% per year (Government of India, Economic Survey 2006-07), this may well prove to be an underestimate. But economic growth does not translate uniformly into poverty alleviation, and there remains a diversity of growth rates within India, both across states and across sectors of the economy. For example, Vinod Ahuja of the Indian Institute of Management in Ahmedabad has estimated that GDP growth during 1995-2005 averaged 3.8% per year in Bihar and 7.1% per year in Karnataka, while agricultural income growth in the two states averaged just 1.8% and 0.4% per year, respectively.

Based on growth in population and income, FAO projects that per-capita calorie consumption in India will increase from about 2,400 kilocalories per day at present to just over 3,000 kilocalories per day in 2050, (Consumption levels in higher-income countries are projected to level off as they approach around 3,300 kilocalories per capita per day.). The number of undernourished people in India is projected to decline to 70 million people (about 5% of the population) by 2050 from 221 million (about 22% of the population) at the turn of the century.

Even as consumption levels in India continue to rise, the composition of diets is changing as incomes rise and preferences change in response to urbanization and globalization (Pingali, 2006). As is typical of countries with rising incomes, the share of calories derived from cereals is declining in India, and is projected to fall below 50% by 2050. Conversely the share of calories derived from higher-value foods like fruits and vegetables, vegetable oils and livestock products is projected to increase, with the most rapid growth in the case of chicken.

Projected growth in demand for cereals is driven primarily by an increase in population. By contrast, demand for higher-value foods is driven by rising incomes and changing preferences in addition to population growth. Demand for potatoes, fruits and vegetables, sugar and meat (except chicken) is projected to double, demand for vegetable oil and milk to triple, and demand for eggs and chicken to increase even more rapidly (Table 2).

How will these levels of increased demand be met? Increases in crop production derive primarily from increases in yield, expansion of arable land, and increases in the frequency with which crops are harvested from a given area (*i.e.*, cropping intensity). Yield improvements accounted for nearly four fifths of South Asian (and global) crop production increases between 1961 and 1999 (FAO 2002). Area expansion accounted for most of the remainder except in South Asia, where increased cropping intensity contributed a greater share. FAO projections suggest these broad trends will continue in the coming decades.

Table 2. Projected food demand in India

Food group	1999-2001	2015	2030	2050
	Million tonnes			
Cereals	159	199	225	243
Potatoes etc.	25	37	46	58
Fruits & vegetables	108	160	208	257
Vegetable oil	11	18	23	29
Sugar	29	40	47	54
Eggs	2	3	6	9
Chicken	1	4	10	18
Milk	66	104	146	196
Beef, mutton & pork	4	5	7	9

Source: FAO, Global Perspective Studies Unit, FAOSTAT.

Note: The figure for cereals includes rice, milled equivalent. The figure for cereal consumption in 1999-01 is the same as the GOI "net availability of cereals figure (e.g. Table 1.17, Econ. Survey 2006-07) except that FAO does not apply a uniform assumption that feed + seed + waste is 12.5% of gross production.

Arable and permanent cropland area in India is projected to grow only slightly, from 170 million hectares in 1999-2001 (of which 58 million were irrigated) to 181 million hectares in 2050 (of which 78 million are projected to be irrigated). Area is projected to grow more rapidly in Brazil, and to decline in China. Over the same period, cropping intensity in India is also projected to increase only slightly, from 101% to 104% in rainfed areas and from 127% to 129% in irrigated areas. By contrast, cereal yields in rainfed areas in India are projected to rise more rapidly, from 1.4 metric tonnes per hectare in 1999-2001 to 2.0 tonnes per hectare in 2050, while irrigated cereal yields are projected to increase from 3.5 to 4.6 tonnes per hectare. Even so, cereal yields in India are projected to remain below those in China, Brazil and other large developing countries.

Combining these trends results in projections of cereal production in India rising from 240 million tonnes (including rice as paddy) in 1999-2001 to 357 million tonnes in 2050, of which 285 million tonnes is projected to come from irrigated areas. In aggregate terms, cereal production in India (including rice in milled equivalents) is projected to increase by 54% (or 0.9% per year) between 1999-2001 and 2050. By contrast, production of potatoes, fruits and vegetables, vegetable oil, sugar and meat (except chicken) is projected to more than double, production of milk to triple, and production of eggs and chicken to increase even faster.

By comparing these figures with those on projected food demand discussed earlier, we can see that food production is projected to exceed demand (for direct human consumption) in India through 2050 for all commodity groups except vegetable oils. The balance is potentially available for use as livestock feed, seed, storage, industrial uses or export.

India ranks in the top tier of producers of fruits and vegetables (with 15 and 11 per cent of the world's total, respectively), but much lower as an exporter, with less than 2 per cent of global trade for fruits and vegetables, in spite of apparent advantages in the cost of production. This is attributed in part to high transportation costs that are often 20 per cent higher than those in competing countries. It also has to do with the presence of a huge domestic market, where food consumption is shifting to higher value foods. Where the contrast between India and other countries, particularly striking is in agricultural marketing, including the low level of supermarket penetration and organization of the supply chain from farm to retail outlet. This situation is changing, but given the low starting point, even the expected high growth rate will not result in a large share of the total market going to supermarkets for a long time.

EMERGING CHALLENGES AND OPPORTUNITIES

Even as farmers in India and around the world respond to increased demand for food, agriculture will become linked ever more closely with the energy sector, both as a consumer and—increasingly—as a producer of fuel. Use of cereals as a feedstock in the production of biofuels (primarily maize for ethanol in the United States) is expected to slow somewhat in 2009 as a result of the economic crisis, but is still projected to reach 125 million tonnes, representing 6 percent of total cereal utilization (FAO, *Food Outlook*, June 2009). Future trends in the use of cereals, sugar, vegetable oils or other agricultural commodities or byproducts in the production of biofuels will depend partly on markets but even more critically on policies, particularly in the United States and the European Union. These remain uncertain, of course, but will play a key role in the evolution of demand for agricultural products, and thus for food security via their effects both on the incomes of agricultural producers and on the prices of food faced by all consumers. Whether these effects translate into opportunities for small farmers and the rural poor more generally depend critically on the ability of farmers in developing countries to respond to new market opportunities. This depends in turn on the quality of research, extension, infrastructure and policies that shape farmers' access to agricultural inputs and outputs.

Most experts agree that the world has sufficient resources and technological potential to meet these demands, if appropriate investments are made and policies introduced. But climate change will complicate the task of increasing agricultural production—improving growing conditions in some areas, making them worse in many others, but changing them virtually everywhere in ways that are still uncertain. Autonomous adaptation by farmers, for example by shifting planting dates, switching crops and varieties, and adopting or expanding the use of technologies like conservation tillage or irrigation where economical, will offset some of the effects of gradual warming, at least initially, but studies suggest that impacts on agriculture due to changes in temperature and precipitation as well as the frequency and intensity of extreme events such as droughts, heat waves or severe storms are likely to become more severe after mid-century. Policies and investments to support adaptation and mitigation will become increasingly important.

Table 3. Projected food production in India, by food group

Food group	1999-2001	2015	2030	2050
	Million tonnes			
Cereals	192	229	262	295
Potatoes etc.	31	46	56	69
Fruits & vegetables	120	175	227	279
Vegetable oil	7	10	13	17
Sugar	34	47	56	68
Eggs	2	4	7	10
Chicken	1	4	10	18
Milk	82	127	178	239
Beef, mutton & pork	4	5	7	9

Source: FAO, Global Perspective Studies Unit, FAOSTAT.
Note: The figure for cereals includes rice, milled equivalent.

WHAT CAN BE DONE?

The food price and economic crises have had a severe impact on millions of people. Those suffering from undernourishment need immediate assistance to prevent a worsening of their situation, and the current crises have rightly focused attention on meeting their needs. The unfortunate reality, however, is that the number of undernourished people was already high and rising even before these crises began, revealing the inadequacies of the present food system and the urgent need for structural changes.

In the short term, safety nets and social protection programmes must be created or improved to reach those most in need. Simultaneously, small-scale farmers need improved access to indispensable means of production and technologies—such as high-quality seeds, fertilizers, feed and farming tools and equipments—that will allow them to boost production. This should also lower food prices for poor consumers, both rural and urban.

In the medium and longer terms, solving the problem of hunger requires investing in agriculture, particularly in low-income food deficit countries. This is important both as a means of increasing food production and, equally, as an engine of economic growth, since most of the world's poor depend at least in part on agriculture for their livelihoods. These countries must be assisted with the necessary technical and financial solutions and policy tools to enhance their agricultural sectors in terms of productivity and resilience in the face of crises. Stable and effective policies, regulatory and institutional mechanisms, and functional market infrastructures that promote investment in the agricultural sector are paramount. Investments in food and agricultural science and technology need to be stepped up. Improved institutions and governance at local, national and international levels are also essential.

In India, the pace of economic growth and national market integration, rising incomes, changing consumer preferences and other factors should add up to significant opportunities for market-driven, more entrepreneurial and higher value-added agriculture. The rapidly expanding market for livestock products and fruits and vegetables should create pro-poor agricultural growth, with increased rural non-farm employment, rising labour productivity and a more competitive small farm sector. An appropriate enabling environment requires investment in research and extension, improved infrastructure for transport, communications, storage, processing and marketing as well as efficient land markets and secure property rights (Pingali, 2006).

While investment and policy improvements in agriculture are necessary to eliminate hunger, they will not be sufficient. Even though most of the world's poor live in rural areas and depend in part on agriculture for their livelihoods, many already earn significant shares of their income from non-agricultural sectors, and that process of transition will continue in the future. For the world's poor, access to food will increasingly depend—as it already does for the vast majority of world's rich—on incomes earned outside agriculture.

In sum, the recent food price and economic crises have reversed earlier progress in reducing hunger, raising the number of undernourised from 848 million before the crises began to 1.02 billion in 2009. Demand on agriculture will only increase in the coming decades, and supply will face new challenges. But most experts agree that the world has the resources and technological potential to produce enough food, if appropriate investments are made and policies put in place. The biggest challenge to reducing hunger will be to go beyond increased agricultural production and improve access to food by the poorest. Now, while public and private attention to agriculture and food security are high, it is critical to maintain and focus that attention on increased investment in agriculture, protection of vulnerable groups, and improved governance of food security at national and global scales. Farmers and investors, scientists and policymakers—the coordinated efforts of all of these stakeholders will be essential. Only a healthy agricultural sector, combined with a growing non-farm economy and effective safety nets, will sustainably eradicate poverty and food insecurity, in line with the Millennium Development Goals and those of the 1996 World Food Summit.

REFERENCES

FAO. 2009. Crop prospects and food situation, No. 3, July 2009. FAO, Rome.

FAO. 2009. The state of food insecurity in the world 2009. FAO, Rome.

FAO. 2009. Food outlook, June 2009. FAO, Rome.

FAO. 2009. Rural income generating activities (RIGA) database. FAO, Rome.

FAO. 2009. The state of agricultural commodity markets 2009. FAO, Rome.

FAO. 2008. The state of food and agriculture 2008, Biofuels: prospects, risks and opportunities. FAO, Rome.

FAO. 2008. The state of food insecurity in the world 2008. FAO, Rome.

FAO. 2008. Crop prospects and food situation, No. 5, December 2008. FAO, Rome.

FAO. 2007. Accelerating agricultural growth in India (prepared in consultation with the Ministry of Agriculture, Government of India, the International Food Policy Research Institute and the M.S. Swaminathan Research Foundation).

FAO. 2006. World agriculture: towards 2030/2050, Interim report. FAO, Rome.

FAO. 2002. World agriculture: towards 2015/2030, Summary report. FAO, Rome.

OECD-FAO. 2009. *OECD-FAO* Agricultural outlook 2009-2018. OECD, Paris, and FAO, Rome

Pingali, Prabhu. 2006. Westernization of Asian diets and the transformation of food systems: Implications for research and policy. *Food Policy*, 12: 281-298.

Sen, Amartya. 1981. Poverty and famines: An essay on entitlement and deprivation. Oxford University Press.

World Bank 2007. Global economic prospects.

World Bank. 2007. A case study of India's horticulture: from competition at home to competing abroad.

7. REGIONAL PATTERN OF ECONOMIC GROWTH AND HORTICULTURAL DEVELOPMENT

Sudha Mysore

Indian Institute of Horticulture Research, Hessaraghatta Lake Post, Bangalore -560 089

ABSTRACT

Growth potential of horticulture production is strong for the under developed and emerging economies, a fact that has been well recognized world over. Horticultural crop development promises these economies the ability to meet domestic food needs along with diversified income sources. India, with its vibrant agrarian economy fits well in to this category and has shown distinct patterns of crop diversification during the last three decades, moving away from cereal base towards commercial horticultural crops. Horticulture in India comprising a variety of crop groups has emerged as a core sector complementing agricultural GDP and an important contributor to the export basket during the last two decades.

During the last decades, economic growth in horticulture has far exceeded than in most agricultural commodities. Since the 1970s, annual growth rates for vegetable supplies have surpassed cereals by 200 percent to 800 percent, with much of this acceleration occurring in the 1990s. The increasing world consumption of horticultural products is driven by rising incomes, urbanization, awareness of health, and changing labour practices.

Keeping in view the changed world trade order in the post WTO regime, much hope is being pinned on this crop group for the much awaited second green revolution in India. Enhanced budgetary allocation from a mere Rs. 34 crores in the IV plan to Rs. 4500 crores in the XI plan, backed by favourable policy and institutional support, horticulture sector is expected to reach a take off stage. Focus is gradually shifting in favour of horticulture as a major source of growth for agriculture in the near future.

While growth in the horticultural sector presents many opportunities for growing rural economies and improving the livelihoods of the rural poor, the development paradigm is not uniform across regions and crops. Analyzing the growth trends across crops and regions provides an insight into the changing pattern of growth and also assist in identifying the critical factors that seem to provide this comparative/competitive advantage triggering growth and development. An attempt has been made in this paper to track down these changes and draw inferences for the future strategy planning.

INTRODUCTION

Growth potential of horticulture is strong for the under developed and emerging economies, a fact that has been well recognized world over. Horticultural crop development promises these economies the ability to meet domestic food needs along with diversified income sources. India, with its vibrant agrarian economy fits well in to this category and has shown distinct patterns of crop diversification during the last three decades, moving away from cereal base towards commercial horticultural crops.

Horticulture sector has emerged as a key sector of the Indian economy, with a contribution of more than 25 percent of the agricultural GDP; provides employment and livelihood support to over 19 percent of the work force; and adds to about 28 percent of the country's foreign exchange earnings. Besides these factors, the sheer tenacity of this set of crops to fit into different production modules and agro climatic zones with definitely higher profitability compared to other crop groups have made crops like fruits, vegetables, flowers, plantation crops to be the most favorite of all crop diversification drive. Having recognized the potential of this crop group, the union government also enhanced budgetary allocation for this group of crops since the IV plan period. The budgetary allocation increased from Rs 34 crores during the IV plan period to over Rs. 10000 crores by the IX plan.

During the last few decades, beginning the IV plan period, economic growth in horticulture has far exceeded than in most agricultural commodities. Since the 1970s, annual growth rates for vegetable supplies have surpassed cereals by 200 percent to 800 percent, with much of this acceleration occurring in the 1990s. The increasing world consumption of horticultural products is driven by rising incomes, urbanization, awareness of health, and changing labour practices. This crop diversification drive is commensurate with the changing policy and institutional framework, in favour of horticulture, there by making it difficult to assess the cause and effect.

While growth in the horticultural sector presents many opportunities for growing rural economies and improving the livelihoods of the rural poor, the development paradigm is not uniform across regions and crops. Some regions and crops seem to have benefited more than the others there by changing the comparative/competitive advantage of regions/crops. What factors triggered such a change, what are the possible implications of such a change for future strategy planning are some of the issues that have been examined in this study. This study based mainly on the panel data on horticultural crops for a period of over 15 years attempts to analyze the emerging regional patterns of horticultural crop growth and development and suggest alternate strategies.

STATUS OF HORTICULTURE IN INDIA

The agro climatic advantage provides India with a unique possibility for the year round production of a number of crops. This is especially true with regard to the highly seasonal and perishable horticultural crops like fruits, vegetable, ornamental crops, medicinal and aromatic crops, spices and plantation crops. According to the latest data of the National Horticulture Board (Anonymous, 2008), horticulture[1] crops occupy an area of 19.3 million ha and contribute to over 185.2 million tones, accounting for 8 percent of the Gross Cropped Area (GCA) and 12% of the total crop output. Though the group comprises wide variety of crops, fruits and vegetables, ornamental and medicinal crops and mushroom put together account for over 67% of the area and 90.4% of this total horticultural crop production. The area and production of horticultural crops have registered a fourfold increase since mid 80s suggesting a tremendous growth potential of this sector.

Trends of Crop Diversification

The Indian economy, which has primarily been agrarian dominated by food grains and cereal crops, started showing signs of crop diversification towards commercial crops since 1983-84. This is largely attributed to the fact that the economy having achieved self sufficiency in food grain production started focusing on other crop groups during this period. Though crop diversification drive was in favour of a number of crops barring cereals, the largest beneficiary of this shift were oilseeds followed by the horticulture sector, lead by spices and condiments and then fruits and vegetables (Annexure 1, Chand *et al.*, 2008). This can be substantiated from the fact that the horticulture sector, especially fruits and vegetables have been the single largest contributor, i.e., 6% in percent share of growth in production during the 90s. This diversification pattern could be attributed to a number of factors including favourable policies and economic reforms adopted by government of India during the early 90s, besides the enhanced household income and increased awareness towards balanced nutrition. In order to study the pattern of crop diversification, the study utilizes the secondary data from published sources from 1991-92 to 2007-08. The study attempts the analysis taking into account two periods, 1991-92 to 2000-01 and 2000-01 to 2007-08.

GROWTH PATTERN OF HORTICULTURAL CROP AREA

In absolute terms, the horticulture sector as a whole gained around 101.81 lakh hectares from 1991-92 to 2007-08, which could be considered a direct result of crop diversification drive away from cereals. The largest share of this area expansion has been accounted for by fruits, vegetables and flowers (85.1%). The gain by the vegetables is of interest as the group registered a decline of around 2,58,000 ha during the period from 1991-92 to 1995-96, but picked up in the later periods. Among the other groups, spices which showed steady gain in the initial years showed a major decline of over 600,000 ha during 2001-2007. Fruits and vegetables have gained nearly 97 percent of the total gain in area expansion drive during 2001 to 2007. It is

[1]comprising fruits, vegetables, flowers, spices and plantation crops

interesting to realize that this gain could be a massive loss for the food grains as a whole and nearly 71% for cereals and 28% for the pulses. Other commercial crops like oilseeds and cotton gained some area but the gain by the horticultural crop group is the highest (Surabhi, 2007).

Table 1. Change in area under horticultural crops (000ha)

	91-92 to 95-96	95-96 to 00-01	01-02 to 2007-08	91-92 to 2007-08	% Change in area
Fruits	483	512	1765	2901	28.49
Vegetables	-258	915	1647	5593	54.94
Spices	211	284	-617	598	5.87
Plantation	435	129	242	928	9.12
Flowers	82	16	55	161	1.58
Total	953	1971	3495	10181	100

GROWTH PATTERN OF HORTICULTURAL CROP PRODUCTION

Horticultural crop production also increased substantially during the study period. It is once again the fruits and vegetables as a group that has contributed substantially towards the increased production. The production gain has been the largest during the period from 2001 to 2007 for fruits, vegetables and plantation crops. During the same period, the other crop groups like cereals and pulses registered declining production trends as well (Table 2).

Table 2. Change in production under horticultural crops (000t)

	91-92 to 95-96	95-96 to 00-01	01-02 to 2007-08	91-92 to 2007-08	% Change in production
Fruits	12875	1631	20502	34871	31.75
Vegetables	-258	22256	37265	67355	61.32
Spices	510	613	338	2203	2.01
Plantation	2132	-172	2587	4547	4.14
Flowers	334	222	335	870	0.79
Total	28921	24714	61227	109846	100

While the changing pattern of area and production suggests the growing significance of fruits and vegetables, it is of interest to study the performance of individual fruits and vegetables as well.

Table 3a and 3b presents the change in area and production of individual fruits and vegetables respectively in two period, i.e., 1991-92 to 2000-01 and 2001-02 to 2007-08. Although the horticultural sector as a whole and that of fruits and vegetables, registered high growth in area, not all crops showed similar trend. The largest gainer has been mango, which registered 50% increase in area, followed by citrus (26%) and banana (13%) in terms of area expansion. This could be attributed to the fact that mango figures in almost all area expansion programmes even under dry land conditions. In terms of production, banana was the largest gainer followed closely by mango and citrus. Pinapple gained least, followed by grapes and guava. In terms of production, banana registered 50 percent gain over the period one, followed by mango and citrus. Production of guava increased the least while that of pine apple declined marginally.

Table 3a: Decadal changes in area and production of fruits (1991-92 to 2007-08)

Fruits	1991-00	2001-08	Percent change	1991-00	2001-08	Percent change
	Area (000 ha)			Production (000t)		
Mango	409.3	686	50.91	1787.9	3735.2	20.37
Banana	106.8	177.3	13.16	8814.5	9067.7	49.45
Citrus	140	346	25.68	1828.9	3190.7	17.40
Guava	56.9	29.8	2.21	615.4	343.5	1.87
Sapota	39.3	77.9	5.78	238.8	497.2	2.71
Papaya	15.3	9.8	0.73	860.9	889.6	4.85
Pine apple	18.4	1.9	0.14	256.9	-5.1	-0.03
Grapes	11.9	18.8	1.40	469.6	620.1	3.38

Table 3b. Decadal changes in area and production of vegetables (1991-92 to 2007-08)

Vegetables	1991-00	2001-08	Percent change	1991-00	2001-08	Percent change
	Area (000 ha)			Production (000t)		
Potato	205.8	574.7	51.35	6805.1	12220.3	50.48
Brinjal	199.6	93.3	8.34	3505	1944.4	8.03
Tomato	167.4	111.9	10.00	3183.4	3018.6	12.47
Onion	161.5	356.1	31.82	193.7	7435.9	30.72
Okra	126.8	-36.4	-3.25	1532.6	-791.8	-3.27
Cabbage	81	19.6	1.75	3138.2	381.5	1.58

In case of vegetables, potato continued to be the most important vegetables with 50 percent increase both in area as well as production during the study period. Onion was the second most important vegetables that gained over 30 percent in area as well as production terms. Tomato and brinjal accounted for 8 to 12 percent gain in area and production, while cabbage gained the least. Okra however, lost around 3.25 percent in area as well as production terms during the study period.

TRENDS IN RATE OF GROWTH IN AREA AND PRODUCTION OF HORTICULTURAL CROPS

Though absolute changes in area and production are important, it is more significant to understand the rate of growth of the sector and its components. The horticulture sector grew at much faster rate than its other counter parts in the period from 1991-92 to 2007-08. The sector as a whole grew at an annual growth rate of 3.14 percent in area and 4.08 percent in production, indicating the significance of productivity for this growth. Within the sector, production of spice crops followed by vegetables shows high rate of growth in the study period. Though all crop groups have contributed significantly towards the over all growth of the sector, the contribution by fruits and vegetables is of primary focus for this study.

Table 4. Growth rate in area and production

COMPOUND GROWTH RATE (%)		
	1991-92 to 2007-08	
	Area	Production
Fruits	4.16	3.82
Vegetables	2.61	4.31
Flowers	6.72	9.37
Plantation Crops	2.11	2.22
Spices	1.48	5.3
Grand Total	3.14	4.08

Although, fruits and vegetables put together accounted for nearly six percent of the growth in area and production terms of horticultural crop sector, the performance of individual fruits and vegetables may be different. While mango followed by banana gained the highest under area expansion, the rate at which this growth took place seems different.

It could be seen from Table 5, among fruits, as against the expected outcome, it was sapota that registered highest rate of growth in area to the tune of 10.34 percent per annum, followed by citrus in the period one. This was further increased to 15 percent in the second period as well. However, the growth rate of productivity has been negative through out, indicating an urgent need to understand the reasons for the same. This is suggestive of the fact that the commercial value of minor fruits seem to be gaining popularity. The low or declining productivity may be due to large number of new plantat-ions that are yet to start yielding. There is a need to revisit the method adopted in estimating the area and production of horticultural crops in general and that of fruits in particular for understanding the reason for low productivity of this group.

The most interesting is the case of mango among fruits, which presents a typical case with large area expansion next to sapota and guava in both the periods; the growth rate in productivity is negative. This is indicative of the fact that as the king of fruits and also being a dry land crop, mango seems to get the top priority in all area expansion drives. Further, the other factors such as biernnial bearing habit, yield fluctuations of the old orchards and the time lag for the new plantations to come for bearing may be responsible for the negative rate of growth of productivity. In contrast to the popular view that the horticultural production is beginning to slump in the post reform period, almost all fruits registered higher growth rate for both area and production in the period from 2001 to 2008 as well. The productivity growth rate however, was lower for almost all fruits in the second period and that of mango and sapota being negative. This trend is suggestive that much of the growth that has been obtained is largely area lead one and the fruit production may show slowing down trends in the near future, if corrective measures are not initiated.

In case of vegetables, brinjal registered the highest growth rate of area as well as production in the first period, followed by tomato. In the post liberalized period, it was onion that registered the highest rate of growth in area as well as production followed by potato. In the over all period, onion followed by tomato showed the highest rate of growth for area and production. This is suggestive of the steady demand and high profitability of these vegetables.

However, okra presents a different case. While the rate of growth in area and production was high in the first period, it has registered negative growth rate in the second period. Although the over all scenario of horticultural crop production is very encouraging, wide variation is clearly visible within the sector.

Table 5. Compound growth rate of area, production and productivity of fruits and vegetables

FRUITS

	Over all			PERIOD 1			PERIOD 2		
CROPS	1991-2008			1991-2000			2001-2008		
	A	P	Y	A	P	Y	A	P	Y
BANANA	2.63	8.55	1.65	2.52	7.32	5.89	5.02	23.74	3.3
CITRUS	5.02	4.94	-0.06	3.53	6.36	2.3	7.24	8.4	1.11
GRAPES	4.43	6.96	2.38	3.47	8.35	4.72	5.68	7.21	1.37
MANGO	4.53	2.31	-2.19	3.85	1.71	-2.33	5.95	4.57	-1.23
PAPAYA	2.53	7.06	4.41	4.53	9.38	4.53	1.48	3.95	2.38
PINEAPPLE	2.23	2.83	0.55	3.47	2.28	-1.13	1.08	1.33	0.15
SAPOTA	11.74	7.25	-4.02	10.34	6.91	-2.99	15.38	9.9	-4.76
GUAVA	3.09	2.89	-0.25	6	6.2	0.13	2.65	1.78	-1.4

VEGETABELS

	PERIOD 1			PERIOD 2			PERIOD 3		
CROPS	1991-2008			1991-2000			2001-2008		
	A	P	Y	A	P	Y	A	P	Y
BRINJAL	2.98	3.94	2.38	7.31	8.75	1.33	2.61	3.18	4.61
CABBAGE	2.16	3.41	2.28	3.67	9.41	5.54	1.09	-1.6	-0.17
OKRA	0.73	0.47	-0.23	3.33	4.87	0.58	-1.32	-3.69	-2.33
ONION	5.08	5.76	0.64	3.76	-0.7	-4.29	9.76	16.45	6.08
TOMATO	4.19	5.44	1.18	6.11	7.93	1.67	3.89	5.64	1.62
POTATO	2.51	4.12	1.07	2.03	3.77	1.71	4.99	5.19	2.78

NOTE: For brinjal 92 & 93 data is not available

It could be inferred from the above analysis that the area under fruits and vegetables has in general increased over the last two decades. While the traditional fruit basket comprising mango, banana and citrus still holds promise, others minor fruits like sapota and guava have shown promise under area expansion lead production growth. The fact that the productivity growth has been negative is indicative of lack of efforts at productivity enhancement in these crops. The results of research effort also need to be disseminated on top priority in these crops. Among vegetables, brinjal and tomato showed high potential for area and production growth up to 2000. The productivity lead growth in the subsequent period in these two crops needs attention. In contrast, cabbage and okra need special attention in terms of R&D efforts as the productivity growth is negative.

CONTRIBUTION OF DIFFERENT STATES FOR THE INCREASED HORTICULTURAL DEVELOPMENT

The contribution of horticulture sector also requires a closer examination from the point of the relative contribution of different growing regions. Hence, this section examines the contribution of different states towards the growth and development of horticultural crop production over the last two decades. Table 6, presents the compound growth rates in area and production of fruits and vegetables by different states for the study period. Almost all states performed well both in areas as well as production of horticultural crops in both the periods. During first period, i.e., 1991-92 to 1998-89, West Bengal registered the highest rate of growth at around 10.54 %, followed by Gujarat and Maharashtra. However, the area growth of fruits and vegetables was significantly higher during the second period, i. e., post reform period especially in case of Jammu and Kashmir, Kerala, Tamil Nadu and Maharashtra. It was however, interesting to note that the traditionally horticultural important states like Karnataka, Orissa and Himachal Pradesh registered very marginal growth in area under horticultural crops. This could be inferred as the increased commercialization and crop diversification drive in favour of horticultural crops by non-traditional horticultural crop regions, because of their profitability. In terms of production also, the post reform period showed significant increase registered by a number of states. Jammu and Kashmir, Andhra Pradesh and Bihar not only retained their high growth rate but also performed well. The most interesting is the case of Gujarat, which had only 2.85 percent growth rate for production in period one showed significant increase to the tune of 15 percent in the second period. However, the traditionally horticultural states like Maharashtra and Karnataka showed very marginal increase at less than 5 percent. The analysis so far is suggestive of the fact that horticultural sector, especially that of fruits and vegetables has been recognized as the most favored crop group for diversification.

Table 6. Trend growth rate in area and output of fruits and vegetables in major states (%)

State/UT	CGR of Area (%)		CGR of Production (%)	
Period	1991-98	2001-08	1991-98	2001-08
Maharashtra	5.71	12.88	4.98	4.32
Andhra Pradesh	3.63	7.25	6.73	12.43
Karnataka	0.98	1.12	6.68	1.06
Tamil Nadu	-4.6	18.36	9.64	7.55
Orissa	4.6	0.58	4.64	0.94
Himachal Pradesh	7.9	0.48	2.05	6.89
Kerala	-1.23	19.72	0.39	8.57
Gujarat	8.34	10.77	2.85	15.08
J & K	-10.49	26.84	-1.83	12.13
West Bengal	10.54	0.97	17.85	8.98
Assam	-2.07	12.93	6.05	6.57
Bihar (old)	6.5	8.45	8.80	10.95
Rajasthan	10.4	4	12.68	6.25
UP (Old)	7.4	6.4	8.25	6.57

Periods: (i) 1991-92 to 1998-99; (ii) 2000-01 to 2007-08

REGIONAL VARIATIONS AND CONTRIBUTION OF DIFFERENT STATES FOR FRUIT AND VEGETABLE PRODUCTION

This section examines the regional variations in growth and development of horticulture taking into account the performance of different states with regard to individual fruit and vegetable production. Data on area and productivity of individual fruits and vegetables during 1991-92 and the triennium value of 2005-2008 have been considered for such an analysis.

Maharashtra followed by Andhra Pradesh emerged as the most important states for fruit production. Andhra Pradesh retained its first position in terms of area and production for mango, grape, citrus and papaya. Maharashtra leads in the area of grape, sapota and citrus, while it is second for mango and banana. In contrast, Karnataka, Tamil nadu and Gujarat appear to hold promise for most of the fruits, though they are not the leading regions.

In case of vegetables, Gujarat and Bihar (old) hold the third and first position; Karnataka leads in case of productivity of tomato and brinjal. Andhra Pradesh and Maharashtra also hold promise holding ranks within fifth place. In order to understand the changing pattern of growth in horticultural crop sector, the state wise analysis was further analyzed taking individual fruits and vegetables.

In case of individual fruits, Gujarat showed the highest area expansion for banana followed by Andhra Pradesh and Tamil Nadu. However, the productivity growth has been very marginal increase or near stagnation. Karnataka has registered negative growth in productivity, despite a 43% increase in area.

Mango is the most significant case, which indicates a negative growth in almost all the states, though area expansion has been significant in number states. Maharashtra followed by Gujarat show the maximum increase in area under mango with the last two decades while the productivity has declined in both the states.

In case of Grapes, barring Maharashtra, which has registered 42 percent increase in productivity, Karnataka, Andhra Pradesh and Tamil Nadu show declining productivity. The area expansion in case of sapota requires a special attention. The states of Karnataka, Maharashtra, Gujarat and Andhra Pradesh have allotted substantial area under sapota in the last two decades. The productivity growth also has been positive and high. However, the compound growth rate of productivity during the same period has been negative which needs a closer examination.

Maharashtra has retained its position as the most important fruit growing state in the last two decades. The state leads in mango, grape and sapota area expansion drive. While the productivity of grape as well as sapota is positive, that of mango is negative. Following Maharashtra closely is Gujarat, which also is showing great potential for fruit production. Area under banana, mango and sapota has increased substantially in the sate. The state also seems to be performing well with respect to the productivity gain.

Table 7. Analysis of changing trend of fruit production across growing regions

Crop	State	1991-92	T.E 2008	% Change in Area	% Change in Productivity
Banana	Tamil Nadu	56886	103633	82.2	0.95
	Maharashtra	56482	75533	33.7	0.68
	Andhra Pradesh	33773	70833	109	0.87
	Karnataka	41342	59033	43	-0.146
	Gujarat	25000	53433	114	0.368
Mango	AP	207596	471533	127	-35
	Bihar	146232	141066	-3.52	-19.6
	Gujarat	32000	102533	220	-17.7
	Karnataka	80803	127533	58	24
	MS	49873	449333	800	-74
	TN	55824	132266	137	-22
Grape	MS	17118	45366	165	42
	Karnataka	6490	11166	72	-16
	AP	2252	2566	14	-16
	TN	2267	2733	21	54
Sapota	Karnataka	13930	25033	80	8
	MS	3824	50553	1221	921
	Gujarat	5500	25033	355	286
	AP	3776	17533	365	280

Similar analysis for vegetables is presented in Table 8. Among the states, Gujarat has emerged as the most significant one for area expansion for vegetables. Gujarat has emerged as the most important vegetable growing state in the last decade with more than doubled area under almost all important vegetables. The fact that the productivity enhancement also is substantial in almost all the vegetables is of significance. Onion seems to be the only vegetable which has registered negative productivity growth in almost all growing regions. Maharashtra state has registered significant expansion in area for tomato and onion but has indicated stagnant or negative rate of change in productivity terms.

Table 8. Analysis of changing trend of vegetable production across growing regions

Crop	State	1991-1992	T.E 2008	% Chane in Area	% Change in Productivity
Brinjal	AP	18897	29467	56	-15
	Maharashtra	24742	29800	20	4
	Karnataka	26050	14666	-25	50
	Gujarat	24450	58767	140	36
	Bihar	68209	54133	-21	1.8
Tomato	AP	41415	82833	100	90
	Bihar	52440	46233	-11.5	85
	Gujarat	5000	29533	490	55
	Karnataka	33446	45933	37	43
	MS	25987	32633	26	-3
	Orissa	42000	100567	140	24
Okra	AP	13352	29367	120	271
	Bihar	70560	56666	-20	0.98
	Gujarat	5500	41833	660	1211
	Orissa	77700	71467	-8.2	-0.3
Onion	AP	20615	35933	74	88
	Karnataka	60000	146267	143	-48
	MS	79124	216967	174	-22
	Gujarat	17900	73766	312	-7
	TN	26745	30933	16	-12
Potato	Bihar	305000	315733	3.5	74
	Gujarat	16800	51966	209	3.86
	Karnataka	35383	61333	73	-57

From this analysis, a pattern of comparative advantage can be drawn for different regions for different fruits and vegetables. While Maharashtra and Andhra Pradesh emerged as the most important states for fruit production,

Gujarat followed by Karnataka have emerged as the most promising states for improving vegetable production.

CONTRIBUTING FACTORS FOR GROWTH IN HORTICULTURE

From the analysis so far it is clear that the emphasis on horticultural crops is increasing over the years. A number of factors seem to contribute to this trend. These factors have been classified into two categories, viz., livelihood based and institutional based.

Livelihood Based Factors

One of the important reasons for increased horticultural crop production is the increased preference for these crops by cultivators. The farmers' choice of a given crop or crop combination is often decided by its profitability and its significance in providing or boosting their livelihood support. Horticultural crops top the list with higher profitability per unit area when compared to any other cereal or pulse group hence fit well into the selection criteria of the farmers. Studies have shown the relative superiority of horticultural crops both in terms of profitability and productivity over a number of other crops (Chand *et al.,* 2008, Mittal, 2007). It is only rational that farmers opt for more profitable crops for sustaining their livelihoods. Hence, it shows beyond doubt that the growing significance of horticultural crops is mainly because they provide the much needed livelihood support to the farmers. This criteria also includes the export lead growth in the recent past where in horticultural crops seem to have better potential for enhanced income through exports as well. The horticulture crop group is also labour intensive there by further adding to the farmers' preference as they also assure employment opportunities whenever grown. However, the growing regional disparities need further investigation. While this could be treated as a basis for achieving regional specialization for the production of a specific crop, the underlying reasons for such a comparative advantage needs to be identified.

Institution Based

The other significant reason for the regional disparities could be attributed to the increased institutional backing that is being provided by the union government or the specific state governments to these set of crops.

Government support: Institutional support through liberal credit and other targets fixed for the government agencies also is responsible for the increased area under a number of horticultural crops. This has resulted in an unprecedented increase in area under horticultural crops, leading to an area lead growth. If the institutions do not take sufficient care to ensure that the package of practices is adopted in total, there is a likelihood that the productivity of the crops would be declining in the near future. States like Maharashtra and Gujarat appear to have taken series of measures towards expanding the growth and productivity of horticultural crops. The Maharashtra state's employment guarantee scheme is an example for state lead support for enhanced area under horticultural crops. There are other initiates as well, such as the National Horticultural Mission and hybrid seeds programmes that aim at district level enhancement in area under horticultural crops. While the traditionally horticultural crop states like Karnataka and Tamil Nadu, do not seem to show a similar drive in enhancing or retaining their position in horticultural crop production in the last decade there by slowing down or losing on the comparative advantage for horticultural crop production.

R&D Support

Yet another important contributor for the enhanced growth and development of horticultural crops is the effective research support. The focused attention provided through enhanced budgetary support is amply visible in the changing technological development in this sector. Technological advancement in fruits mainly stems from the improved package of practices and new production technologies. While in case of vegetables, the yield enhancing hybrid technology has been the major contributor. The role of private sector is substantial in vegetables like tomato and cabbage, while the public sector efforts are equally important in all other crops as well.

CONCLUSIONS AND POLICY IMPLICATIONS

Crop diversification in favour of horticulture seems to have started from the early 80s, but has accelerated from the mid 90s. The pattern of diversification has been significantly in favour of fruits and vegetables. The performance of individual crops and regions has been changing showing signs of comparative advantage. While the underlying reason for such a diversification is primarily the higher profitability in comparison with other crop groups, the pattern of development needs attention. While some of the traditionally horticultural crop growing states retained their significance and consolidated their lead in horticultural crop production, some non traditional crops and regions also have shown huge potential in the recent past. Gujarat has emerged as one such state, while Maharashra and Andhra Pradesh represent the traditional belt. The changing pattern of growth and development suggest the growing significance of livelihood and institutional parameters for the enhanced horticultural crop growth. An analysis of this nature not only highlights the regional pattern of horticultural crop development, it also helps identify the critical gaps but also help prioritize institutional support as well.

REFERENCES

Birthal, P.S, P.K. Joshi and Ashok Gulati 2007. Vertical Coordination in High Value Food Commodities: Implications for Smallholders. In Agricultural Diversification and Smallholders in South Asia, edited by P.K. Joshi, Ashok Gulati and Ralph Cummings Jr. Academic Foundation, pp. 405-440.

Cold Chain Summit 2007, organized jointly by Ministries of Agriculture and Food Processing Industries and the CII, March 20-21, 2007, CII, New Delhi.

Coulter, Hugh, 2004. Wholesale Market Management Manual - Support for Agricultural Producers - Support to the Vertical Structure of Market. European Union, April 2004.

Kumar, Praduman and Surabhi Mittal 2003. Crop Diversification in India: Analysis by State and Farm Size Group. Agricultural Situation in India. Special Number, August 2003. Directorate of Economics and Statistics, Ministry of Agriculture, Government of India, pp 273-280.

Mittal, Surabhi 2007, Can horticulture be a success story for India? ICRIER Working Paper No 187, August 2007.

NHB 2008, Indian Horticulture Database. National Horticultural Board. Ministry of Agriculture, Government of India.

Raghunath, S, D. Ashok, Puneet Prakash Mathur, Thomas Joseph, 2005, Indian Agricultural Produce Distribution System -- Towards an Integrated Agri Produce Flow, Indian Institute of Management, Bangalore.

R Chand, S.S. Raju and L.M. Pandey, 2008, Progress and potential of horticulture in India, *Indian J of Agricultural Economics*, **63**(3): 237-245

Sudha, M., 2008, Triggering agricultural development through horticultural crops, *Indian Journal of Agricultural Economics,* **63**(3): 540-536.

8. CONTRIBUTION OF VEGETABLE CROPS TO ADDRESS POVERTY, HUNGER, AND MALNUTRITION IN SOUTH ASIA

[1]M.L. Chadha, [2]Peter Hanson, [3]Satish K. Sain and [4]Roohani Pal

[1] Director, AVRDC-RCSA, ICRISAT Campus, Patancheru, Andhra Pradesh, India
[2] Plant Breeder and Global Theme Leader-Breeding, AVRDC-The World Vegetable Center, Shanhua, Tainan, Taiwan
[3&4] Visiting Scientists, AVRDC-RCSA, ICRISAT Campus, Patancheru, Andhra Pradesh, India

ABSTRACT

The "Green Revolution" contributed greatly to the reduction of hunger in South Asia. Today, the region claims to have a buffer stock of a million tonnes of food grains. Despite the grain surplus, South Asia continues to struggle with food insecurity, poverty, and a high incidence of malnutrition. Out of 963 million undernourished people in developing countries, 524 million live in South Asia, where per capita annual income is only one-third of the average of US$970 for all developing countries. World Bank estimates reveal that around 35% of the population in Bangladesh is chronically undernourished, followed by 25% in India and Sri Lanka, and 20% in Nepal and Pakistan. Just three crops— wheat, rice, and maize—meet more than 50% of the region's daily requirement of protein and calories. Iron and vitamin A deficiencies devastate the lives of millions in South Asia. Although the area and production of vegetable crops have increased from 3.43 million ha and 2.24 million t in 1961 to 7.47 million ha and 10.03 million t in 2007, respectively, the per capita consumption of vegetables in the region remains very low (159 g compared with the global average of 321 g). South Asia accounts for 11.23% of global vegetable production, yet many people in the region lack adequate access to vegetables, which are essential for good nutrition and health. Vegetables can balance staple-based diets, providing protein, vitamins, fiber, minerals, micronutrients and phytochemicals associated with health and the prevention of chronic diseases. To reduce dependence on a handful of crops and to meet growing demand, more and higher quality food needs to be produced. Vegetable production offers a means for impoverished farmers and households to generate income. Since its founding in 1971, AVRDC–The World Vegetable Center has sought to alleviate poverty and malnutrition in the developing world through the increased production and consumption of safe vegetables. The Center, headquartered in Taiwan, has three regional offices, including the Regional Center for South Asia, established in Hyderabad, India in 2006. The Center collects, characterizes, conserves, and improves vegetable germplasm resources for worldwide use. The Center's improved lines help meet the nutritional and economic needs of the poor; its technologies empower farmers and provide globally accessible, user-friendly, science-based approaches to vegetable production. This paper highlights the Center's role in addressing poverty and hunger, and in improving the nutritional security of South Asia by developing and promoting high yielding, abiotic and biotic resistant, nutritionally improved lines, good agricultural practices (GAPs), and technologies for production, postharvest, and marketing of vegetables.

INTRODUCTION

The Green Revolution in the 1960s and '70s, transformed agriculture around the world and increased grain production by 250% (http://en.wikipedia.org/wiki/Green_Revolution). Rice and wheat now constitute 60% of the total area devoted to food grain production in South Asia. Despite this fact, South Asia is home to the greatest concentration of the world's 2.1 billion poor living on less than US$2 a day (CGIAR, 2008). About 55% of the world's undernourished people live in South Asia, where per capita annual income is only one-third of the average of US$ 970 for all developing countries. The region struggles with food insecurity, poverty, and a very high incidence of malnutrition; according to estimates in 2005, approximately 850 million people suffer from chronic malnutrition. Undernutrition in South Asian countries ranges from 38% to 51%. About 47% of the children of ages 0-5 years are either moderately or severely underweight and the rate for moderate or severe wasting is 18%. Additionally, 29% of infants are born underweight in South Asia (FAO, 2006). South Asia accounts for 74 million of the world's total 178 million of under-fives who are chronically malnourished. The prevalence of underweight children in the region is due to the low status of women in South Asian countries, their lack of nutritional knowledge, and inadequate feeding and caring practices for young children. Natural disasters occur frequently in South Asia. Cyclones and floods threaten Bangladesh and the coastal parts of India, while drought is common in the arid and semi-arid parts of India and Pakistan. The effect of natural calamities is most severe on food-insecure households. Most of the South Asian population subsists on a staple-based diet. Food security has become increasingly dependent on rice, wheat, and maize, which constitute more than 50% of the daily requirement of proteins and calories (FAO, 1996).

To reduce dependence on a handful of crops, more food and higher quality food needs to be produced. Vegetables are rich in vitamins, minerals, plant proteins, phytochemicals, and fiber. Vegetables can enhance nutrition and provide balance to starchy, staple-heavy diets. Despite the advantages of consuming vegetables, the per capita consumption of vegetables in South Asia remains low (73 kg/year) compared with the recommended 200 kg/year (WHO, 2002). Consumption of adequate quantities of vegetables, especially green leafy vegetables, is essential to meet the dietary requirement of vital micronutrients. Vegetables can address the "hidden hunger" of micronutrient malnutrition, which can have serious consequences for children and pregnant or lactating women.

A strong vegetable sector is essential for human and economic development. Vegetables help combat malnutrition, diversify diets, and alleviate poverty by creating jobs and new sources of income for smallholder farmers and landless laborers. Vegetable production supports agribusiness, creating new economic and entrepreneurial opportunities, especially for women. Vegetable production diversifies and generates farm income, usually to a greater degree than other agricultural products. Processing vegetable crops for value addition creates additional job opportunities, especially for women in the rural sector. Vegetable production develops management and leadership skills among farmers. When carried out following good agricultural practices, vegetable production enhances agricultural sustainability by naturally enriching soils, diversifying cropping systems, and reducing cereal pest populations. Intensive cultivation and improved productivity per unit area is the key to rural economic and social development. Since 1971, AVRDC – The World Vegetable Center has sought to alleviate poverty and malnutrition in the developing world through increased production and consumption of safe vegetables. To enhance food and nutritional security, agricultural output must keep pace with population growth and changing dietary demands. AVRDC-The World Vegetable Center effectively mobilizes resources from the private and public sectors to promote safe production of nutrient-rich vegetables that nourish people, generate income, and nurture the earth.

EXPANDING VEGETABLE R&D IN SOUTH ASIA FOR GREATER IMPACT

AVRDC – The World Vegetable Center is the world's leading international institute dedicated to vegetable research and development. It is based in Taiwan, and has regional centers in Thailand (Asian Regional Center), Tanzania (Regional Center for Africa), and India (Regional Center for South Asia), with offices in Cameroon, Indonesia, Laos, Madagascar, Mali, the Solomon Islands, South Korea, Uzbekistan, and Vietnam. The center develops vegetable lines and other technologies to increase vegetable production and consumption in developing countries, leading to more income opportunities and healthier diets for the poor. Since 1990, the center has had an active R&D program through the South Asian Vegetable Research Network (SAVERNET), a collaboration with South Asian National Agriculture Research Systems (NARS) and a bilateral project in Bangladesh. In April 2006, the center established the Regional Center for South Asia (RCSA), at the International Crops Research Institute for the Semi-Arid Tropics (ICRISAT) campus in Hyderabad, India. This was an important step forward to initiate R&D programs aimed at building a vegetable-based agricultural system to benefit small-scale farmers and provide a basket of safe vegetables to consumers. Vegetables must be integrated into the region's cereal-based farming systems to improve the nutrition of rural families and to meet the growing food demand from the urban poor living in megacities. Vegetable production creates more jobs than cereal-based crops per hectare; the economic opportunities generated by vegetable production and related industries particularly empower women—an economically disadvantaged segment of the South Asian population.

The center currently focuses its activity in South Asia on five research themes: germplasm conservation, evaluation, and gene discovery; genetic enhancement and varietal development of vegetables; safe and sustainable vegetable production systems; postharvest management and market opportunities; and nutritional security, diet diversification, and human health. The center collaborates closely with the National Agricultural Research Systems (NARS) on projects of mutual benefit.

GERMPLASM TO BREED BETTER VARIETIES

AVRDC – The World Vegetable Center collects and preserves the genetic heritage of its mandate vegetable crops for future varietal improvement efforts. In 1972 the center's genebank contained 590 accessions; today, it is the world's largest public collection of vegetable germplasm, with more than 56,000 accessions and many wild relatives of major vegetable species, including more than 10,000 accessions of indigenous vegetables. The current collection has 44,563 accessions of the center's principal crops and 11,952 accessions of other crops, comprising 404 species of 161 genera from 152 countries.

The center characterizes vegetable germplasm resources and shares useful germplasm material with NARS, and the public and private sectors worldwide. Promising lines of tomato, sweet and hot pepper, okra, eggplant, garlic, onion, cucurbits, soybean, mungbean, and other vegetables with disease, insect, drought and high temperature resistance, high yield, and improved nutrition have been developed at the center. The gene bank's wild germplasm accessions provide the genetic resources breeders use to develop improved inbred lines with desired traits. To date, 325 AVRDC breeding lines of various vegetables have been released officially by NARS in 91 countries.

Since the center's founding, it has distributed more than 571,000 seed samples to researchers in at least 180 countries. The center's improved vegetable lines, along with improved management practices, have raised the yield potential and promoted year-round production of vegetables in the tropics. The AVRDC Vegetable Genetic Resources Information System (AVGRIS), an online database, provides easy assess to documentation on collected and characterized material of mandate crops and indigenous vegetables. The database is part of the CGIAR System-wide Information Network for Genetic Resources (SINGER).

IMPROVED LINES FOR YEAR-ROUND SUPPLY

The center bred the world's first heat-tolerant tropical tomato lines, creating a new industry in the tropical lowlands; has benefited smallholder farmers and consumers. These lines possess improved horticultural characteristics such as determinate habit, firmness, deep red internal and external color, 80-120 g fruit weight, heat tolerance, and resistance to bacterial wilt, fusarium wilt, *Tomato mosaic virus* (ToMV), and grey leaf spot. For highland areas, lines are developed for high yield, with semi-determinate to indeterminate habits; round, deep globe fruits with >100 g fruit weight; and resistance to late blight, early blight, and fusarium wilt. More than 120 of the center's improved tomato lines have been released worldwide. Approximately 75% of the vegetable seed companies in Asia use improved inbred lines from AVRDC as revealed by a recent survey of 29 Asian seed companies. About 33% of tomato cultivars to be released in the near future have been developed from AVRDC germplasm material. The combination of resistance to both bacterial wilt (*Ralstonia solanacearum*) and geminivirus diseases has enabled farmers in South India to double their tomato yields from 19 t/ha to 37 t/ha in on-farm trials conducted under high geminivirus disease pressure. Geminivirus-resistant lines also allow farmers to reduce pesticide applications to control whitefly (*Bemisia tabaci*), the insect vector that transmits geminiviruses. These improved tomato lines help fulfill the daily requirement of vitamin A, and serve as an important source of income to smallholder farmers.

The center's heat-tolerant Chinese cabbage lines are well-known in the region. Approximately 16% of Chilli Pepper cultivars to be released in South Asia in the near future have been developed from AVRDC germplasm material. In 2007, 88 sets of the International Chilli Pepper Nursery and 67 sets of the International Sweet Pepper Nursery were distributed to co-operators in more than 30 countries, including South Asian countries.

Red onion lines developed by the center produce 50% higher yields (20 to 34 t/ha), and have longer self life, resistance to onion smudge, and 70% larger bulbs than the most widely grown variety, 'Red Creole,' thus offering greater marketing opportunities for farmers. The center also conducts breeding research on cucurbits, okra, and indigenous vegetables; more than 5000 accessions of indigenous vegetables are in the genebank.

Through the South Asian Vegetable Research Network (SAVERNET), 170 varieties of more than 22 different vegetable crops have been shared among the six participant countries (India, Sri Lanka, Nepal, Bangladesh, Pakistan, and Bhutan). Several varieties were released and promoted on a large scale in these countries. This program generated high internal rates of return (more than 90%) and high levels of economic surplus (more than US$500 million), which accrued more to producers than consumers (AVRDC, 1996).

The center's *Vigna* germplasm collection is the largest in the world, with about 11,000 accessions of 12 *Vigna* spp. Intensive and extensive, formal and informal collaboration with SAVERNET and others led to the release of 112 improved mungbean varieties based on AVRDC breeding lines in 27 countries around the globe, including 35 lines in South Asia. These lines are well-adapted to spring, summer, and autumn seasons in different agro-ecosystems of the Indo-Gangetic plains. For the past three decades, the center has made significant progress in developing high yielding (up to 2.5 t/ha), bold-seeded (5 to 6 gm for 100 grains) mungbean breeding lines with early (55 days) and synchronous maturity. The lines are resistant to *Mungbean yellow mosaic virus* (MYMV), Cercospora leaf spot, and powdery mildew. Mungbean has been introduced successfully into rice-wheat, rice-potato, and other cropping systems in the Indo-Gangetic Plains; it can be grown during the fallow period between cereal crops. The economic value of improved mungbean varieties and technologies in Pakistan alone has been estimated to be US$20 million. Under the Seed Village Program and through demonstrations, 16,000 t of mungbean seed were produced. With partner NARS, AVRDC organized training for extension specialists and farmers in mungbean cultivation, and for rural women and girls in mungbean value addition and nutritional improvement of diets through mungbean-based recipes.

By 2005, the estimated area planted with AVRDC's improved mungbean cultivars was more than 2.5 million hectares in Pakistan, Bangladesh, India, and Nepal (Shanmugasundaram, 2007). In Punjab the improved variety SML668 was planted on 95% (summer) and 60% (rainy season) of the mungbean area in 2006. By the end of 2008, the area planted to SML668 in Rajasthan, Haryana, and Himachal Pradesh was estimated to be around 60,000, 70,000, and 50,000 ha respectively. In the states of Bihar and Jharkhand, Haryana, Himachal Pradesh, and Tamil Nadu, which recently have been involved in AVRDC mungbean popularization program, SML668 is performing better than local varieties.

Since 1981, the center has developed vegetable soybean lines adapted to the tropics and subtropics. Fifteen countries, including Bangladesh (GC 83005-9), India 'Swarna Vasundhara' (GC 98009-1-1-2), Pakistan (AGS 190), and Sri Lanka (AGS-190) have released more than 39 vegetable soybean cultivars from AVRDC breeding materials (Shanmugasundaram and Yan, 2004). These lines are resistant to downy mildew, and are early maturing, high yielding, and suitable for mechanical harvesting. Vegetable soybean is delicious and nutritious, providing a rich source of protein, cholesterol-free fat, vitamins, and minerals to the diet.

INTEGRATED PEST MANAGEMENT FOR A SAFER FOOD SUPPLY AND A HEATHIER ENVIRONMENT

The center has developed and adapted integrated crop management systems for safe vegetable production. Component technologies include high quality seed of superior varieties with disease and pest resistance and tolerance to abiotic stress; starter solution fertilization to promote strong seedling growth; drip irrigation for managing water resources; protected cultivation in nethouses, net tunnels, or greenhouses to deter pests; sex pheromone traps, biofumigation, and colored sticky traps to catch pests; grafting onto flood-resistant rootstocks; and crop management techniques such as staking, mulching, raised beds, pruning and binding branches. Together, these strategies can help farmers safely produce more and better quality vegetables. It has been shown that year-round tomato production is possible in South Asia's uplands and lowlands using integrated technologies including lines tolerant of biotic and abiotic stress, tomatotone, rain shelters, plastic mulch, and tunnels. Year-round production of tomatoes increases the availability of fresh, nutritious food for consumers and generates income for farmers. Grafting to control bacterial wilt, Fusarium wilt, and root-knot nematode in tomato and eggplant is conferring better yield in most areas. In India, polythene sheet mulch has improved growth and yield for several crops such as okra, eggplant, and a range of cucurbits (Pandita and Singh, 1992). The center is working to adapt low-cost drip irrigation technologies for vegetable production in dry areas and in areas where acceptable quality water is scarce.

IPM strategies have been developed for eggplant and leafy vegetables, two of the most heavily sprayed food crops in South Asia. The center's IPM technologies to control diamondback moth and eggplant fruit and shoot borer (EFSB), the most destructive insect pests of crucifers and eggplant worldwide, are being adopted by farmers. The IPM program for diamondback moth began in Southeast Asia and has been introduced throughout Asia. In India it has been estimated that adoption of IPM to control diamondback moth would result in cost savings of at least Rs. 3,000 – 4,000/ha. In some parts of South Asia, farmers controlled EFSB by spraying pesticides up to 84 times during the 6-7 month cropping season. Since 2000, the Center's IPM program for smallholder farmers in Bangladesh and India reduced pesticide applications for eggplant fruit and shoot borer by 65–75%, cut production costs by 30%, increased farmers' incomes by 60%, and reduced farmers' exposure to pesticides. Less pesticide residue remained on eggplant sold to consumers. Farmers adopting the IPM technologies increased their production area by approximately 20%.

The center has developed and integrated molecular marker and screening techniques into traditional breeding process to facilitate inbred line development and germplasm enhancement. Researchers apply phenotypic and molecular criteria to identify lines resistant to biotic and abiotic stress, and lines with desirable nutritional qualities.

ENHANCING NUTRITIONAL SECURITY AND QUALITY

Quality, not only yield, has always mattered at AVRDC. Since 1971, the center's research has focused on legumes as a supplemental protein to cereals and vegetables as a more efficient means of delivering vitamin A and making dietary iron more available to bodies starved of micronutrients. The center screens vegetables for key vitamins A and C, iron, and phytonutrients such as beta-carotene, methionine, and flavonoids, and has developed several vegetable lines rich in nutrients. These lines include "golden tomatoes" that provide 3-6 times more vitamin A than the standard types; improved, tropically adapted chilli lines with high levels of antioxidants; sweet pepper lines rich in provitamin A, C and E; vegetable soybean lines rich in vitamins, protein, fat, isoflavones, and free from cholesterol; and mungbean lines with higher iron content to increase the iron available to the body. AVRDC broccoli lines are good-tasting, nutritious, and have the potential to protect against cancer. High vitamin sweet pepper lines could contribute significantly to improved nutrition. Within the past three decades, breeding of principal crops for higher contents of nutrients and health promoting factors such as beta-carotene, other carotenoids, antioxidants in pepper, lycopene, flavonoids in tomato, methionine and iron in mungbean, isoflavones in vegetable soybean, glucosinolates in leafy crucifers, carotenoids in pumpkin, and anti-insulin resistance in bitter gourd is being done to combat micronutrient malnutrition.

A good portion of the iron contained in cereals, legumes, and vegetables is not available to the body unless those foods are prepared properly. More than 70% of pregnant women in South Asia are deficient in iron; the importance of ensuring food nutrients can be used by the body is obvious. AVRDC conducted a survey of iron bioavailability in more than 20 vegetables and learned that cooking vegetables significantly increases the availability of iron to the body. Cooking increases the iron availability from 5-12% in amaranth, 16-32% in sweet pepper and from 3-8% in vegetable soybean. Cooked vegetables also increased the availability of iron from other foods in the same meal. Sprouting a bean can improve its food value. Research has shown that sprouting and then cooking mungbean increases the amount of available iron. AVRDC developed low-cost, easy-to-prepare mungbean dishes with high iron bioavailability (AVRDC, 1998; AVRDC, 2000, 2001; Amirthaveni and Yang, 1998) and published mungbean recipe books based on these principles, which have made a significant impact in the region for reducing iron deficiencies (Amirthaveni and Yang, 1998, Bains *et al.,* 2003). In Punjab, a sensory evaluation of iron-rich mungbean recipes revealed that recipes combining mungbean with other vegetables attained the maximum overall acceptability score and also enhanced iron availability (Bains et al., 2006). In Pakistan, the release of new mungbean varieties led to an increase in consumption. Weinberger (2003) studied the impact of enhanced iron consumption through mungbean and its effect on the productivity of anemic female workers in Pakistan. The study showed the economic benefit to the country was US$3.51 to 4.21 million in additional annual income primarily from women whose health improved from mungbean consumption.

Since 1980, the center's garden programs have sought to increase the nutritional status and income of the rural and urban poor through school, home, and market garden technologies. The center has designed school and home gardens using a variety of vegetable crop species to provide a year-round vegetable supply. Growing vegetables in school and home gardens is the most direct way for many poor families to improve their access to a variety of micronutrient-rich food. From 1991 to 1990 the center, in collaboration with NARS and NGOs, promoted a specially designed 16-square-meter home garden for year-round production in Bangladesh. An evaluation of the home gardening program found an increase in vegetable and energy consumption, a small reduction of night blindness, and an increase in income compared to controls during the study period (Cohen *et al.* 1985; HKI/AVRDC, 1993; AVRDC, 1994; Yusuf and Islam 1994; Greiner and

Mitra 1995; Talukder *et al.,* 2000). The center also conducts research on nutritional and commercial aspects of indigenous vegetables, especially *Moringa*, ivy gourd and amaranth and has selected varieties with improved nutrition. Improved varieties, growing techniques, and cooking methods developed by the center for many crops are fueling resurgence in their popularity. Selected high yielding varieties suited to local tastes are being promoted to farmers with the help of local partners. Supermarket displays and innovative promotions have raised urban consumer awareness.

Home garden kits developed and distributed by AVRDC to the survivors of tsunami-affected areas in Sri Lanka and Indonesia (2005) and Bangladesh (1998) included nutritious and marketable vegetables (*e.g.* water convolvulus, peppers, and pak-choi), fertilizer, and hoe heads to increase the availability of vegetables in a very short period. This effort helped people grow food quickly to prevent hunger and also generate some income from the sale of surplus produce.

Indigenous vegetables (IVs) are underutilized local species. Most are rich in nutrients, well adapted to the locations where they grow, have greater resistance to diseases and pests, and can withstand adverse climates and environments. Tropical IVs can help alleviate nutritional deficiencies. With the center's help, selected superior IV lines have been identified and released by national partners as varieties, or have been included in public and private sector breeding programs, especially in Bangladesh and the coastal parts of India. Amaranth, Malabar spinach, kangkong, jute mallow, ivy gourd, cowpea, *Moringa*, and others now supplement diets throughout the region. Amaranth, Malabar spinach, kale, and *Moringa* are rich in vitamin A (more than 2000 IU); legumes and taro are rich in vitamin B (above 17 mg); amaranth, kangkong, bitter gourd, and *Moringa* contain vitamin C (above 20 mg). Amaranth, kangkong, mustard, spinach, beans, onion, and *Moringa* are a good source of calcium (20 mg); amaranth, kangkong, spinach, and *Moringa* are rich in iron (above 3 mg); spinach, beans, onion contain phosphorus; and onion, okra and asparagus supply iodine. By extending the production area of indigenous vegetables with improved varieties, these species can generate income for smallholder farmers and increase the supply of nutrients to their families and communities.

Many vegetables are highly perishable and up to one-third of the world's harvest is lost before it ever reaches the consumer. High postharvest losses make vegetable production and marketing a risky business for smallholders. The center researches and develops low-cost postharvest technologies to reduce losses for two important vegetables—tomato and chilli pepper—with benefits for farmers and consumers. Recently, AVRDC identified lines suited for long-term storage and developed simple practices, such as using limestone to absorb moisture, which extends the storability of soybean seeds for two years or even longer. Reducing postharvest losses through research and training is the first step in creating value-adding industries based on packaging and processing. Women play a significant role in value addition after harvest. Their participation in vegetable value chains has implications for family income and nutritional status. Employment in postharvest activities can empower women within the household and at the community level.

CONCLUSION

Vegetables are vital for health and wealth. Vegetables diversify the human diet and provide essential nutrients, minerals, and vitamins. Vegetable production generates income for small-scale farmers, supports agribusiness and entrepreneurial opportunities, diversifies farm income, and develops management and leadership skills among farmers. In sum, a strong vegetable sector can become an engine for economic growth in South Asia.

AVRDC's Regional Center for South Asia works to bring about improved rural livelihoods and increased food and nutritional security through sustainable safe vegetable production. With appropriate postharvest techniques, enhanced value addition, and stronger public-private sector collaboration, market channels in the region can be linked for year-round production and consumption. Due to the broad range of agroclimates in South Asia, genetic resources for vegetable crops need to be collected for biodiversity security. Improved vegetable lines and varieties need to be exchanged and shared through partnerships among scientists in the region for further dissemination to NGOs and the private sector to enhance crop productivity and soil health, reduce cost of inputs for vegetable farmers, and increase their incomes. Overcoming micronutrient deficiencies and reducing child malnutrition must move higher on the agenda; food and nutritional security is cross-sectoral, and involves public and private sectors and actors. To help reduce poverty and food insecurity, priorities for implementation should include scaling-up vegetable cultivation in the smallholder sector, especially with high value vegetable products; investments in local, national and regional infrastructure; and educational programs through schools, health centers, hospitals, and communities to raise awareness of the value of consuming safe, nutritious vegetables.

ACKNOWLEDGEMENT

Thanks to Ms. Maureen Mecozzi, Editor, AVRDC - The World Vegetable Center, for her valuable contribution in editing this paper.

REFERENCES

Amirthaveni S. and Yang RY. 1998. High-iron mungbean recipes from South Asia. Shanhua, Taiwan: Asian Vegetable Research and Development Center. AVRDC Publication No. 98-480. 34 p.

AVRDC. 1994. In: *AVRDC Progress report* 1993. AVRDC, Shanhua, Tainan, Taiwan.

AVRDC. 1996. *Vegetable Research Networking in South Asia: SAVERNET Phase I final report.* AVRDC, PO Box 42, Shanhua, Tainan, Taiwan 741, ROC. AVRDC Publication No. 97-465. 451 p.

AVRDC. 1996. Increasing iron bioavailability by cooking. In: *AVRDC Progress report 1995*. p. 130–131. Shanhua, Taiwan: Asian Vegetable Research and Development Center. AVRDC Publication No. 96-449.

AVRDC. 1998. Mungbean as a better dietary iron source for South Asia. In: *AVRDC Progress report 1997*. p. 129-131. Shanhua, Taiwan: Asian Vegetable Research and Development Center. AVRDC Publication No. 98-481.

AVRDC. 2000. Iron bioavailability of vegetables. In: *AVRDC Progress report 1999*. p. 70–72. Shanhua, Taiwan: Asian Vegetable Research and Development Center. AVRDC Publication No. 00-503.

AVRDC. 2001. Role of vegetables in food and nutrient consumption in rural India. In: *AVRDC Progress report 2000*. p. 69–70. Shanhua, Taiwan: AVRDC Publication No. 01-520.

Bains, K., Yang, R.Y. Shanmugasundaram, S. 2003. *High-Iron Mungbean Recipes for North India*. Shanhua, Taiwan, AVRDC-The World Vegetable Center. AVRDC Publication 03-562. 34 p.

Bains, K., Agarwal, R. Barakoti, L. 2006. Development and impact of iron-rich mungbean recipes. p. 140-154, S. Shanmugasundaram (ed.). In: *Improving income and nutrition by incorporating mungbean in cereal fallows in the Indo-Gangetic Plains of South Asia DFID Mungbean Project for 2002-2004*. Proceedings of the final, workshop and planning meeting, Punjab Agricultural University, Ludhiana, Punjab, India, 27-31 May 2004.

CGIAR. 2008. Story of the month: May, 2008. http://www.cgiar.org/monthlystory/may2008.html

Chadha, M.L. 2007. Vegetable research and development strategies for alleviating, hunger, malnutrition and poverty in South Asia. *Proceedings of South Asia Regional Planning Meeting held at AVRDC-RCSA, ICRISAT campus*, Hyderabad, India. p. 139-144. AVRDC Publication No.: 07-699.

Cohen, N., Jalil, M.A., Rahman, H., Matin, M.A., Sprague, J. Islam, J., Davison, J., Leemhuis De Regt, E. and Mitra, M. 1985. Land-holding, wealth and risk of blinding malnutrition in rural Bangladeshi households. *Social Sciences Medicine*, **21**(11): 1269-72.

FAO. 1996. Report on the state of the world's plant genetic resources for food and agriculture. Rome, Italy, 511 p.

FAO. 2006. The State of Food Insecurity in the World 2006. Washington D.C.

Greiner, T. Mitra, S.N. 1995. Evaluation of the impact of a food-based approach to solving vitamin A deficiency in Bangladesh. *Food and Nutrition Bulletin*, **16**(3):193-205.

HKI and AVRDC. 1993. *Homegardening in Bangladesh. Evaluation report*. Helen Keller International, Dhaka, Asian Vegetable Research and Development Center, Taiwan.

Pandita M.L. and Singh, N. 1992 Vegetable production under water stress conditions in rainfed areas. In: Kuo C.G. (Ed.) *Adaptation of food crops to temperature and water stress*. AVRDC, Shanhua, Taiwan pp 467-472.

Shanmugasundaram, S. 2007. Exploit mungbean with value-added products. *Acta Hort*. **752**: 99-102.

Talukder, A., Kiess, L., Huq, N., de-Pee, S., Darnton-Hill, I. and Bloem, M.W. 2000. Increasing the production and consumption of vitamin A rich fruits and vegetables: lessons learned in taking the Bangladesh homestead gardening program to a national scale. *Food and Nutrition Bulletin*, **21**(2): 165-172.

U.S. Census Bureau-World POP Clock Projection. 2009.

Weinberger, K. 2003. The Impact of Iron Bio-availability-enhanced Diets on the Health and Productivity of School Children: Evidence from a Mungbean Feeding Trial in Tamil Nadu, India. In: D.J. Watson, (Ed). *International Conference on Impacts of Agricultural Research and Development:* Why has impact assessment research not made more of a difference? Proc. of a conference organized by the Standing Panel on Impact Assessment (SPIA) of the Interim Science Council, Consultative Group on International Agricultural Research (CGIAR), and the Economics Program, the International Maize and Wheat Improvement Center (CIMMYT), San José, Costa Rica. Mexico D.F.: CIMMYT. 4-7 February 2002.

Yusuf, H.K.M. Islam, N.M. 1994. Improvement of night blindness situation in children through simple nutrition education intervention with the parents. *Ecology Food and Nutrition*, **31**(3/4): 247-256.

World Health Organization, 2002. "The world health report 2002 – Reducing risks, promoting healthy life". Geneva: WHO.

9. HORTICULTURE AND FOOD SECURITY

Surabhi Mittal

Senior Fellow, Indian Council for Research on International Economic Relations (ICRIER), Core-6A, 4th Floor, India Habitat Centre, Lodi Road, New Delhi – 110 003

ABSTRACT

Access to food is one of the main pillars of food security. At the household level and for individuals it is the access to food, both in terms of physical access and the capacity to purchase the food that defines the household's food security status. Horticulture can be promoted as a means of agro-diversification, providing the much needed impetus to the growth of agricul-tural sector, through increase in income and employment. The country's main policy focus until recently was only on grains and cereals, but it has changed with the launch of the National Horticulture Mission, 2005-06. Also the ongoing liberalization and the emergence of an integrated global market have opened new vistas for the Indian horticulture. The diversification into the production of high value commodities also provides an increasing role to small holding farmers. The paper highlights the shift in cropping pattern in favour of horticulture and also analysis of the economic feasibility of this diversification. There are evidences that show that the shift towards fruits and vegetable is economically viable and beneficial to farmers, but this diversification needs to be planned in a systematic manner. Certain strategies and policies are also suggested in this regards.

With changing consumption pattern, it is important to mobilize the small farmers to diversify to meet the increasing domestic demand of horticulture products. In spite of being one of the largest producers of fruits and vegetables in the world, several factors like low crop productivity, limited irrigation facilities and underdeveloped infrastructure facilities like cold storages, markets, roads and transportation, are stumbling blocks which prevent the horticulture potential from being fully exploited. This paper advocates for improving the facilities in the horticulture sector so that the impact could be felt after its true potential is exploited and the new practices of distribution and management needs are introduced. These measures will be successful only if necessary reforms are implemented to integrate the domestic market both vertically and horizontally and provide farmers with the freedom to sell their produce wherever it's more profitable for them. While technological up-gradation and associated institutional changes are identified as thrust areas for future development of the horticulture sector, exports are considered to be most important for the growth of the sector. This development will also help in overall growth of the economy through generation of extra foreign exchange, creating employment opportunities and also upliftment of the small and marginal farmers, with definite positive implications on income and employment.

INTRODUCTION

Food security has two main pillars, one is food availability and other is access to food. Availability of food is usually discussed in relation to total production, food stocks and trade of food at the country level. At the household level and for individuals it is the access to food both in terms of physical access and the capacity to purchase the food that defines the household's food security status. In the present times when the economy is facing high food inflation the purchasing power of the poor and vulnerable population is depleting fast. The bad monsoon this year has further created situations of low levels of production, which further impacts the farmer's income level. The high food prices also do not prove to be an opportunity for farmers as these prices do not actually get passed on to them due to lop sided price policies, poor infrastructure and institutional mechanism. In this situation the farmers and urban poor population are seemingly struggling with the dire problems of food insecurity. In such a situation, can the high value cash crop sector like horticulture acts as the source of diversified incomes to the rural poor?

With overall higher economic growth and urbanization, changing lifestyles, increasing market integration and trade liberalization there is an increase in demand for horticultural products. Thus horticulture can be promoted as a means of agro-diversification for the second Green Revolution, providing the much needed impetus to the growth of agricultural sector, which can further lead to increase in income and employment.

The country's main policy focus until recently was only on grains and cereals, but it has changed in recent times with the launch of the National Horticulture Mission (NHM), 2005-06. In the period 1948-80 not much planned efforts had been made for horticultural development, except for some technical support and development efforts for specific commodities like spices, coconut and potato. During 1980-92 there was consolidation of institutional support and a planned process for the development of horticulture. It was only in the post-1993 period that a focused attention was given to horticulture development through an enhancement of plan allocation and knowledge-based technology. Although this decade was called a "golden revolution" in horticultural production, the productivity of horticultural crops has increased only marginally from 7.5 tonnes per hectare in 1991-92 to 10.3 tonnes per hectare in 2007-08 (NHB, 2008). Government of India launched the NHM with a mandate to promote integrated development in horticulture, to help in coordinating, stimulating and sustaining the production and processing of fruits and vegetables and to establish a sound infrastructure in the field of production, processing and marketing with a focus on post-harvest management to reduce losses.

There are evidences that Indian agriculture is diversifying into the production of high value commodities, also providing an increasing role to small holding farmers. But on the other hand Indian rural economy has also been facing the challenge of inability to manage the problems involved with this transition. This paper thus highlights the transitions and the challenges that are arising along with. The paper also suggests measures to overcome the constraints to ensure a better income security for the rural poor and nation on a whole.

In 2007-08 the total area under fruits and vegetables was 13.58 million hectares and the aggregate production stood at 189.39 million tonnes (NHB, 2008). As a result of this huge spurt in horticulture produce, India has become the second largest producer of fruits and vegetables in the world, next only to China. But India is still lagging behind in actual exports of these produce. The Indian horticulture sector is facing severe constrains such as low crop productivity, limited irrigation facilities and underdeveloped infrastructure support like cold storages, markets, roads, transportation facilities, etc. There are heavy post-harvest and handling losses, resulting in low productivity per unit area and high cost of production. However, on the other hand, India's long growing-season, diverse soil and climatic conditions comprising several agro-ecological regions provide ample opportunity to grow a variety of horticulture crops. Thus, efforts are needed in the direction to capitalize on our strengths and remove constrains to meet the goal of moving towards a formidable horticultural growth in India. The foreign trade policy in 2004-09 emphasized the need to boost agricultural exports, growth and promotion of exports of horticultural products. Horticulture contributes nearly 28 per cent of GDP in agriculture and 37 per cent of export share in agriculture. Further, the foreign trade policy 2009-2014 has introduced a single window system to facilitate export of perishable agricultural produce that will help to reduce transaction and handling costs.

PAST AND PRESENT TRENDS IN AREA, PRODUCTION, PRODUCTIVITY AND CONSUMPTION

Cropping Pattern Change

Agricultural diversification is an important instrument for economic growth. Diversification largely depends upon the opportunities and responsiveness of farmers to technological breakthrough, consumer demand, government policy, trade arrangements and development of irrigation, roads and other infrastructure (Kumar and Mittal, 2003). Changes in cropping patterns are responsive to these factors. The aggregate cropping patterns of the country are represented by the gross cropped area allocation among different crops and commodity groups. India has experienced a considerable degree of crop diversification in term of changes in the area under various crops since the Green Revolution which was largely in favor of food grains to meet the objective of self-sufficiency and country's food security. In past one decade, the changes in cropping pattern are more towards the horticulture sector and commercial crops like cotton. The change in cropping pattern in last two decades from food grains to horticulture[1] is illustrated in Figure 1. The decline in area under food grain in the period 1990-2000 had mainly been because of shift in areas away from coarse cereals and pulses towards fruits and vegetables. In the period of 2000-2007/08 the area under paddy also declined along with coarse cereals.

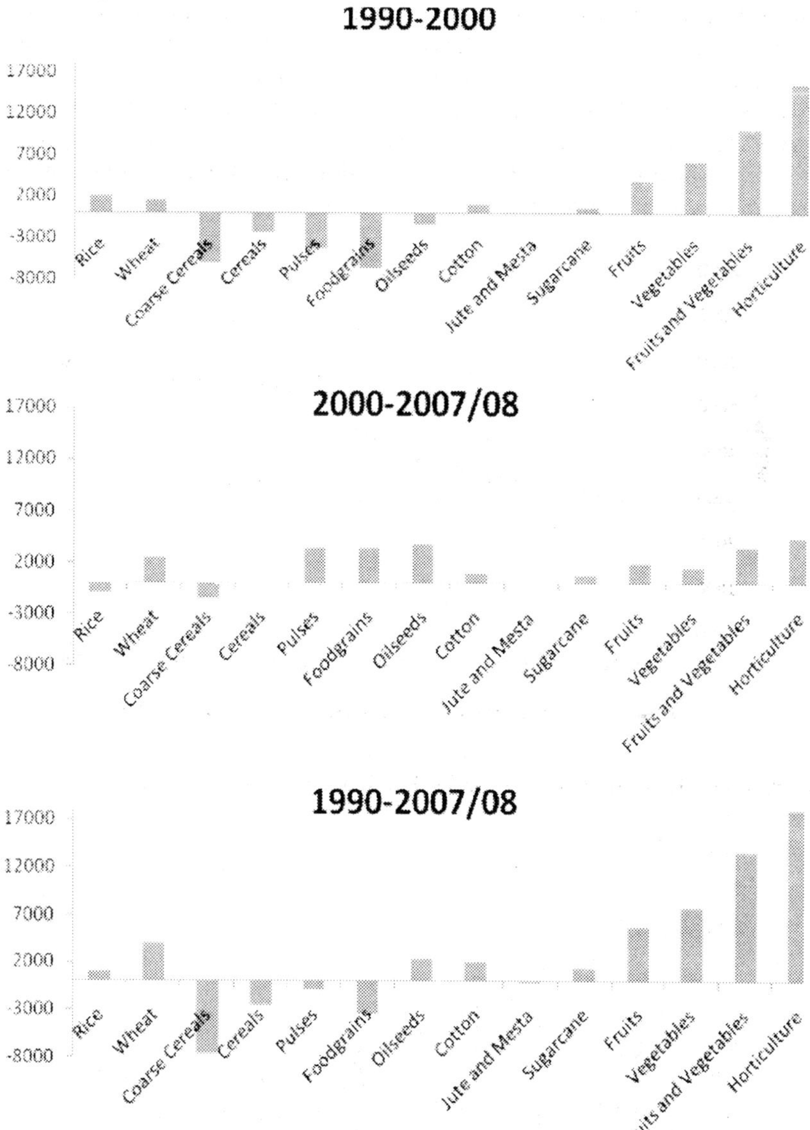

Figure 1. Changing cropping pattern, 1990-2007/08 (in thousand hectares)

1. In this context horticulture includes fruits, vegetables, plantation crops, spices and flower.

The increase in area under fruits and vegetables had been of a greater magnitude in the decade of 1990's than in 2000's. Overall a major reallocation of land towards horticultural crops was visible in the period 1990-2000 and in the next decade the shift was only marginal. But overall in the last two decades, cropping pattern has shown a shift towards horticultural commodities. Within the horticulture sector, the change in area under cultivation for different groups is presented in Figure 2. Within the horticulture sector, it is the fruits and vegetables that have shown the maximum gain in the area allocation. Figure 3 show that almost all the major states in India have followed this trend. This gain in area under horticulture and mainly under fruits and vegetables is a collective impact of the diversification of production pattern of the producer and the increased demand of consumers due to shift in their consumption pattern.

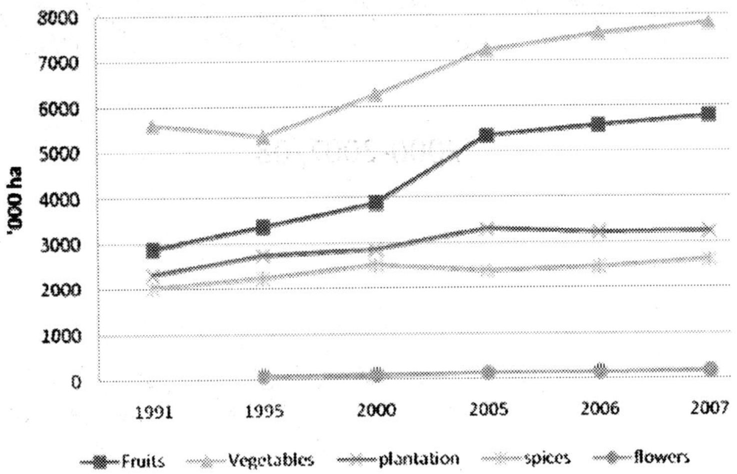

Figure 2. Trend in area allocation under horticulture sector in India

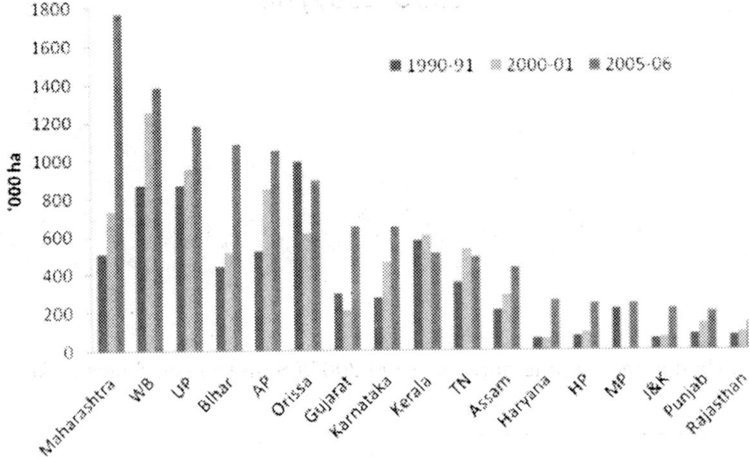

Figure 3. Change in area under fruits and vegetables in major states of India

PRODUCTION AND PRODUCTIVITY TRENDS

During 1990's within the horticulture sector, the production of vegetables is the highest followed by fruits. Fruits and vegetables combined form the major contributor to the total horticulture production. A constant trend is observed in production of plantation, spices and flowers also. Table 1 presents the growth rate in area and production of fruits and vegetables over the period. For fruits the area growth had been 3.28 per cent in 1990-95 which increased to 7.07 per cent in 2000-04 and 4.55 per cent in 2004-07. The production in fruits grew at the rate of 9.43 per cent in the initial period but later the growth had been declining which again picked up in 2004-07 period. The area and production growths for vegetables had been high at 3.15 per cent and 6.22 percent respectively in 1995-2000 which later declined but again improved in the post NHM launch era.

Table 1. Average annual rate of growth in area and production

(Unit: per cent)

Period	Fruits		Vegetables	
	Area	Production	Area	Production
1990-95	3.28	9.43	-1.00	4.67
1995-2000	2.58	1.62	3.15	6.22
2000-04	7.07	4.04	1.41	1.50
2004-07	4.55	7.67	5.00	7.09

In the world India is the biggest producer of banana and mango is the second largest producer of lime. China is the other biggest producer of these produce in the world. European countries are the leaders in grapes production with India ranking sixteenth in the world. Brazil leads in the production of orange and papaya and Thailand in pineapple. India and China are the world leaders in the major vegetable production. For brinjal, cabbage, cauliflower, onion and pumpkins China is the biggest producer followed by India. India ranks first for green peas. In spite of potato being the number one among the vegetable production in the country, India ranks third in the world production. For tomato India ranks sixth in world production (Mittal, 2007).

Along with huge diversification in area allocated to production of fruits and vegetables and high production levels, the sector growth is slow (Economic Survey 2008-09). The most important factor impacting the horticulture sector's growth is the low and declining productivity. As presented in Table 2 for fresh fruits and Table 3 for fresh vegetables, the decline in productivity as well as the low productivity rates as compared to the world's high-yields is quite visible. For the fresh fruits, India yield is the highest in grapes but has been stagnant or declining for citrus fruits and mango.

Mango has the highest share in our exports, and with its low yield the horticulture sector would be hit. Also the area under mango plantation remained static because of land constraint. The existing mango plantation requires rejuvenation or replanting. Rejuvenation, a short-term measure, has helped to retain or increase the productivity by a few years but the orchard owners feel that the need to replant the orchard. As per the existing forest rules, cutting of trees is not allowed and hence replanting the orchard can't take place, which is why increasing the productivity of mango is facing several hindrances. Except for cassava and tapioca where India has the highest yield in world's vegetable production, for most of the vegetables the yields are stagnant. For all the fresh fruits and vegetables, the potential yield possible is higher than the existing yields. For cassava, the higher yields can be explained by the fact that mostly cassava is cultivated under contract farming on commercial scale for exports.

Table 2. Trends in yield for major fruits in India

(Unit: tonnes / hectare)

Commodity	1993-94	2007-08	Gains in Yield	Potential Yield (Highest in the World) 2007-08
Apple	6.3	7.6	1.3	France (39)
Banana	27.6	35.9	8.3	Costa Rica (52)
Citrus Fruit	9.0	9.0	0.0	Spain (51.2)
Lime/Lemon	10.1	8.5	-1.6	Turkey (30.5)
Orange	8.9	6.7	-2.2	USA (35.4)
Grapes	18.1	26.0	7.9	India
Guava	10.8	11.1	0.3	Guatemala (26.8)
Litchi	5.6	6.0	0.4	-
Mango	8.3	6.3	-2.0	Brazil (17)
Papaya	22.7	33.4	10.7	Indonesia (81)
Pineapple	16.2	15.1	-1.1	Kenya (44)
Sapota	13.8	8.2	-5.6	-
Total Fruits	11.7	11.00	-0.7	-

Note: World yields are from FAO, 2006. Figures in parenthesis are the world highest yield.

Table 3. Trends in yield for major vegetables in India

(Unit: tonnes / hectare)

Crop	1993-94	2007-08	Gains in Yield	Potential Yield (Highest in the World) 2007-08
Brinjal	15.3	17.0	1.6	Japan (31)
Cabbage	15.5	22.2	6.7	Korea (63)
Cauliflower	15.2	18.1	3.9	Italy (24)
Okra	10.3	10.3	0.0	-
Onion	10.9	15.1	4.2	USA (56)
Peas	8.4	8.2	-0.3	France (12)
Tomato	14.3	17.9	3.7	USA (66)

Potato	16.6	19.3	2.7	Netherlands (45)
Sweet Potato	8.5	9.1	0.6	Japan (23.8)
Tapioca	24.5	33.5	9.0	India
Lettuce	6.6	6.6	0.0	USA (36.2)
Pumpkin & Gourd	9.7	9.7	0.0	Spain (42)
Beans	2.7	2.8	0.1	Morocco (15)
Cassava	24.5	27.9	3.4	India
Total Vegetables	13.5	14.8	1.3	-

Note: World yields are from FAO, 2006. Figures in parenthesis are the world highest yield.

If the information about the production undertaken on commercial and non-commercial basis is looked into, then it is quite possible that the average yield level may appear far satisfactory. This further implies that if the farming is taken up in an organized manner with the use of inputs, their application, and harvesting techniques imparted to the producers then the yield level can be raised.

CONSUMPTION TRENDS

Diversification in both the production and consumption pattern is observed in India. The economy is moving from being a supply-driven economy to a demand-driven economy thus with the change in consumption pattern of the population the production diversification is also evident. This section deals with the current demand and supply situation and also gives projections for fresh fruits and vegetables production and demand.

According to Engel's law with the economic growth and increase in per capita income, the consumption would shift from the staple food to high value commodities like fruits and vegetables, milk and milk products, fish and eggs. The NSS consumer expenditure data shows that the annual per capita consumption of vegetables has increased from 47.9 kg per person per annum in 1983 to 67.65 kg per person per annum in 2004-05 (Table 4). Fruits consumption has also increased from 3.3 kg per person per annum to 10.7 kg per person per annum during this period. The magnitude and trend of increase in consumption levels of vegetables and fruits has been quite similar in both the rural and urban areas.

Table 4. Annual per capita consumption of fruits and vegetables in India

(Unit: kg/person/annum)

Groups	1983	1987-88	1993-94	1999-00	2004-05
Vegetables					
Rural	46.30	50.17	60.33	74.01	67.34
Urban	50.95	56.89	63.86	78.16	68.26
All India	47.93	52.37	61.63	75.57	67.65

Fruits					
Rural	2.85	9.73	7.50	9.80	9.45
Urban	4.16	15.70	13.41	16.07	13.05
All India	3.31	11.68	9.68	12.15	10.67

Fruits are expected to increase to 17.43 million tonnes by 2010 and 25.47 million tonnes by 2020. For vegetables, this demand is expected to be 103.16 million tonnes by 2010 and 137.25 million tonnes by 2020. Demand for both fruits and a vegetable is expected to rise at the rate of 4-5 per cent per annum in the next 15 years. The per capita demand is also expected to rise at the same rate (Mittal, 2006).

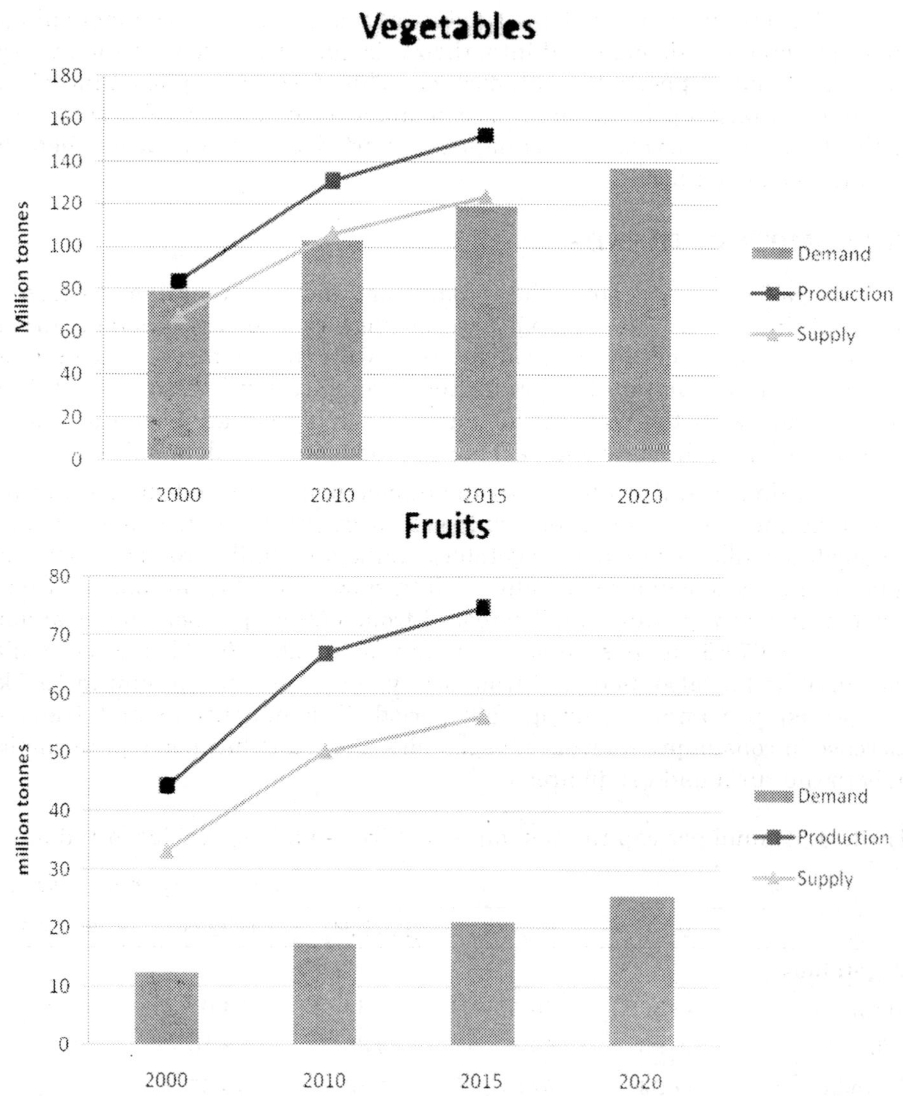

Figure 4. Projected demand, production and supply of fruits and vegetables in India

This huge increase in the demand of fruits and vegetables for domestic consumption is a challenge to be met by the country. Along with this is the goal to produce horticultural products for exports, which can act as an engine of growth to the agricultural sector. Due to food security concerns, lack of credit with farmers to diversify, and lack of the risk- taking attitude of farmers, it is difficult to allocate large areas of land for horticulture. Diversifying land away from cereals to horticulture remains a constraint in spite of it being more profitable for farmers to produce horticulture products than cereals. The way out is to increase our productivity and remove the other supply constraints.

The National Horticulture Mission aims to double the production of fruits and vegetables by 2010. The Figure 4 also presents projected future production and supply of fruits and vegetables in India for 2015-16. Based on 1999 as the base year, it is estimated that the production of fruits and vegetables would increase to 66.9 million tonnes and 131 million tonnes respectively by 2010 and almost double by 2015 (Kumar and Kumar, 2003). Due to huge post-harvest losses the actual supply of fruits and vegetables would be 20-25 per cent less than the production. This will help in meeting the domestic demand on the margin leaving no scope for increasing our export potential. Thus it is required that efforts should be diverted to minimize and remove the supply constrains and make the supply chain efficient in order to improve the horticultural and agricultural growth rates. This increasing trend is likely to follow in the future also and projected demand and supply scenario as presented in figure 4 show a scenario of likely shortfalls.

SCOPE OF HORTICULTURE TO IMPROVE INCOMES AND ACHIEVE FOOD SECURITY

A benefit-cost analysis helps in determining the profitability of farmers to diversify their land to horticultural production from the conventional practices of cultivation. Benefit cost ratio (BCR) is computed as an indicator of economics of investment criterion. This ratio helps in judging the feasibility of investing in the proposal. This ratio is a simple calculation that depicts the total financial return for each rupee invested in cultivation[1]. The BCR is computed for cereals and fruits and vegetables[2] and presented in Table 5. The results show that for wheat and paddy the gross returns are

---

1. If BCR is 3.56, it means that for every rupee invested in one hectare of land under cultivation for a given produce, the return is about Rs. 3.56 per hectare after the sale of the produce.
2. For horticulture produce, the establishment cost is an important component. For fruits the total cost of establishing and managing an orchard is spread over a period to have the average annual estimate of the cost of cultivation. For fruit orchards, all the costs are net of the returns from these intercropping. The average annual costs are worked out taking into account the number of years of the gestation period.

only on the margin and the BCR is mostly less than one. This implies that the farmers are not able to meet the cost of production. The reason why farmers continue to cultivate the staple food grains is that part of the produce is for self-consumption. If the implicit cost of family labour and land is not included in these calculations then there is some positive income also to meet household expenses. The increasing cost of inputs and declining yields are making it unprofitable for farmers to cultivate the staple food.

Table 5. Cost-benefit ratio for major cereals, fruits and vegetables for selected states

Region	Benefit- Cost Ratio				
Northern States	Mango	Litchi	Aonla	Wheat	Paddy
Uttar Pradesh	4.60	2.61	2.65	1.06	0.92
Bihar	-	1.69	-	0.91	0.80
Southern States	Bhindi	Brinjal	Tomato	Jowar	Paddy
Karnataka	1.71	2.14	2.01	0.62	0.95
Andhra Pradesh	1.53	1.92	-	0.50	0.94

The BCR for selected fruits and vegetable[1] is much better than that of food grains. For mango, after five years of plantation the fruit is ready for marketing. The maximum production is derived only after 10 years. Based on the economics of 10 years of mango plantation in Uttar Pradesh, with 100 plants per hectare, the annual BCR is computed at 4.6. Aonla and litchi are the intercrops in this region but in the calculation of BCR the returns from these intercrops are discounted. Litchi in Uttar Pradesh, tomato and brinjal in Karnataka, aonla in Uttar Pradesh have a BCR of more than two. This indicates that the gross returns are double the cost of cultivation of the produce. None of the horticultural crops presented in the table shows BCR less than one or even on its margin. This is indicative of the economic benefit of crop diversification towards fruits and vegetables. The marketing costs are also included in the calculations. If the markets are brought close to the farm gate or in the supply chain the produce is directly procured from the farm gate then it will be more beneficial for farmers, as they will be getting a better price for the fresh produce. In addition, the cold storage facility provided in the transportation would preserve the quality further, so that the consumers get better quality. This would further reduce the post-harvest losses and thus the quantity saved in the process will be an addition to the net availability.

Since the BCR of horticulture is more than the BCR of cereals, this implies that it is profitable to cultivate fruits and vegetables rather than cereals. The economic feasibility of shift of land away from cereals to

---

1 The information used for the computation pertains to the year 2005. It is only after the launch of the National Horticulture Mission that information on cost of cultivation of horticulture produce is being recorded. But no data was published at the time of the analysis.

horticulture produce is justified. But this feasibility does not imply shifting all the land away from cereals. The need is to re-allocate land in a manner that the optimal output and income can be generated, keeping in mind the domestic demand, export targets and improving the economic conditions of the farmers. Keeping food security of the country as the prime concern, the poor, the small and marginal farmers should try to diversify and increase the income levels. Regions where productivity of rice, wheat and other basic cereals has declined, or is not profitable for farmers to produce, the farmer needs to diversify his cultivation portfolio. In year-round cultivation, even if small amount of land is diversified towards horticulture commodities, and more importantly towards vegetables, then the income level of the farmer can improve.

Constraints

Besides the issue of diversification for improving incomes, the shift is also needed to meet the increasing domestic demand of fruits and vegetables. There are various supply constraints which are keeping the per unit productivity and per unit availability low. The supply chain efficiency is an important issue which needs to be taken into account along with the production diversification to reap the maximum benefit from the changed scenario.

Major constraints in production and marketing in fresh fruits and vegetables as listed in the literature (Kumar, Sant, P.K. Joshi and Suresh Pal, 2004) are non-availability of quality seeds, inadequate irrigation facilities, lack of soil testing facility and extension staff. Inefficiency in pest management, credit availability constraint, high cost of production, lack of information, huge post-harvest losses, lack of roads, cold storage, inadequate space, poor market network and high transportation cost. A lot of research initiatives and investments have already gone in, which have resulted in the growth in the supply of fresh fruits and vegetables in the past decade. But a lot more needs to be done especially in the field of research and investment to improve the infrastructure conditions and reduce the post-harvest losses in the sector so that the per-unit productivity and per capita availability of fruits and vegetables is increased. The general constraints faced by this sector are the timely delivery, grading, packaging, good quality, poor market infrastructure, agro-processing plants, marketing credit, proper pricing, uniform grading and standardization of weights and measures; inadequate and poor dissemination of market information, poor post-harvest handling, low and declining productivity.

Supply Chain Management

One of the weaknesses of the supply chain is that it is a multi-layered marketing channel lacking in infrastructure. Efficient supply chain requires strengthening all the levels of infrastructure such as the inputs delivery, credit, irrigation, improved procurement, minimizing post-harvest loses, cold storage chains, better and efficient processing and marketing techniques, efficient storage, warehouse and also efficient and competitive retailing. The infrastructure to improve efficiency and the linkages of the supply chain is very poor, which is affecting the growth potential of the horticulture sector. Timely availability of inputs, development of organized input market and infrastructure for its storage and distribution will add to the productivity of the sector. Development of cold chain network will help in reducing the post-harvest losses of fruits and vegetables. Improving the post-harvest management means an overall improvement in the per unit productivity. Cold chain infrastructure will require a huge sum of Rs. 18,000 crore to Rs. 20,000 crore investments in the next five years.[1] Investment is required to build the cold storage chain from the refrigerated trucks which procure the produce from the farm gate, then pre-cooling chambers, cooled area for sorting and grading the produce, refrigerated trucks for transferring the produce to the market of distribution or part of delivery. The development of this chain requires huge investment in technology, infrastructure and maintenance. A study by Raghunath *et al.* (2005) has estimated that with strengthening the supply chain the benefits to consumers and producers can increase by 20-25 per cent in the most perishable commodity like tomato. Due to inefficiency in the supply chain the price received by the farmers is only about 24 to 58 per cent of the retail price the consumer pays.

Post-harvest Management

Another issue in the supply chain is the inefficient post-harvest management. The proper integration of post-harvest technology into marketing supply-chain is crucial. Cultivable waste of about 24-40 per cent is reported because of the inefficiency. Post-harvest management does not only mean reducing waste but also maintaining the quality. The post-harvest technologies aim to address some of these issues by optimizing quality, safety and by reducing waste. Poor handling of produce lowers market quality and can substantially reduce producer returns. Our difference in prices between the farm and the retail is highest in the world. Improved handling methods and the resolution of regulatory requirements allow access to more distant domestic markets and also international markets. Appropriate production practices, careful harvesting, and proper packaging, storage, and transport contribute to good produce quality. Post-harvest management needs to focus on enhancing value through the application of cost-efficient and environmentally sound post-harvest techniques, to reduce losses and increase efficiency. Changes in production methods can also affect post-harvest product quality. Post-harvest technology of fresh fruits and vegetables in recent years gained enormous momentum to save losses during harvesting, handling, storage and transpiration quality and quantity. The extent of losses of fruits and vegetables in India is estimated at about Rs. 10,000 crore to 12,000 crore per annum, and the loss of quantity ranges between 10 per cent and 80 per cent in the most perishable fruits and vegetables. The major causes of loss are improper handling, poor packing, improper storage, uncontrolled

---

1. CII in Cold Chain Summit, 2007

temperature, etc. Thirty per cent of India's fruit and vegetable produce goes waste because of the lack of cold storage chains. Presently, there are about 1,300 cold storage facilities in India, of which 50 per cent is used only for potatoes and a large percentage of these are either underutilized or non-utilized for most of the year. A huge investment is required to meet the gap in the cold chains. There is the need to manage the available resources efficiently and minimize the quality and quantity losses. Proper post-harvest handling requires proper cooling and packing facility, clean and fast transportation, careful handling, proper management of environment, temperature, relative humidity, ventilation and sanitation. Value can be added to the produce by improving the product presentation through grading, packaging and labeling. Farmers should be encouraged to adopt the 'Good Agricultural Practice' (GAP) scheme to improve quality.

WAY FORWARD

Development of agriculture requires collaboration among various stake-holders along with efficient vertical and horizontal integration. The horticulture sector in particular has to prioritize development of research in the issues of genetics, biotechnology, integrated and sustainable production systems, post-harvest handling, storage, marketing and consumer education. Diversification offers an attractive option and a major source of pushing up growth of agricultural sector. Diversification towards horticulture products creates opportunity for raising farm incomes significantly, along with creating employment opportunities while technological up-gradation and associated institutional changes are identified as thrust areas for future development of the horticulture sector, exports are considered to be most important for the growth of the sector.

The two golden rules for successful development of the horticulture sector are to ensure consistency in supply and provide recorded and demonstrated traceability of products. Thus, production strategies are the most crucial in strategy development. The basic production factors required for high value horticulture production are non-restrictive policies in favor of horticultural development; suitable and controllable climate conditions, availability of labour and horticultural growing skills, basic local general infrastructure for transportation and marketing, i.e. access to road, train, sea, air transport, telecommunications, power and water; and basic local horticultural infrastructure, i.e. access to horticultural inputs and services. Producer strategy should be designed based on financial resources, managerial skills and entrepreneurial capacity. The development strategy should be based on innovation. Production innovations initially focused on efficiency and effectiveness in order to increase yields and lower costs. Now it is important that the production innovations should focus on developing sustainable production techniques, and also focus on adding value in terms of packaging and processing.

India being the land of small and marginal farmers and studies have been advocating the fact that small farmers are going to feed India, it is important to mobilize and help them to diversify to meet the increasing domestic demand of horticulture products. As identified from African nations the small farmers are the key to initiate the horticultural revolution and with technical change and increase in international competitiveness large scale operations and vertical integration takes place. Thus to sustain the growth and development of the small land holder, farmers should be monitored to identify, select and support horticultural and entrepreneurial smallholders. Linking small farmers with high value urban and export markets would lead to the development of the rural sector. An uncontrolled expansion of horticultural production should not be encouraged. Horticultural crop diversification should be encouraged by intercropping horticultural with non-horticultural crops. This will yield more food, more income and better soil health. To increase the production and productivity of fruits and vegetables, introduction of vegetables in the crop rotation and adoption of recommended practices is very important. The use of vegetables in intercropping also helps in increasing the incomes of farmers during the period when the fruit orchard has not become commercially viable. The diversification plan of the horticulture sector needs to identify potential crops area wise and the area under low yielding vegetables and fruits should be shifted to more productive and profitable one.

There is a strong need to strengthen the research on horticultural crops to develop demand-driven technology by improved variety, pest management, etc., in both public and private sectors. These technologies should be quickly disseminated through government institutions, NGOs and even private participants by encouraging farmers' participation and upgrading their technical capabilities. A small targeted group of talented stakeholders and small entrepreneurs should be identified and assisted in obtaining access to funding, technical assistance and supply chain partners. This small group should be actively encouraged to develop business plans and feasibility studies. The horticultural development requires a minimum set of basic production factors, an optimal crop management infrastructure, post-harvest infrastructure, entrepreneurial management and horticultural expertise, logistical infrastructure and supporting financial infrastructure

Development of horticultural sector should be accompanied by the growth of the agro-processing industry. The opportunity exists to promote the industry by intensifying production of a required, appropriate variety of tomatoes, cucumbers, mangoes, pineapples, lemons, etc., for the products like ketchup, sauce, juice and pickles. Thus the production strategy should target not only meeting the domestic and export demand of fresh products but also of the processed products. There is the need to improve post-harvest operations related to handling, storage and marketing of fresh and processed produce. Volumes saved in post-harvest losses are actually the surpluses generated, without additional cost. The horticulture sector has an immense potential of generating employment. Additional employment can be generated by developing the horticulture-based agro processing units. This sector needs to be developed as an organized industry and has to be run collectively by all the stakeholders with farmers as entrepreneurs.

The sale of fruits is generally through pre-harvest contractors, so that the farmer gets an advanced payment and covers his risk. Vegetables are usually sold through commission agents and very little of pre-harvest contracting is done. Due to this the net returns are generally low. Farmers devote more time to their field crops rather than to the orchards. If the farmer does the marketing of his produce himself then the net returns to him would double. The marketing cost of fruits and vegetables is almost 50 per cent of the total cost of production, thus, there is a need to set up institutional agencies that can advance credit to farmer and motivate them to market the produce themselves. Post-harvest losses in horticultural crops range from 15-50 per cent. At micro level these losses increase the marketing cost of the product and at macro level they also reduce the per capita availability. Thus there is need to develop technologies, methods and mechanics to reduce these losses. There is a need to remove the distortions in the present supply chain and create better integration between the different links of the supply chain and reduce the losses. This will result in a net gain to producers, consumers and to the nation. Food processing also needs to be strengthened to create demand for agricultural produce, cut down or eliminate post-production losses and provide value added products. Investment in food processing, especially in rural areas would lead to a higher realization to farmers as this would generate additional demand farm products as well as create employment opportunities in the non-farm sector crating a virtuous cycle in the process.

Farmers usually procure inputs from the retail market and end up selling their produce in the wholesale market. Buying at retail price and selling at wholesale price is the most uneconomic way doing the business. Thus the involvement of an institutional structure in coordinating the demand of individual farmers can reduce the total cost of inputs to them. The market needs to be demand driven rather than supply driven. The price of the produce should not be based on the prevailing wholesale price but on the basis of cost of cultivation of that produce. Farmers should be their own price setters rather than price followers. There is also an immediate need to integrate the production, marketing and processing processes of the produce to get maximum benefits from fruits cultivation. Contract farming, cluster farming and cooperative farming are mechanism whereby the corporate sector can also help in establish linkages between farmers and markets. This needs to be backed by ensuring effective dispute settlement mechanism along with adequate awareness, information and support to small farmers

India horticulture sector needs to exploit its export potential by planning its production (reducing fluctuations in supply), improving the tractability and packaging mechanism and by targeting the right market. The window of international demand for the horticultural products is very small. Thus a planned strategy is to be made to target the markets during that period. India can lead in the export of non-traditional fruits like cassava, sapota, litchi and guava. Further, successful implementing of the core marketing strategies will help in future expansion of the domestic and international markets. In this regard the potential states have been identified by NHM, but the exports face certain non tariff barriers too. To enhance exports there is a need to develop air transport cargo system specialized for fresh fruits and vegetables, along with the airports, roads and rail connectivity with the area of procurements. The country's capability to generate surpluses for exports depends on its ability to tab the potential of small farmers. Thus there is a need to develop the credit structure and farmer export cooperatives. Awareness should be created among the farmers to understand the requirements of the importing countries. These non-tariff barriers need to be understood to get a breakthrough in the market access. For this, assistance from APEDA and exporters' association as well as training to the farmers is necessary. Quality control and longer shelf life are crucial for exports. Organic production of fresh fruits and vegetable is important to capture markets in Europe.

India can look forward to emerge as a major producer of horticultural products and thus secure reasonable market access for its agro exports, which are largely dependent on the competitive technologies that will help in enhancing export potential. This development will also help in overall growth of the economy through generation of extra foreign exchange, creating employment opportunities and also upliftment of the small and marginal farmers, with definite positive implications on income and employment. Thus, it is important that the government should create a positive environment that will ensure a mutually beneficial relationship between farmers and organized sector. Along with investment in infrastructure, development of extension activities and linkages with farmers is also an important area where the government can play an influential role.

REFERENCES

Kumar, Praduman and Surabhi Mittal. 2003. Crop diversification in India: Analysis by state and farm size group. *Agricultural Situation in India*. Special Number, August 2003. Directorate of Economics and Statistics, Ministry of Agriculture, Government of India, pp 273-280.

Kumar, Praduman and Promod Kumar. 2003. Demand, supply and trade perspective of vegetables and fruits in India, *Indian J. Agril. Marketing*, **17**(3): 121-130.

Kumar, Sant, Joshi, P.K. and Suresh Pal, 2004. Impact of Vegetable Research in India. *NCAP Workshop Proceedings No. 13*. Published by National Centre for Agricultural Economics and Policy Research, ICAR, India.

Mittal, Surabhi, 2006. Structural shift in demand for food: Projections for 2020. *ICRIER Working Paper* No **184**, August 2006.

Mittal, Surabhi, 2007. Can horticulture be a success story for India? *ICRIER Working paper No. 197*. August, 2007.

NHB, 2008. *Indian horticulture database 2008*. http://nhb.gov.in/database2008.pdf

Raghunath, S, Ashok, D. Puneet Prakash Mathur, Thomas Joseph. 2005. *Indian agricultural produce distribution system -- Towards an integrated agri produce flow*. Indian Institute of Management. Bangalore.

10. INTEGRATED HORTICULTURE AND NUTRITION DEVELOPMENT: LESSONS LEARNED FROM FAO PROJECTS

Biplab K. Nandi[1] and Lalita Bhattacharjee[2]

[1] *Former Senior Food and Nutrition Officer,*
FAO Regional Office for Asia and the Pacific, Bangkok
[2] *Nutritionist, National Food Policy Capacity Strengthening Programme,*
Food and Agriculture Organization of the United Nations, Bangladesh

ABSTRACT

Food insecurity and malnutrition result in serious public health problems and lost human potential developing countries. The major problems faced by the poor include, low productivity, food insecurity and malnutrition, despite the productivity gains good food production the world is still some way from meeting the Millennium Development Goals. This is because the focus has been on reducing hunger by supplying more calories rather than a more balanced diet. Rightly it had been noted that the 'right to food needs to become right to good food". Adopting a nutrition orientation to food production offers grater possibilities for long-term sustainability of food resources in the communities.

The fallowing recommendations are suggested for consideration:

- Crop diversification in addition to horticulture, small livestock and fisheries production should be strengthened in all food policy and action.
- National food policy should strengthen achievement of balanced production and availability of a range of diverse foods.
- Augmenting the production of pulses, vegetables and fruits should be strongly considered given the decrease in pulse production world wide.
- There is need to widen reliance on variety of foods, including traditional, underutilized, indigenous plants, fruits and vegetables and promote dietary diversification.
- Agro-based activities at household and community levels (nutrition food preparations, processing, preservation and storage of foods) would need to be expended and strengthened.
- Safety and hygienic handling, wise purchase should be an integral part of fruit and vegetable promotion programmes.

- Strengthening the implementation of nutrition education through Food Based Dietary Guidelines (FBDGs), School Nutrition Programmes, Information Technology at household and community levels should be considered as a policy tool by agriculture, health and food sectors.
- Intra familial food distribution issues-strengthen nutrition education strategies in national and community level programmes should be considered as strategic policy priorities.
- There is need to have in place early warning systems –Food Insecurity and Vulnerability Information Mapping Systems (FIVIMS), vulnerability, food and nutrition indicators for taking timely practical action and measures.
- There is need to promote urban agriculture activities and ensure food availability within urban household, slum communities, urban labour sectors, etc,.

1. INTRODUCTION

Achieving food security and nutritional well-being is central to the achievement of the Millennium Development Goals (MDGs) and the World Food Summit (WFS) targets. Commitment towards and achievement of these goals and targets remain a major challenge for the world's community and Asia in particular. Amidst these challenges, it is encouraging that analysis of recent trends in FAO's flagship publication *The State of Food Insecurity in the World 2008* shows a reduction in the number of undernourished in Asia and the Pacific, although at a pace too slow to meet the target of halving hunger by 2015.

FAO recognizes that healthy, well-nourished people are both the outcome of successful social and economic development and constitute an essential input into the development process. Achieving the WFS nutrition-related goals requires that national and sectoral development policies and programmes are complemented by effective local action aimed at improving household food security and promoting the year-round consumption of nutritionally adequate diets. These actions need to occur within the framework of promoting sustainable livelihoods and need to address variety of relevant local issues leading to various forms of poverty and malnutrition, including problems of chronic and seasonal food shortages, lack of dietary diversity, inadequate family care and feeding practices, and poor living conditions. Emphasis is also being placed on interdisciplinary activities to assist national and district level stakeholders and communities to identify and implement intersectoral strategies and actions that lead to sustainable reductions in malnutrition.

These activities are being actively pursued by the FAO Regional Office for Asia and the Pacific as part of its regular field programme. FAO is providing technical assistance to national governments and related stakeholders to better plan, target and monitor food security and programmes. In collaboration with the FAO Representations in the countries such as Bangladesh, India, Indonesia, Lao People's Democratic Republic, Sri Lanka and Viet Nam the FAO Regional Office has undertaken initiatives to promote integrated home gardening in selected villages of the countries targeting vulnerable groups of rural farmer households. The paper provides an account of projects implemented in selected Asian countries by the Ministries of Agriculture and related partner ministries with technical support from FAO.

2. PROGRESS IN FOOD SECURITY AND NUTRITION

Trends in undernourishment show the number of undernourished people in the world at 848 million in 2003-2005. 832 million in the developing countries, 541 in Asia and 313 in South Asia (Table 1)

Table 1. Prevalence of undernourishment

Region	No. undernourished (millions)	% undernourished (millions)
World	848	13
Developing countries	832	16
Asia	541	16
South East Asia	87	16
South Asia	313	22
India	230	21

Source: FAO, SOFI, 2008[1]

Over the past two decades, progress has been made in reducing undernourishment in the developing countries. The incidence of undernourishment has declined from 28 percent of the population two decades ago to 17 percent at present. However, as a result of population growth, the decline in absolute numbers has been slower. In addition, the decline was far more pronounced in the course of the 1980s, but appears to have slowed down in the 1990s. It must be pointed out that most of the improvement has been concentrated in Asia and the Pacific, which halved the incidence of undernourishment over the past two decades. Recent estimates show that the current global food crisis and economic situation has pushed the number of hungry and undernourished to over a billion.

Food Consumption Pattern and Nutrition

While there has been a rapid increase in food production in Asia, this has not been sufficient to address nutrition needs, in particular micronutrient needs of the populations. The emphasis has largely been on the augmentation of production of wheat and rice and this has resulted in substantial increase in the per capita availability of these major cereals. Pulses and legumes, which contribute to the nutrient quality of cereal-based diets has lagged behind in production, resulting in a substantial reduction in the per capita availability of these latter food grains.

1 FAO (2008) State of Food Insecurity in the World, FAO, Rome.

Figure 1 gives the share of various food groups as percentage of dietary energy supplies (kcal/capita/day) in selected countries of South Asia.

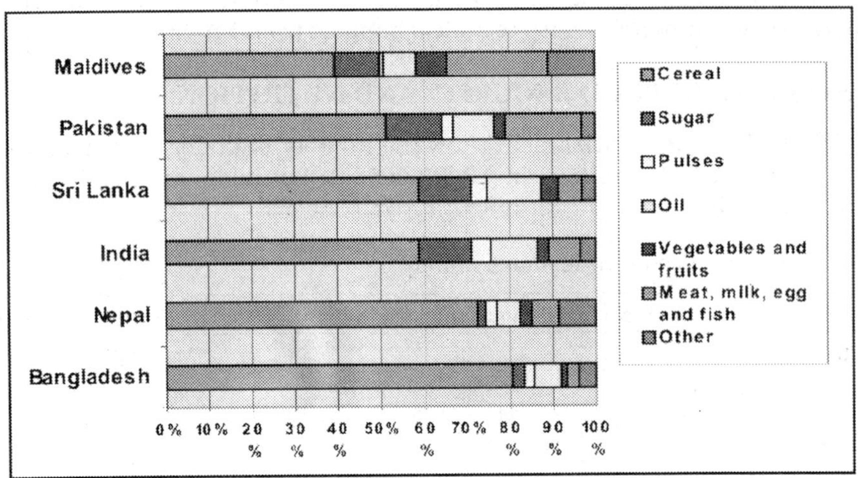

Fig. 1. Share of food groups as percentage of total dietary calorie supply in South Asian countries, 2005 (Source FBS, FAO 2005)

South Asia is the poorest region in the world after Sub-Saharan Africa. The following is an analysis of the dietary calorie supply[1] and nutrition status of South Asian countries, drawing data from FAO Food Balance Sheets[2]. This was done using the FAO method involving the estimation of the distribution function of dietary energy consumption on a per-person basis. During the decade from 1995 to 2005, all countries of the region, except Pakistan, improved their food calorie supply, but the average per capita food calorie supply of the region is still the lowest compared to the other regions of Asia and the Pacific. Within the region, Bangladesh, the second poorest country after Nepal, has always the lowest per capita calorie supply, with highest dietary energy supply from cereals (DES Cer %, an indicator of poverty), to the tune of 80%, followed by Nepal, 72%. Among the countries, Maldives has the highest per capita Gross National Income (GNI) (US$2680), the highest per capita calorie supply (2600 Kcal) and the lowest DES Cer % (39%). Maldives and Pakistan are the highest animal products consuming countries, but most of it is fish (marine) in Maldives and beef and milk in Pakistan. The diet in Maldives is also rich in fruits and vegetables (7.3% of total calories), compared to only 2.6% in Pakistan and <2% in Bangladesh. The data reflect poor diversification in the diets of Bangladesh and Nepal and

1. Calculated as dietary energy supplies by FAO

2. Yusuf, HKM, Bhattacharjee L, Nandi BK (2008) Trends and patterns of dietary energy supply and nutrition status in South Asian countries, 1995 -2005, S Asian J of Pop and Health, 1 (1): 1-12; for the trend and pattern of dietary intake (calorie supply from different food groups) in the 6 countries of the region, data from the FAO Food Balance Sheets were taken for three year points during the decade of 1995-2005 (1995, 2000 and 2005, respectively) (FAOSTAT, 2007).

Pakistan. The diets of other countries, particularly Maldives, are relatively more diversified.

Prevalence of undernutrition in total population is the highest in Bangladesh (30%) and lowest in Maldives (10%). Bangladesh, India, Nepal and Pakistan experience high under-5 stunting and underweight rates in the range of 43% to 51% and 47-48%, respectively. Sri Lanka has the lowest child stunting rate (about one-third that of Bangladesh, India or Nepal and one-half that in Maldives). One-way ANOVA test of the data shows that total per capita caloric supply is significantly (r^2 = 0.937, p = 0.001) negatively associated with % undernutrition in total population. There is also a positive, but insignificant (r^2 = 0.511, p = 0.110), association between national income and total per capita calorie supply (data not shown). Thus, national income largely, but not wholly, determines food consumption and nutritional status of a country.

3. IMPLICATIONS FOR FOOD AND NUTRITION SECURITY

Promoting dietary diversity and the use of locally-available, nutrient-rich foods has been recommended as an important component of food based strategies for sustainable dietary improvement. Decreases in dietary diversity, linked to increased availability and thus consumption of cereals and cereal products in developing countries, have in turn been linked to poorer nutrition. Cereals have also tended to replace indigenous and traditional food crops that supply macro and micronutrients in the diets of rural and poor communities. Many of these traditional staple food crops contain more iron and other micronutrients than the wheat, rice and maize that have replaced them. Further, urbanization and changes in the employment patterns of populations have affected the dietary patterns and lifestyles. Along with this dependence on too few plants, many plants have been either forgotten or neglected, while others still require research on their potential use and value. There is an obvious and crucial link between the maintenance of the diversity of food crops and effective strategies that ensure optimum nutritional status.

4. AGRICULTURAL AND FOOD POLICIES

Current agricultural policies largely emphasize primary agricultural production and do not explicitly have micronutrient outcomes among its goals. Accordingly, the thrust is mainly on staples which provide energy and protein with much less emphasis on fruits and vegetables which are rich sources of micronutrients. There is also lack of integration at the farm level between plant and animal production. This is particularly true for small livestock and fish that are particularly important in improving the micronutrient value of diets of rural communities in developing countries. Effort also needs to be directed toward the cultivation and use of edible, indigenous plants as sources of micronutrients. Ample evidence exists on the availability of wild foods in communities, particularly local green leafy vegetables and fruits that are rich sources of pro-vitamin A[1] .

Governments need to acknowledge that the micronutrient deficiency problem exists and they should take the lead in sensitizing the public on the magnitude of the problem, its causes and effects, and about possible strategies. A well developed policy environment could facilitate and support small scale and semi commercial farming systems that maximize outputs of micronutrient-rich foods. Nutrition considerations need to be seriously integrated into agriculture, livestock, aquaculture and related programmes. Alongside, multi sectoral approaches with greater government support for district and community leadership are critical. Recent exemplary efforts have been made in some Asian countries towards implementing food policies that are comprehensive, pro poor and encompassing food security in all its dimensions including food production and availability, access to food and utilization of food for nutrition[2]. A plan of action has also been developed under the leadership of the Ministry of Food and Disaster Management in collaboration with 11 partner ministries and departments to monitor the implementation of the National Food Policy Plan of Action. Technical support is being provided by the FAO National Food Policy Capacity Strengthening Programme.[3]

5. SELECTED FAO PROJECTS IN ASIA

FAO case studies[4] from Asian countries have demonstrated nutrition improvement outcomes through integrated food based strategies as being sustainable mechanisms for overcoming problems of food security and nutrition. Plans of action for food and nutrition policies have incorporated dietary diversity and related nutrition indicators for suitable guidance in agriculture and poverty reduction strategies

5.1. FAO–GoB-UNDP Integrated horticulture and nutrition development project: Bangladesh

An effective case of increasing the production and consumption of micronutrient rich vegetables and fruits and in increasing the diversity of the diet is demonstrated through an integrated horticulture and nutrition development project (IHNDP) in Bangladesh. As part of the IHNDP's food

---

1. Tontisirin K, Nantel G, Bhattacharjee L (2002) Food based strategies to meet challenges of micronutrient malnutrition in the developing world. Proceedings of the Nutrition Society, 61: 243-250.

2. Bangladesh National Food Policy (2006)

3. FAO National Food Policy Capacity Strengthening Programme with funding support from EC and USAID and implemented in collaboration with the Food Planning and Monitoring Unit, Ministry of Food and Disaster Management, Bangladesh

4. Bhattacharjee L, Saha S and Nandi BK (2007) Food based nutrition strategies in Bangladesh: Experience of horticulture and nutrition development, FAO Regional Office Publication, Bangkok.

based strategy, activities were implemented to include the total food chain activities ranging from production in the farm, group garden and individual home stead garden, including procurement from local markets, to processing domestically in preparation of meals to consumption within the home.

IHNDP illustrated the links between different horticulture components that prevent micronutrient malnutrition and promote food security, health and livelihoods. It demonstrated the effectiveness of household production in addressing undernutrition and micronutrient deficiencies in Bangladesh.

A nutrition strategy based on household production is more sustainable than external intervention in the form of supply of drugs and centralised fortification facilities outside the control and supervision of the beneficiaries. Strengthening it with institutional support should be a key government concern. Dietary diversification through horticultural food intake and supported by nutrition education is, therefore, seen as a sustainable approach to fighting micronutrient malnutrition.

The overall objective of IHNDP was to improve the efficiency of the horticultural system and associated support services in the project areas using modern technologies to ensure food and nutritional security. The central aim was to create conditions for diversified food production for consumption by providing rural farmers with necessary knowledge, technology and skills to make nutritious food available to their community and improve household dietary quality. With an immediate objective of "increasing nutritional awareness among the beneficiaries and develop a comprehensive food-based nutrition programme to reduce malnutrition in the target groups comprising mainly women and children", the nutrition component of the project aimed to diversify the food habits of the target groups by promoting consumption of horticulture-based food as a sustainable solution to the problem of micronutrient malnutrition.

The project's Nutrition Education Strategy set up community-based nutrition education programmes to create nutritional awareness among various groups such as women, farmers and school children. Mass media educational messages and programmes on the advantages of consuming vegetables and fruit were developed and disseminated.

Strengthening links between horticulture and nutrition

As part of strengthening links between horticulture and nutrition a variety of micronutrient rich vegetables and fruits are being promoted through the project and include different types of leafy vegetables (Indian spinach, stem amaranth), carrot, country beans, red pumpkin, tomato, broccoli, garden peas, okra, onions and green chillies.

Vegetable production and nutrition supply

Assessment on the yield of vegetables in the demonstration plots at different locations during the two seasons, summer and winter provided valuable insights on the nutritional contribution of the vegetables yielded on the farm and home stead gardens. Based on the average yield of horticultural produce including tomato, garden peas, carrot, Indian spinach, okra and stem amaranth from the farm or group garden, the available yield per household and per capita was calculated (Figure 2).

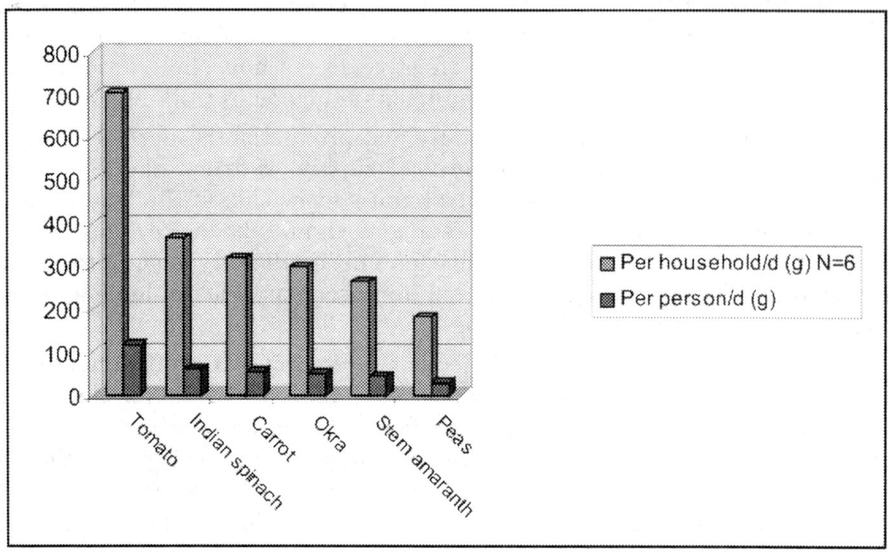

Figure 2. Average vegetable yield (g) per HH and person/day

It was noted that about 201 g of vegetables /day was available per person /day from the group garden. This is encouraging since it is close to the recommended figure of 215 g/day for Bangladesh. The National Plan of Action for Nutrition (NPAN) for Bangladesh had recommended that the daily consumption of vegetables from the present level of 72 g per capita per day be increased to the recommended amount of 215 g. Approximate nationally recommended requirements were vegetables 200 g/person/day, fruits 70 g/ person/day, potato 125 g/person/day and sweet potato 25 g/person/day. The present production of horticultural crops is far below than the requirement[1]. The demand of the crops will increase further by 2010 with the increase of population. At the national level, special efforts have reportedly been made in the horticulture sector to produce major fruits and vegetables on a year round basis through application of appropriate technology. Accordingly, the development of the horticulture sector has received governmental priority in addressing micronutrient malnutrition through the Integrated Horticulture and Nutrition Development Project.

Promoting year round fruit supply

In subsistence agriculture country like Bangladesh, the nutritional status, and to a large extent health, depends on access to the quantity and

1. Hossain SMM (2000) Paper on "Horticulture Research in Bangladesh", Horticulture Research Centre, BARI, Bangladesh

quality (including diversity) of food supplied by agriculture. Expanding and improving household integrated horticulture production systems to help in promoting nutrition and address changes in rural livelihoods are strategies that can be promoted. IHNDP demonstrated models of high density mixed fruit orchards and the multi bed vegetable systems that can be promoted to ensure a regular and substantial supply of micronutrients to the household (Table 2).

Table 2. Estimated production of fruits over a ten year period/day/household (g)

Species	No. trees	Year 1	Year 2	Year 3	Year 4	Year 5	Year 6	Year 7	Year 8	Year 9	Year 10	Total
Mango	2	0	0	40	160	300	438	570	570	570	570	3.24
Jackfruit	2	0	0	0	0	0	164	300	438	570	570	2.21
Drumstick	5	60	130	200	270	270	270	270	270	270	270	2.32
Guava	6	40	410	410	270	190	136	80	80	80	80	1.79
Papaya	8	240	190	0	0	0	0	0	0	0	0	438
Banana	6	330	410	130	50	50	50	50	50	50	50	1.28
Hog Plum	6	100	270	410	270	240	210	191	164	164	164	2.21
Lime	4	50	210	210	160	130	100	109	109	109	109	1.34
Lemon	2	20	210	210	160	130	100	109	109	109	109	1.31
Custard apple	2	0	0	100	100	80	50	20	0	0	0	383
Carambola	4	40	410	410	540	680	820	958	1.09	1.09	1.09	7.0
Acerola	2	50	80	100	136	136	136	136	136	136	136	1.20
Pineapple	204	0	750	270	0	0	0	0	0	0	0	1.0
Total		934	2.83	2.34	1.89	1.97	2.24	2.54	2.76	2.90	3.06	23.5

The horticulture technology models showed that a wide range and quantity of fruits can be produced over a 10 year period at the household level. All these fruits are rich in vitamins and minerals (particularly vitamin A, vitamin C, potassium, iron) commonly known as micronutrients. Accordingly, they will contribute to both micronutrient food security and dietary diversity. Right in the first year, one can estimate a total availability of 155 g of fruit /person/day. Even if ½ of the quantity is consumed, it would serve to provide about 77 g of fruit which is much higher than the current consumption of 34 g/day and it would greatly help bridge the micronutrient gaps in the diet.

Figure 3 above shows the vitamin A supply[1] (as ß-carotene) from fruits/person/day which is noted to be adequate in providing a substantial part of the household's micronutrient requirements, right from the 1st year

1. In terms vitamin A availability , the bioavailability factor used for beta carotene is : 1 mcg beta carotene = 0.25 mcg retinol

itself. The recommended dietary allowances[1] for β-carotene for an adult are 2400 μg. Even if only papaya and drumstick leaves are taken, it would serve to furnish 97% of the RDA for β-carotene. In the 2nd year it is noted that these two foods would provide nearly twice the RDA for Vitamin A. For subsequent years the contributions from mango and drumstick leaves would be more than three times the RDA.

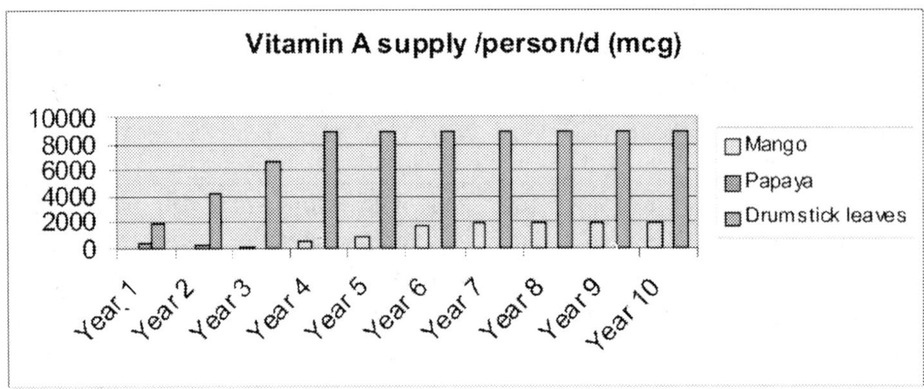

Figure 3.

Nutrition education, food preparation and processing technologies

The nutrition education programme played an important role in supporting the vegetable production programme. Through the food preparation technologies transferred through the demonstrations, women farmers were encouraged to prepare and consume horticulture based nutritious recipes in the household and promote their intake among their families. The excess yield was sold to provide some income. Accordingly, the community and home garden produce made substantial contribution to meeting the day's nutrient allowances. Tomato are a good source of vitamin C providing 80% of the recommended dietary allowance (RDA) and a little over ¼ the RDA for vitamin A as beta carotene. Carrot is an excellent source of vitamin A and contributes to twice the amount of the RDA (215 %). Indian spinach provides nearly ¾ of the RDA for vitamin A, and above the RDA for vitamin C (130 %). Stem amaranth also provides adequate vitamin contributing to 92 % of the RDA for beta carotene and about ¼ the RDA for calcium.

Vegetable and fruit consumption in project areas increased while programmes on food intake methods and young child feeding promoted changes in nutritional behaviour among children. Nutritional improvement also resulted from simple agro-processing technologies such as dehydration, pickling, bottling, pulping and preparing preserves and relishes from a variety of vegetables and fruits. This built synergies between nutrition and

1. Recommended daily allowances for Indians (2004), Indian Council of Medical Research (ICMR).

food processing by reducing micronutrient losses, increasing shelf life, supplementing daily nutrient intake and adding value to some products.

Techniques to preserve micronutrient (vitamin A) rich foods using drying and pickling were developed and strengthened and were scaled up for application at field level. Programmes worldwide have shown advantages of simple methods of processing wherein examples of solar drying and production of leaf concentrates which have demonstrated the advantage of reducing the volume of the leaves and increasing the concentration of pro vitamin A carotenoids. Lack of proper processing and preservation technologies force farmers to sell the produce immediately after the harvest when prices are low. A number of options have to be developed for product processing that increases longevity, shelf life and nutrition so that products could be marketed at higher prices as well as consumed at home.

School nutrition programme

The school nutrition programme implemented in 30 schools and covering 1800 adolescent girls showed encouraging trends in consumption of vegetables and fruits and provided diversity in the habitual diet. Evaluation of food frequencies at a quarterly interval of the programme showed that that over ¾ of the children were seen to be taking leafy vegetables at least every other day (Figure 4).

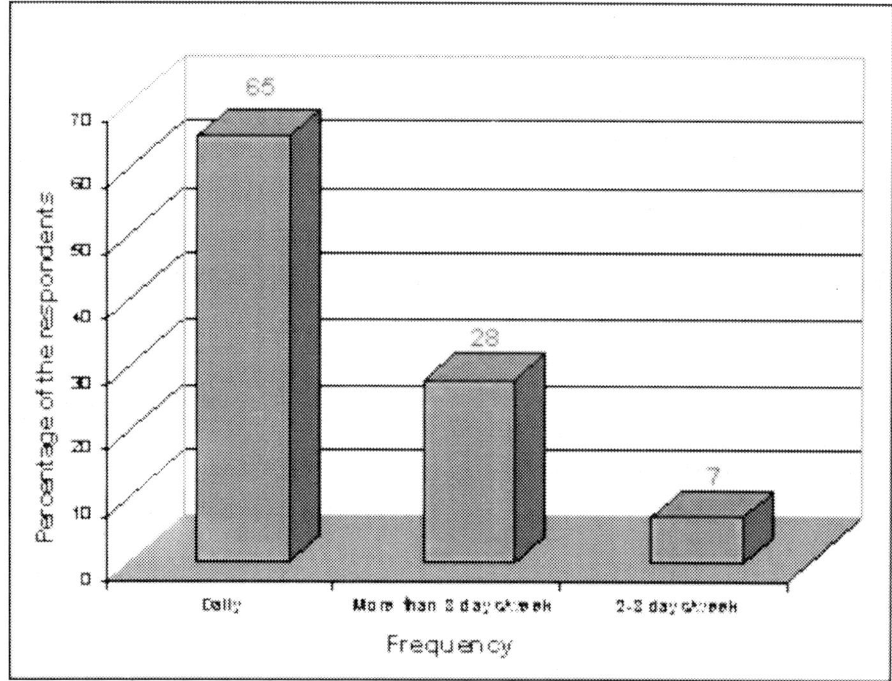

Figure 4. Frequency of consumption of vegetables by students.

Figure 5 shows the fruit consumption which was much lower compared to the consumption of vegetables with around ½ of the children taking fruits between 2-3 times/week.

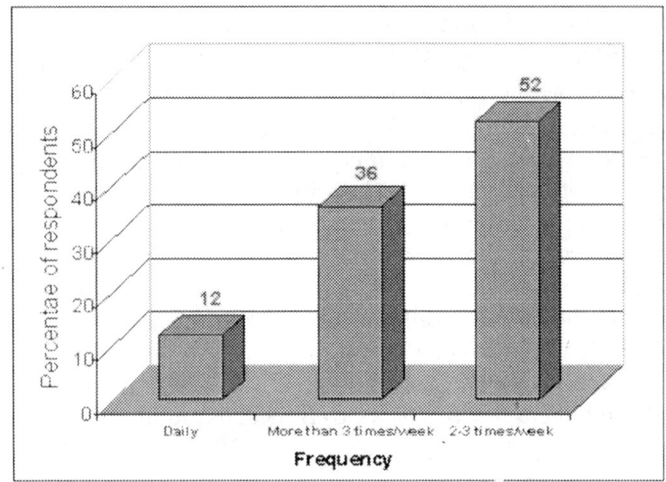

Figure 5. Frequency of conumption of fruits by students

Fruits are generally more expensive than leafy vegetables and as many of the households did not have home gardens, they did not have ready availability of fruits on a regular basis. Students particularly mentioned the need for setting up a home garden in their households. The school garden programme is rightly promoting the growing of local fruits such as guava, mango and papaya which are rich sources of micronutrients such as vitamin C and vitamin A.

Dietary improvement outcomes

Results of the food consumption and dietary assessment[1] among IHNDP covered households showed improvement in food consumption and dietary pattern as compared to non IHNDP covered households. (Table 3).

Table 3. Summary of food consumption and dietary pattern

(% households)

Indicator	Project households	Non-project households
Leafy vegetables frequency (4-7 times/week)	52	23
Yellow orange vegetables and yellow orange fruits frequency (1 time/week)	85	45
Vegetables and fruits consumed	46 -64%	22 -54%
Vegetables sold	9-20%	5-10%
Given to relatives, friends	10-20%	5-10
Meat, fish, poultry, milk frequency (1-3 times/week)	96	86
Initiation of complementary feeding at 6-7 months	62	38

1. Conducted in collaboration with Institute of Nutrition and Food Science, Dhaka University, (2004)

Few significant findings were:

- Project covered households consumed between 46 to 64% of vegetables and fruits produced in their home gardens and sold between 9 to 20 % of the produce which was significantly higher than that of non-project covered households.
- Consumption of leafy vegetables, yellow and orange vegetables as well as vitamin C rich fruits increased in the project households.
- Over 60% of project covered households started giving complementary foods to infants between the ages 5 to 7 months as compared to only a third of non project households who did so. Continuation and practice of breast feeding alongside was also being seen.
- The project households brought significantly more area under rice and vegetables and produced significantly more number of summer and winter fruits, leafy and non-leafy vegetables and spice.
- The project households consumed and sold significantly more summer and winter leafy and non-leafy vegetables and spice, and significantly fewer project households suffered from food insecurity.
- Significantly more project households consumed leafy vegetables, yellow and orange vegetables, yellow and orange fruits, sour fruits and vegetables and vitamin C rich fruits.
- Significantly more project households had rice, pulse soup and vegetables, and rice, fish and dark leafy vegetable as habitual diet.
- Significantly more project household children were initiated complementary feeding on time and more project household children relished leafy vegetables, potato, pumpkin, tomato and okra.

5.2. Integrated home gardening for nutrition improvement in Lao PDR

Promotion of home gardens through community based approaches showed that it could be an efficient way of tackling problems of malnutrition in Lao PDR. Growing vegetables in the home gardens is a traditional practice in Lao PDR, with gardens varying in size, biodiversity and seasonal produce being adapted to local resources and preferences. Vegetables are grown scattered in the courtyard, in pots or on roof tops. These traditional gardens do not use improved technology, grow only limited number of vegetables and fruit varieties and are generally maintained during the course of the year with limited means. Most houses also raise chicken while some keep pigs. While these serve as a source of food for the household they also serve as an important source of income. Most rural households produce vegetables mainly for home consumption and the yield produced is so low that often does not even meet the household requirements. It was within this context that FAO in collaboration with the Government of Lao PDR executed the "Pilot Project on Promoting Home Gardens for Improved Nutritional well being" in selected villages of Central Lao PDR. An integrated home gardening programme that included the production of horticulture crops (vegetables and fruits), small livestock (poultry and pig) and fish supported by a nutrition education component was implemented.

Identifying and promoting nutritious foods

The home gardening programme planned in collaboration with the Ministry of Agriculture and Forestry, the community and the FAO team promoted a number of different types of foods rich in vitamins and minerals (micronutrients). The gardens were planned to provide year round supply of nutrient rich foods that were compatible with local taste preferences. These included a variety of leafy vegetables, beans, gourds and other vegetables such as hot pepper, pumpkin, tomato and herbs like lemon grass, coriander, etc. Fruits included mango, custard apple, jackfruit, pomelo, sweet tamarind, lemon, longan, litchi, guava and sapodilla were also planted in the community. Poultry including chicken and duck while local species of fish namely tilapia, silver carp and common carp were also promoted.

Improving home gardening

Prior to project commencement of the FAO project, vegetables were grown in 50% of households within a limited area. However, due to poor techniques and poor care, the yield and quality of the garden produce were low and reportedly met just 10-15% of the household consumption needs. Almost all vegetables were also being grown only in the dry season. The project thus focused on improving the home gardening through the use of a net house technique for year round production of vegetables. Within one year, the production of vegetables grown in the net house reached 73.480 tons, amounting to a per capita production of 96 kg/person/year. From August 2003 when the seeds were provided to farmers until August 2004, each household produced 8-9 crops and the total vegetable production was 73.480 tons working out to 245g of vegetables per capita. As per the present national production levels, vegetable crops are grown in about 31,000 ha, which constitutes about 3.9% of the total area under all crops, producing a total output of 117,300 t amounting to an annual per capita availability 23.5 kg compared to the minimum required level of 73 kg. The production and yield output through the project was therefore much higher (245 g vegetables/day vs present per capita daily availability of 64.3 g).

Nutrition education strategies

A Nutrition Education Strategy was implemented that aimed at improving the nutrition and health practices of pregnant and lactating women and children aged 0-5 years. The key and desired behaviour changes related to breast feeding, complementary feeding and management of childhood diseases, improving nutritional status of mothers by improving their food and nutrient intakes and health seeking behaviours. To this end, nutrition education activities included food preparation and complementary food demonstrations and growth monitoring and promotion.

Increase in home garden food base

There was a slight shift in the source of vegetables to the households at the end of 18 months after the implementation of the project. The ratio of vegetables from the home gardens as compared to that from the forest and markets increased after the commencement of the project. The ratio of vegetables from the home garden increased by 35%, while the ratio of vegetables from forest and market sources reduced by 17 and 18% respectively as seen in Table 4.

Table 4. Sources of vegetables for households in target village (%)

TCP/LAO/2902

Villages	From forest		From market		From home garden	
	before	during	before	during	before	during
Ban Cheng	43.0	32.5	25.3	12.8	31.7	54.7
Ban Nakhoung	48.0	29.8	36.0	8.9	16.0	61.3
Ban Na tham	60.0	33.0	23.5	8.0	16.5	59.0
Ban Somsa ath	46.1	34.0	30.5	13.8	23.4	52.3
Average	49.3	32.3	28.8	10.8	21.9	56.8
Increase (+), Decrease (-)		-17		-18		+35.0

The production of vegetables and fruits provided the household with direct access to important nutrients (micronutrients) that were not either available to them or were not within their economic reach. The increased food production outcomes of the project showed that advancing the dietary diversification as an integrated strategy was vital in addressing nutritional deficiencies and related conditions.

Vegetables for consumption and income

A nutrition education programme was linked with the home gardening activities where nutrition messages were regularly given to the farmers. Awareness about diversity of vegetable sources as well as health contributions was raised. After one year of project implementation, it was noted that not only had the area for growing vegetables increased but the kinds of vegetables being grown had also become varied. Of the 29 kinds of vegetables recommended by the project, 26 were grown in home gardens. Before commencement of project, only a fourth of the households were growing these vegetables, but after the project period it was almost ¾ with an increase by half over a one year period. There was a shift in the source of vegetables after the implementation of the project. The ratio of vegetables from the home gardens as compared to that from the forest and markets increased. The production of vegetables and fruits provided the household with direct access to important nutrients (micronutrients) that were not either available to them or were not within their economic reach.

Food consumption

Consumption of leafy and yellow orange vegetables increased over the period of observation from 17.5% to 55.0% of households in the first semester and up to 80.0% in the post intervention period (p : 0.001). Additionally, consumption of wild fruits, leaves, and berries also increased over time (20.0%, 27.5%, and 60.0% respectively, p: 0.01). Households consumed fish more frequently, while other sources such as meat and chicken were added 4 to 6 times per week. Findings at baseline, first semester (6 months) and at the end of project completion showed a progressive improvement in the consumption of micronutrient rich foods among selected target households. The findings suggest that exposure to nutrition education led the target households to increase their consumption of fruits and vegetables. The main reasons cited by the households for increasing the consumption of micro-nutrient rich foods were the nutrition advice given by the project. The households also mentioned that they were expecting more production of food and that there would be more money from the sale of surplus.

Impacts on nutrition

Weight and height measurements of 196 children under five years were followed up beginning with a baseline conducted in April 2003, a semester measurement in October 2003, and post-intervention period measurement in July 2004. Children identified as undernourished were provided special dietary rehabilitation and their parents were given individualized dietary guidance on the preparation of nutrient dense recipes for their children. The baseline survey found that the prevalence of underweight (WFA < - 2SD from the median of the reference population" as moderate malnutrition) was 23.2% and 9.5% (WFA < - 3 SD from the median of the reference population" as severe malnutrition). After six months of implementation along with nutri-tion education, a semester anthropometric measurement showed that the prevalence of underweight declined to 20.4% of <-2SD and to 1.9% of <-3SD. The final anthropometric measurement was taken 9 months later and it was found that the prevalence of underweight declined to 15.9%. The rate of severe underweight declined from the baseline to post intervention period (p: 0.02) (Table 5).

Table 5. Percentage of moderate and severe underweight children in three periods of project implementation

Time	Number of children measured	Weight for age (Underweight)	
		<-2SD	<-3SD
Baseline (April 2003)	95	23.2	9.5
First semester (October 2003)	103	20.4	1.9
Post intervention (July 2004)	88	15.9	2.3
		p: 0.31	p: 0.02

The results of the project in demonstrating a decline in the prevalence of both severe and moderate undernutrition among children under five years, were considered important and note worthy indicative of the projects impacts on nutrition.

Monitoring and evaluation

A Participatory Monitoring and Information System (PMIS) provided the necessary base for evaluating the project implementation on a periodic basis. From a list of core indicators, a functional list of simple and practical indicators was used for monitoring each of the home gardening components. Accordingly, the village development committee was guided to combine the information obtained from each of the community collaborator groups responsible for each of the home gardening activities and combine it into an overall monitoring system.

Sustainability initiatives

Sustainability initiatives have commenced by the community and government wherein inter sectoral collaboration fostered by the project at district and provincial levels are serving to promote implementation of the home gardening activities on a broader scale. The improvement of the community's self supporting mechanisms through the model home gardens are seen to pave the way for sustainability. Training of community collaborators has been seen as important in sustaining the interest of the community in carrying on the activities. Group savings activities and small scale community based ventures have been initiated with every target household contributing between $ US 0.50 to $ US 1.00 towards savings each month. The targeting of households and communities also help to reduce costs, effectively reach the number of households at one time and enable selection and graduation of households into and out from the programme. The input supply of seeds, fruit tree saplings, fish and small livestock for home gardens supported with nutrition education has implications for food based nutrition improvement and serves as a cost effective method for promoting household food security and nutrition. The produce obtained from the community garden is being sold in the market and are serving as a source of income for the community. It is encouraging to note that the needs for developing and promoting community gardens have been determined by the villagers themselves and the detailed logistics /mechanism for this process have been discussed at the community development committee.

5.3. Integrated home gardening for nutrition improvement in Vietnam

The FAO/Government Cooperative Programme in Vietnam demonstrated a case of the community based approach in nutrition improvement and household food security specifically in the prevention of vitamin A deficiency in 1998-2000. The project used a preventative, educative, food based approach for reducing child malnutrition in rural Vietnam where 45% of the children were malnourished[1]. The approach combined targeting of households vulnerable to household food insecurity with home gardening for food and income and nutrition education. The project established that with low levels of inputs at commune level (18 US$ per household), rural households would improve the level of their children's nutritional status using diverse variety of foods produced in their own home gardens. Implemented in 100 communes in four provinces in Vietnam, some 10,500 households with malnourished children have benefited. A major impact of the project was the reduction of malnu-trition by 12 percent compared with non-project areas in the same provinces. The successful improvement of food production in 80 percent of households in the same two year period was a major achievement and indicates the feasibility of the household food–based strategy (Table 6).

Table 6. Impact on malnutrition and home gardens

Province	Rate of malnutrition in project sites (%)			Rate of malnutrition in other areas (%)			HHs with "improved home garden status" %
	1997	1999	Change	1997	1999	Change	
Phu Tho	53.1	42.8	10.3	35.9	39.8	-3.9	82
Nam Dinh	40.2	27.3	12.9	48.4	39.3	9.1	81
Binh Dinh	42.2	32.9	9.3	39.8	39.9	-0.1	78
Binh Thuan	39.4	20.8	18.6	35.9	39.1	-3.2	80
Average	44.1	31.9	12.4	40.0	39.5	0.5	80

Knowledge, attitude and practices (KAP) assessment provided insights in the typical food insecure families targeted by the project. Results indicate that the project has been able to improve nutrition and household food availability in a relatively short time, in particular nutrition knowledge and food diversity (Table 7).

Table 7.

Indicator	With project for 7 months	Without project
Improvement in child nutrition status	37%	0%
Mother with good knowledge of nutrition	65%	15%
Increased daily fruit and vegetables consumption	25%	5%
Consume more than 5 different foods daily	63%	10%

The impact of the project on food consumption and nutritional status of women and children showed statistically significant decreases in the prevalence of underweight for the three project communes surveyed and, for stunting, in two out of the three. These improvements were in excess of trends reported for national and regional trends and control communes.

1. Terminal Report, FAO Project, 2000

5.4. Prevention and elimination of micronutrient malnutrition, Tamil Nadu, India

As part of preventing and controlling micronutrient malnutrition in Pennagram block of Dharmapuri district in Tamil Nadu, India, the food based approach was promoted and testified in ultra poor communities. Expertise was provided in community and household horticulture along with nutrition education, training and practical skills development[1] .

By the end of 18 months, community nurseries were established in the project villages. Nutrition linked horticultural activities were imprinted in the minds of the community, with the message, that consumption of fruits and vegetables is important to improve health. Free distribution of seeds and seedlings served as incentives and appeared to have created considerable interest among the communities. Support as fencing, reclamation of land and provision of seedlings has been extended through the project. Specifically, participatory development and targeting of women through self help groups and labour intensive entrepreneurial initiatives have helped to reach the unreached in the community.

6. HYGIENE AND SAFETY OF FRUITS AND VEGETABLES SAFE AT HOUSEHOLD AND COMMUNITY LEVELS

Elements of hygienic and safe handling and processing of fruits and vegetables have been essential components of nutrition and consumer education in the FAO projects mentioned above.

Fruits and vegetables like all other foods need to be properly kept, handled and cooked. Though fruits and vegetables are generally considered safe, they are reportedly the second most cause of food poisoning after sea food, and cause a similar number of food poisoning to meat. Poisoning can be caused by contamination, damage or substances naturally present in fruits and vegetables[2].

Fruits and vegetables can get contaminated with bacteria, moulds or parasites anywhere along the food chain (Box 1)

Box 1: When vegetables and fruits can get contaminated

- While they are growing, for example if animal manure and pesticides are used.
- When they are picked, if the picker has unclean hands.
- During transport and packing.
- Where they are sold, from cross contamination or fom people handling them; or
- In the home, from other foods, dirty hands or dirty food preparation surfaces.

---

1. M. S. Swaminathan Research Foundation and FAO (2001). Promoting Household Nutrition Security, Baseline Survey Report in "Workshop on Alleviating Micronutrient Deficiency: Role of Horticulture and Home Gardens, Chennai, India.
2. FAO/WHO 1998. Surface decontamination of fruits and vegetables eaten raw :a review. WHO.

The only way to ensure that fruits and vegetables are as safe as possible is to take care at every part of the food chain. Even in home gardens there is need to look at the growing practices in the garden and kitchen.

Often consumers believe that there is great danger from pesticides and other chemicals used during production of fruits and vegetables. However, for all substances approved for use on foods, safe intake levels have been assessed. Each substance usually has a recommended maximum level per person and often maximum levels for individual foods. The Codex Alimentarius Commission, the International Body for Food Standards and Food Safety, recommends maximum residue limits take into consideration the amount of pesticides in food that can consumed daily over a lifetime without appreciable health risk to the consumer. These limits also have a built-in safety factor to account for differences in sensitivity.

Some fruits and vegetables contain natural toxins that can cause illness. Generally methods of food preparation will reduce the level of toxin to make it safe to eat. For example, cyanogenic glycosides are found in raw chaya leaves and also raw cassava. Cooking destroys the toxin and so cooking should make it safe to eat. Neither should be eaten raw, though, as they can cause serious health problems. Some toxins may be present when a fruit or vegetable is not ripe or when it is not stored properly. Green areas on potatoes contain high amounts of glycoalkaloids. These cause stomach pain and diarrhoea, although it is rarely fatal. It is best not to eat any damaged, green or sprouted parts of a potato. Bruised *apples* are often seen to be soft and brown in the bruised area. They contain a substance called patulin which slowly develops and can cause health problems. Any bruised areas on apples should therefore be avoided.

7. POLICY IMPLICATIONS FOR NUTRITION AND HEALTH

Low fruit and vegetable intake is a major contributing factor to such micronutrient deficiencies. According to the FAO/WHO[1] people are not consuming enough[2] vegetables and fruits and promoting the consumption of vegetables and fruits are becoming a global priority. Recent research has shown that a regular and increased consumption of a variety of vegetables and fruits provide substantial benefits of risk reduction for a number of non-communicable diseases such as cardiac disease, diabetes mellitus, degenerative eye disease as well as certain types of cancer. Nearly three million deaths and more than 26 million disability adjusted life years were attributable to low fruit and vegetable intake globally, according to previous data from 2000. Further, 31% of ischemic heart disease, 19% of ischemic stroke, 19% of gastric cancer, and 12% of lung cancer worldwide could be prevented by increasing dietary intake of fruits and vegetables to the minimum recommended five daily servings.

---

1. FAO/WHO (2003) Report of the Expert Consultation on Diet, Nutrition and the Prevention of Chronic Disease, Geneva.
2. FAO/WHO recommends a daily intake of 400 g of vegetables and fruits

Because food based approaches involve the production, marketing and consumption of indigenous and locally –available micronutrient rich foods, and the promotion of dietary diversity, they can assist economic development. It is vital therefore that agriculture related programmes being implemented at the national and community levels, intensify the nutrition component to their programme. An effective Nutrition Information, Education, Communication Strategy and Programme should be promoted using an integrated and multi sectoral approach. For example, in horticulture, there is need to examine the effects of the vegetable and fruit production programmes more closely on household level production. Generally such programmes have long gestation periods. Effects of food based strategies take longer to manifest, because such strategies "treat the root of the problem, which is insufficient nutrient rich foods, and involve attitudes and behaviour change"[1]

Inadequate evaluation, weak protocols and statistical analyses have been cited as impediments that slow progress in understanding the real potentials of interventions based on food production and nutrition education in controlling micronutrient deficiencies[2]. The bioavailability of micronutrients at the host level and food level are also aspects that need to be further studied. Although evaluations are often weak and do not allow firm conclusions about impact, there is a consistent trend indicating a positive association between food based approaches, food and nutrient intake and nutrient status in some cases.

While, these projects offer valuable lessons in using horticulture-based nutrition development strategies to improve food security and nutritional standards in other developing countries, establishing regional, national, and international food and nutrition policies is essential to ensure that all people at all times have equitable and dignified access to sufficient, safe, healthy, and culturally appropriate food including adequate fruits and vegetables.

8. CONCLUSIONS AND RECOMMENDATIONS

Food insecurity and malnutrition result in serious public health problems and a lost human potential in developing countries. The major problems faced by the poor include, low productivity, food insecurity and malnutrition. Despite the productivity gains of food production the world is still some way from meeting the Millennium Development Goals. This is because the focus has been largely been on reducing hunger by supplying more calories rather than a more balanced diet. Rightly it has been noted that the 'right to food needs to become right to good food".[3] Adopting a

---

[1] Kuhnlein, HV (2003) Micronutrient nutrition and traditional food systems of indigenous peoples.Food, Nutr. & Agri. (33): 33-37.

[2] Ruel MT (2001) Can Food Based Strategies help reduce Vitamin A and Iron Deficiencies? A Review of Recent Evidence, IFPRI, Washington, D.C.

[3] Swaminathan MS and Pedro Medrano (Eds) (2004). Towards a Hunger Free India from Vision to Action: Ethical Dimensions –Right to food.East West Books (Madras) Pvt.Ltd, 470 pp

nutrition orientation to food production offers greater possibilities for long-term sustainability of food resources in the communities.

The following recommendations are suggested for consideration:

- Crop diversification in addition to horticulture, small livestock and fisheries production should be strengthened in all food policy and plans of action
- National food policy should strengthen achievement of balanced production and availability of a range of diverse foods.
- Augmenting the production of pulses, vegetables and fruits should be strongly considered given the decrease in pulse production world wide.
- There is need to widen reliance on variety of foods, including traditional, underutilized, indigenous plants, fruits and vegetables and promote dietary diversification.
- Agro-based activities at household and community levels (nutritious food preparations, processing, preservation and storage of foods) would need to be expanded and strengthened.
- Safety and hygienic handling, wise purchase should be an integral part of fruit and vegetable promotion programmes.
- Strengthening the implementation of nutrition education through Food Based Dietary Guidelines (FBDGs), School Nutrition Programmes, Information Technology at household & community levels should be considered as a policy tool by agriculture, health and food sectors.
- Intra familial food distribution issues -strengthen nutrition education strategies in national and community level programmes, should be considered as strategic policy priorities.
- There is need to have in place early warning systems –Food Insecurity and Vulnerability Information Mapping Systems (FIVIMS), vulnerability, food and nutrition indicators for taking timely practical action and measures.

There is need to promote urban agriculture activities and ensure food availability within urban households, slum communities, urban labour sectors, etc.

11. CONTRIBUTION OF HORTICULTURE AND HOMESTEAD FOOD PRODUCTION TO HOUSEHOLD FOOD AND NUTRITION SECURITY AND DECREASED ANEMIA AMONG YOUNG CHILDREN - LESSONS LEARNED FROM SCALING-UP PROGRAMS IN ASIA (BANGLADESH, CAMBODIA, NEPAL AND PHILIPPINES)

Aminuzzaman Talukder[3, 5], Nancy Haselow[5], John Deidrick[5], David Spiro[2], Ellen Villate[4], Hou Kroeun[3], Amin Uddin[1], Jennifer Nielsen[6], and Victoria Quinn[6]

Helen Keller International [1]Bangladesh, [2]Nepal, [3]Cambodia, [4]Philippines, [5]Asia-Pacific Regional Office, Cambodia, [6]Head Quarter, USA

ABSTRACT

Micronutrient (MN) deficiency, particularly anemia, is a serious public health problem among young children and women throughout Asia. HKI and partners have been implementing homestead food production programs (HFPP) in Asia to increase and ensure year round availability and intake of MN foods by women and children. To review the impact of horticulture and HFPP on food security and nutrition and identify lessons learned for further scaling up in new areas of existing countries and to other countries. Baseline (BL) and end line (EL) data from HFPP programs in four countries covering 28,000 households (HH) between 2003-2007 were reviewed. Data, using pre-coded questionnaires, were collected from program HH participating in HFPP and from control HH (in similar communities with no HFPP) in non-HFPP villages in all countries. Blood samples were collected from ~1200 non-pregnant women and ~1000 children < 5y at BL and EL to measure change in hemoglobin (Hb) level, using hemocue. A review of the M&E process and obstacles to scaling up was also undertaken.

Horticulture and HFPP intervention significantly improved diet diversification [2-3 at BL to 8-9 at EL ($P<0.001$)] and increased micronutrient rich foods consumption among program HH compared to the control group in all countries. Program HH consuming vegetables, fruits and animal foods (egg, liver) significantly increased from 18 to 42% ($p<0.001$) from BL to EL compared to the control group (12-18%) at EL. The median income from HFPP produce significantly increased ($p<0.001$) among the target HH (US$ 1 to 7) compared to the controls. The money was mainly spent on foods such as meat ($p<0.005$), cooking oil and fish ($p<0.05$). Anemia prevalence among children < 5y decreased in all countries, however, the change was only significant in Bangladesh and the Philippines [Bangladesh 63.9-45.2% ($P<0.001$), Nepal 65.0-57.5% ($P=0.209$), Philippines 42.9-16.6% ($P<0.001$), Cambodia 78.0%-70.2% ($P=0.153$)]. Anemia prevalence remained unchanged in control children in Nepal and slightly decreased in the other 3 countries. Horticulture and HFPP intervention reduced anemia among children <5 y. Program HH diversified their diet by increasing intake of MN-rich foods and using the income generated to buy other food, thus contributing to improved HH food and nutrition security. Considering the current global food crisis, programs such as horticulture and HFPP should be scaled up quickly to mitigate the adverse affects of rising food prices and reduced access to food in the marketplace. To scale-up HFPP further: Monitoring and evaluation of HFPP should be strengthened to see impact on health, nutrition and economics. HFPP should be integrated with synergistic programs to accelerate scale-up. Essential nutrition actions (ENA) should be integrated into horticulture and HFPP for better impact on health and nutrition.

INTRODUCTION

There is significant evidence that malnutrition, particularly micronutrient malnutrition, is a contributing factor in up to 35% of under-five childhood mortality cases and growing evidence that malnutrition plays a similar role in maternal mortality (1). At least one half of preschool-aged children and pregnant women in Bangladesh, Cambodia, Nepal and the Philippines are affected by micronutrient malnutrition, including deficiencies of vitamin A and iron (2-5). Micronutrient deficiencies are highly prevalent in these countries because the typical diet lacks diversity and is low in quality. In addition, micronutrient deficiencies are highly prevalent among infants and young children because of their high nutritional needs relative to energy intake and the effects of frequent infection (including sub-clinical infection) on appetite, nutrient absorption, and nutrient loss. Malnutrition during early life leads to permanent growth stunting (6), and adversely affects brain development and other functional outcomes. Micronutrient malnutrition in particular leads to sub-standard physical and cognitive functioning and inhibits work productivity. Among children, anemia has negative psychological and behavioral effects, reduces physical activity and increases the risk of child morbidity. Not only are these effects harmful to an individual's health and well-being, the high prevalence of anemia among women and children also negatively impacts social and economic development. Economic losses due to iron deficiency in South Asia have been estimated at 5% of GDP annually.

Inadequate dietary intake is the main cause of micronutrient deficiencies, and thus it seems logical that food and horticulture activities could contribute to improvements in nutrition and micronutrient status (7). Dietary diversification involves improving dietary intake by having a sufficient, affordable and diverse supply of micronutrient-rich vegetables and fruits throughout the year, and their adequate consumption. While plant foods are an important source of vitamin A and other micronutrients, it is now well known that the bioavailability of vitamin A from plants is lower than was originally thought (12). Therefore, it is crucial to increase the consumption of animal foods. For this reason, Helen Keller International (HKI) initiated pilot projects in Bangladesh, Cambodia and Nepal to integrate animal production into on-going horticulture and home gardening programs. This integrated approach is referred to as homestead food production (HFP) that includes horticulture intervention and small animal and poultry production.

As a means of addressing dietary diversification, horticulture intervention under homestead food production is the main strategy that HKI has used in Bangladesh, Cambodia, Nepal and the Philippines. The HFP program has been shown to improve household food security and nutritional status of women and children by empowering families to produce nutritious foods year-round, while providing nutrition education to promote improved nutrition practices. The benefits of horticulture intervention and HFP include increased food production and availability, increased consumption of nutritious foods, improved food diversification, and income generation (8). In addition, food-based approaches contribute to poverty reduction in a variety of ways, which in turn adds to improved nutritional status (10, 11). Increa-sing availability and consumption of micronutrient-rich foods as part of food-based strategies is considered a sustainable approach because the process empowers individuals and households to take ultimate responsibility over the quality of their diet through their own production of nutrient-rich foods and educated consumption choices (9).

GOALS OF THE STUDY

The overall aim of the project was to improve the intake of micronutrient rich food among women and young children. The goal of the study is to review the impact of horticulture and homestead food production program and identify lessons learned for scaling to new areas of existing countries and to other countries.

METHODOLOGY

Data were collected between 2002 – 2007 from different HFP programs in four countries. Selected HH from communities that participated in HFP (received nutrition education, training, inputs and were exposed to best practices for HFP) were identified as 'Intervention' households'. 'Control' households were geographically separate from the intervention HH in different villages and didn't participate in HFP but had similar demographic profiles as the intervention group. The inclusion of control group enabled HKI to take into account other social and economic factors outside of the HFP program that may have influenced income and nutritional status during the program period by comparing the change in intervention households from baseline to endline with the change in control households during the same period.

Cross-sectional data were collected from program households participating in the HKI's HFP program and from control (non-program) households in four countries. Data were collected at baseline (BL, 2002-2003) and end line (EL, 2006) after 3-4 years of intervention. A pre-coded-structured questionnaire was used to collect data on food production, food consumption, and diversity of vegetable and fruits, income and utilization, and anthropometry measurements, along with health, nutrition and socio-economic indicators.

Capillary hemoglobin (Hb) was measured using the Hemo Cue device (Angelholm, Sweden). Anemia was defined as Hb<120g/L (non-pregnant women) or Hb<110g/L (children 6-59 months of age). Blood samples were collected from ~1200 non-pregnant women and ~1000 children < 5year age from intervention group and same number from control group at BL and EL to measure change in hemoglobin level. Vitamin A intake was assessed among sub-samples by using the 24–VASQ (HKI's 24-vitamin A semi-quantitative) food frequency method. A review of the monitoring and evaluation process and obstacles to scaling up was also undertaken. Confidentiality of all information has been maintained. Ethical consent for these studies was granted by the National Ethics Committee in all the four countries

DESCRIPTION OF HKI HORTICULTURE AND HFP PROGRAM IN ASIA

The HKI Asia-Pacific horticulture and food production program, integrated with nutrition education, works by providing technical assistance, training, agricultural supplies and management support to partner NGOs for a period of three years. The first pilot project was initiated in Bangladesh 20 years ago and based on this pilot experience; it was refined and scaled up throughout the country. A similar model was adapted to the local contexts in Cambodia, Nepal and the Philippines. Altogether HKI has reached over one million households that represent approximately 5 million beneficiaries. The HFP model provides support to each NGO partner to establish Village Model Farms (VMF) in the target communities, which are then a good source of seeds, seedlings and saplings and chicks for household production. The VMF also serves as a focus for community support, to demonstrate agricultural methods, and to provide practical training and inputs for targeted household (HH) production and for women's groups of HH producers/farmers. Each VMF supports two groups, and each group consists of about 20 households represented by women. Accordingly, one VMF supports approximately 40 HHs. HH producers are encouraged to improve their gardening practices by growing several varieties of vegetables and fruits throughout the year. In this way, the program increases the availability and diversity of food throughout the year, (Fig. 1).

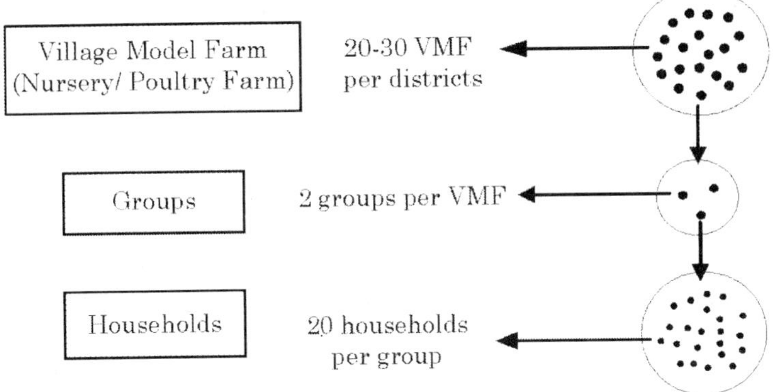

Figure 1. Model for implementation of homestead gardening program

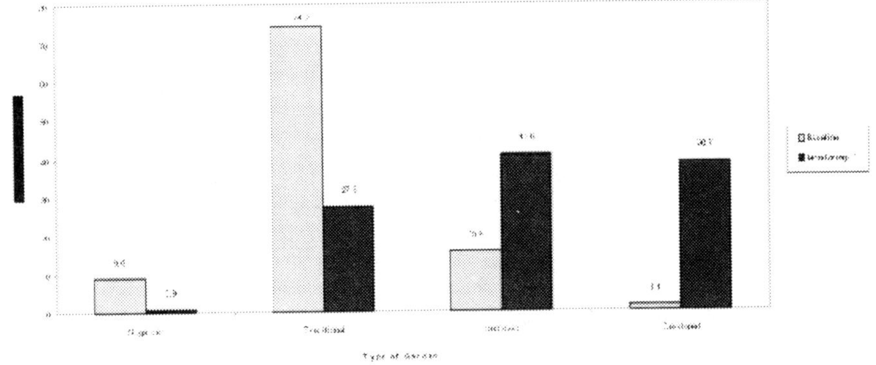

Figure 2. Shifting of gardening practices after 18 months of program participation (n=1949)

RESULTS

Changes in Gardening Practices

The observed changes in gardening practices after 18 months of horticulture intervention are presented in (Figure 2). Based on their seasonal cultivation and the variety of crops raised in Cambodia, homestead gardens can be classified into three categories: traditional, improved, and developed.[a] Approximately 90% of the households practiced homestead gardening but 74% had a traditional garden at the start of the program. After 18 months of participation, the percentage of households who practiced year-round gardening ('developed gardening') had increased significantly from 1.5% to 39% (χ^2

a. Traditional gardens are scattered plots that produce on a seasonal basis only gourds and a small selection of traditional vegetables. Improved gardens refer to those that produce a greater variety of crops on a fixed plot, but still only during certain seasons. Developed gardens are defined as fixed plot gardens that produce a wide variety of fruits and vegetables year-round.

test, p<.001). In addition, the percentage of households without a home garden decreased from 9% at baseline to less than 1% after one year (χ^2 test, p< .01).

Household Consumption of Vegetables Produced in the Homestead Gardens

Figure 3 shows the quantity of vegetables from homestead gardens consumed during the three-month prior to data collection. There are three phases of the program[1]. Households in the active phase consumed a median of 85 kg of vegetables, and households in the completed phase consumed a median of 70 kg of vegetables. These quantities were found to be significantly higher than the median of 38kg of vegetables consumed by control households during the same period.

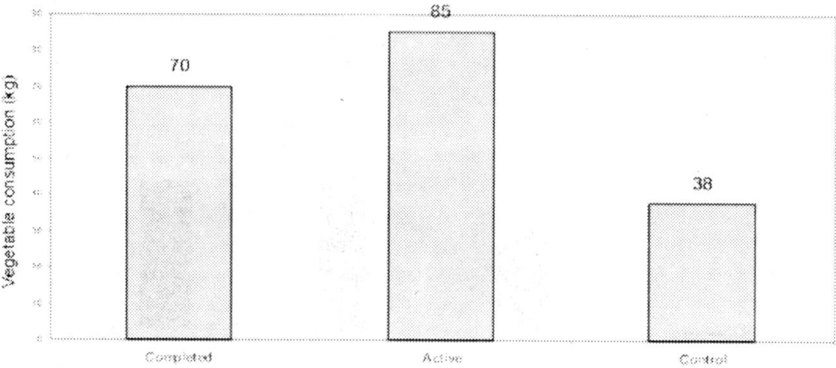

Figure 3. Vegetables consumed by household as produced in own garden in last 3 month by program participants (n=1423).

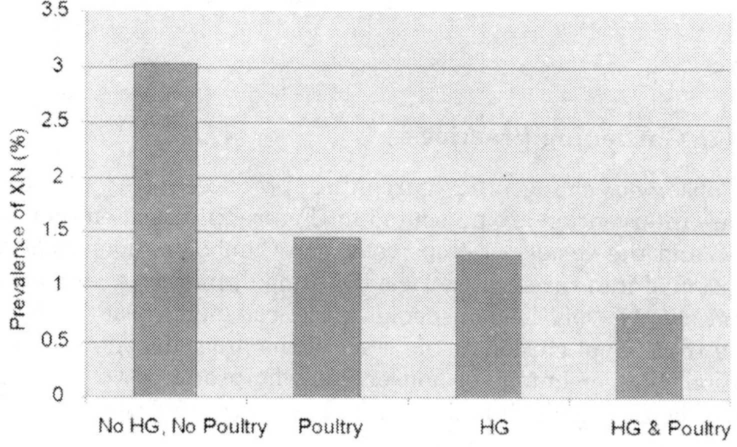

Figure 4. Prevalence of night blindness (XN) among children under five (12-59 months) that had not received VAC by home garden and poultry ownership (n=4296)

---

1. Completed phase is program already ended 2-3 years back. Active phase is still going on and Control phase is we don't do any intervention

Impact on Night Blindness Among Children Under Five Years of Age

HKI re-established the importance of HFP including vegetable and fruits in the control of night blindness based on the findings of the last national vitamin A survey in rural Bangladesh (15). HFP has an important role to play to control night blindness considering the rate of night blindness (clinical vitamin A deficiency) among children from households with a garden and poultry was lower than for those from households without a garden and poultry, particularly when they had not received a vitamin A capsule in the 6 months prior to the survey (Fig. 4).

Anemia Prevalence Among Children 6-59 Months

Anemia prevalence among children 6-59 months decreased in all countries, however, the change was only significant in Bangladesh and the Philippines. Figure 5 shows that in Bangladesh 63.9-45.2% (P<0.001), Nepal 65.0-57.5% (*P*=0.209), Philippines 42.9-16.6% (P<0.001). Cambodia 78.0%-70.2% (*P*=0.153). Anemia prevalence remained unchanged in control children in Nepal and slightly decreased in the other three countries.

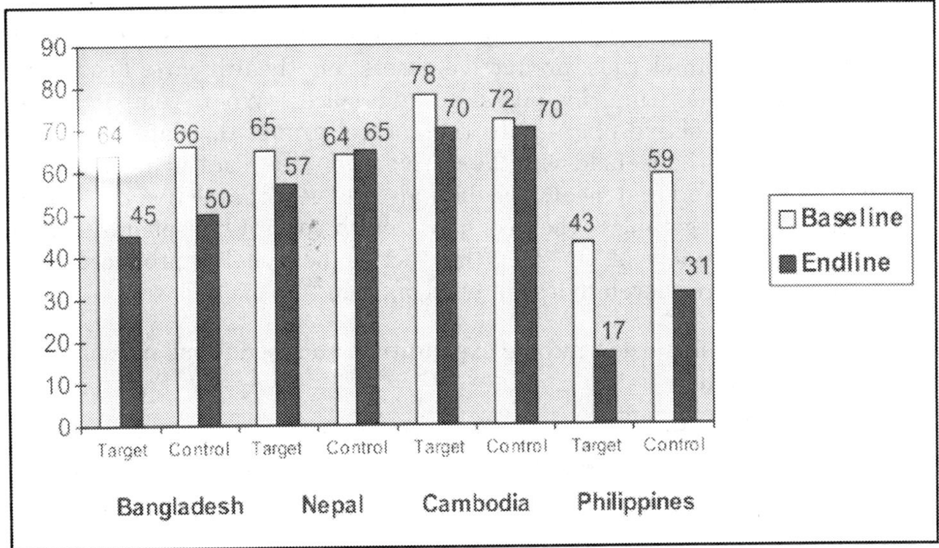

Figure 5. Anemia prevalence among children aged 6-59 mo from program and control households in Bangladesh, Cambodia, Nepal and Philippines at BL and EL.

Anemia Prevalence Among Non-Pregnant Women

Among target women in Nepal and Bangladesh, anemia prevalence declined by 26% (p=0.009) and by 12% (p=0.075), respectively, whereas it remained unchanged among control women in both countries (Figure 6). In Cambodia, there was a non-significant reduction among control women only. Blood sample were not collected from the Philippines HFP program area.

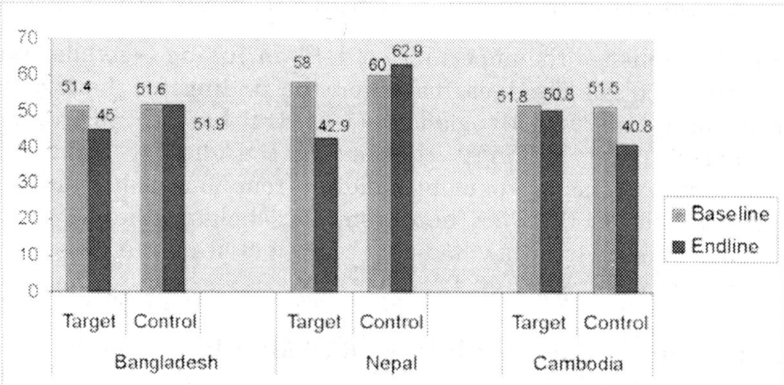

Figure 6. Anemia prevalence among non-pregnant women from program and control HHs in Bangladesh, Cambodia and Nepal at BL and EL.

Main Use of Income Earned by Selling Garden Produce

The results presented in Table 1 show that food was the most frequently purchased item by household type by using the income generated through homestead gardening. Other items that income was invested in were children's education, clothes, productive assets, and health care. Food items purchased included primarily oil, salt, and spices. When comparing the households in the active phase with the control group, the data show that significantly more households in the active than the control used garden money to purchase oil and meat. The data also suggest that active households purchased fewer vegetables when compared with control households. This is not surprising in view of the fact that active households produced more vegetables and were therefore less in need of them.

Table 1. Homestead garden income expenditure and food items purchased by households

Income Expenditure	Household		Sig. L	Food Item	Household		Sig. L
	Active (n=466)	Control (n=176)			Active (n=250)	Control (n=108)	
Food	54.9	62.5	NS	Oil	83.2	74.1	***
Education	48.4	25.6	***	Salt/Spices	77.1	70.4	NS
Clothing	40.9	29.5	**	Rice	37.6	47.2	NS
Productive Assets	28.2	26.1	NS	Fish	44.0	39.2	NS
Health Care	22.6	17.6	NS	Pulses	24.4	28.7	NS
Housing	7.1	5.1	NS	Meat	32.4	19.6	**
Social Activities	4.4	6.2	NS	Vegetables	2.0	9.3	**

, * = Significant at $p< 0.01$ and $p< 0.001$ probability levels respectively

NS = Not significant. Data on income expenditure represent only households that generated income, and food items represent only households that purchased food using garden income.

Explaining a Better Economic Situation of Homestead Gardeners

The reasons behind an improvement in households' economic situation were investigated. Table 2 presents the main reasons given by beneficiaries whose economic situation had improved. The data show that 65.2% households in the completed phase and 84.1% of households in the active phase consider their garden as one of the main reasons. On the other hand, only 17.7% of control households regarded their garden as a main reason for the improvement.

Table 2. Main reasons for better economic situation of homestead garden beneficiary and control households.

Reason	Households (%)			Sig. L
	Completed (n=598, 83%)	Active (n=607, 84%)	Control (n=315, 44%)	
Homestead garden	65.2	84.1	17.7	***
Increase in other incomes	66.4	51.2	65.2	***
More members earning	43.3	30.1	44.3	***
More productive assets	43.6	33.4	26.3	***

DISCUSSION

A year-round supply of food to households through home gardening illustrates that the program has been instrumental in augmenting food availability to the households, thus contributing substantially to household food security. A sufficient availability of food, however, does not necessarily guarantee sufficient access to it. Access to food is ensured when all households and all individuals within those households are able to obtain appropriate food for a nutritious diet.

In rural areas of all the countries in the study, food from homestead production, including vegetables, fruits and poultry, are often the only source of micronutrients in the family diet (13). The production of fruits, vegetables, and meat provides the rural household with accessibility to the nutrients required, which otherwise may not be readily available or within their economic means. Equally important, the HFP produce can be sold, generating additional income for the family. The nutrition education component of the HFP program promotes informed food purchasing and consumption choices. Based on HKI's monitoring and evaluation findings, the additional income is often used to buy supplementary food items such as meat, fish, and cooking oil, which further increase the diversification of the family's diet (14). In this review, 60% of the intervention respondents mentioned that they used income earned from HFP to purchase food, with 70% of these buying high quality items such as fish, chicken, beef or pork. In many cases, this income also covered school education, health-care services, and/or productive activities like reinvesting in HFP or other productive assets like buying poultry, land, livestock etc. HKI has also found more recently that this investment in productive assets helps with being able to withstand times of economic difficulty or natural disaster (16). Thus, homestead food production has the proven potential to improve accessibility to those foods that increase year-round household food security and improve the nutritional status of family members.

Additionally, increased vegetable crop diversification in homestead gardens has been found to be associated with an increase in nutritional quality of garden produce in terms of iron, vitamin A, vitamin C, and fiber content. The observed increase in the number of vegetable and fruit crops, especially vitamin A-rich crops in this program, suggests a potential improvement in the diet quality of garden produce accessible to households.

Horticulture intervention and HFP also has a positive affect on micronutrient deficiency status, including vitamin A deficiency and anemia. Horticulture intervention and HFP has been shown to be an important way to improve the intake of vitamin A and other micronutrient rich foods, particularly for poor households in countries like Bangladesh, Nepal, Cambodia and Philippines.

CONCLUSION

This paper has shown that the HKI homestead food production program has had an impact on reducing micronutrient malnutrition through increasing the availability and consumption of micronutrient rich foods and by improving the economic status of families. Homestead food production has been shown to be an important way to improve the intake of vitamin A and other micronutrient-rich foods, particularly for poor households in countries like Bangladesh, Nepal and Cambodia, thus lowering the risk of vitamin A deficiency. Furthermore, HFPP was found to contribute somewhat to reducing anemia. This was particularly observed among children under five in all the countries and among non-pregnant women in Bangladesh and Nepal. A reduction in anemia levels due to an increased consumption of animal and plant source foods is a marker for a reduction of other micronutrient deficiencies, such as vitamin A and zinc.

Food availability, including the availability of animal foods, and food security are major components of poverty alleviation. Homestead food production increases the availability of plant and animal foods, as well as their year-round consumption. Both the increased availability and the increased choice of foods enhance dietary diversity and vegetable and animal food consumption. In addition, homestead food production provides additional income for poor households, which is mainly used for purchasing foods of higher nutritional quality, including animal products. Significantly, participation in the homestead food production program, together with the income that the activity generates, has been found to empower women. Their participation in household decision-making increases and that, amongst others, has a positive impact on food preparation practices and care seeking behavior (17).

Integration of homestead food production into existing home gardening activities should be used as a means to address food security and improve nutrition among other vulnerable households, including HIV/AIDS affected households. Households with PLWHA require additional food and have added health care costs, therefore resources become even scarcer within these families. When HKI included households with PLWHA in its HFP program in Cambodia, all food produced through the program was consumed. In addition, the income earned from HFP was used to purchase other food items to further improve food security and nutrition as diversity in food is important for the increased intake of micronutrients. Therefore, this kind of intervention has been shown to be important for most vulnerable groups such as HIV/AIDS affected households.

Horticulture intervention and HFP increases the availability of plant and animal foods, as well as their year-round consumption. This has been shown to be an important way to improve the intake of vitamin A and other micronutrient-rich foods, particularly for poor households in countries like Bangladesh, Nepal, Cambodia and Philippines. Equally important, the HFP produce can be sold, generating additional income for the family. The nutrition education component of the HFP program promotes informed food purchasing and consumption choices, and the added income allows the purchase of additional high quality foods such as fish, cooking oil, chicken, beef or pork. HFP also has an affect on micronutrient deficiency status. Nutrition education plays an important role in changing behaviour and improving IYCF and nutrition practices among household members as the nutritional status of children becomes more strongly tied to increased food availability and dietary intake as a result of the HFP program.

As noted above, although the study found that anemia among children improved in all four countries from baseline to endline, further research would allow HKI to assess more fully the impact of HFP on other health and nutrition indicators of the beneficiary population. Considering the current global food crisis, what can be demonstrated immediately, however, is that programs such as horticulture and HFPP should be scaled up quickly to mitigate the adverse affects of rising food prices and reduced access to food in the marketplace.

REFERENCES

Black, R.E. *et al.* 2008 Maternal and child undernutrition: global and regional exposures and health consequences. Lancet: 371:243-60

HKI/IPHN, 2002. Anemia is a severe public health problem in pre-school children and pregnant women in rural Bangladesh. *Nutritional Surveillance Project Bulletin* **10.** HKI, Dhaka.

HKI/Cambodia, 2001. Initial finding from the 200 Cambodia National Survey Supporting document to the Micronutrient Workshop help on February 20, 2001 in Phnom Penh, Kingdom of Cambodia. Helen Keller International, Phnom Penh.

Nepal Micronutrient Status Survey, 1998. Kathmandu, Nepal. Ministry of Health, Child Health Division, HMG/N, New ERA, Micronutrient Initiatives, UNICEF Nepal and WHO.

Food and Nutrition Research Institute, Department of Science and Technology. Philippine Nutrition Facts and Figures, 1998 and 2003.

Martorell, R., Schroeder, D.G. Rivera, J.A. and Kaplowitz, H.J. Patterns of linear growth in rural Guatemalan adolescents and children. *J. Nutr.* **125** (4 Suppl): 1060S-1067

Kiess, L., Moench-Pfanner, R. and Bloem, M.W. 2001, *Food-based strategies*: Can they play a role in Poverty alleviation? Food Nutr Bull: 2001

Talukder, A. *et al.*, "Increasing the production and consumption of vitamin A-rich fruits and vegetables: Lessons learned in taking the Bangladesh homestead gardening programme to a national scale," *Food and Nutrition Bulletin* (United Nations University Press, 2000).

Ruel, M.T., Levin, C.E., 2000. Assessing the potential for food-based strategies to reduce vitamin A and iron deficiencies: A review of recent evidence. Food Consumption and Nutrition Division Discussion Paper no 92. International Food Policy Research Institute, Washington DC, 2000. See also Food Nutr Bull 2001; 22: 94-95.

De Pee, S., Bloem, M.W. and Kiess, L., 2000. Evaluating food-based programmes for their reduction of vitamin A deficiency and its consequences. *Food Nutr Bull* **2000**; 21: 232-238.

Bloem, M.W., Moench-Pfanner, R. and Kiess, L., 2001. Combating Micronutrient Deficiencies – An important component of poverty reduction. Biomedical and Environmental Sciences 2001; 14: 92-97.

West, C.E. *et al.*, 2002. Consequences of revised estimates of carotenoid bioefficacy for dietary control of vitamin A deficiency in developing countries. *J. Nutr.* **32**, 2920S-2926S

Helen Keller International/Cambodia, 2004. Homestead Food Production Activities in Cambodia – A Mapping Review Report. Phnom Penh, Cambodia, 2004.

Helen Keller International/Cambodia (HKI), 2007. "Homestead Food Production Program improves food and nutrition security by increasing consumption of micronutrient-rich foods and family income in households with HIV/AIDS and other chronic diseases." *Cambodia Nutr. Bull.* **7**(1)

Kiess, L., Bloem, M.W., de Pee, S. Hye, A., Khan, T., Talukder, A. and Huq, N. Bangladesh: Xeropthalmia free. The results of an effective vitamin A capsule program and homestead gardening. In: American Public Health Association 126th Annual Meeting Report, Washington, CD: American Public Health Association, 1998:361 (Abstract).

Homestead Food Production Program in Char area in Bangladesh. The final evaluation of the project, 2008. Helen Keller International, Bangladesh.

Homestead Food Production Program improves food and nutrition security by increasing consumption of micronutrient-rich foods and family income in households with HIV/AIDS and other chronic diseases. *Nutr. Bull. Helen Keller Int. Cambodia,* **7**, (1) October 2007.

12. IMPROVING LIVELIHOODS OF SMALL FARMERS THROUGH THE PRODUCTION OF HIGH-VALUE HORTICULTURAL CROPS IN PROTECTED AGRICULTURE IN DRY AREAS

Mahmoud El-Solh[1], Ahmed T. Moustafa[2] and Aden Aw – Hassan

Director General, International Center for Agricultural Research in Dry Areas (ICARDA), Aleppo, Syria

ABSTRACT

Land and water resources in the world's dry areas are under great pressure, as a result of rapid population growth, climate change and other factors such as increase in salinity. There is a critical need for sustainable and nature-friendly production techniques or systems to increase water productivity in agriculture, especially in arid areas, to sustain and improve the livelihood of those who depend on farming. One important technique is Protected Agriculture (PA) for the production of high-value horticultural crops. This paper illustrates the impact of ICARDA's PA work in different countries. Water productivity under PA is significantly higher than in open fields. ICARDA and its partners in arid areas in Afghanistan, Yemen, Pakistan and Oman have tested and demonstrated the potential of PA to increase farm incomes. An Integrated Production and Protection Management (IPPM) program for PA was developed and implemented in Arabian Peninsula countries aiming at reducing pesticide use and enhancing water and land productivity. The use of IPPM in Yemen has reduced chemical spraying in cucumber production from 18-20 times to only 2-3 times. Various hydroponics systems have proved highly successful under farmers' conditions. The Soiless Vertical Growing System in Kuwait and Oman increased farmers' returns on investment and saved water, fertilizers and labor.

ICARDA research focus on protected agriculture has been targeting resource-poor small farmers. In Afghanistan, pilot growers who adopted PA techniques increased their annual income by up to 138 percent. Similarly, in Yemen, cucumber producers using PA produced 2-3 crops per year instead of one crop, with higher productivity, better quality, reduced water requirements and lower cost of production. As a result, growers' income increased up to four-fold. In Pakistan, adoption of PA substantially increased the income earned per m^3 of water. PA reduces seasonality, enabling farmers to supply local markets with high quality food out of season and creating more jobs. It is also suitable for women and women-headed families. ICARDA's research on horticulture includes value chain analysis and linking small farmers to markets. Studies have identified different actors and their functions, costs, returns, and marketing margins along the value chain of several commodities in Egypt, Jordan and Morocco. Post-harvest processing opportunities, contractual arrangements with small farmers, and production, institutional and policy constraints were also studied. The results have important policy and development implications to enhance the production and marketing efficiency of horticultural commodities and improve smallholder farmers' access to markets. It is important to note that protected agriculture requires initial investment, which small-scale poor farmers find difficult. Policy and institutional innovations are needed to make appropriate financial instruments (e.g. microfinance) accessible for small-scale producers. Dissemination of knowledge on optimal production practices, water use and pest management is also critical.

1. INTRODUCTION

Land and water resources in the world's dry areas are under great pressure as a result of rapid population growth, climate change and many other factors, including increases in soil salinity. There is a critical need for sustainable production techniques that increase water productivity in agriculture and improve the livelihoods of those who depend on farming. Water productivity under Protected Agriculture (PA) is significantly higher than in open fields. ICARDA and its partners in arid areas in Afghanistan, Yemen, Pakistan and Oman have tested and demonstrated the potential of protected agriculture to increase farm incomes. In Afghanistan, pilot growers increased their annual income by up to 138%. Similarly, in Yemen, cucumber producers using protected agriculture produced 2-3 crops per year instead of one crop, and as a result, income increased up to four-fold. In Pakistan, adoption of protected agriculture substantially increased the income earned per m^3 of water. An Integrated Production and Protection Management (IPPM) program for protected agriculture has also been implemented in Yemen which has reduced chemical spraying in cucumber production from 18-20 times to only 2-3 times. In addition, various hydroponics systems have proved highly successful under farmers' conditions in Kuwait, Oman, and the UAE.

Protected agriculture reduces seasonality, enabling farmers to supply local markets with high quality food out of season, and creates more job opportunities. It is also suitable for women and women-headed families.

ICARDA's research on horticulture includes value chain analysis and linking small farmers to markets. Studies have identified different actors and their functions, costs, returns, and marketing margins along the value chain of several commodities in Egypt, Jordan, Morocco and Tunisia. Post-harvest processing opportunities, contractual arrangements with small farmers, and production, institutional and policy constraints were also studied. The results have important policy and development implications to enhance the production and marketing efficiency of horticultural commodities and improve smallholder farmers' access to markets. It is important to note that protected agriculture requires initial investment, which poor small-scale farmers find difficult. Policy and institutional innovations are needed to make appropriate financial instruments (e.g. microfinance) accessible for small-scale producers. Dissemination of knowledge on optimal production practices, water use and pest management is also critical.

2. HORTICULTURE TO IMPROVE RURAL LIVELIHOOD

The massive increase of world population and their needs for clean water and healthy food during last decades have put a tremendous pressure on limited natural resources, especially in the dry areas. Demand for water has risen twice as quickly as the increase in population in the last century (FAO, 2009), therefore more regions are chronically short of water. It is estimated that by 2025, 1,800 million people will be living in countries or regions with absolute water scarcity, and two-thirds of the world population could be living under water-stress conditions (FAO, 2009). By 2025, countries such as Pakistan, South Africa and large parts of China and India will join countries in the Middle East and North Africa which are classified as having absolute water scarcity today. It is projected that these countries will shift water from agriculture into other sectors such as industry and household use, and will therefore be more dependent on imported food.

The scarcity of water and land has been intensified by increases in soil salinity and climate change, seriously affecting poor rural people in arid regions whose livelihoods depend on conventional agriculture. As a result, poverty increases further, and it is probable that in rural areas, water scarcity affects most people. In large parts of the developing world, irrigation remains the backbone of rural economies. Smallholder and farmers with limited resources who make up the majority of the world's rural poor (IFAD, 2008) occupy marginal and low productive land and depend mainly on limited natural water resources. Their livelihood largely depends on income from agricultural production, and they are therefore not only highly sensitive to changes in water availability but also to changes in market prices and the demand for better quality food products.

Low productivity of agricultural activities and the low resulting income has complex effects on rural livelihoods. For example in Sudan, low income from agricultural activities forces young people to migrate to urban areas. Family health is also affected with an increasing pressure on the family labor force, including women and children (IFAD, 2007). Similar situation exists in the rain fed terraces of Yemen which are the major producers of food crops but farmers' earnings are very low. As a result, the terraces suffer from neglect due to lack of investment and scarcity of labor due to migration to cities (Moustafa *et al.*, 2003).

Horticulture and production of high value vegetable and fruits crops tide with simple but effective techniques for improving productivity per unit of water and land could have a great impact on poor rural families. Horticulture is a well-established and economically important industry that makes a significant contribution to livelihoods in the dry land, rainfed and irrigated farming systems of the NENA region. Many of the major species are indigenous and particularly well adapted (IFAD, 2008b). Horticulture and production of high value fruit and vegetable crops has great potential to provide poor rural families with possibilities for income diversification and better livelihood alternatives. Studies consistently show that the ratio of benefits to costs for fruit and vegetable crops is two times higher than the corresponding ratio for predominant cereals and pulses (IFAD, 2008). Generally, small growers working in the production of fruits and vegetables generate better income than farmers that are engaged in the production of cereal crops alone. Studies from developing countries frequently show higher average net farm incomes per household member among horticultural producers. For example, in Kenya it is calculated that difference in farm incomes between horticulture growers and cereal growers is about five time (McCulloch and Ota, 2002) or in Colombia the horticulture farm generate income significantly higher than cereal by about 117% (Weinberger and Lumpkin, 2005).

The high value of fruits and vegetables not only benefits growers by increasing their incomes but also plays an important strategic role in improving the diets of people around the world. The World Health Organization (WHO) estimates that low fruit and vegetable intake contributes to approximately 2.7 million deaths a year from chronic diseases and causes about 31 %of ischemic heart diseases and 11 %of strokes worldwide. It ranks low fruit and vegetable intake as the 6th main risk factor for mortality in the world (WHO, 2003).

3. PROTECTED AGRICULTURE

Protected agriculture (PA) where crops are growing under protective housing (plastic houses or greenhouses), is an intensive and dynamic form of crop production. In PA the environment and therefore, timing of the production can be controlled and yields can be substantially improved. Protected agriculture and its associated production techniques can significantly improve productivity per unit of area and land for the production of high value cash crops. Figure 1 illustrates and compares the yield of tomato, cucumber and pepper under protected agriculture and open field where the yields in greenhouses are significantly higher for all three crops.

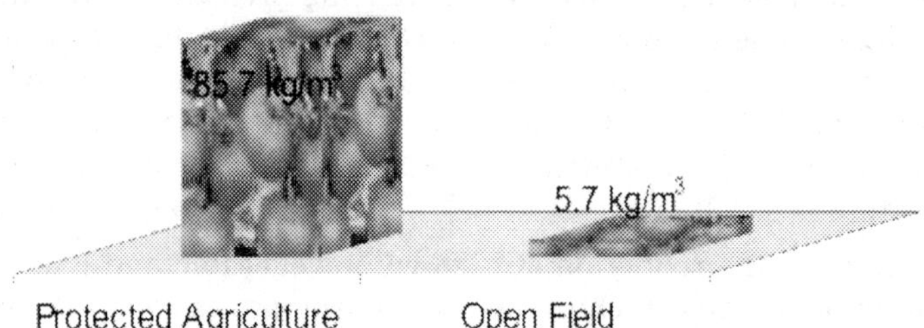

Figure 1. Productivity of 1 m^3 of water protected agriculture vs. Open Field *(Guma, Y.T.and A. Al-Masoum, 1996)*

Furthermore, protected agriculture would significantly reduce the amount of water and chemicals used in producing high value fresh produce compared with open field production (Moustafa *et al.,* 2006b). Figure (2) compares the productivity per unit of water for tomato in open field and protected agriculture. Water productivity would be even higher when soilless production techniques are implemented under protected agriculture.

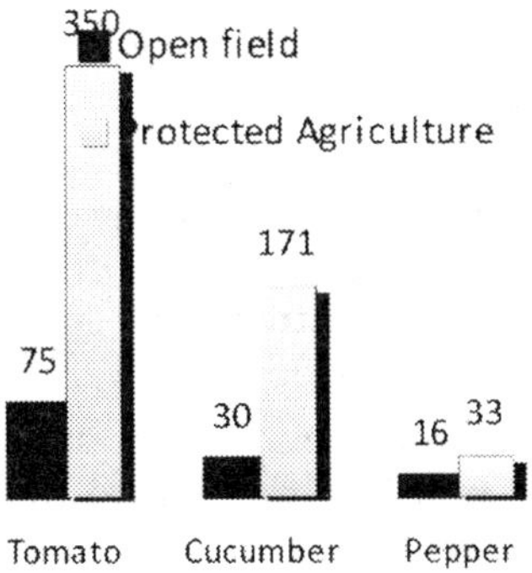

Figure 2. Production levels (t/ha) of open field and protected agriculture *(Jensen and Malter 1995)*

There is a wide range of vegetable crops are grown over an extended period during the year under protective housing. The technology is suitable for use on marginal or other non-productive lands and, if simple structures are used, the technology is both affordable and sustainable. It is, therefore, particularly suited to marginal and water deficient areas where farming communities are highly vulnerable and are struggling to improve their livelihoods (Moustafa, 2007a). Protected agriculture has the potential to improve rural livelihood and contribute significantly to both the development of rural communities and to the national economy. It can play an important role in supplying local markets with fresh produce that in many cases could not be grown otherwise. Furthermore, PA and its associated industry would improve employment within rural communities and create opportunities for the disadvantaged, particularly women, and the disabled. It also offers potential for development of a private service sector in the construction and supply of protected agriculture equipment. Ultimately, high yielding quality produce from protected agriculture could be expanded to serve the export market and generate a valuable source of foreign revenue.

In general, protected agriculture and its associated techniques could help smallholders in rural and peri urban areas to enhance their livelihoods and increase their incomes by different means including:

1. Significantly increase the yield and quality of products which could generate better revenue;
2. Utilize marginal and non-productive lands in relatively small areas which other crops can not be cultivated;
3. High water productivity compared to open field which save significant amount of water to be used for expansion of farming systems to generate more incomes.

4. With good management, PA significantly reduce the amount of agro-chemical used thus produce safe food and reduce cost of production;
5. Offer year round production regardless of season, so multiple cropping on the same unit of land is possible. The flexibility of the system means that farmers can take advantage of market seasonality and higher prices.
6. Protected agriculture can increase the level of diversification through producing high quality vegetable crops which can not be produced in rural areas otherwise
7. Production of high value crops under greenhouses can be easily managed and handled by rural women with limited land and resources therefore Protected Agriculture has good potential for supporting rural family headed by women.
8. On a small scale, the system can be managed by household labor alone, or on a larger scale can generate additional employment opportunities in the intensive management and care of crops. It particularly favors the involvement of women, as it does not require heavy labor, and provides an enclosed and conservative environment in which women can work more satisfactory.
9. Offers opportunities for the development of peri-urban agricultural production close to major markets, reducing transport and packaging costs.

4. HIGHLIGHT OF ICARDA'S RESEARCH FOR DEVELOPMENT ACTIVITIES ON PROTECTED AGRICULTURE

ICARDA's research for development activities on protected agriculture, which mainly were carried out in the Arabian Peninsula, have fruited in two technology packages namely Integrated Production and Protection Management (IPPM) and soilless (hydroponics) production systems.

4.1. Integrated Production and Protection Management (IPPM)

The protected, warm and humid greenhouse environment is suitable for crop production but also attracting the presence of a number of pests and diseases. Overuse, misuse, and careless application of chemical pesticides can result in the accumulation of toxic substances on plants. The extensive use of chemicals to control diseases and pests in greenhouses results in complex problems of insect resistance, and health and environmental hazards. Crops can be protected by control measures that avoid heavy reliance on pesticides, thereby reducing use of hazardous chemicals. These control measures are part of an Integrated Production and Protection Management (IPPM) program developed by ICARDA and implemented in number of research stations and in pilot private farms. IPPM components are including:

- GH climate management
- Irrigation and fertilization management

- Agro-management and practices
- Mechanical protection
- Biological control
- Chemical control

To investigate the impact of IPPM on production of cash crops under protected agriculture a study conducted in Yemen, where two greenhouses each 500 m^2 were used. One of the greenhouses used as control unit while IPPM techniques were implemented in the second greenhouse. Both houses planted with cucumber during the period between 1 August and 31 December 2003. The percentages of infection with different pest and diseases and number of sprays are illustrated in Figure 3 and 4. The greenhouse under IPPM only sprayed two times while the control greenhouse which was under normal growers management sprayed 19 times (Kirshi *et al.*, 2003).

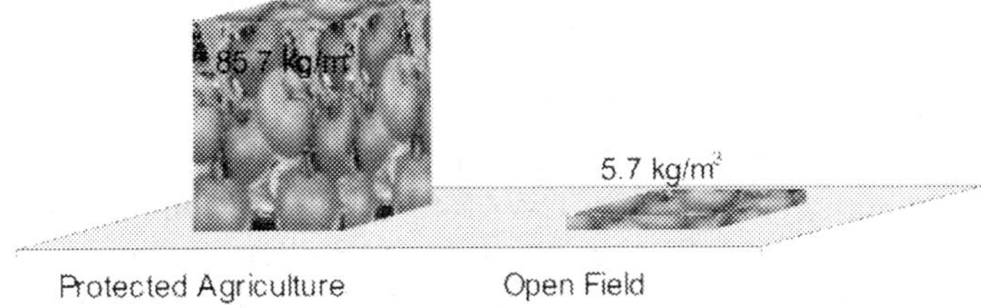

Figure 3. Productivity of 1 m3 of water protected agriculture vs. open field *(Guma, Y.T. and A. Al-Masoum, 1996)*

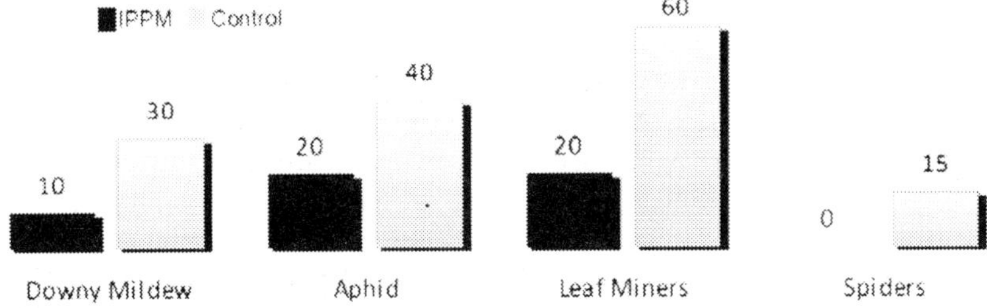

Figure 4. Percentage of infection under IPPM and normal management in Yemen (1 Aug to 31 Dec 2003)

As a part of the IPPM program, soil solarization techniques were developed and simplified for growers and a technical booklet published by ICARDA. The technique was implemented on research stations and private farms with excellent results, avoiding the use of hazardous chemicals (Abu Gharbia, 2002).

4.2. Soilless (hydroponics) Growing Techniques

Greenhouses, although extend production season but losing their productivity due to soil related constrains especially salt accumulation and soil born disease and nematodes. Nematodes are one of the world's major agricultural pests, causing an estimated $80 billion in worldwide crop damage annually

(Divergence 2009). In some cases; such constrains forced growers to stop production, abandon their farm and lose investments.

On the other hand, lack of good quality water for agriculture, especially in CWANA region, put the improvement of water productivity particularly for cash crops on top of the researchable issues list. Traditional techniques in protected agriculture may be highly productive but their relative use of water may be high due to run off and infiltration; thus, the water-use efficiency may be relatively low. In arid countries, rapid evaporation from the soil surface may also lead to salinity problems.

Soilless techniques offer a way of improving water-use efficiency and obtaining better water management in crop production. Soilless cultivation is intensively used in protected agriculture to improve control over the growing environment and to avoid uncertainties in the water and nutrient status of the soil and to improve water productivity. Especially in soilless closed systems water-use efficiency is maximized by recycling all the water and nutrients not used by the plants. It also overcomes the problem of salinity and the accumulation of pests and diseases.

ICARDA through research 4 development program in close collaboration with NARES of many countries in CWANA, developed, enhanced and simplified number of soilless production techniques. Soilless culture (Hydroponics) in large scale could be very complicated and needs strong management and sensitive measurement tools. ICARDA has succeeded to enhance and simplify such techniques to be applicable at the small growers' farms. Various hydroponics systems for growing cash crops, including cucumber, tomato, pepper, musk melon, green bean and strawberry, were tested in research stations. Soilless culture not only shows a practical solution for soil problems but also manages to significantly increase water use efficiency and produce more crops with less water. After proved successful at the research stations, the techniques are being transferred to growers in many countries (ICARDA Annual Report, 2007c).

5. EFFECT OF TRANSFERRING THE PROTECTED AGRICULTURE AND ITS ASSOCIATED TECHNOLOGY ON GROWERS LIVELIHOOD

ICARDA in collaboration with National Agricultural Research and Extension Systems (NARES) transferred the enhanced protected agriculture technology packages to number of growers in different countries. In all cases, capacity building and development of NARES human resources were received special attention to improve the sustainability of the projects achievements. Furthermore, technical backstopping was provided to the pilot growers by ICARDA scientists and consultants.

5.1. Improve Income of the Resource-poor Prowers in Yemen Mountain Terraces

In Yemen, the rain fed terraces are major producers of food crops, but the money they earn is low. The terraces suffer from neglect as a result of lack of investment and a scarcity of labor due to migration of young men to the cities. Terraces are supported by an intensive network of water diversion which collects the water harvested in upstream and delivers the surplus water to downstream terraces and the valleys. Neglect in proper maintenance of the upstream terraces could destructively affect the farms on downstream and valleys especially during the rainy seasons and floods.

To save the terraces and to provide growers with suitable livelihood alternative as a means to boost their incomes ICARDA in collaboration with NARS introduced protected cultivation into mountain terraces. In 2000 three pilot growers in mountain terraces of Al Mahweet, Yareem and Taez were selected and greenhouses were constructed at their sites. Greenhouses were made from locally available materials. Harvested rain was the main source of water used in the three locations through drip irrigation system. Pilot growers were visited regularly by ICARDA scientists and NARES researchers and extension agents to provide technical backstopping. The overall yield of greenhouses for the first season was quit encouraging where they produced about the same yield as greenhouses in NARS research station (12.54 kg/m^2) (Table 1).

The Table 1. Illustrates some of the data from the three pilot growers' site (Mukred *et al.,* 2003).

IPPM techniques also were introduced and successfully practiced by pilot growers. Plants were strong and healthy, and had low incidences of pests and disease.

Table 1. Cucumber production records under protected agriculture in Yemen terraces (First season 2000)

Parameters	Sites		
Location	Al Mahweet	Yareem	Taez
GH area (m^2)	216	132	252
Plant density	2.2	2.2	2.2
Days to first harvest	52	57	55
Duration of harvest	56	75	65
Yield (Kg/m^2)	11.0	9.3	12.4
Average yield per plant (KG)	5.0	4.2	6

The cost benefit analysis also showed that the cost of materials and construction can be covered within three growing seasons. The charted Fig. 5 illustrates the cost of materials, construction and income reported by growers in each pilot site.

The success story of protected agriculture in Yemen terraces engorges ICARDA and NARS to reach other donors for expansion of the activities. In 2005, the French-Yemen Food Aid Program funded new project for ICARDA to introduce protected agriculture technology to 35 growers in Taiz (ICARDA. 2005). The 35 pilot growers were selected based on number of characteristics including access to water and rood as well as their interest to participates and follow up the project instructions. The project provided growers with greenhouse structures and production facilities while growers provided the water resources and labor force. Each pilot growers received on the job training and technical backstopping on greenhouse production and management by ICARDA scientists and local project implementation team for about 3-4 seasons. Cucumbers produced from the greenhouses generated higher revenue for the growers (ICARDA 2005) (Fig. 6).

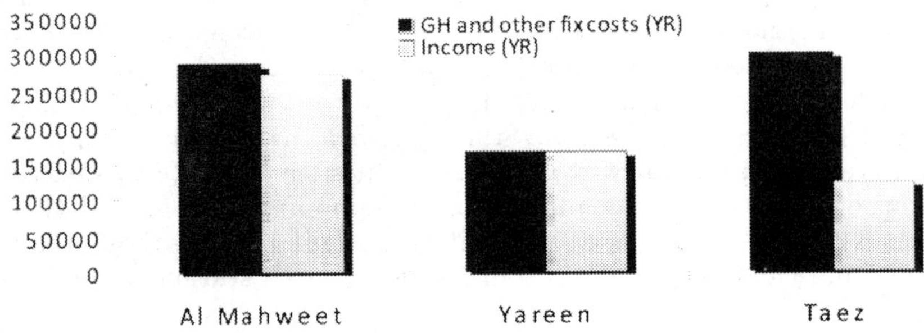

Figure 5. Cost of production and income of pilot growers in Yemen mountain terraces for cucumber production (first seasons 2000)

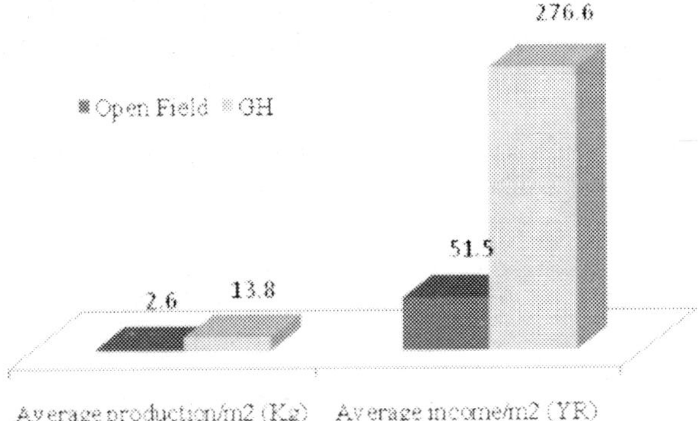

Figure 6. Production and income of cucumber from open field and protected agriculture in Yemen (Data from pilot growers, 2005)

Cucumber productions from the greenhouses were increased by 420% comparing to that of the open field (Fig. 2). Data for greenhouse production and selling prices were collected from actual figures of 25 growers who adopted the technology in Taiz. The open field data also belong to same growers prior to adaption of the technology (ICARDA 2006).

5.2. Promotion of Sustainable Production System for Cash Crops under PA in Afghanistan to Enhance Farmers Income and as Alternative Livelihood

In Afghanistan, ICARDA in collaboration with National Agricultural Research and Extension Systems and financial support of USAID introduced and transferred protected agriculture techniques to 65 growers. A protected agriculture center (PAC) was established by the project in Kabul to support the research and training activities. During the implementation phase of the project, that started on 2004 and lasted for two years, ICARDA organized 18 training course for local growers and NARES in addition to number of workshops and Farmers Field Schools.

The impact of the greenhouses technology on Afghan grower's income was assessed. All growers did not change their cropping patterns, the number of crop or the size of land they use to cultivate prior to their introduction to protected agriculture. Therefore, for these farmers the greenhouses were "add-ons" to the portfolio of crops they used to grow and incomes. Results show that protected agriculture contributed between 8% and 138% increase in income generated from crop production in 2005 (Moustafa, 2007a). This economic success has somehow catalyzed the interest of new farmers to install greenhouses in their farms. Table 2 demonstrates and compared the net benefit of cucumber production under protected agriculture and open field in Afghanistan (Moustafa *et al.,* 2006a).

Table 2. Comparison of profitability of cucumber production under greenhouse and in open field (1 Jerib=2000m^2)

Item description	Unit	GH Production[1]	Open field Produ-ction	Net change GH over open field (%)
Production [2]	Kg	19500	3700	427
Producer price [3]	Af/Kg	24	18	33
Material input cost	Af	118200	4650	2442
Labour cost	Af	76800	18000	184
Depreciation/season	Af	60000	-	
Total production cost	Af	255000	22650	913
Total income	Af	468000	66600	603
Net income/return to land	Af	213000	43950	443
Yield per unit of land	MT/Jerib	20	3.7	427
Water efficiency [4]	Kg/m^3	18	2.1	778
Net return to labour & Management	Af/manday	1166	688	69
Net return per unit of water	Af/m3	221	24	805

Data source: information obtained on spring production of cucumbers from the grower in Parwan.

1. For adequate spacing and ventilation 6 greenhouses could fit on One Jerib of land.
2. Open field cucumbers are longer and bigger than those being produced in greenhouses due to variety difference; accordingly 6 open field cucumbers weigh 1 Kg while 8 greenhouse cucumbers weigh 1 Kg.
3. Cucumbers are sold per unit not per kilogram; greenhouse cucumber carries a price premium for quality.
4. Assume open field uses 10 times more water than greenhouse cucumber production.

Productivity of cucumber per unit of area under greenhouses was five times more than that from open field. Similarly, water productivity also shows significant increase of Nine folds. Pilot growers managed to get the higher market prices for greenhouse cucumber compare to the local varieties that produced in open field (Moustafa 2007b). Overall, the net income from one jerib of cucumber under greenhouse was about five times more than the same in open field.

There are number of success story in Afghanistan where protected agriculture techniques were adopted by small growers with limited resources. For example, Mr Mula Dad, a 28 years old grower from Parwan province in Afghanistan who inhereated one jerib of land and cultivated wheat and turnip in rotation before his exposure to the greenhouse technology. Married with 2 children, Mr Mula Dad had been working as a labourer in the construction sector; his income was not only volatile but also very low (about US$40 per month). During his first season on 2005, he produced about 26,000 cucumbers from his 270 m^2 greenhouse and sold them at Afs 3 per cucumber on average. He remembered that he earned at least Afs 60,000 (US$1200) cash from this activity (some cucumbers were given to friends and visitors) which he spent on his family and had a small saving (Moustafa, 2007a).

Moreover, protected agriculture provides good livelihood alternative for poppy growers. While the ban on poppy cultivation is a positive development, there are serious implications for the rural economy. Poppy farmers and workers engaged in harvesting poppy, a highly labor intensive process; have lost a major source of income. If the ban on poppy cultivation is to be maintained, assistance is urgently needed in the creation of alternative income earning activities. Interviews with growers shows that the net benefit of each Jerib (2000 m^2) of poppy for growers is about US$500-600. Poppy cultivation is very difficult and has only one crop per year. The experience of ICARDA in Afghanistan shows that a grower with good management and by technical support could get a benefit of US$700-1200 per year (two season) from a 270 m^2 under the GH (Moustafa, *et al.*, 2006) (Fig. 7).

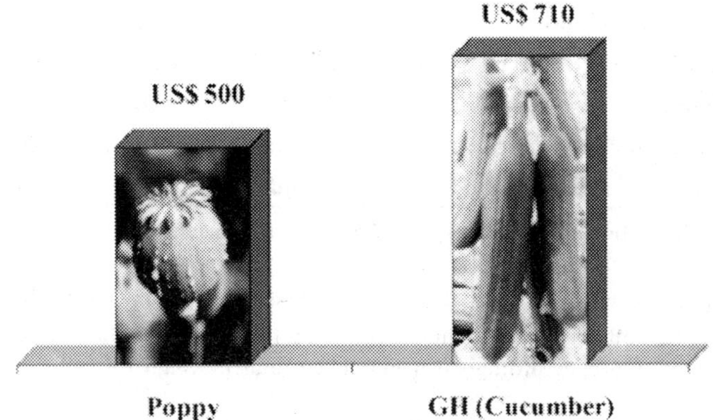

Based on interview with farmers and socio-economic study

Figure 7. Comparision between the net income of one jerib of poppy and one GH (270 m^2) of cucumber (one year production)

5.3. Enhance Rural Livelihood by Increasing Farm Income; Protected Agriculture for Cash Crops in Balochistan, Pakistan

Protected agriculture provides poor rural families and small growers with excellent opportunities for crop diversification and intensification. Growers can produce high quality vegetable crops under protected agriculture which cannot be grown otherwise.

In Baluchistan, ICARDA and its local partners promoted production diversification through protected agriculture. Three pilot growers were selected for the adoption of protected agriculture to produce high value vegetable crop beside their traditional crop which is wheat (ICARDA, 2007a). The production of cucumber from the greenhouses helped growers to generate better income per m^3 of their valuable water by 10 folds as compared to the wheat production (Shomo *et al.*, 2008).

The feasibility study which was conducted by ICARDA revealed that the cucumber produced in greenhouses was clearly distinguished by the buyers and sold at premium prices during the whole production cycle as compared to those produced in the open field. The average premium received per crate (18 kg) was around \$2 which is 30-40% higher prices (Shomo *et al.*, 2008) (Table 3 and 4).

Table 3. Total gross revenue obtained from cucumber production from PA in Quetta (2005-06 first growing seasons)

Month	Yield (kg)	Price (\$/kg)	Gross revenue (\$)
August (9-31)	685	0.30	204.80
September (1-30)	456	0.40	182.90
October (1-12)	42	0.46	19.40
Total	1183	0.35	407.10
Per square meter	4.38		1.50

Table 4. Net returns from cucumber production under PA

	Total value	Proportion of revenue
Total gross revenue	\$407.12	100%
Variable cost	\$151.20	37%
Fixed costs	\$156.70	39%
Total costs	\$307.90	76%
Net returns	\$99.20	24%
Net return/m^2	\$0.37	
Net return per m^3 water	\$2.88	

5.4. Soilless (Hydroponics) Techniques to Improve Water Productivity and Combat Soil-Borne Diseases and Salinity Buildup for Greenhouse Growers

The arid climate and environmental conditions of the Arabian Peninsula necessitate the use of protected agriculture for the production of high value vegetable crops. Extensive water scarcity, salinity build-up and soil-borne diseases are the major constraints for greenhouse growers in the region. Despite all efforts including soil replacements and heavy chemical treatments but the loss were increasing and many growers abandoned their greenhouses.

To tackle such challenges, hydroponics would be the correct action. However, the complexity of the hydroponics systems' in terms of management is a major risk factor for growers to adopt such technology therefore, the research activities focused on simplifying the system and use local available materials to cut down on cost. Ready made fertilizers which are available in the market were used. Yield and quality of hydroponics' production were superior compared to production from the traditional soil techniques. The techniques were transferred to different growers for the production of cucumber, tomato, strawberry, green beans, peppers, musk melon and lettuce (Moustafa *et al.*, 2006).

CONCLUSION

Horticulture in general has a great potential to improve growers' livelihoods, especially those in poor countries where most rural people with limited resources. They produce only to meet their own food needs rather than market-oriented crop production, and the adoption of horticulture, especially protected agriculture, would help a shift from conventional farming to market orientation production systems.

Protected agriculture can provide rural people with nutritious vegetable crops to help address food security and alleviate poverty. The increased consumption of vegetables together with the additional food that rural families could purchase through the increased income generated by protected agriculture, will contribute to improved food security and nutrition of the most vulnerable, such as children, pregnant women and the elderly.

Protected agriculture reduces seasonality, enabling farmers to supply local markets with high quality food out of season and creating more permanent jobs. Protected agriculture is labor intensive therefore it could provide rural society with number of job opportunity for the disadvantaged, particularly women. In addition, supporting the production facilities required number of off-farm technicians that can be occupied by local youth. Furthermore, it also offers potential for development of a private service sector in the construction and supply of protected agriculture equipment. High yielding quality produce from protected agriculture could be expanded to serve the export market and generate a valuable source of foreign revenue.

Ultimately, protected agriculture with its associated techniques could increase the quality and quantity of cash crop production in rural areas with limited resources and utilizing marginal lands, therefore it would be a good answer for helping resource-poor growers to come out of poverty.

REFERENCES

Abu Gharbieh, W. and A.T. Moustafa., 2002. Soil solarization for protected agriculture, *Al-Morshed Magazine*, February 2002. pp.16-23.

CGIAR.2007. Challenge program: High value crops - Fruits and vegetables; plugging the income and nutrition gap in food security development. Concept Note - Challenge Program on High Value Crops - Fruits and Vegetables. http://www.cgiar.org/pdf/cp_cn_highvaluecrops.pdf

Divergence, 2009. Crop Protection, Plant Parasitic Nematodes. Divergence Inc. St. Louis. USA. http://www.divergence.com/ma_crop_protection.php

FAO., 2009. Water & poverty, an issue of life & livelihood. FAO, Rome, Italy. http://www.fao.org/nr/water/issues/scarcity.html

Guma Y.T. and A. Al-Masoum, 1996. Production and marketing of vegtables in the United Arab Emirates with refrence to greenhouse industry, Faculty of Agricultural Scince, UAE Unversity, *Acta Hort.* 434, ISHS 1996.

ICARDA., 2005. Supporting rural development and food security in terraces of Yemen. Adoption of sustainable protected agriculture techniques for production of cash crops in Taiz. ICARDA, Aleppo, Syria. http://www.icarda.org/aprp/APRP-AE/taez/index.htm

ICARDA., 2006. Greenhouses increase farmers' incomes in Yemen and Afghanistan. ICARDA Annual Report 2006. ICARDA, Aleppo, Syria. Pages 67-68.

ICARDA., 2007a. Enhancing rural livelihoods by increasing farm income: Protected agriculture for high-value crops in Balochistan, Pakistan. Leaflet. ICARDA, Dubai, UAE.

ICARDA., 2007b. ICARDA in the Arabian Peninsula, Ties that bind; Twenty years of collaboration in scientific agricultural research for development between the National Agricultural Research Systems of Arabian Peninsula countries and ICARDA. ICARDA, Aleppo, Syria. 52pp.

ICARDA., 2007c. Better livelihoods and more efficient water use, from greenhouses. ICARDA Annual Report 2007. page 30-31

ICARDA-APRP., 2007. Sustainable Management of Natural Resources and Improvement of the Major Production Systems in the Arabian Peninsula. Final report, 2000-2005 ICARDA - Arabian Peninsula Regional Program. ICARDA, Aleppo, Syria. 69pp.

IFAD., 2007. Rural poverty in the Sudan. http://www.ruralpovertyportal.org/web/guest/country/home/tags/sudan [Accessed on 6 July 2009]

IFAD., 2008a. Rural Poverty Report 2001 - The Challenge of Ending Rural Poverty. http://www.ifad.org/poverty/

IFAD., 2008b. The role of high-value crops in rural poverty reduction in the Near East and North Africa. IFAD, Rome, Italy? http://www.ifad.org/pub/pn/crop.pdf

Jensen M.H., A. J. Malter., 1995. Protected Agriculture a Global Review, Washington DC. World Bank Technical Paper Number 253.

Kirshi A., N. Al Ragihi and A.A. Al Thoris., 2003. Dissemination of Integrated Production and Protection Management Techniques of Cucumber Pests in Protected Agriculture under Farmers Condition in Dhamar, ICARDA-APRP Annual Report 2002/03, Dubai, UAE. Page 175-177

McCulloch, N. and M. Ota., 2002. Export horticulture and poverty in Kenya. IDS Working Paper 174. Institute of Development Studies, Brighton, UK.

Moustafa, A.T., 2001. Integrated Production and Protection Management for the Green Houses. Al Morshed, November 2001, Abu Dhabi, UAE.

Moustafa, A.T., 2006. Economic Perspective of Protected Cultivation in the Arabian Peninsula, Proceeding of Investment in Protected Cultivation in GCC Countries workshop,15 May 2006, Abu Dhabi, UAE, http://www.icarda.org/aprp/APRP-AE/pawsabu/ATM.htm

Moustafa, A.T., 2007a. Improving family incomes and livelihood in rural Afghanistan through promotion of sustainable production systems for high value crops with less water. ICARDA, Aleppo, Syria. 24pp.

Moustafa, A.T., 2007b. Protected Agriculture: A Different Greenhouse Effect, ICARDA Caravan ISSUE NO: 24, June 2007. Page 29-3

Moustafa, A.T. and A.W. Mukred, 2002. Protected Agriculture Earns More Income from Less Water for Terrace Farmers in Yemen, ICARDA Caravan: ISSUE NO 16, June 2002. Page 22-23

Moustafa, A.T., A. Al Bakry, M. Al Rawahi, A.R. Al Mohammadi and R. Ahmed, 2003. Development of Soilless Growing Techniques for the Production of High Quality Cash Crops using Less Water in the Arabian Peninsula, More Crops per Drop. Proceeding of 7th International Conference on Dryland Development (IDDC) 14-17 Sep 2003, Tehran, Iran. Pages 485-488.

Moustafa, A.T., K. Amegbeto, M. Wadid. S. El-Abd and A. Nejatian, 2006. Improving rural livelihood in Afghanistan through the promotion of sustainable production technologies for high value crops. Proceeding of 8th IDDC, Peking, China, 25-28 Feb 2006. Pages 676-681.

Moustafa, A.T., S. Oraifan, A. Al Bakry, and A. Nejatian, 2006. High density cropping system for cash crop production in marginal land with less water. Proceeding of 8th IDDC, Peking, China, 25-28 Feb 2006. Pages 839-842.

Mukred, A.W., A T. Moustafa, A. Al Kirshi and M. Al Dhbhani, 2003. Introduction of protected agriculture into mountaine terraces of Yemen. Proceeding of 7th International Conference on Dryland Development (IDDC) 14-17 Sep 2003, Tehran, Iran. Pages 504-511

Shomo, F., A. Khan, K. Shideed, A. Mustafa and M. Islam, 2008. Feasibility of protected agriculture for cash crop production in Balochistan. 9th International Conference on Dryland Development 7-10 November 2008, Alexandria, Egypt. Proceeding in Press.

Weinberger, K. and T.A. Lumpkin, 2005. Horticulture for poverty alleviation—the unfunded revolution, AVRDC Publication No. 05-613, Working Paper No. 15. Shanhua, Taiwan: AVRDC – The World Vegetable Center, 20 pp.

WHO, 2003. Fruit, Vegetables and NCD Disease Prevention, Global Strategy on Diet, Physical Activity and Health, http://www.who.int/dietphysicalactivity/ publications/ facts/ fruit /en/index.html.

III
Contribution of Horticulture Technology

13. DYNAMICS OF TECHNOLOGIES IN HORTICULTURE FOR IMPROVING PRODUCTION, PRODUCTIVITY AND PROFITABILITY[1]

H.P. Singh

Deputy Director General (Horticulture), Indian Council of Agricultural Research, Krishi Anusandhan Bhawan-II, Pusa Campus, New Delhi 110 012

ABSTRACT

In the changing global scenario, demand for healthy food is increasing due to rising population with higher expandable income. More diversified products with ensured food safety are needed. Thus production system management and cultivars have to change to ensure better marketing of the produce for higher returns. Horticulture, which includes fruits, vegetables, spices, flowers and medicinal and aromatic plants have proved beyond doubt its potentiality for gainful diversification. In the last two decades, production of horticultural crops has increased manifold and horticulture has moved from rural confine to commercial production and this changing scenario has encouraged private sector investment in production system management. Since growing of horticultural crops is rewarding to the farmers in terms of returns per unit area, the sector is expected to contribute significantly for food and nutritional security, employment opportunity and poverty alleviation. However, with declining land and water and with a threat of climate change, there is a rising concern for meeting the growing requirement. Therefore, the technology needs a change and has to be dynamic to address the emerging issues which provide both challenges and opportunities. Although horticulture in India has been moving dynamically and there is an impressive growth in production owing to development and adoption of technologies, yet there are issues which would need appropriate attention for dynamics of technologies for gainful utilization of emerging opportunities.

1

Paper for presentation in Horticulture for livelihood security & economic growth. International Conference on Horticulture (ICH 2009) to be held from November, 9-12, 2009 at PNASF Bangalore.

IMPACT OF TECHNOLOGICAL CHANGE AND CHALLENGES AHEAD

In the past, Government and other stakeholders have taken initiatives in development of technology, its dissemination and adoption by farmers which has the growth impacted in terms of increased production, productivity and also availability of horticultural produce. Production of horticultural produce has jumped 8 times from the production level in 1950-51. However, challenges ahead are to have sustainability and competitiveness, and to achieve the targeted production of 300 million tons at the end of XI plan from the current level of 214 million tons in the environment of dwindling land and water resources (Table 1). Despite of numerous challenges and shortcomings the horticulture has impressive growth. Among the horticulture crops, fruit crops recorded a two fold increase in area and production in the last decade. India is the second largest producer of fruits (68.5 mt) from 6.10 m ha area and contributes 11.2 per cent share in global fruit production. India occupies first place in the production of mango, banana, papaya, pomegranate, sapota and aonla. The productivity of grape is the highest in the world. India has higher national average productivity in banana and sapota as compared to world average productivity. However, productivity in citrus, mango, apple, guava and pineapple continue to be lower than world average.

Table 1. Projected demand of horticultural crops

Commodity	Production (mt)			
	1991-92	2008-09	2011-12	2020-2021
Fruits	28.63	68.50	81.00	98.00
Vegetables	58.53	129.11	185.00	220.00
Spices	1.90	4. 15	5.50	6.50
Coconut	7.49	10.15	20.00	24.00
Cashew nut	0.40	0.70	1.70	2.00
Cocoa, other	-	4.81	6.80	9.50
Total	**96.95**	**214.72**	**300.00**	**360.00**

[*Source*: Singh, HP, 2007, Indian Horticulture, 52(4): 3-9 and NHB Database, 2009]

Vegetable crops, which occupy 8.0 mha has the production of 129 mt. In the last one and half decade, country's vegetable production has almost doubled and gross vegetable productivity of the country has increased by one and half times. Major contribution to vegetables comes from potato, tomato, brinjal, okra, beans and cucurbits. Among the cucurbits besides melon, gourds, cucumber and gherkins are important. Cucurbits like pointed gourd, spine gourd are also gaining importance of commerce, which have much more value for export for its medicinal and therapeutic uses.

Commercial floriculture sector has recorded fast pace of growth during the last decade. Intensive floriculture has contributed considerably for generation of employment. During the last decade export has grown manifold and area has expanded to 1,67,000 ha with production of 9,87,000 mt of loose flowers and 4.8 million cut flowers. Floriculture provides ample opportunity both for domestic market and export, which includes cut flowers, loose flowers, potted plants, bedding plants, foliage and dry flowers. Medicinal and aromatic plants, which have immense potential have got due recognition in the decade. Its cultivation and use is promoted across the country in regionally differentiated manner. Among the aromatic plants, mentha has become important for commercial cultivation.

India is the largest producer, consumer and exporter of spices and spice products in the world and produces more than 50 spices. The spice production in India is in the order of 4.14 mt from an area of about 2.63 mha. Area and production of spices in the country have registered substantial increase over the last fifteen years with average annual growth rate of 2.7% and 6.2% in area and production, respectively. Chilli is the major spice crop occupying about 29% of area under cultivation having contribution of 31% of total spices production in the country. Turmeric accounts for 20% of production and 7% of area, while garlic accounts for 20% of production and 6% of area. Seed spices contribute 14% of production and occupy 32% of area while black pepper contributes 1.1% of production and occupies 9% of area of the total spices in the country. India is leading producer of plantation crops in the world with contribution of 22.34% in coconut, 25% in cashewnut and 56% in arecanut.

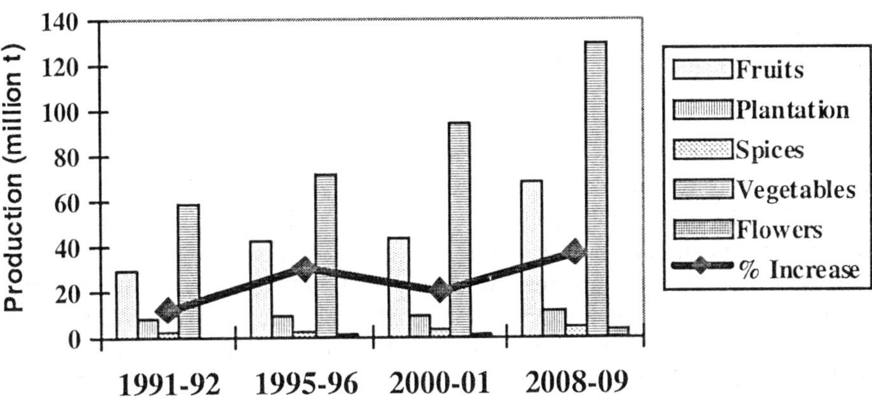

Fig. 1. Increase in production of horticultural crops over decades
(Source, Singh, H.P. 2009)

Past trend in development has been satisfying in terms of technological adoptions, production, availability and export of horticultural produce, and this trend has been marked as "Golden Revolution". The expansion of horticulture in nontraditional areas resulted in increase in area under fruits from 4.01 million hectares (mha) in 2001-02 to 6.10 mha in 2008-09 with an increased production of about 25 million tons (mt). The area under vegetable cultivation also showed a similar increase of 2.1 mha from 2000-01 to 2008-09 with an increased production of 21 mt (Fig. 2). India has emerged as second largest producer of fruits and vegetables and has first position in several horticultural crops. Production and export of flowers have increased manifold and the country has a major stake in global trade of spices and cashew nuts. Export of medicinal plants, fruits and vegetables have also exhibited rising trend.

Though India is leading in the productivity of some horticultural crops like grapes, cassava, ginger, turmeric, still there is a scope to increase the productivity in other horticultural crops in comparison to other countries. The Government has addressed many of these constraints with added thrust in IX and X Plan in regionally differentiated manner. Technology Mission for Integrated Horticulture Development in North-East Region including Sikkim and National Horticulture Mission has created awareness across the region which provided an insight for horticulture-led transformation and subsequently National Horticultural Mission, and micro irrigation projects have been launched in X Plan and are persuaded in XI Plan to achieve the transformation of horticulture utilizing the technology. It is also proposed to have technology mission on cashew and also micro-irrigation, since mission approach has been effective in effective dissemination of knowledge and adoption of improved practices leading to desired growth.

INSTITUTIONAL SUPPORT FOR RESEARCH

Horticultural research in the country is being carried out at 10 ICAR institutes (with 24 regional stations) six Directorates and seven National Research Centers (on major crops). Area specific, multi-disciplinary research is also being conducted under 14 All India Co-ordinated Research Projects each on Tropical, Sub-Tropical, Arid Fruits, Vegetables, Potato, Tuber Crops, Mushroom; Floriculture, Medicinal and Aromatic crops, Betelvine, Palms, Cashew, and Spices at 251 centers located at various research institutes and State Agricultural Universities. In addition, five network projects have been approved in XI plan to address specific issues. Research on horticulture is also undertaken in multi-disciplinary institutes, departments of horticulture in 34 Agricultural Universities, two Deemed Universities and three Universities of Horticulture and Forestry. Besides the above, traditional Universities, and other Central organizations like the Council of Scientific and Industrial Research, Department of Biotechnology, Atomic Energy Research Centre, Defense Research Development Organization and others also undertake research projects in basic/strategic area. The Horticulture Division of ICAR is responsible for national level planning and promotion of major research program in relation to horticultural crops. The main focus of program is to safeguard the genetic resources, develop suitable cultivar of different crops, augment the production and protection technologies and also technologies for enhanced shelf-life. Outreach research programs to combat the serious diseases caused by Phytophthora, Fusarium, Ralstonia, Alternaria and viruses have initiated which is cutting across the discipline. Five National Research Centers on Cashew, Onion and Garlic, Mushroom, Medicinal and Aromatic Plants and Oil palm have been upgraded to function in directorate mode having all India program and AICRP on floriculture has been strengthened and upgraded to Directorate of Floriculture. Two genomic network projects are being operated in India at various ICAR institutes on horticultural crops namely, International Solanaceae Genome Network program and International Network on Improvement of Banana and Plantain. Several international programs have been taken up which address the issues.

TECHNOLOGICAL CHANGES

Most important changes being noted in agriculture are the diversification, modernization, market orientation and commercialization, involving introduction of new crops and varieties, increased share of horticulture in the cropping pattern, diversion into processing and export oriented production of a large number of crops in a fairly large proportion of the total cultivated area. Several technologies have been advanced in the complete value chain involving technology for orchard establishment, availability of true to type planting material, plant architecture engineering and management, mulching, fruit thinning, integrated nutrient management, water management, integrated pest and disease management, post harvest technology, processing, marketing etc. The dynamics in horticulture sector have occurred because it has received lot of importance from all the stakeholders from public sector, private sector and farmers during last decade. This is primarily the result of realization that diversification to horticultural crops is now the major option to improve livelihood security to small farmers, employment and nutritional security. Keeping in view the dynamic needs of diverse stakeholders under the National Agriculture Research System, the R&D on horticulture has been undertaken in several multi-crop and multi-disciplinary institutes and the several technologies have emerged.

Adoption of High Yielding Cultivars

Many high yielding varieties and hybrids of different horticultural crops have been developed for different regions and have been adopted for cultivation by many farmers. Till date, 1596 improved high yielding, high quality coupled with disease and pest resistant varieties and hybrids have been released by various institutes/universities for cultivation in diverse agro-climatic conditions of the country (Table 2 and 3). In Fruits–197; Vegetables–519; Floriculture-203, Tuber crops–158; Spices and Plantation crops–388; Medicinal and Aromatic plants–126; Mushrooms–5 varieties have been released. Regular bearing mango hybrid, export quality grapes, multiple disease resistant vegetable hybrids, high value spices and tuber crops of industrial use have been developed. Improved varieties have revolutionized the horticultural sectors. High-yielding Gauri Sankar and Sree Bhadra sweet potatoes have focus in minimizing malnutrition and improving nutritional security.

Besides breeding for disease and pest resistance, quality also assumes great importance in horticulture. A large number of cultivars have been developed which carry resistant genes. Use of biotechnological tool to develop resistance to fruit borer in brinjal is in the final stage of release. Similarly, breeding to develop grape cultivars suitable for vine making, black pepper cultivars rich in aroma compound 'Caryophyllene', development of processing tomatoes etc. are some of the research programmes being carried out in various horticultural institutes. Varieties are being bred for processing qualities such as Kufri Chipsona in potato for chips making, high TSS white onion NRCOG W448, grape varieties suitable for wine making, papaya varieties for table and papain production are some of the successful research attempts being carried out at various ICAR institutes.

Table 2. List of high yielding varieties/hybrids in fruit crops

Fruit Crop	Varieties/Hybrids
Mango	Mallika, Amrapali, CISH M2, Ratna, Arka Puneet, Arka Arunna, Arka Anmol, Arka Neelkiran, Sindhu (pulp making), Dashehari 51 (regular-bearing, high yielding)
Banana	FHIA -01, FHIA -03, Co-2, H-1 and H-2
Grape	Arkavati, Arka Kanchan, Arka Hans, Arka Shyam, Arka Neelmani, Tas-e-Ganesh, Sharad Seedless, Sonaka and Dilkush
Acid lime	Pramalini, Vikdram, PKM -1, Sai Sarbati and Jai Devi
Guava	CISH-G-1, CISH-G-@, ISH-G-3, Arka Mridula, Arka Amulya, Lalit (Pink peel), Safed Jam, Kohir Safeda.
Papaya	H-39, Cp-81, Co-1-7, Surya, Pusa Delicious, Pusa Majesty, Pusa Giant, Pusa, Dwarf, Pusa Nanha
Litchi	Swarna Rupa, Sabour Madhu, Sabour Priya
Pomergranate	Hybrid Ruby, Ganesh, Hyti (red arils), Arakta and Bhagwa
Pineapple	PKM -1
Ber	Gomah Kirti, Umran, Gola and Kaithalli
Aonla	Chakaiya, Krishna, Kanchn, NA-6, NA-7 and NA-10
Beal	NB-5, NB-9, CISH B-1, CISH B-2, Pant Aparna, Pant Shivani
Apple	Sunehari, Lal Ambri, Chaubatia Princes, Amrich, Amred and Royal Ambri
Custard apple	Arka Sahan
Sapota	Kalipatti, Gupthi, Cricket Ball, PKM-2 and DHS
Kiwifruit	Abbott, Allison, Bruno, Hayward and Monty
Peach	Shan-e-Punjab and Flordasun (low-chilling)

Table 3. Vegetables and spices varieties resistant to pests and diseases

Crop	Name of variety	Resistant against
Tomato	Vaishali, Rupali, Rashmi	Fusarium & Verticillium wilt
	Arka Vardan	Root knot nematode
	Arka Alok, Arka Abha, Shreshta	Bacterial wilt
	Arka Ashish	Powdery mildew
Brinjal	Pusa Purple Long	Shoot borer and little leaf

	Pusa Purple Cluster	Little leaf
	Arka Nidhi, Arka Keshav, Arka Neelakanth	Bacterial wilt
Chilli	NP46A	Tolerant to thrips
	PMR 57/88K	Powdery mildew
	Pant C1	Tolerant to mosaic and leaf curl
	Musalwadi	Tolerant to die back and powdery mildew
	Jawahar 218	Tolerant to leaf curl and fruit rot
	Punjab Lal, Pusa Sadabahar	TMV, CMV and leaf curl virus
	Ujjwala	Bacterial wilt
Cabbage	Pusa Mukta, K1	Black rot
	Pusa Drumhead	Black leg
Cucumber	Poinsett	Downy mildew, powdery mildew, anthracnose, angular leaf spot
Bottle gourd	Punjab Komal	CMV
Watermelon	Arka Manik	Powdery mildew, Downy mildew and anthracnose
Muskmelon	Arka Rajhans	Powdery mildew
Okra	Punjab Padmini, Parbhani Kranti, Arka Abhay, Arka Anamika	Yellow vein mosaic
Pumpkin	Arka Suryamukh	Fruit fly
French bean	Arka Bold	Rust
Pea	FC1, Jawahar Matar 5, Jawahar Peas 83, JP4 (JM6), Arka Ajit	Powder mildew
Onion	Arka Kalyan, Arka Kirthiman	Purple blotch, basla rot and thrips
Amaranthus	Arka Suguna	White rust
Black pepper	IISR Shakthi	Phytophthroa capcisi
	IISR Pournami	*Meloidogyne* spp
Cardamom	IISR Vijetha	Mosaic
	IISR Avinash	Rhizome rot
Ginger	IISR Mahima	*Meloidogyne* spp
	Maran	Soft rot
Cumin	GC-4	Wilt

Hybrid Technology

Appreciable progress has been made in the development of hybrids in several vegetable crops and hybrid seeds of tomato, chilli, cucumber and muskmelon are being produced at several locations. Besides this, imported seeds of most of cole crops (especially cabbage) are available to the Indian farmers. All India Co-ordinated Vegetable Improvement Project (AICVIP) has so far recommended for cultivation of more than 35 hybrids. Besides, many hybrids of vegetable crops developed and marketed by private sector are also available to the farmers. At present, area under vegetable hybrids is 0.5 million ha in India, about 10% of total area is under vegetables. Hybrid tomatoes are cultivated on commercial scale in Karnataka, Maharashtra, Gujarat and Andhra Pradesh. Entire cabbage growing in southern part of Maharashtra and West Bengal are mainly F_1 hybrids. Area under high yielding F_1 hybrids in important vegetable crops ranges 17.8-31.5% in tomato (31.5%), cabbage (31.39%) and brinjal (17.8%) and areas under capsicum and chilli are also under expansion. High production, earliness, superior quality, uniform produce and resistance to biotic and abiotic stresses are the main advantages of F_1 hybrids.

Keeping in view the dynamic needs, the research efforts in various institutes has focus on development of hybrids with multiple disease resistance, early maturity and utilizing male sterility system. Advanced lines of tomato (TLBR-1, 2, 3, 4, 5, 6) with combined resistance of bacterial wilt were evolved at IIHR, Bangalore which could now be utilized for the develop-ment of markers and stacking of resistance genes against ToLCV, early blight in the parental inbreds. CMS lines have successfully been utilized to produce potential experimental crosses of onion and commercial hybrids of chilli. Nuclear male sterile lines (GMS) of tomato, brinjal and chilli were introduced/ established and are being successfully utilized for developing a number of cost effective experimental crosses at various centers. Considerable success has also been achieved for adoption of hybrids. Hybrid seeds of chilli (CH-1, CH-3, CCH-3), brinjal (Kashi Sandesh), tomato (TH-1) are being now produced by the farmers by up scaling their skill. The parental lines of a number of hybrids developed have been sold on non-exclusive basis to the seed companies with the aim to promote these hybrids among the farmers.

Utilization of Biotechnological Tools

Somatic hybridization and anther/embryo culture

One of the successful attempts of protoplast fusion to produce somatic hybrids for developing good rootstocks is in citrus. Use of meristem culture and micro grafting, which involves excision of extreme tip of shoot doom (0.1-0.5 mm) and extensively grafting onto *in vitro* raised rootstock. It is very successful in citrus for elimination of viruses. Anthers of capsicum variety Arka Gaurav and tomato hybrid Avinash 2 responded to culture with an embryogenic like response without an intervening callus phase. Androgenesis has been successfully used for brinjal, pepper, cabbage, cauliflower, potato, asparagus and carrot, whereas gynogenesis has been successful in onion. Embryo rescue has been successfully employed in production of hybrids of *Musa accuminata* x *Musa balbisiana, Carica papaya* x *Carica cauliflora,* interspecific crosses in pineapple and seedless x seedless grape varieties. This technique facilitates breeding of early maturing fruit cultivars in which fruits ripens before seeds reach germinable stages. Triploid progenies from immature fruits/seeds of 4n x 2n crosses of watermelon could be salvaged through embryo culture.

Use of marker technology

It is one of the most important technologies of immediate concern in heterozygous species like fruit crops where genepool is rapidly shrinking. Genetic characterization of available native and introduced germplasm in different fruit crops is necessary as it has several immediate implications. The outcome may help breeders in understanding the genetic diversity, evolutionary pattern, characterization of genotypes into groups/classes, identification of mutants and synonyms, confirmation of hybridity, identification of useful gene sources and their tagging, settling of disputes over claims on genetic material under patent regime, precision breeding, marker aided selection (MAS), etc. Efforts are under way to fingerprint mango, banana, cashewnut, kiwifruit, walnut, grape, citrus, etc. by different research centers. In root-knot nematode resistance (Mi) gene in tomato, DNA sequence has been isolated that overlap with parts of desired gene and, therefore are more certain to be inherited with it serving marker. These sequences are also being used to isolate Mi gene to facilitate breeding this valued trait into new varieties and even other species. QTL mapping are in progress in many crops such as brinjal, tomato, capsicum while association mapping (linkage disequilibrium) is used in case of perennials such as black pepper, cardamom and coconut. Gene pyramiding to converge useful genes in one back ground variety of commerce is the mainstay of biotechnological research and is in progress in solanaceous vegetables.

Transgenics in horticulture

Efforts are in progress at various institutions in India to tackle issues of managing disease resistance, resistance to insect pests, nutritional quality improvement and to extend self life of fruits and vegetables through development of transgenics. A network project is in operation which involves 6 horticultural crops at ICAR laboratories as a part of mega network project on transgenics (Table 4).

Table 4. ICAR network projects on transgenics

Crop	Traits
Banana	Resistance to Sigatoka leaf spot, *Fusarium wilt*, Bunchy top virus, Banana streak virus
Brinjal	Resistance to fruit and shoot borer
Cassava	Resistance to mosaic virus
Papaya	Resistance to ring spot virus and leaf curl virus
Potato	Resistance to potato viruses, PVY, PALCV
Tomato	Resistance to early blight, leaf curl virus, fruit borer, tolerance to high temperature, drought and extended shelf life

Bt produce one or more proteins referred to as δ-endotoxins. These crystal proteinms or δ-endotoxins are characterized by their toxicity to the larveae of distinct range of host insects. The most damaging insects are usually Lepidopteran followed by Coleopteran and Dipterans. A large number of transgenics with *Cry-l* AB gene have been produced. ELISA tests have shown high proportion of Bt protein in total soluble proteins of these transgenics. The presence of the *Cry* gene has also been confirmed by Southern blotting. Another gene, which produces osmotin, has been transferred to a variety, which is under evaluation for resistance to late blight.

Improved shelf life and quality

Post harvest losses take a heavy toll in vegetable crops. Biotechnology can help remove the plant components that cause early deterioration after harvesting. For instance, a technique to reduce the presence of a normal tomato enzyme is involved in the softening of ripe tomato fruits. This technique involves engineering plants with and antisense gene so that production of enzyme is significantly reduced. Improved storage and shelf life of genetically modified vegetables with quality traits would increase the availability and improve the possibilities of balanced diet. At low temperatures, potatoes accumulate reducing sugars making them unfit for processing. This is because of vascular invertase activity. An invertase inhibitor (inhh) gene (accession AY145781) from tobacco has been cloned and integrated in T-DNA of pBinAR. This construct is used for transformation of both table and processing potatoes to reducing sugars accumulation during storage. Nutritionally improved transgenic potatoes have been obtained by transferring the amaranth seed albumin gene (AmA1) from *Amaranthus hypochondriacus* in to potato.

Bioactive components

Active components such as caroteinoids from tomatoes, glucosinolates from Brassica vegetables, phytoestrogenes from soybean and phenolics and antioxidants from various plants protect and prevent numerous disease such as cancer and cardiovascular disease. These antioxidants are also associated with slowing the ageing process and improving overall health. Phytoestrogens from *Dioscorea* species such as wild yam (*Dioscorea villosa)* also possess numerous physiological benefits. Some sweet potato varieties have anti-diabetic potential. Most of these bioactive components are plant secondary metabolites. The genes used for transformation in development of transgenics of vegetable crops is given in Table 5 and 6.

Development and utilization of detection and diagnostic methods based on ELISA, PCR and Microarray are gaining importance. Biotechnological tools have augmented conventional approaches in identification and isolation of resistant sources/ genes and development of crops resistant to specific viruses through transgenic approaches. Virus resistant transgenic plants are obtained by inserting segments of viral nucleic acid into plant genomes that leads to silencing of genes of the virus that have homologous sequences, thereby making plants resistant. There are several cases of successful transformation of a susceptible into a resistant crop through genetic engineering with parts of the genome of the virus such as coat protein, movement protein, replicase protein and non-coding regions (Table 6). Some of the nonviral genes also have resulted in virus resistance include the gene for ds RNase from *Schizoa sccharomyces* pombe, tobacco resistant gene 'N', mouse protein kinase, systemic acquired resistance gene etc. Combining a host gene for resistance with pathogen derived genes or with genes coding for antimicrobial compounds provide for a broad and effective resistance in many host – pathogen combinations.

Table 5. Transgenic developmental activities in vegetable crops

Institute	Vegetable crops	Transgenic	Target
CPRI, Shimla	Potato	Bt cry 1 Ab	Developed palnts resistance to potato tuber moth
		Osmotin	Developed water stress tolerant plants
		Coat protein	Developed plants resistance to potato virus Y (PVY)
Delhi University, Delhi	Tomato	Ctx-B and Tcp antigens of Vibrio cholerae	Edible vaccine development
	Brinjal	Chitinase, glucanase and thaumatin encoding genes	Create disease resistant plant
IARI, New Delhi	Brinjal	Cry 1Ab	Lepidopteran resistant plants
	Tomato	Cry 1Ab	Lepidopteran resistant plants
	Cauliflower	Cry 1Ab	Lepidopteran resistant plants
	Cabbage	Cry 1 Ab	Lepidopteran resistant plants
	Tomato	Acc synthase	Controlling fruit ripening
		Replicase gene	Develop tomato leaf curl virus resistant plants
IIHR, Bangalore	Muskmelon	Rabies flycoprotien gene	Develop edible vaccines
	Tomato	Leaf curl virus sequence	General leaf curl virus resistant plant
	Tomato	Chitinase and glucanase	Develop plant resistant to fungal diseases
JNU, New Delhi	Tomato	OXDC	Develop fungal disease resistance plant
	Potato	Ama 1	Develop protein rich potatoes
IAHS, Bangalore	Tomato	Alfalfa glucanase and tomato leaf curl virus genes	Generate plants resistant to viral and fungal diseases

Institute	Vegetable crops	Transgenic	Target
Proagro PG-S (India) Ltd, Gurgaon	Tomato	Cry 1 Ab	Lepidopteran resistant plants
	Brinjal	Cry 1 Ab	Lepidopteran resistant plants
	Cauliflower	Cry 1H/Cry 9C	Lepidopteran resistant plants
	Cauliflower	Bar, barnase, barstar	Superior hybrid cultivars
	Cabbage	Cry 1H/Cry 9C	Lepidopteran resistant plants

Table 6. Transgenic vegetable crops and genes used for their transformation

Vegetable crops	Modification	Gene
Cantaloupe	Modified fruit ripening cantaloupe	S-adenosylmenthionine hydrolase gene from *E. coli* bacteriophage T3
Potato	Insect and virus protected potato	The cry IIIA gene from Bacillus thuringie-nsis sp. tenebrionis and the potato virus Y coat protein gene
	Insect protected potato	The cry IIIA gene from Bacillus thuringiensis sp. Tenebrionis
	Insect protected potato	The cry IIIA gene from Bacillus thuringiensis
Radicchio rosso	Male sterile radicchio rosso	The barnase gene for Bacillus amyloliquefaciens
Sqush	Virus – resistant squash Freedom II	Coat protein genes of cucumber mosaic virus, Zucchini yellow mosaic virus and watermelon mosaic virus 2
	Glufosinate tolerant sugarbeet	Phosphinothricin acetyltransferase gene from Streptomyces viridochrome genes
	Gluphosinate tolerant sugarbeet	The enolpyruvylshikimate-3-phosphate synthase gene from Agrobacterium sps. Strain CP4 and a truncated glyphosate oxidoreductase gene from Ochrobactrum anthropi
Tomato	Insect protected	The Cry Ia (c) gene from Bacillus thuringiensis subsp. Kurstaki
	Modified ripening	S-adenosylmethionine hydrolase gene from E. coli bacteriophage T3
	Flavoured tomato	Antisense polygalacturonase gene from tomato
	Improved ripening	A fragment of aminocyclopropane carboxylic acid synthase gene from tomato
	Delayed softening	A fragment of polygalacturonase gene from tomate

Rootstocks for Enhanced Profitability

Several rootstocks with immense potential have become integral part of the fruit industry. The strong horticulture industry today based on grapes, citrus, apple and few other fruit crops would have never flourished in absence of better rootstocks. There is need to use efficient rootstocks for boosting fruit production and productivity in the country. The success of root stock is determined by its tolerance to prevailing conditions of soils, climate and disease, while still producing high yields of good quality fruit. Most important place among citrus rootstocks is occupied by Rough lemon, Karna Khatta, Cleoptra mandarin and Rangpur lime for yield and tolerance to disease and salinity in several states both for sweet orange and mandarins. The use of rootstock in grape cultivation has gained popularity and almost all newer vineyards are being planted on rootstocks only. The popular root stock for grape are Dogridge B and 110R. Selective root stocks have become an integral part of the commercial cultivation of grape and citrus. Mango is still commonly propagated on variable seedling rootstocks of unknown parentage. Most of the Indian varieties are monoembryonic but about a dozen varieties from South India are polyembryonic. The exploitation of polyembryonic is under evaluation. Several components of future commercial fruit growing will depend on successful use of rootstocks for better scion compatibility, canopy architecture, fruit quality, nutrient absorption, water use efficiency, biotic and abiotic stress tolerance, adaptation under the influence of climate change. There is need to breed newer rootstocks for fruit growers of citrus, grapes, mango, apple, aonla etc.

High Density Orcharding

High density planting technology has been standardized for many crops and also adopted by many fruit growers in India. It has become a success story in many fruit crops like banana, pine apple, citrus, papaya, coconut, mango, cashew and few arid fruit crops. High density orchards not only have provided higher yield and net economic returns per unit area in the initial years, but also facilitated more efficient use of inputs. Banana cultivars in the high density planting with spacing between 1.2 m x 1.2m and 2.0 m x 2.0 m having 6,944 to 7,500 plants ha^{-1} produced the yield range of 86.60 to 174.39 t ha^{-1}. In cashew, high density planting with plants of 312 to 625 or more grafts ha^{-1} are planted. Closer spacing of 4 m x 4 m or 6 m x 6 m or 5 m x 5 m in the beginning, and thinning out in stages and thereby maintaining a spacing of 8 m x 8 m by eleventh year has been followed. This enables higher returns during initial years and as trees grow in volume, the final thinning is done. Four fold yield increase up to six years and two to five fold increase up to 12 years was observed.

In mango, 3 m x 3 m spacing is prepared for high density planting. In guava, 10' x 10' (1,076 plants ha^{-1}) in a square system of planting has been found highly successful. Papaya panted at 2.5 m x 2.5 m spacing. Spacing Nagpur Orange and Acid limes at 6 m x 3 m (555 plants ha^{-1}) was found ideal for higher yield of at Akola and Rahuri, respectively, but they had more incidence of canker compared to plants spaced at 6 m x 6 m. Spacing of 5 m x 5m (400 plants ha^{-1}) was ideal for *Khasi* mandarin. High density orchards start fruiting in the third year of the planting and remain productive for 10 to 12 years. Alternate rows can be removed when plants canopies are intermingling. Technologies for meadow orcharding in guava, high density planting, canopy management and rejuvenation of old and senile orchards of mango, aonla and guava are developed and demonstrated. Coconut based high density multi-species cropping system helps to improve soil properties, realize the better stability for the farm net income and generates additional employment.

Efficient Nutrient Management

Among various inputs, fertilizers alone account for 20-30% of the total cost of production in horticultural crops. Moreover, the efficacy of fertilizers applied in soil is low due to various losses and soil fixation. Fertilizer efficiency studies using isotope labeled fertilizers have shown nutrient derived from fertilizer (NDFF) values of 0.5 to 30%. There is need of a system which can determine the need of nutrients for all stages of growth and also understand the soil ability to supply those needed nutrients more efficiently. Monitoring soil health and leaf analysis in fruit crops (Table 7) will be useful to ensure a balanced nutrient management and safeguard against hidden hunger as well as luxury consumption by horticultural crops. Concerted research work for attainment of high nutrient use efficiency is required to evolve plant types that are highly responsive to copious supply of nutrients and tolerant to marginal soil fertility conditions with assured yields. Besides, use of liquid form of fertilizers through fertigation appears to be promising for deeper application with sizeable input saving. The use of bio-fertilizers, VAM fungi, biological nitrogen fixers and other beneficial microbial agents also need to be optimally used to attain an efficient management. There is also an immense scope of harnessing geographical information system (GIS) using remote sensing (RS) to address farmer/site specific farming including site specific nutrient management under precision farming.

Table 7. Optimum nutrient norm for fruit crops

Nutrient	Unit	Anab-e-Shahi	Thompson Seedless	Alphonso	Totapuri	Guava	Acid Lime
Nitrogen	%	0.50-1.25	0.87-1.61	0.78-1.65	0.84-1.53	1.69 – 2.20	1.53-2.10
Phosphorous	%	0.46-0.68	0.29-0.65	0.02-0.33	0.064-0.147	0.16 – 0.23	0.10-0.15
Potasium	%	1.16-2.42	2.00-3.02	0.77-1.73	0.52-1.10	1.20 – 1.67	0.96-1.66
Calcium	%	1.40-2.07	0.98-1.36	0.76-1.63	1.97-3.20	0.60 – 1.27	3.05-3.43
Magnesium	%	0.24-0.58	0.63-1.10	0.40-0.65	0.40-0.65	0.35 – 0.50	0.40-0.60
Sulphur	%	0.08-0.18	0.09-0.13	0.03 -0.13	0.014–0.21	0.29 – 0.47	0.25-0.29
Iron	ppm	38-107	54-80	657-963	48 - 86	114 – 178	117-194
Manganese	ppm	18-86	42-209	13-408	57-174	34 – 77	21-63
Zinc	ppm	25-94	30-88	7.8-18.3	25-33	29 – 41	25-50
Copper	ppm	10-30	5-10	14.3-17.8	3.10-8.00	6 – 12	8.7-14.8

Organic Horticulture

Demand for green food is on increase and harnessing the potential of organic farming which address soil health, human health and environmental health is considered of greater significance. In last few years, organic farming has attracted many farmers across the country especially combined with ecotourism and have experimented successfully. India is best known as an exporter of organic tea and also has niche market for spices and fruits and vegetables (Table 8). Fresh tropical fruit that are having high potential in the world market are banana, date, fig, pineapple, avocado, guava, mango and mangosteen, melon and papaya; other fruits like kiwi, durian, litchi, jackfruit and carambola also have scope for production. Large scale investment in the sector will overcome the marketing constraints of these produce and hence will be a highly remunerative venture in the days to come.

Table 8. Major horticultural products produced in India by organic farming

Groups	Products
Commodity	Tea, Coffee
Spices	Cardamom, Black pepper, White pepper, Ginger, Turmeric, Vanilla, Tamarind, Clove, Cinnamon, Nutmeg, Mace, Chili
Fruits	Mango, Banana, Pineapple, Passion fruit, Sugarcane, Orange, Cashew nut, Walnut
Vegetables	Okra, Brinjal, Garlic, Onion, Tomato, Potato

Source: Garibay and Jyoti (2003)

Efficient Water Use Technology

Water is important for growing of crops and with all our efforts 40% of the total area is irrigated. It is also believed that even if all the resources of water are utilized, whole area cannot be irrigated. Irrigation efficiency is only to the tune of 30-35%, therefore only means to provide the water is effective management and enhancing the water use efficiency. Therefore lots of initiatives were taken to apply water in such a manner which can provide maximum output. When, where and how water should be applied has been worked out. Among various methods tried drip irrigation has proved successful in exhibiting high water productivity by saving irrigation water from 25-60% in various orchard crops and vegetables from along with 10-60% increase in yield as compared to conventional method of irrigation. It is one of the latest methods of irrigation which is becoming popular in areas with water scarcity and salt problems. The fertigation has become the state of the art in orchard crops and vegetables because nutrients can be applied to plants in the correct dosages and at the time appropriate for the specific stage of plant growth. Fertigation requirement in fruits (mango, banana, grapes, papaya, and pomegranate, citrus and strawberry), vegetables (tomato, chillies, brinjal, okra, potato, muskmelon, cucumber) and ornamental crops (rose, carnation, gerbera) and plantation crops (coconut, arecanut and coffee) have been standardized to improve the nutrient and water use efficiency which increased from 120 to 290% (Table 9). Few progressive farmers are growing high value fruit crops of grapes, pomegranate and banana using fertigation. Similarly, efficient water management practices like zero tillage and natural/synthetic mulches have also been developed to tackle the problem of water scarcity. Another measure could be the development of drought-tolerant and low water-requiring varieties for which the work has already been started and tangible results are expected in few years.

Table 9. Water productivity enhancement in horticulture crops

No.	Crop	Efficient technique for enhancing water productivity
1	Arid fruits	• Drip irrigation – 25 % saving in irrigation water • Double ring method of irrigating anola, water saving 50-60 percent
2	Banana	• Drip irrigation, 30% water saving & 40 % increase in yield
3	Cashew	• Crescent bund treatment of terrace increased the soil moisture content by 14.8% and reduced the run-off by 4.6 l/s/ha. • Drip irrigation, requirement of water reduced by 66%
4	Citrus	• Drip irrigation, requirement of water reduced by 40%
5	Grapes	• Tas-A-Ganesh grafted on rootstock for better WUE – 25% water saving • Mulching and spray of anti-stress combination reduced 25% irrigation requirement
6	Oil Palm	• Drip irrigation better than hose irrigation
7	Onion	• Drip irrigation, 40 % saving in water
8	Orchid	• Misting in green house
9	Plantation crops	• Drip irrigation, 34% water saving in coconut, arecanut & cocoa
10	Pomegranate	• Drip irrigation, saving of water is 50%
11	Potato	• Scheduling irrigation at 15-25mm Cumulative Pan Evaporation is optimum for different soil type
12	Spices	• Drip irrigation 7 litre/day/vine during summer increased the yield of black pepper
13	Temperate fruits	• Irrigation at 80% ET increased fruit set in apple • Use of black polythene / dry grass as mulch saved water by 40-70% & increased yield of apple
14	Tuber crops	• Irrigation at 25% available soil moisture is ideal
15	Vegetables	• Genotypes for better WUE was identified in Capsicum (Arka Mohini), Onion (Arka Kalyan), tomato (Arka Lohit) • Irrigation at 100% Cumulative Pan Evaporation for tomato was better • Highest WUE (108 q/ha-cm) was with drip irrigation at 40% Pan Evaporation.

Horticulture Based Cropping Systems

The horticulture is the best option for crop diversification and maximizing system productivity. There are several success stories of cropping system based on horticulture, which have proved meaningful in providing high net returns to farmers by making sustainable use of farm resources and environment on a long term basis. During the initial stage of establishment the widely spaced orchard crops are commonly utilized by farmers for intercropping with vegetable crops, spices, flower and medicinal and aromatic crops. In order to ensure a yield continuum and regular income, it is imperative that cropping system is managed in this way that light, nutrient, water, space etc. are utilized effectively.

The successful intercrops for young mango orchard are lady's finger-carrot, chilli-onion, brinjal-tomato, bottle gourd-radish and sponge gourd-tomato in Bihar. Cucumber and amaranth are profitable intercrops with Nendran banana in Kerala. Nagpur mandarin + cotton + soybean intercrop gave the highest net return. Vegetables successfully intercropped with papaya. Shade loving medicinal and aromatic crops like patchouli, rose geranium, long pepper, *sarpgandha, kacholam* etc., are successfully grown under coconut and areca nut. The choice of crop selection is mainly based on farmer's need. Potato and cabbage are the most suitable crops in rotation under normal monsoon conditions in Nilgiri hills. Intercropping tomato with drumstick earned the net profit of Rs 10,026 ha^{-1} in Andhra Pradesh. The elephant foot yam is widely grown as intercrops in banana, papaya and other orchards in Andhra Pradesh. It has been also reported that elephant foot yam and grater yam crops were more profitable intercrops in coconut garden. Cassava can be profitable intercrop during initial establishment period of coconut. Sweet potato fit into rice based cropping system for rotation. Spices like black pepper, ginger, turmeric, vanilla, nutmeg, clove and some medicinal plants are the ideal intercrops for coconut.

Horticulture Based Farming Systems

Indian farmer cannot take risk with small land holding size. Therefore for ensuring sustainability and economic viability there is need to have farming system model which may include fruit crops, vegetables, plantation crops, tuber crops, spices, livestock, tanks for fish farming. The need is increasing to produce more and more from less and less resources. Enterprise mix will diffuse the risk to farmers. Horti-silviculture and Agri-horti-silvipastoral systems with animals provide sustainable systems particularly small and medium farmers. A model developed showed that fodder from terrace risers, horticultural and silvi-pastoral crops can sustain an unit of 10 goats with reproduction efficiency of 170% and the pigs can be met part of their nutrient requirement through succulent grasses, grains and radish produced in watershed. Comparison of the labour productivity of farming systems revealed that agri-horti-silvipastoral land use system provided 374% incremental gain over the agricultural land use system that gave only 80%.

A system with mushroom as component showed that on an average of Rs. 11,755 per year was obtained in rice-poultry-fish-mushroom multi-enterprise system compared to Rs. 6334 per year from conventional system from one acre holding. The per day net income was worked out Rs. 32.3 in case of integrated system compared to Rs. 17.35 per day in conventional system. The feasibility and risk in pulses and oilseeds production *vis-à-vis cereals* and vegetable production in Kangra district (Himachal Pradesh) showed that risk could be notably reduced if farmers diversified their cropping pattern and found vegetable + dairying as the most appropriate risk efficient choice for farmers. They could increase the farm income as much as 47.76% (from Rs 70, 450 to Rs. 1,04,100). They found that pure vegetable farming was more risky and less remunerative. The grass-legume fodder from coconut garden in the ratio of 3:1 (30 to 40 kg) could support four milch animals per hectare. The incremental benefit of coconut + dairy over coconut alone was Rs. 22, 763 per year per ha. The total output from 1.04 ha mixed farming model was 19, 125 coconuts, 9275 liters milk, 526 kg poultry (live weight) 50 numbers of Japanese quail bird and 40 kg fish. The net income was Rs. 66, 678 per year.

Hi-tech Horticulture and Precision Farming

Hi-tech horticulture is the deployment of modern technology which is capital intensive, less environment dependent, having capacity to improve the productivity and quality of produce. On the other hand, Precision Farming involves the application of technologies and principles to manage spatial and temporal variability associated with all the aspects of horticultural production for improving crop performance and environment quality. Precision farming calls for efficient management of resources through location-specific hi-tech interventions. Hi-tech horticulture encompasses a variety of interventions such as micro irrigation, fertigation, protected/greenhouse cultivation, soil and leaf nutrient based fertilizer management, mulching for *in-situ* moisture conservation, micro propagation, biotechnology for germplasm, genetically modified crops, use of biofertilizers, vermiculture, high-density planting, hi-tech mechanization, green food, soil-less culture, biological control etc.

A Management Information System (MIS) was developed for coconut plantations comprising monthly weather data, harvest, cultural operations, pest and disease incidence and treatment of pests data base. From the analysis it was observed that application of variable nitrogen and phosphorous fertilizers will increase the farm profit by 23% in coconut. For precision farming of cassava, modified QUEFTS (Quantitative Evaluation of Fertility of Tropical Soils) model for fertilizer calculation, leaf colour chart (LCC) for real time N management and Chlorophyll Meter (SPAD-502) have been standardized. The Site Specific Nutrient Management (SSNM) approach compared to farmers' practice gave good response. A fertilizer calculation chart for a particular yield goal based on yield in nutrient omission trials has also been developed for cassava. Precision farming application of fertilizers has been proved profitable than the recommendations based on package of practices. About 17 Precision Farming Development Centers (PFDC) have been established in different agro-climatic regions under National Horticulture Mission (NHM). Activities like green house construction, mulching, shade net and plastic tunnels are also being promoted and 195.2 ha have been already brought under protected cultivation. The precision farming technology can be adopted in horticulture production by simulating computer based fertilizer, water, growing media application, light and temperature controlling and spraying of pesticides as and when the crop requires to improve the production. Utilization of these interventions orchestrated together having aim of achieving higher output in given time period leads to precision farming, which is largely knowledge driven.

It has been recognized that hi-tech horticulture would play a major role in horticulture sector in the coming years to improve production and productivity of horticultural crops. Therefore, adoption of hi-tech horticulture and precision farming assume greater significance. The crops where some of the components of precision farming have been practiced are banana, grape, pomegranate, capsicum, tomato, chilli, cashew and selected flowers. The dissemination of technology on hi-tech horticulture and precision farming has to be addressed through training programmes for the benefit of farmers as well as field functionaries.

Biological Control

Biological control has been the most significant approach to plant health management. It promises through modern biotechnology to be even more significant in the twenty-first century (Table 10). Although the number of biocontrol products is increasing, these products still represent only about 1% of agricultural chemical sales. Yet these are important contributions because biocontrol agents offer disease management alternatives with different mechanisms of action than chemical pesticides. Trends in research include the increased use of biorational screening processes to identify microorganisms with potential for biocontrol, increased testing under semi commercial and commercial production conditions, increased emphasis on combining biocontrol strains with each other and with other control methods, integrating biocontrol into an overall system. Plant health management in horticultural crops involves not only pre-harvest but also post harvest health management strategies such as production of pest and disease-free planting materials, use of bio-inoculants and other growth enhancing soil amendments, indexing for major pathogens and certification of planting materials, seed plot technique and mother garden technique and other such measures.

In Banana, cut-pseudostem trap smeared with bio-agents like *Beauveria bassiana* or *Heterorhabditis indica* are effective in controlling both stem and corm weevils. Growing of sunnhemp as inter crop and application of bio control agents like *Trychoderma viridea, Pseudomonos fluorescence, Paecilomyces lilacinus* and VAM are effective in controlling the nematodes in banana. Effective biocontrol agent and fungicides application have been developed for the control of wilt in banana. Molecular diagnostic kit has been developed for the identification of the virus plants and its eradication for maximum profitability. In citrus, the three important parasitioids viz., *Tamarixia radiata* against psylla, *Citrortichus phyllocnistoides* and *Cirrospilus quadristriatus* against leaf miner were identified and protocol for production of biocontrol agents developed.

Table 10. Biocontrol agents used against pathogens and pests

Biocontrol agents	Pathogen/Pest and crop
Beauveria bassiana, Heterorhabditis indica	Corn and stem weevil of banana
Trichoderma viride, Pseudomonos fluorescense, Paecilomyces lilacinus and VAM	Nematodes of banana
Tamarixia radiata	Psylla of citrus
Citrortichus phyllocnistoides, Cirrospilus quadristriatus	leaf miner of citrus
Mallada boninensis	blackfly, psylla, leaf miner, mealy bug, aphids and mites of citrus
Cheilomenus sexmaculatus	aphids, psylla and blackfly of citrus
Encarsia (Prospaltella) perniciosi, Aphiitis sp., Proclia sp., Chiloneurus sp., Chilocrus bijugas, Coccinella septumpunctata	San Jose scale of apple, pear, peach, almond and plum
Aphelinus mali and predators coccinella Septem punctata and Hippodamia variegare	Woolly aphid of apple, pear, peach, almond and plum
Typhlodromous sp., Amblyseius sp., and Zetzellia sp.,	Eurpoean red mite of apple, pear, peach, almond and plum
Trichoderma spp., Aspergillus niger, Pseudomonas fluorescens	Phytophthora and Pythium infecting many crops, anthracnose and grunaria leaf spot in grapes; Nematodes infecting different crops, date palm sucker rot, fruit rot and stem blight in ber
Metarrhizium anisopliae, Oryctes baculovirus	coconut rhinoceros beetle
Apanteles taragamae Viereck, Bracon brevicornis Wesmael, Goniozus (Perisierola) nephantidis Mues, the pupal parasitoids Brachymeria nosatoi Habu	Coconut leaf eating caterpillar
Bacillus amyloliquifaciens	Bud rot of coconut
HaNPV and SlNPV	Heliothis armigera and Spodoptera litura infecting vegetable crops

PCR-based diagnostic protocol has been developed for rapid detection of CTV and citrus greening bacterium. In grapes, Dogridge rootstock has been recommended for cultivation under the soil / water salinity and drought conditions and accepted very well by the grape growers all over the country. About more than 1000 ha area has come under this rootstock and all upcoming areas are covered with this rootstock. Forecasting based disease management (for downy mildew and powdery mildew) resulted in considerable saving of sprays during one year of production cycle. In pomegranate, fruit cracking has been reduced with 1-2 sprays of boron (0.2%) resulting in better gains. Strains of nitrogen fixing bacteria *Azospirillum bresilense,* VAM inoculations @ 200g/100kg mixture in potting material have resulted in better growth and production of pomegranate in field with superior nutritive quality.

In temperate fruit crops, insect pests such as sanjose scale, aphids, woolly aphid, peach leaf curling aphids, stem and root borer, shot hole borer, tent cater pillar, codling moth and European red mite and diseases like scab, powdery mildew, leaf spot, brown rot, gummosis, canker are the important problems. A comprehensive approach involving cultural, biological, and chemical measures have been evolved for the integrated management of all these pests. Similarly, IPM methods with emphasis on biological control or use plant based pesticides have been developed for the management of pests and diseases affecting arid horticulture crops.

Bio-intensive IPM Technologies

Bio-intensive technologies based on IPM approach has been developed to manage two major pests of coconut viz. rhinoceros beetle *Oryctes rhinoceros, Aceria guerreronis* and leaf eating caterpillar *Opisina arenosella.* For the management of coconut rhinoceros beetle, the bioagents are, the green muscardine fungus *Metarhizium anisopliae* and *Oryctes baculovirus*. Methods to mass multiply the larval parasitoids of *O. arenosella* viz., *Apanteles taragamae* Viereck, *Bracon brevicornis* Wesmael, *Goniozus (Perisierola) nephantidis* Mues, the pupal parasitoids *Brachymeria nosatoi* Habu, have been developed. The red palm weevil pheromone lure is produced and supplied to farmers and developmental agencies across the country. The integrated disease management has been developed for root wilt of coconut and yellow leaf disease of arecanut. In cashew, phytosanitation measures combined with swabbing the pest infested portion after removal of internally occurring pest stages followed by drenching the root zone with chlorpyriphos (0.2%) as post extraction prophylaxis (PEP) could effectively reduce reinfestation of stem and root borers

Effective management practices for viruses, late blight, bacterial wilt, tuber borne diseases, potato apical leaf curl virus and potato stem necrosis have been developed. Disease management for cassava mosaic disease includes production of disease free planting materials, multiplication of planting material in vector free zones and mass scale multiplication of healthy planting materials through the involvement of farmers. An IPM package was developed for the control of sweet potato weevil, with synthetic sex pheromone as the main component. Integrated disease management package involving tolerant variety, biocontrol agent (*Trichoderma viride*) and minimum use of fungicide was effective in reducing the *Phytophthora* leaf blight of taro.

Future thrusts should include identifying and cloning resistance genes and improving the efficiency of conventional and molecular breeding, studies on mode of parasitism to develop resistant lines against nematodes, use of biotechnological interventions wherever resistance source is lacking in wild and cultivated types, development of a good certification system for planting material production, keeping vigil and addressing new emerging diseases and pests, identification, characterization and efficacy especially multiple efficacy of biocontrol organisms in controlling more than one pathogen and their commercialization, residue analysis in produce, development of integrated package for management of diseases and pests under organic cultivation system.

Post Harvest Technology

In order to make horticulture a viable enterprise, value addition is essential. Processing is a fast growing sector in the world economy. Presently, less than 2% of the fruits and vegetables are processed in the country as compared to 30% in Thailand, 70% in Brazil, 78% in Philippines and 80% in Malaysia. India already enjoys an enviable position in the world horticulture. Harvest indices, grading, packaging, storage techniques have been developed/ standardized for major horticultural crops. The demand for some of the processed fruits and vegetables has grown enormously in country. Value addition through dehydration of fruits and vegetables including freeze drying, dried and processed fruits, vegetables and spices and fermented products play an important role in horticultural crops. Development of new products like juice punches, banana chips and fingers, mango nectar and fruit kernel derived cocoa substitute, essential oils from citrus, fruit wines, dehydrated products from grape, pomegranate, mango, apricot and coconut, grape and fruit wines, value-added coconut products like snowball tender coconut, coconut milk powder and pouched tender coconut water (Cocojal) etc. are getting popular day by day. Improved blending/ packaging of tea and coffee have opened new markets.

Consumer friendly products like frozen green peas, ready to use salad mixes, vegetable sprouts, ready-to-cook fresh cut vegetables are major retail items, which have already started peeping out of retail windows. In order to reduce dependence on refrigerated storage, low cost eco-friendly cool chamber for on farm storage of fruits and vegetables has been developed. For preventing the post harvest losses proper storage, cold preservation, packaging and transport methods with Hazard Analysis Critical Control Point (HACCP) norms have to be given more thrust. As food consumption patterns are changing towards more convenient foods, the demand for products like pre-packed salads, frozen vegetables etc. are increasing. The future research areas may be oriented towards developing varieties with prolonged shelf life with better processing qualities, standardization of Modified Atmosphere Packaging and Storage systems with greater emphasis on safety (pesticide free), nutrition and quality (CODEX standards).

Mechanization in Horticulture

India is having largest number of tractors in the world, (about 3.2 million) producing about 0.25 million per year. Still most of the horticultural operations in India are done manually or with animal power. Wherever, the farming operations are mechanized the crop productivity is high. The machinery developed by different SAU's and agricultural research institutes need to be popularized. The tractor drawn implements have been found wide acceptability in potato cultivation, from sowing to harvesting. Suitable varieties of fruits, vegetable, spices and plantation crops are to be selected and the cultivation practices be standardized for mechanization. In fruit crops, the tractor operated pit hole digger and bucket excavators (JCB) have been developed by different firms and they need adoption for planting fruit trees. The fruit nurseries need the adoption of media siever, media mixer and plastic bag filler. The automatic grafting machines developed in advanced countries are yet to be adopted. The commercially available rotoslasher is to be adopted for weeding. The electrically / pneumatically / hydraulically operated secateurs, loppers need adoption for pruning. The self propelled pruner for tree crops and grapes orchards need adoption. The harvesting aids for mango, guava, sapota, orange, straw berry developed by different institutes need adoption by farmers. The transplanters for vegetable seedlings developed at IIHR, PAU and CIAE needs adoption. The commercially available boom sprayer, air assisted sprayer can only be used in raised bed cultivated crops. The combine harvesters for digging potato, detopping and bagging of onions and garlic are urgently required. The mushroom spawn and cultivation machinery developed at IIHR, Bangalore and NRCM, Solan may be adopted to mechanize the Indian mushroom industry.

EMERGING ISSUES

Horticultural crops are highly dynamic, especially the flowers, vegetables and many fruits wherein consumer preference determines the economics of the scale. How to make the sustainable development in production of fruits, vegetables, tubers, plantations and tuber crops for meeting the growing demand of rising population and nutritionally rich horticulture produce and how to achieve the high production levels keeping in view the present and future needs are some of the issues needs to be addressed in a systematic manner. The challenges ahead are much greater than before, as more has to be produced with declining land and water. The issues are dynamic but they move around the sustainability of production and competitiveness.

Climate Change

The climate change has far reaching impact on horticultural crops. A study conducted using GIS at IISR has indicated that many areas presently suitable for spices would become unsuitable in another 25 years. There would be new areas which will become suitable for cultivation. But the major threat would be to the endemic gene pool that are native to these locations. They would become extinct leaving us poor of valuable genes. Other studies have indicated that lack of chilling requirement during winter and warmer summers in lower elevations lead to the shifting of apple production to higher elevation (2700 m msl). High temperature and moisture stress increased sun burn and cracking in apples, apricot and cherries and increase in temperature at fruit maturity lead to fruit cracking and burning in litchi. Low fruit set in mango was due to sudden rise in temperature (Tmax 35°C) during peak flowering. Low temperature (4 to11.5°C) in the month of January, accompanied by high humidity (>80%) and cloudy weather delayed panicle emergence. In citrus untimely winter rains promote vegetative instead of flowering flushes. Increase in temperature with low RH (less than 60%) reduced leaf spot diseases in banana especially in sub humid region. In vine grapes anthocyanin development is influenced by difference between day and night temperatures with high variation (15-20°C) promoting colour development.

In potato, reduction in marketable grade tuber yield to the extent of 10-20% was due to high temperature. Advancement in appearance of aphids by two weeks with increase in 1°C reduced growing period for seed crop in potato. Elevated CO_2 (550 ppm) enhanced tomato, onion and cassava production and enhanced the quality of sweet potato tubers. In coconut, impact of climate change related events like consecutive droughts and cyclones adversely affected nut yields. Elevated CO_2 increased growth of coconut seedlings by ~20-30%. Annual C sequestration in coconut above ground biomass varied from 15 t CO_2/ha/year to 35t CO_2 /ha/year depending on cultivar, agro-climatic zone, soil type and management. Ten year old irrigated oil palm indicated carbon sequestration potential up to 30-35 t C/ha. The tea mosquito bug populations and damage score in cashew were negatively correlated with minimum temperature, RH (85-90%) and positively with bright sunshine hours (9-10 h). Bright sunshine hours less than 2 h per day triggers incidence of inflorescence blight in cashew. In black pepper, increase in rainfall during December-January decreased the productivity, while March-April rainfall increased productivity. Increase in temperature by 1-2 degrees reduced the growth parameters and photosynthesis by about 5-10% and could enhance the concentration of beta caryophyllene, the main quality component.

Quality Assurance

Acceptance of the horticultural produce in the high-quality conscious era require the produce to be of high standards free from toxins and other elements of health hazards, which could be achieved through, popularization of improved agro-technique, emphasis on cultivation of specific varieties for table, processing and export, emphasis on harvesting horticultural produce at optimum maturity, grading, packing and transport, setting up of quality control laboratories, wide dissemination of international quality standards for various commodities and, ensuring pesticide residues within permissible limits. With the opening up of the economy and emergence of free trade there shall be increasing competition to capture the markets, which is equally true for horticultural products. The strategies, to overcome the implication on Indian horticulture, would include emphasis on creating awareness about implications of WTO regime among horticultural entrepreneurs, identification of horticultural cropping zones, harmonization of sanitary and phytosanitary standards and strengthening of mechanism to meet the requirement and, strengthening of quality assurance.

Efficient Regeneration System

Seed and planting material assume much more significance to achieve the goal of accelerated production of quality horticultural produce. However, due to presence of a large number of viruses and other pathogens, there are all possibilities for transmission of diseases through clonal propagation, many of which will be lethal. Further, many horticultural crops are perennial in nature and any mistake done in the choice of cultivar, and quality of planting material and seed will invite more risks to the farmers. The quality planting material production has been one of our focus areas. Micro-propagation technique in various fruits and spices has proved very effective in production of healthy planting material. Banana and potato are the success stories for quality planting material production on large scale. For cleaning of on infected materials in citrus, shoot tip grafting technique has been standardized and serpentine method of quality planting material production has been developed in black pepper. For mass production of disease free planting material through micro-propagation, protocols are available for few crops and require to be developed for many other horticultural crops. The standards developed for various crops are to be effectively implemented to ensure the quality production of planting material. The development of sensitive detection techniques like ELISA, ISEM, NASH and PCR have been of great help in eliminating quarantine material even with low concentration of pathogen and practically zeroing their incidence in basic seed and vegetatively propagated planting material. To ensure the quality of planting material there is need for quality and health management system for seed and planting material, so that increasing demand projected for 2011-12 can be met (Table 11).

Table 11. Horticultural planting material requirement for XI plan period

Crops	Targeted Seed/ Planting Material requirement by 2011-12
Vegetables	80,000 MT seeds
Fruits	1.46 billion saplings
Potato and other tuber crops	7,62,113 MT + 1.32 million setts
Plantation crops	10.74 million plantlets
Spices	18.81 million Nos. plus 9,165 MT
Medicinal and Aromatic plants	26,869 MT + 44,741 suckers

Water Productivity Enhancement

With increasing demand for food, water scarcity is going to be a serious problem in near future. Among the various impacts of climate change, erratic rainfall and reduced available water are very critical. If natural resources like water are not managed properly, it will be the single major limiting factor in agriculture sector. It would be better to adopt / develop technologies and take measures so as to improve the water productivity through existing water resources. This problem could be addressed through judicious use of water and minimizing the losses in conveyance and application. Water harvesting and use of drip/ micro irrigation has been standardized, which would save a lot of ground water for future.

Dominance of Small Farmers

Indian agriculture is characterized by dominance of small and marginal farmers having very small land holdings. The average size of the land holding has reduced to 1.32 ha in 2000-01 from 2.30 ha in 1970-71. The absolute number of operational holdings has increased from about 70 million in 1970-71 to about 121 million in 2000-01. Due to division of farm holdings, the number of operational holdings has gone up (by about 51 million), but the average size of holding has declined steadily. Share of farm holding having less than 2 ha has increased from 70% in 1970-71 to 82% in 2000-01. Further, more than 60% of the farmers in the country are operating and depending for their livelihood on less than one ha. Development strategy for agriculture must focus on small farms. Ensuring sustainability and economic viability of small holders and improving their competitiveness in production and marketing by facilitating better access to improved technology, inputs, credit and markets should be accorded priority for higher and inclusive agricultural growth. Horticulture is best option for small farmers, they can grow high value crops, flowers, medicinal and aromatic plants under protected cultivation, mushrooms, bee farming and post harvest processing and preparation of value added items.

FOCUS OF RESEARCH INITIATIVES IN XI PLAN AND BEYOND

The focus of research has been completely reoriented in our 11th plan proposals, where the utilization of genetic resources gets the priority and to work along with the partners especially with the private partners. Some of the focus areas of research are;

- Increasing the value of production by reducing variability in yield, quality, reducing crop losses and increasing marketability
- Develop system for productive use of water to get 'more crop per drop' by increasing the water and nutrient use efficiency
- Develop alternative to chemicals and fertilizers, integrated management and decision support systems to reduce the use of chemicals for the control of pests and diseases

- Develop the production systems that minimize wastes and maximize recycling.
- Improve the understanding of interactions between native ecosystems and production systems and develop best practices to conserve biodiversity
- Develop production and post-harvest technologies to improve product quality and minimize environmental impacts
- Development of new innovative diagnostic techniques for rapid, accurate and cost effective detection of high impact pests and diseases
- Understand social needs of communities and build the capabilities for practice change, and for effective utilization of resources and adoption of technology
- Facilitate accelerated adoption of improved technologies and best practices and respond to needs including bio-security threats
- Upgradation of NRC as directorate for effective outreach programmes.

New Research Initiative in XI Five Year Plan

Several research initiatives are being undertaken, for example, in the network mode we have the Phytophthora, Fusarium and Ralstonia which are the serious problems affecting fruits and vegetables. Some of the other research initiatives are :

- 41 new centres to cover agro-climatic zones
- Network programme on Phytophthora, wilt and Ralstonia
- Network programme on leafspot diseases
- Network programme on viral diseases
- Genomics in horticulture crops
- Outreach programme on sucking pests
- Integrated approach cutting across the barrier of discipline/crop
- NAIP projects on horticultural crops
- Network programme on onion & garlic
- Network programme on temperate horticulture

CONCLUSION

The decade has witnessed the emergence of horticulture as commercial enterprise moving from rural confine, owning to technological advancement and policy environment provided by the Government. Past investment has been rewarding in terms of increased production, productivity and export of horticultural produce. Despite achievements in horticulture sector the challenges confronting are still many. Although India is ranked as second in the world production of fruits as well as vegetables; the availability of fruits and vegetables still continues to be much below the dietary requirements.

With increase in per capita income and accelerated growth of health conscious population, demand for horticultural produce is on the increase which is expected to further accelerate. But there is a growing competition in open economy, which demands competitive price of standard qualities produce. This opens up opportunities and challenges. Hence, the potentialities which exist need to be harnessed and gains have to be sustained. India is required to enhance producing high quality produce to face the stiff competition under the liberalized economy. Attention is therefore required to tackle the issues of low productivity, poor quality and high costs of horticultural produce. Development of improved cultivars with high quality characteristics, productivity, resistance to pest and disease and tolerant to abiotic stresses and technologies for improving the efficiency of water and nutrients by reducing variability in yield, quality and reducing pre and post harvest crop losses are the needs of the hour. The crop monitoring mechanism would need strengthening to enable the forecast of crop prospects. Efforts will also be needed to ensure timely availability of quality seed and planting material to farmers. Horticultural development has to be seen with an integrated approach, addressing important gaps, in harnessing the potential for horticultural development. However, it will require sustained research efforts to improve competitiveness, enhancing efficiency of production system, assessing quality and safety and enhancing the capability of community to manage the change. Consequently, technology driven horticulture will contribute significantly for economic development.

REFERENCES

Singh, H.P., Kaushish, S.P., Kumar, A, Murthy, T.S. and Samuel, J.C. 2001. Microirrigation. Central Board of Irrigation and Power. P1-794.

Singh, H.P. 2004. Approaches for accelerated development of horticulture –thrust during tenth plan. **In** Horticulture in India. Sigh H. P. and Samuel, J.C. (eds.), DAC, Ministry of Agriculture, New Delhi. NHB, Gurgaon.1-23p.

Singh, H.P. 2008. Can India surpass China in fruit production? The Hindu Survey of Indian Agriculture 2008. pp 59-63.

Singh, H.P. 2008. Technological interventions in Indian horticulture – Way ahead. *Indian Horticulture* **53**(1):3-9.

Singh, H.P. 2008. Mitigating global warming... Development in horticulture crop production with climate change. *Indian Horticulture.* **53** (5):39-45.

Singh, H.P. 2008. Horticulture Crops : Research initiatives. **In** Recent initiatives in Horticulture. Eds. Chadha, K.L., Singh, A.K. and Patel, A.B. The Hort. Society of India, New Delhi, pp1-36.

Singh, H.P., Parthasarthy, V.A. and Prasath, D. 2009. Horticulture crops- Varietal wealth. Studium Press (India) Pvt. Ltd.

Singh, H.P., Rai, Mathur, Pandey, S., Kumar, S.2009. Vegetable varieties of India. Studium Press (India) Pvt. Ltd.

Singh, H.P. 2009. Impact of horticulture in conservation agriculture. Indian Farming **58**(11):3-8.

14. IMPROVING PRODUCTIVITY IN HORTICULTURAL CROPS

Michael Böhme, Ina Pinker, Idha Arsanti, Huynh V. Chuong, Nabarath Baniya and Yaroslav Shevchenko

Humboldt University of Berlin, Dept. Horticultural Plant Systems, Lentzeallee 75, 14195 Berlin, Germany

ABSTRACT

The productivity of horticultural crops is influenced by different factors in particular by the potential of the cultivar used, the cultivation technology, and the regulation of climatic and edaphically growing factors as well as the use of treatments in order to reduce influences of stresses. In this regard, the 'Land Suitability Analysis and Evaluation' for the horticultural crops are very important but so far rarely used in order to identify the appropriate land for the different crops. In accordence with the results of the land analysis and evaluation limitations and input requirements for the sustainable production, can be determined. Examples for 'Land Suitability Analysis and Evaluation' were carried out for vegetables in Indonesia, for seasonings and vegetables in Nepal, and fruits in Vietnam. Recommendations for the farmers to improve the productivity as well for the authorities in the regions for political supports could be given for various crops. In some cases there is a lack of suitable land for cultivation of those horticultural crops with high economical value or the climatic growing conditions are not stable. Under such conditions the use of protected cultivation and additionally of bioregulators can increase the vitality and therewith the productivity of horticultural crops. In several experiments the positive effect of different bioregulators (Humate, Lactate, *Bacillus subtilis*) could be proved if one of the growth factors was at a critical level. A combined application of bioregulators could be beneficial in complex stress situations in order to improve the productivity of the horticultural crops.

INTRODUCTION

Productivity is a measure of output from a production process often in relation to the corresponding input. Following this definition we do have several possibilities to consider the productivity in horticultural crops.

- High yield and sustainable production of vegetables, legumes and fruits

- High quality following international standards e.g. GLOBALGAP (EUREPGAP)
- High marketability and high commercial value to raise economic level
- To produce crops with high nutritional value; not only 'modern' but healthy
- To secure high biodiversity in particular cultivation of indigenous crops

There are different factors influencing the productivity in horticultural crops, as cultivars, cultivation technology and the growing conditions. But first of all information has to be collected about:

- appropriate farming system and evaluation of the land suitability
- appropriate and advanced cultivation methods
- use of growth regulator in order to stabilise the productivity in stress situation

This paper summarizes some of our researches concerning and connecting all these aspects and possibilities to ensure high productivity in horticultural plant systems.

LAND SUITABILITY ANALYSIS AND EVALUATION FOR PRODUCTION OF HORTICULTURAL CROPS

In order to have high yield and quality of horticultural crops it is very important to select the appropriate farming system (Dixon *et al.,* 2001) and to estimate the suitable land for certain crops (Huyen, 2007, Huyen and Böhme, 2007). Vegetables and legumes are included in many different 'Farming Systems' in South Asia that means in regions with similar natural resources and traditionally dominant pattern of farm activities and household livelihood. These farming systems for horticultural crops, in particular for vegetables have often strengths but also constraints (Arsanti and Böhme, 2008). The strengths of Vegetable Farming System (VFS), e.g. in Indonesia are: Successful technology adoption by farmers, relatively low labour cost, extensive source of information from organization and stake holders and good mechanism of the market.

The constraints are mainly due to the lack of suitable land for the crops. In this regard the 'Land Suitability Analysis and Evaluation' using (Global Information System (GIS) is very important in order to determine the appropriate land for various horticultural crops. Other constraints are monoculture of crops and land erosion often as result of mismanagement of conservation of natural land resources.

Land suitability analysis and evaluation of farming systems were carried out for vegetables in Indonesia (Arsanti and Böhme, 2008), for seasonings and vegetables in Nepal (Baniya *et al.,* 2009; Baniya, 2008) and fruits in Vietnam (Huynh, 2007). Based on this, recommendations for the farmers to improve the productivity as well for the authorities in the regions could be given for different crops.

Many researchers worked to establish a methodology for useful and comprehensible classification of land suitability. This aim is completely fulfilled with the suitability categories published by the FAO (1976, 1986). In general these categories are divided in suitable and non-suitable, furthermore in classes as shown in Table 1. Such classification can be used for a certain crop, but also for a group of crops, e.g. solanaceous plants.

Table 1. Structure of land suitability classes and subclasses (FAO, 1976)

Order	Class	Description
Suitable (S)	S1 (Highly suitable)	Land having no, or insignificant limitations to the given type of use
	S2 (Moderately suitable)	Land having minor limitations to the given type of use
	S3 (Marginally suitable)	Land having moderate limitations to the given type of use
Non-suitable (N)	N1 (Currently not suitable)	Land having severe limitations that precl-ude the given type of use, but can be improved by specific management
	N2 (Permanently not suitable)	Land that have so severe limitations that are very difficult to be overcome

Limitation factors leading to the classification as shown in Table 1 are often additional indicated with small letters, e.g. S2g indicating limitations by inappropriate soil conditions. The information about such limitations can be useful in order to organize the cultivation schedule in an appropriate way.

Such limitations can be:

d: Limitations by rainfed cultivation time,

f: Limitations by flooding in rainy season,

g: Limitations by inappropriate soil conditions,

i: Limitations by irrigation conditions,

r: Limitations by low rainfall,

s: Limitation by salty infiltration in dry season,

Evaluation of land suitability needs (1) well developed Land Information System (LIS), whereas basic data can be provided by GIS in combination with a multi-criteria approach, (2) information about growth requirements of horticultural crops, (3) geographical information and infrastructure, (4) marketing and national market system (Huynh and Böhme, 2007). That means very detailed information related to the physical, climate and social-economic conditions are necessary for complex recommendations for growing of the right crop on the right land as follow:

- Land use structure- farming system (monoculture or crop rotation, intensive vegetable growing or, irrigated or non-irrigated land);
- Soil classification;
- Soil sloping, soil depth, texture, fertility;

- Climate information (air and soil temperature in average and min. and max., precipitation, humidity, wind)
- Crop requirements (regarding soil and climate condition, cultivation time, mix cropping)
- Infrastructures (roads for transporting the commodities, distance to the market, supply of seeds, plant material, fertiliser, pesticides)
- social-economic information

The current physical conditions in particular the soil quality description is of high importance for the land suitability evaluation. The GIS data e.g. soil fertility level have to be evaluated in the region investigated in relation to the crop requirements. In the end a database is available with the evaluation for every soil characteristic and the available are (Fig. 1).

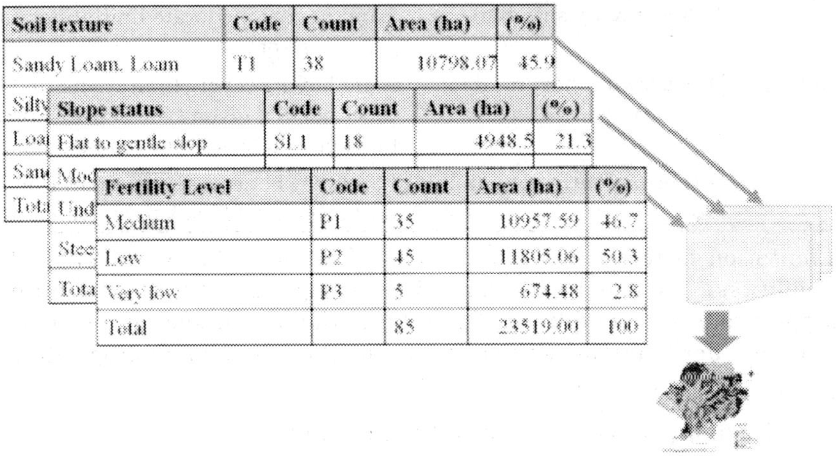

Soil texture	Code	Count	Area (ha)	(%)
Sandy Loam, Loam	T1	38	10798.07	45.9

Slope status	Code	Count	Area (ha)	(%)
Flat to gentle slop	SL1	18	4948.5	21.3

Fertility Level	Code	Count	Area (ha)	(%)
Medium	P1	35	10957.59	46.7
Low	P2	45	11805.06	50.3
Very low	P3	5	674.48	2.8
Total		85	23519.00	100

Figure 1. Data collected in Kathmandu valley, for land suitability analyse, example for cabbages or mustards

After collection of all data it has to be decided for which crop and under what for conditions the specific land is suitable. Analytic Hierarchy Process (AHP) is a widely used method in decision-making. AHP is introduced by Saaty (1977), with the basic assumption that comparison of two elements is derived from their real-time importance. It is developed to select the best from a number of alternative conditions with respect to several criteria. AHP allows for both the inconsistency in the decision and provide the means to improve the consistency. Land suitability analysis consists of multiple criteria and alternatives which must be evaluated by a decision-maker in order to achieve a goal. The AHP provides a systematic method for comparison and weighting of these multiple criteria and alternatives by decision-makers.

To find the right decision this method was used for physical, economical and social criteria. As an example the criteria for physical conditions were selected as follow:

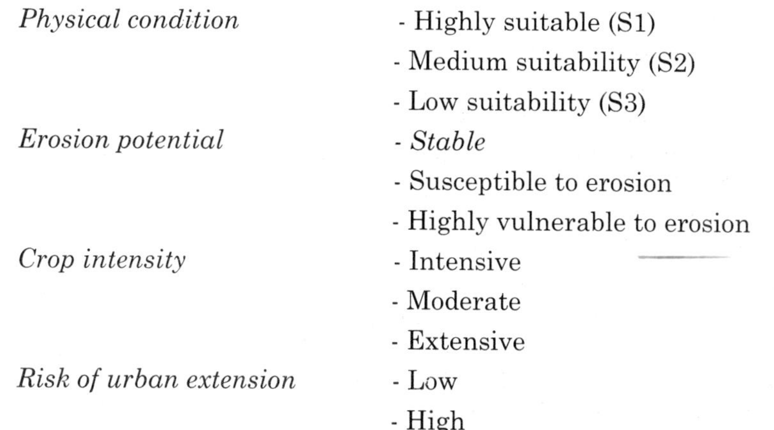

Physical condition	- Highly suitable (S1)
	- Medium suitability (S2)
	- Low suitability (S3)
Erosion potential	- *Stable*
	- Susceptible to erosion
	- Highly vulnerable to erosion
Crop intensity	- Intensive
	- Moderate
	- Extensive
Risk of urban extension	- Low
	- High

Based on such criteria the suitability index was then calculated after weighting of each criterion (Fig. 2).

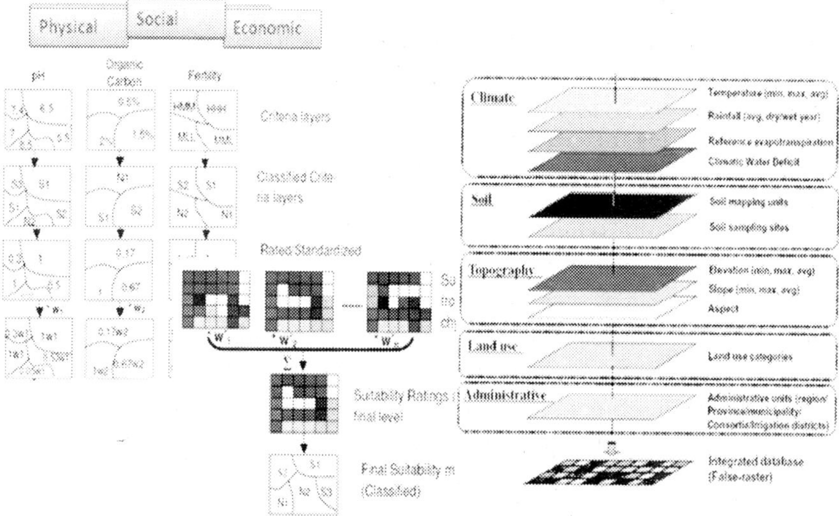

Figure 2. Final land suitability evaluation using the AHP method for vegetable and use of GIS for overlapping the different suitability levels.

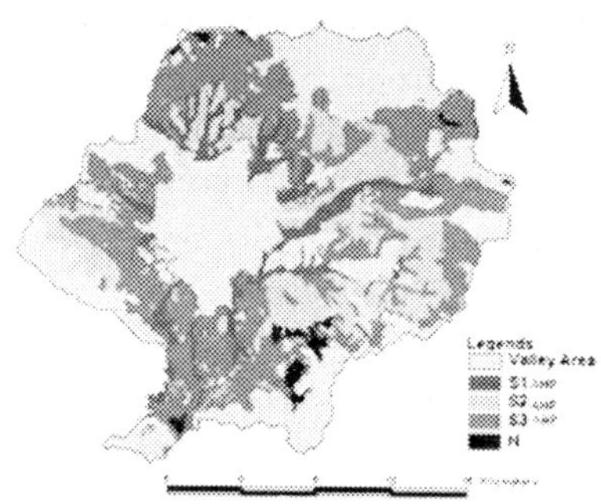

Figure 3. Final map for suitable land for cabbage/mustard cultivation in Kathmandu valley, Nepal

The final land suitability decision is a result of such evaluation considering all limitations from physical, economical and social origin. After overlapping of all data a map will be drawn where the suitability for the investigated region are visible (Figure 3).

This map and the information about the existing limitations can be used for recommendations to the farmer in order to cultivate the right vegetable in the right time on the right field. Therewith a tool is given to estimate suitable land for high productivity in horticultural crops.

Similar studies regarding preparation of final land suitability evaluation were done in Vietnam, in particular for citrus, lemon, mandarin and mangosteen orchards. In this research also different level of land suitability (S) and non-suitability (N) were determined; several limitations for high productivity can be identified in the region studied. Some of them are dominant and are very difficult to transform for example soil type, soil layer depth, and slope. Other limitations like soil fertility (pH, organic matter (OM), N, P, and K) and soil texture can be improved by technical methods and intensive cultivation (Huynh, 2007).

Studies in Indonesia were focused on economical competitiveness for cultivating vegetables by improving of the farming systems (Arsanti, 2008). Vegetable Farming System (VFS) is the most extensive human land use and soil characteristics are a critical determinant of vegetable productivity, including the effective production, high yield and quality of the vegetable products. Each VFS has distinct soil and soil management conditions for optimal production. It has been suggested that soil productivity is the net resultant of soil degradation processes and soil conversation practices (Kavetskiy *et al.*, 2003).

In order to increase the productivity in vegetable crops the activities in Vegetable Farming Systems should be focused as follow:

- Regular technical assistance for vegetable development.
- Increase production of vegetables through appropriate cultivation methods in relation with the quality assurance system and food safety.
- Improving vegetable cropping pattern to control the production and price of vegetables; transition from monoculture to multi-culture-system with vegetable and agricultural crop rotation.
- Development of competitive agribusiness area, especially for export commodities and import substitution products based on the agro-eco-system advantage or infrastructure in each region.
- Application of scientific technologies to increase the quantity and improve the quality of products.

USE OF APPROPRIATE AND ADVANCED CULTIVATION METHODS

The productivity in horticulture crops is influenced not only by the suitable land and the appropriate farming system, but also from cultivation techniques and method. If the land (soil) and climate conditions are not or less suitable, cultivation of vegetables in greenhouses is often the only possibility. In such cases when the soil is not suitable for vegetable cultivation, e.g. deserts, marginal soil in hilly areas, salty soil, urban regions etc., then the use of soilless (hydroponically) systems are recommended. There are many different systems known, the main groups are 'substrate culture', 'water culture' and 'aeroponics'. Additionally these systems can be divided in simple and advanced techniques, 'open' and 'closed' systems and low and high input systems. In our research we used all these different techniques cultivating cucumbers, tomatoes, bell pepper, eggplants and several cucurbits, but also for herbs and leafy vegetables. Beside all advantages of soilless culture in greenhouses the growers have sometimes to react on different stress situation for the crops.

USE OF GROWTH REGULATORS IN ORDER TO ENHANCE THE PRODUCTIVITY IN STRESS SITUATIONS

The presented research was focused on the sustainability of the cultivation through minimizing of stress influences on plant growth. Even in case of optimal farming system, suitable land as well as advanced cultivation technology because of several reasons, stress situation can reduce the productivity of horticultural crops in the field and in the greenhouse as well. Cultivation of horticultural plants in greenhouse occurs under controlled conditions. Nevertheless, situations with suboptimal growing factors cannot be excluded. Suboptimal values of pH, EC, temperature are abiotic stress factors often limiting the productivity of the horticultural crops (Böhme *et al.,* 2008), however, the stress situation are often not as pronounced to be recognized. In this regard the research was focused on use of plant growth regulators or bioregulators because they can be applied as stress relieving agents.

The background of the research was the appearance of different stress factors or suboptimal conditions as follow:

- Suboptimal supply with nutrients
- Extreme pH- or EC-values
- Missing buffer effect of the soil therefore :
    - fluctuation of the nutrient supply is more dangerous
    - injuries of pathogens are more intensive and higher risk for fast spreading
- Lack of oxygen can be harmful for root growth (in particular in water culture as NFT)
- Fluctuation of the temperature in the air or in the rhizosphere
- Plant growth inhibition due to pests

In the literature there are different definitions for substances regulation or influencing the plant growth.

Plant Growth Regulators

A plant growth regulator is defined as any substance or mixture of substances intended to accelerate or retard the rate of growth or maturation, or otherwise to change the behaviour of plants or their produce.

Synthetic compounds – Paclobutrazol, Ethephon

Natural compounds – Phytohormones, Humic acid, Lactic acid

Plant Growth Strengtheners

In Germany more the 200 such strengtheners are registered and often used in organic farming.

There are three groups of substances classified for:

- enhancing plant resistance against pathogens
- protecting plants from abiotic stress
- application on cut flowers for longer vase life

Such plant growth strengtheners are available on/as:

- inorganic basis - SiO_2 and silicates, $CaCO_3$, Al_2O_3, $NaHCO_3$ and
- organic basis - Compost extracts, alga extracts, plant extracts or -oil, humic acids;
- homeopathic substances
- microbial basis as Trichoderma spp., Talaromyces flavus, Pythium oligandrum, Bacillus spp., Pseudomonas spp., Streptomyces spp.

Bioregulators or Biostimulators for Horticultural Crops

In general the effect of such bioregulators is maximizing the net photosynthesis and minimizing of physiological disorders. Substances belonging to this group are all of organic origin or microorganisms. These regulators are mostly treated if the plant growth is non-uniform as result of (1) environmental stresses (temperature fluctuations, frosts, moisture levels, salt levels) (2) intermittent stress (nutrient disorder, attacks of pest and diseases).

Table 2. Previous research with biostimulators at the Dept. Horticultural Plant Systems

Content and conditions	Substances used	Vegetables
Salt stress (EC)	Humates	Tomato
Sub-optimal nutrient supply	Humates, Fulvic acid	Tomato, Water spinach
Temperature stress	Humates	Tomato, Water spinach
Mineral & organic substrates	Different Humates	Tomato
EC & pH stress	Humates, Lactates, Bacillus subtilis	Tomato
Pest control –white fly	Humate	Tomato
Comparison of nutrient solution	Lactate & mineral fertilizer	cucumber

With the aim to improve the productivity in horticultural crops also biostimulators can be used especially if the growing conditions are not appropriate or satisfying. In several experiments (David *et al.*, 1994, Böhme *et al.*, 2005, Böhme *et al.*, 2007) the positive effect of different bioregulators used could be discovered if one of the growth factors was at a critical level Table 2.

This research showed that Humates are effective in nutrient and salt control as well as suboptimal temperature. Lactates have a good effect if the pH-value is unstable.

Research with Mixed Biostimulators

As example for the effect of mixed biostimulators are presented results of research with cucumber cultivation in greenhouse. The experiment was conducted with *Cucumis sativus* L. cv. Indira RZ. The growing conditions can be described with a mean temperature in the greenhouse during the day 27.8°C, night 24.5°C, relative air humidity 75.2%. Cucumbers were grown in a substrate culture with trickle irrigation in containers with eight liters perlite. The biostimulating mixtures used in the experiments consists on K-Humate (Humintech), LACTOFOL'O' (lactate based mixture) and *Bacillus subtilis FZB* 24. Different combinations of these substances were tested (control variant without treatment) LACTOFOL 'O'+ *B. subtilis* FZB24; K-Humate + *B. subtilis* FZB 24; LACTOFOL 'O'+ *B. subtilis* FZB 24 + K-Humate). Mixtures consisted of 0.2% *B. subtilis* FZB 24, 0.1% K-Humate and 0.2% Lactofol'O' respectively.

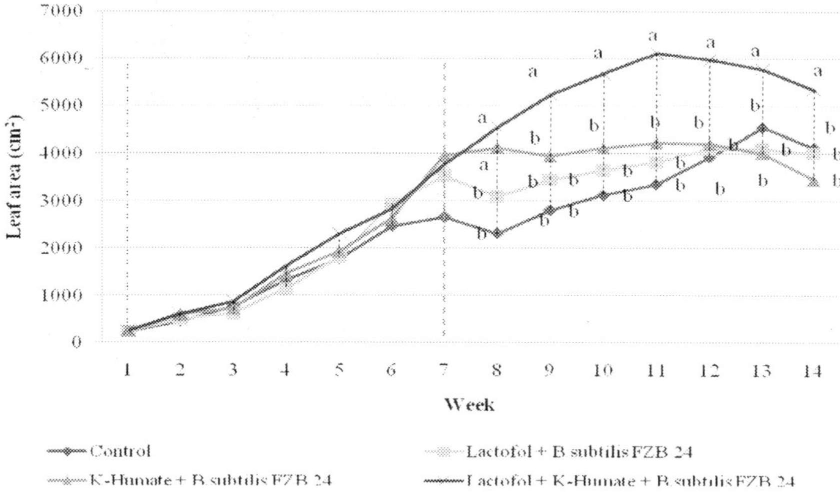

Figure 4. Influence of different biostimulating mixtures on evolution of leaf area of cucumber plant. (One-way ANOVA, Tukey's HSD $p<0.05$. Comparison only within measurement date. Different letters mean statistical significant difference).

At the end of last treatment with the mixed biostimulator, that means after the four week, stress factor was applied. In experiment with pH as stress factor – pH values was adjusted to suboptimal level (pH 3.2) by adding

H_3PO_4 to nutrient solution. This pH stress was maintained through one week. For temperature stress, temperature in the climate chamber was reduced from 25°C to 6°C with duration of three hours.

Application of biostimulating substances influenced dynamic of leaves formation on all treated variants (Fig. 4) with the largest leaf area finally after treatment with the LACTOFOL'O'+ K-Humate + *B. subtilis* FZB 24 – biostimulating mixture. The differences between the variants and the control were not statistically significant. Differentiation in leaf area dynamic can be observed beginning from 8-th week. However, treatments were conducted three times on -1st, 2-d and third week, but positive effects of it were observed only five weeks later (on 8th).

Application of a biostimulating mixture containing *Bacillus subtilis* FZB24 (0.2%), K-humate (0.01%) and LACTOFOL 'O' (0.1%) proved to be useful in reducing stress influence on the growth of cucumber plants. Application of either pH- or temperature stress markedly reduced the leaf area if no biostimulator was applied (Figure 5).

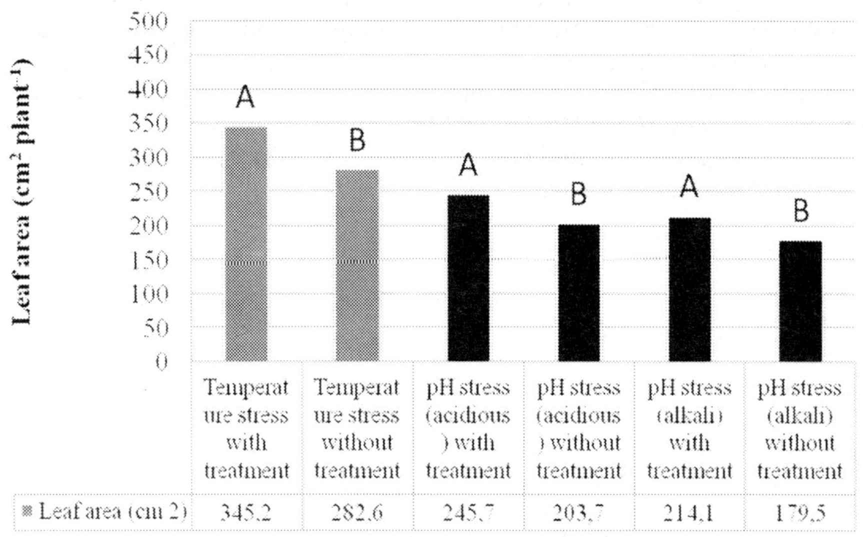

Figure 5. Effects of biostimulators under abiotic stress conditions, suboptimal temperature and pH value (t-test for pairs of variants with the same suboptimal growing factor).

As early stress indicator, chlorophyll fluorescence Fv/Fm-value showed a positive effect of the curative biostimulator treatments for stress counteraction in plants. Cucumber plants with no biostimulator treatment showed very low Fv/Fm values implying that they were not able to recover from temperature- as well as pH- stress (Figure 6).

The plants treated with the biostimulating mixture develop larger assimilation apparatus in comparison to all other variants resulting obviously in higher yield (Fig. 7). This variant developed more flowers and also more marketable fruits. The gap between the number of flowers and marketable fruits on this variant is the narrowest.

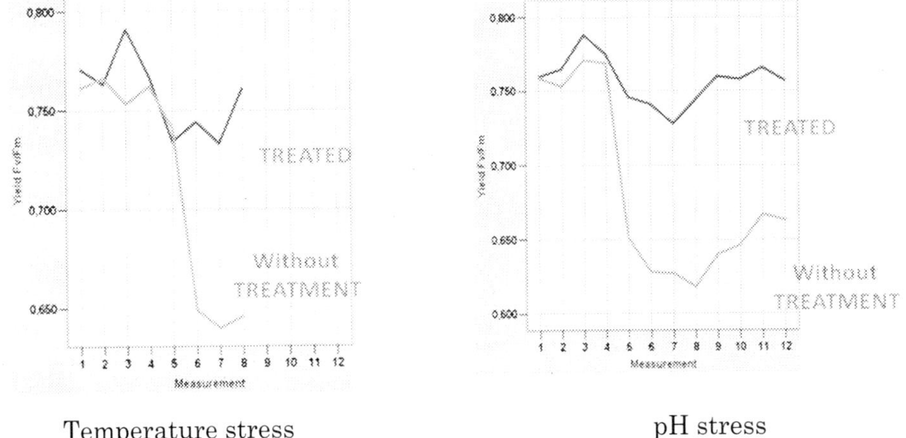

Temperature stress pH stress

Figure 6. Effect of biostimulators on electron efficiency of Photosystem II of cucumber plants under temperature or pH stress.

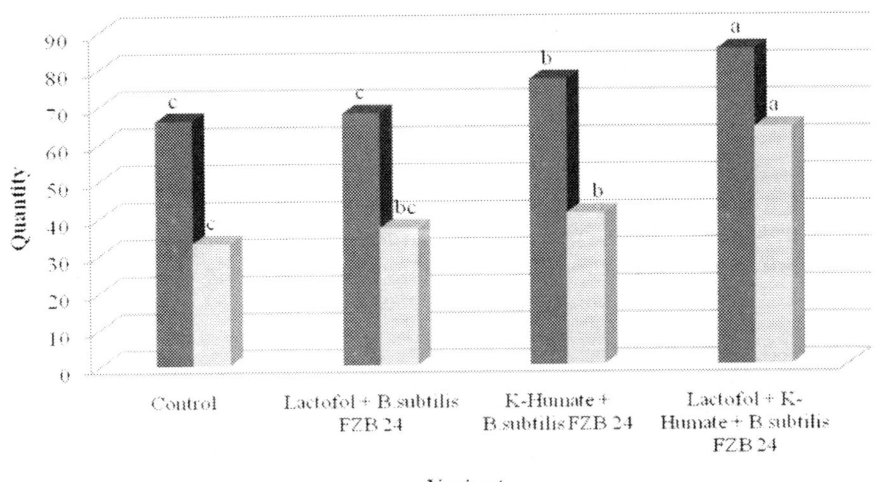

Figure 7. Influence of different biostimulating mixtures on flowers and marketable fruits quantity of cucumber plant. (One-way ANOVA, Tukey's HSD $p<0.05$. Different letters mean statistical significant difference).

In the control and the variants only about 50% of the flowers developed into marketable fruits. This positive effect of biostimulators on fruit quality was already reported earlier, however in this experiment the single compounds have been compared with the mixture of all three.

CONCLUSION

The productivity of horticultural crops is influenced by climatic and edaphically growing factors. The yield potential of selected crops and of cultivars can be exploited if these factors are optimal or the cultivation technology is well adapted. In certain extent inappropriate growing conditions can be regulated with special methods or treatments in order to reduce influences of stresses.

- To improve the productivity in horticultural crops we need first of all evaluation of suitable land in the different regions.
- It is important to evaluate the land use in this farming systems. In this regard database for Land Information System (LIS) is necessary. Tools of Global Information Systems (GIS) can be used in order to collect data of different layers and to develop a map which meets the requirements of the horticultural crops.
- With help of a multi-criteria analysis the final land suitability map for the different groups of vegetables or fruits can be developed.
- In regions with arable land for agricultural crops it is not easy to find land for horticultural crops belong to the classification 'high suitability'
- In several experiments the positive effect of different biostimulators used could be discovered if one of the growth factors was at a critical level.
- The biostimulators tested in the experiments have different effects-Humates are effective in nutrient and salt control as well as sub-optimal temperature. Lactates have a good effect if the pH-value is unstable. Microorganisms like *Bacillus subtilis* are effective under different stress conditions they have an unspecific effect; mainly they are effective if the nutritional status is suboptimal.

REFERENCES

Arsanti I. W. 2008. Evaluation of Vegetable Farming Systems in Upland Areas of Java and Sumatra, Indonesia, PhD Dissertation, Humboldt-University of Berlin, Faculty of Agriculture and Horticulture, Germany

Arsanti I. W. and Böhme M. H. 2008. Evaluation of Profitability and Competitiveness of Vegetable Farming Systems in Upland Areas of Indonesia, Proc. IInd on Supply Chains in Transit. Econ. Acta Hort. 774: 49-54

Baniya N. 2008. Land Suitability Evaluation Using GIS For Vegetable Crops in Kathmandu Valley /Nepal, PhD Dissertation, Humboldt-University of Berlin, Faculty of Agriculture and Horticulture, Germany

Baniya N., Böhme M., Baniya S. 2009. Physical Land Suitability Assessment for the Large Cardamom Amomum subulatum Roxb. Cultivation in Hills of Kathmandu Valley, The Chinese Journal of Population, Resources and Environment, Vol. 7 No 4: 59-63.

Böhme, M., J Schevtschenko, Pinker, I. 2005. Effect of biostimulators on growth of vegetables in hydroponical systems. *Acta Horticulturae (ISHS)* **697**: 337-344.

Böhme, M., I. Pinker, J. Shevchenko: Effects of biostimulators on nutrient supply in vegetables. Procc. Gent 2007, pp 6.

Böhme, M., J. Shevchenko, I. Pinker. 2008. Use of biostimulators to reduce abiotic stress in cucumber plants. Proc. XXVII IHC–S9 Endogenous and Exogenous Plant Bioregulators, *Acta Hort.* 774: 339-344.

David, P.P., Nelson, P.V. and Sanders, D.C. 1994. A humic acid improves growth of tomato seedling in solution culture. *J. Plant Nutrition* **17**:173-184.

Dixon J., Gulliver A., Gibbon D. 2001. Farming Systems and Poverty – Improving Farmers' livelihoods in changing world – FAO and World Bank, Rome, Italy; Principal Editor Malcolm Hall.

FAO, 1976. A Framework for Land Evaluation: Soils Bulletin 32. Rome, Italy: Food and Agriculture Organization of the United Nations.

FAO, 1986. Land evaluation for development: Food and Agriculture Organization of the United Nations, Rome, Italy: 4-59 (74p).

Huynh, V. C. 2007. Multi-criteria Land Suitability Evaluation for Selected Fruit Crops in Hilly Region of Central Vietnam, PhD Dissertation Humboldt University of Berlin, Faculty of Agriculture and Horticulture, Germany.

Huynh V.C. and Böhme M. H. 2007. Land suitability analysis for selected crops using GIS and multi-criteria approach in Central Vietnam Case study of Huong Binh commune in Thua Thien Hue province, Biannual Conference of Commission 1.5 Pedometrics, Division 1 of the International Union of Soil Science, Tübingen, 43.

Kavetskiy S., Böhme M., Kraynyuk M., Hofman G., and Cleemput O. 2003. Soil Quality and fertilization. Tempus Tacis, NAUU, Kiev, ISBN 966-8006-20-8.

Saaty, T.L. 2000. Fundamentals of decision making and priority theory with the analytic hierarchy process: RWS Publications, Pittsburgh, Vol. 6: 21-28.

15. SIMPLIFIED SOILLESS PRODUCTION SYSTEM (HYDROPONICS) ADOPTED BY GROWERS IN UAE FOR THE PRODUCTION OF HIGH QUALITY CASH CROPS

Ahmed T. Moustafa[1], Mansour I. Mansour[2], Saeed H. Al Bagham[2], Ali H. Al Hamoudi[2], Arash Nejatian[1], Saleh Akrot[2] and Hamlan Al Kabishi[2]

1. International Center for Agricultural Research in the Dry Areas (ICARDA); 2. Arabian Peninsula Regional Program (APRP).

ABSTRACT

Especially in arid areas such as the United Arab Emirates (UAE), protected cultivation in greenhouses is an important production system, supplying local markets with fresh, high-value vegetables. However, salinity build-up and soil-borne diseases are major constraints in greenhouse systems in the UAE. Despite all efforts, including soil replacements and heavy chemical treatments, growers continue to face heavy losses. A soilless (hydroponics) production system was introduced to small-scale greenhouse growers through a joint research program between ICARDA and the Ministry of Environment and Water (MEW) in UAE. The system was developed using locally available materials and simplified to suit conditions on the pilot farms. It is suitable for a range of high-value crops including cucumber, tomato, melon and peppers. Thirteen pilot growers in Ajman, Fujairah, Ras Al Khaimeh, and Sharjah have adopted a closed hydroponics system using perlite as a growing medium. In addition, three different closed systems (each in a 240m^2 greenhouse) have been established at the Research Station in Hamranieh for research, demonstration and training for growers and MEW staff. In the first season (Feb to Jun 2009), greenhouses in private farms were planted with cucumber and muskmelon. All costs (initial investment as well as operations) were borne by the growers. ICARDA and MEW scientists and extension agents provided technical backstopping and training; Integrated Production and Protection Management (IPPM) practices were successfully implemented. The hydroponics system produced crops with 30-50 percent higher yield and superior quality compared to soil based greenhouse production. Water and fertilizer requirements were reduced by 30 percent and 15 percent respectively. This success has encouraged further adoption of hydroponics systems. Similar successes have been achieved in other countries in the

Arabian Peninsula. The number of greenhouses with hydroponics system in Oman has increased from four in 2005 to 69 in 2009. Preliminary analysis in UAE shows that the typical cost of installing a hydroponics system in a 240m^2 greenhouse is US$ 1000; this can be fully recovered in a single year. ICARDA and MEW are now working with pilot growers to conduct a detailed impact assessment.

INTRODUCTION

The major constraints for vegetable production in UAE could be summarized in the followings: (1) harsh environment, (2) water scarcity and (3) lack of arable land and suitable soil.

1. Harsh Environment: UAE is characterized by an arid and semi-arid climate. Rainfall is highly erratic in space and time with annual precipitation generally less than 50 mm. Temperatures are generally high, reaching 50°C at times in summer-the main problem is prolonged hot periods (over 35°C) through the summer, when the relative humidity is often high. The average minimum and maximum temperatures in Abu Dhabi are illustrated in the following Figure 1.

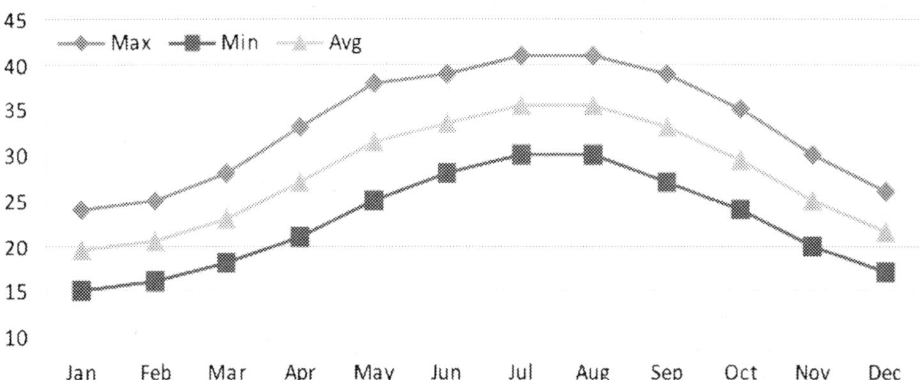

Figure 1: Historical temperature data for UAE by months.
Source: Abu Dhabi International, United Arab Emirates

The temperature in UAE, like other countries of the region, limits the production periods of the vegetables and other field crops. Research outputs and growers' experience in UAE show that majority of vegetable crops can only be cultivated between Septembers to May in open field.

2. Water Quality and Scarcity: Water scarcity is another major constraint. In 2001 the renewable water resources per capita in UAE was about only 56m^3 while the same for Middle East and North Africa region was about 1223m^3 (Earth Trends 2003). The quality of the water also is a major issue. Ground water quality is described as brackish with an average salinity of 4000 ppm (Brook, 2006). At the moment 99% of drinking water in UAE is coming from high tech desalination units. Seawater intrusion, which is a common problem along the Arabian Peninsulas, is also reported in UAE (Mohsen *et al.*, 2007).

Shortage and high cost of quality water, which makes the agricultural activity not economically feasible, is forcing the growers to stop production or to install desalination units in their farms.

3. Lack of Arable Land and Suitable Soil: Land availability and soil quality is also a major limiting factor for agriculture production. The total arable land and area under permanent crops in UAE based on CIA factsheet in 2005, was only 3.04% of total state area (CIA, 2005).

Without irrigation, agriculture in UAE is not possible. Ground water reserves are being depleted faster than they are being replenished and meanwhile desalination, the main source of drinking water, has a high environmental cost (UAE Interact 2008). Sustainability in agricultural production highly depends on unconventional water resources such as desalination water. Utilizing such expensive water for production of vegetable and agricultural products required a production system with higher productivity per unit of water and land.

Furthermore, to meet the increasing demand for food in UAE and achieve the food security the agricultural production system should be enhanced to increase the productivity per unit of valuable water and land while tolerating the extreme heat and conserving the fragile environment.

WHY SOILLESS CULTURE (HYDROPONICS)?

Soil is the most available growing media and plants normally grow on it. Main functions of soil are to provide anchorage, nutrients, air, and water to the plants rooting system. However, soils do pose serious limitations for plant growth. Presence of disease causing by soil organisms and nematodes, unsuitable soil reaction, salt accumulation due to irrigation, unfavorable soil compaction, poor drainage, degradation due to erosion, etc. are some of them.

In greenhouses, agricultural practices in soil will increase the soil-borne pathogens such as nematodes and accumulate salinity. Moreover, the present conventional growing system in soil, wastes a lot of fresh water due to run off and deep percolation. In arid countries, rapid evaporation from the soil surface may also lead to salinity problems. Soilless culture (Hydroponics) would be the answer to many of such constraints and to successfully produce high quality yield.

In the absence of soil in hydroponics systems, soil related constraints are likely disappeared which prevent the expensive soil sterilization process and application of hazardous agro-chemical in greenhouses.

Hydroponics offer opportunities to provide optimal conditions for plant growth and therefore, higher yields can be obtained compared to conventional soil based agriculture.

Soilless techniques also offer a way of improving water-use efficiency and obtaining better water and fertilizer management in crop production.

There are two main types of soilless cultivation:

1. Open systems, where plants are grown in containers or channels filled with an inert medium. Water and nutrients are supplied

through drip irrigation and the surplus nutrients and water is allowed to run off as waste.

2. Closed systems, where plants are grown in containers or bare roots placed in pipes or channels lined with polyethylene. A film of nutrient solution circulating through the roots or dripped into the containers and back to a catchments tank. This provides good contact between the solution and air, sufficient to maintain the oxygen level required by the roots without additional aeration of the solution.

Different hydroponics production systems are implemented by commercial growers while research is continued to further enhance the technology.

Soilless techniques offer a way of improving water-use efficiency and obtaining better water management in crop production. Soilless cultivation is intensively used in protected agriculture to improve control over the growing environment and to avoid uncertainties in the water and nutrient status of the soil and to improve water productivity. Especially in soilless closed systems water-use efficiency is maximized by recycling the water and nutrients. It also overcomes the problem of salt accumulation and increasing the pathogens in soil.

SIMPLIFYING THE TECHNOLOGY TO ENHANCE THE ADOPTION PROCESS

ICARDA through research for development program in close collaboration with NARES of many countries in CWANA, developed, enhanced, and simplified number of soilless production techniques. Soilless culture (Hydroponics) in large scale could be very complicated and needs trained management and sensitive measurement tools. ICARDA has succeeded to enhance and simplify such techniques to be applicable at the small growers' farms. Various simplified hydroponics systems for growing cash crops, including cucumber, tomato, pepper, muskmelon, green bean and strawberry, were tested in research stations in collaboration with NARES specially in Arabian Peninsula countries and after proved successful started to be transfer to private farms and growers (ICARDA Annual Report 2007) (Figure 2).

Different crops required different hydroponics production system. For production of cash crops such as strawberries and beans, the vertical soilless production system was adopted to maximize growing space by growing the crops vertically. Such technique for strawberry has been investigated as joint research activities between ICARDA and NARES in Kuwait, Oman, Bahrain, and Saudi Arabia and after proved promising in terms of land and water productivity started to be transferred to growers. Agro-economic comparison between vertical soilless and traditional soil bed systems was conducted in Kuwait. The study was based on cost and production data reported by a private farm in Al Wafra, Kuwait (Moustafa, *et al.*, 2006). The total cost of production in the vertical soil-less system was 40% less than soil beds, while production per unit area increased by four times and water productivity raised by 70% (Moustafa *et al.*, 2003).

Figure 2: Hydroponics production systems for strawberry and tomato

Horizontal hydroponics production systems are required for the production of number of vegetable crops such as tomato, cucumber, eggplant, muskmelon and pepper. ICARDA through joint and collaborative research activities with NARES in the number of CWANA countries succeeded to enhance and develop such systems for the production of high quality cash crops in the dry area.

The main structure of both vertical and horizontal systems is the same and designed in a way that could be easily changed and shifted to each other with the minimum costs. Furthermore, a production system can accommodate both types (vertical and horizontal) which provide more management tools and facilities for growers to get the maximum befit out of his/her investment. The hydroponics production systems could be installed and operated also in open field. Utilizing such system under net houses in UAE is under investigation where the primary results are promising.

INTRODUCING TECHNOLOGIES TO GROWERS AND NARES IN EMIRATES

To enhance the capacity of NARS in UAE for further assessment and the adoption of soilless culture techniques, ICARDA-APRP established three soilless production systems at Hamranieh Research Station. Similarly, five systems established at five pilot growers sites which increased to 13 sites after the first growing seasons.

The study objective was:

- To demonstrate the potential of soilless production systems for increasing yield and water productivity and reduce the use of agrochemicals;
- To compare different soilless techniques for cash crops productions;
- To enhance the soilless production system for further adoption in emirates;
- To enhance and develop the MEW capacity and human recourse for further research and development activities on soilless production systems.

- To develop guideline and recommendations for growers to adopt and manage the soilless production systems based on collated data;

Pilot growers whom adopted the technology were provided with technical backstopping by ICARDA scientists in collaboration with NARES through number of visit, field days, workshops and farmers field school.

RESEARCH AND TECHNOLOGY TRANSFER ACTIVITIES

Three non-cooled greenhouses (8m width and 30m length each) at Hamranieh Research Station (HRS) were selected for the establishment of three soilless production systems. During system installation, researchers, extension agents and technicians received on the job training. Three different growing canals' materials and design *i.e.,* (1) cement block, (2) polypropylene plastic, and (3) ditch covered with polyethylene sheet were used. Perlite was utilized as growing media in all systems. All three greenhouses were connected to one catchment tank. Water and nutrients were delivered through drip irrigation system to the plants and then to growing canals connected to the catchment tank using PVC drain lines (Figure 3).

Figure 3: Establishment of soilless culture system, Hamranieh RS, UAE

Another GH was used for comparison where cucumber was planted in soil using the typical management techniques and fertigation for conventional soil production system.

Readymade compound fertilizers were selected carefully from the market to make the stock solution. The PH was maintained at 6 and EC at 2500-3000 µS during the production period. A written guideline was developed by ICARDA scientist for nutrition solution managements.

The production process were monitored and followed by MEW and ICARDA scientists. Data on fertilizer application, PH and EC and production were recorded daily.

Similar arrangement was carried out at farmers' fields. All systems were equipped with automatic irrigation controller. Pilot growers received on-the-job training and technical backstopping during the production period through regular visits by ICARDA scientists and MEW extension agents.

In all greenhouses at research station and growers site, Integrated Production and Protection Management (IPPM) techniques were implemented. IPPM consists of a number of control measures which promoted by ICARDA and implemented in number of research stations and pilot private farms (Moustafa, 2001). IPPM components are including: GH climate management, irrigation and fertilization management, agro-management and practices, mechanical protection, biological control and chemical control, ICARDA & NARES previous research activities on IPPM application in UAE resulted in reduction of agro-chemical utilization for production of cash crops under greenhouses. Similarly, Implementing IPPM in Yemen reduced agrochemical spraying from 19 to only 2 times (Mukred *et al.,* 2003). The three Greenhouses at Hamranieh RS sprayed only once against powdery mildew. No other pesticides were applied during the production periods.

RESULTS AND DISCUSSIONS

The primary production data for cucumber under soilless culture and its comparison with the normal soil bed at the research station revealed the superiority of hydroponics. Productivity per unit of area of soilless culture was about six times more than that in soil. Figure 4 illustrates the production in soil and soilless culture in HRS and growers field. The highest production and water productivity obtained at HRS with the cement block canal greenhouse. The cucumber production at the greenhouse with plastic canal was reduced by 36% comparing to other system. This was mainly due to difficulties in water management in plastic canals.

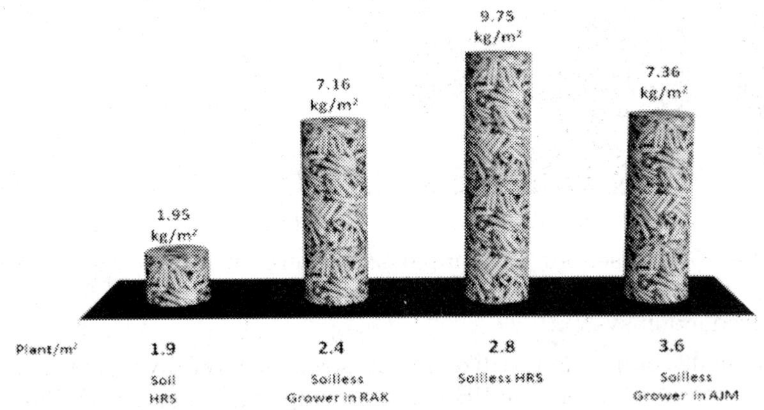

Figure 4. Cucumber production from soil and soilless (Hydroponics) in Hamranieh RS and pilot growers in Ajman and Ras Al Khaimeh (February to May 2009)

Soilless system not only increases yield but also speed up the production by 20 days earlier. Figure 5 illustrates that the cucumber in soilless started production about 3-4 weeks earlier than the same in soil. This is due to the better growing situation that soilless production provided to plants. In soilless systems, plants were provided with optimum requirements of water and nutrients.

Total water used during the production period for each GH was about 26.4m^3. Water productivity of cucumber in soilless culture was significantly higher than the same in the soil (Figure 6). The water productivity in soilless reached to 81kg/m^3 while in soil this figure was only 6kg/m^3.

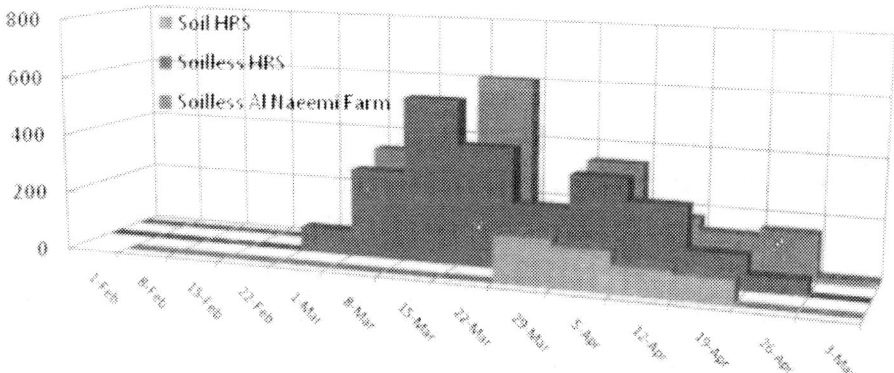

Figure 5. Weekly harvest of cucumber (kg) from soil and soilless (February-May 2008)

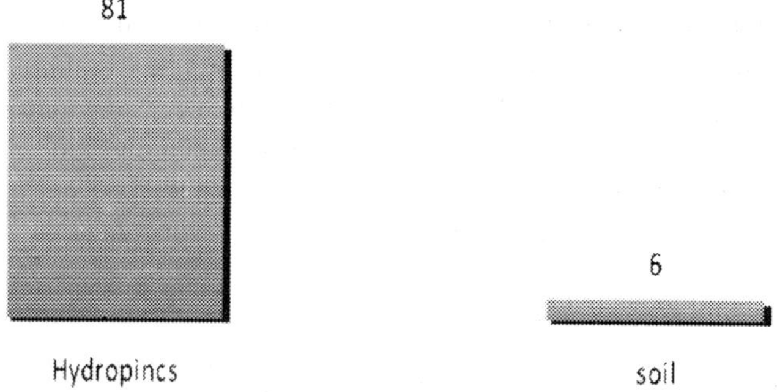

Figure 6. Cucumber water productivity (kg/m3) in soil and soilless production at Hamranieh RS

IPPM application significantly reduced number of agro-chemical spray. In all GH at research station and growers filed only one spray for downy mildew were carried out.

Cost of the production, income and net benefit of cucumber production in soilless culture and conventional soil base system under cooled GHs in UAE are compared and illustrated in Figure 7. The primary cost for establishment of hydroponics is higher than soil based system. However, this would be easily recovered with saving on production materials, water and labor. Saving on production materials in hydroponics is mainly due to less seedling welding, optimum use of fertilizer and skipping over the costly soil sterilization process. The better quality of cucumber produced in hydroponics provided growers with higher premium in the market resulted in better income and net profit comparing the soil based system.

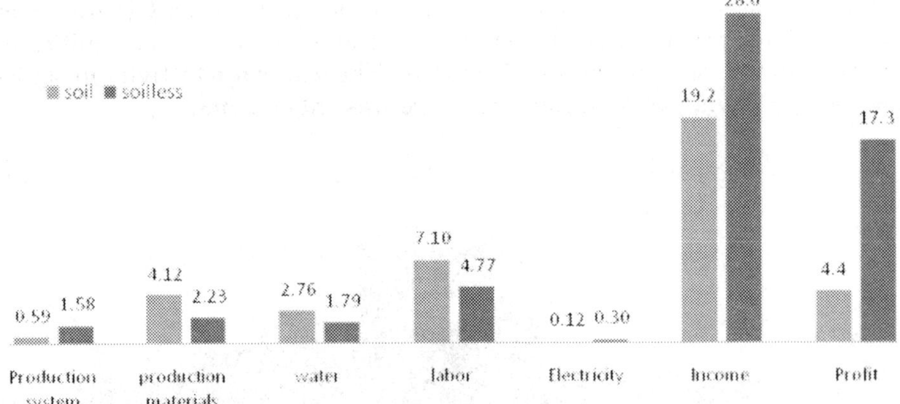

Figure 7. Cost of production, income and net benefit for cucumber production in soil and soilless

Source: Data collected during the study and interview with growers and researcher in Emirates

REFERENCES

Abu Dhabi Int'l, United Arab Emirates: Climate, Global Warming, and Daylight Charts and Data, http://www.climate-charts.com/Locations/u/UE41217.php;

Brook, M. 2006. Water Resources ABU Dhabi Emirate U.A.E. Retrieved from Abu Dhabi global Environmental Data Inicitive: http://www.agedi.ae/Resources/SOE/ Abu% 20Dhabi%20 Emirate%20Water%20Sector%20paper.pdf

CIA., 2005. United Arab Emirates. Retrieved Jan 14, 2010, from The World Fact Book: https://www.cia.gov/library/publications/the-world-factbook/geos/ae.html;

Earth Trends., 2003. Water Resources and Freshwater Ecosystems- United Arab Emirates, http://earthtrends.wri.org/pdf_library/country_profiles/ wat_cou_784. pdf

Mohsen, S., and Anvar, K. 2007. Seawater intrusion in the coastal aquifer of Wadi Ham, UAE. IAHS-AISH publication ISSN 0144-7815 CODEN IAPUEP. http://cat.inist.fr/?aModele=afficheN&cpsidt=19688750

Moustafa, A.T., 2001. Integrated Production and Protection Management for the Green Houses. Al Morshed, November 2001, Abu Dhabi, UAE.

Moustafa, A.T., 2006. Economic Perspective of Protected Cultivation in the Arabian Peninsula, Proceeding of Investment in Protected Cultivation in GCC Countries workshop,15 May 2006, Abu Dhabi, UAE, http://www.icarda.org/aprp/APRP-AE/ pawsabu/ATM.htm

Moustafa, A.T., Al Bakry, A. Al Rawahi, M. Al Mohammadi, A.R. and Ahmed, R. 2003. Development of Soilless Growing Techniques for the Production of High Quality Cash Crops using Less Water in the Arabian Peninsula, More Crops per Drop. Proceeding of 7th International Conference on Dryland Development (IDDC) 14-17 Sep 2003, Tehran, Iran. Pages 485-488

Moustafa, A.T., Abdin, F. Al Zeir, K. Al Qurain, F. H., 1999. Production of Cucumber in Plastic Green Houses at National Agriculture and Water Research Center with ICARDA Under IPPM. ICARDA-APRP Strengthening Agricultural Research and Human Resource Development in the Arabian Peninsula, Final Report 1995-2000, Aleppo, Syria.

Moustafa, A.T., Oraifan, S. Al Bakry, A. and Nejatian. A., 2006. High density cropping system for cash crop production in marginal land with less water. Proceeding of 8th IDDC, Peking, China, 25-28 Feb 2006. Pages 839-842.

Mukred, A.W., Moustafa, A. T. Al Kirshi, A. and Al Dhbhani, M., 2003. Introduction of protected agriculture into mountain terraces of Yemen. Proceeding of 7th International Conference on Dryland Development (IDDC) 14-17 Sep 2003, Tehran, Iran. Page 504-511

UAE Interact., 2008. Abu Dhabi's water conservation plans, http://www.uaeinteract.com/ docs/Abu_Dhabi%E2%80%99s_water_conservation_plans/30105.htm, access Nov 2009.

16. ASIAN SMALLHOLDER VEGETABLE FARMERS AS IPM EXPERTS: EXPERIENCES FROM THE FAO REGIONAL VEGETABLE IPM PROGRAMME

Jan Willem Ketelaar[1] and Prabhat Kumar[2]

[1]*Chief Technical Advisor, FAO Regional IPM Programme, Bangkok, Thailand*
[2]*PhD (Hort. Entomology), Affiliated faculty and Senior Research Scientis, Bangkok, Thailand*

ABSTRACT

Pesticide use in vegetable production among smallholder farmers in Asia remains unnecessary high. Enduring concerns over farmer health and environmental pollution caused by indiscriminate use of pesticides call for better crop protection strategies. New concerns over food safety in relation to pesticide residues on fresh and processed vegetables intended for domestic and export markets provide momentum for further development and application of Integrated Pest Management as part of Good Agricultural Practices promotion efforts. Climate change is facilitating the spread of new invasive pest species, often prompting farmers to use more pesticides. Sucking pests, many of them vectoring important and new plant diseases, are now dominating the complex of pest problems that vegetable farmers face on a day-to-day basis. FAO has been working with Asian governments, civil society organizations and the private sector to develop robust IPM strategies for a range of economically important vegetable crops during the last decade. Case studies of successfully employed vegetable IPM strategies, resulting into major reductions in pesticide use, will be detailed in this paper. Apart from assistance to governments to strengthen pesticide regulatory systems, FAO also continues to assist National IPM Programmes in the Asia region to implement farmer IPM training, using the innovative and successful Farmers Field School approach. This paper will therefore also outline training achievements to date and results from this important training work, aimed at turning vegetable *farmers* into IPM experts. For further details on FAO's Regional IPM Programme, see website: www.vegetableipmasia.org

INTRODUCTION

Vegetable Production in Tropical Asia

Vegetables are an important part of a complete diet of Asians, adding valuable nutrients that would otherwise be insufficiently available in food staple crops such as rice, wheat or maize. Often at the expense of rice land, farmers increasingly switch to more rewarding cash crops as to meet the growing demands for such vegetables as crucifers, carrots, potatoes, tomatoes and beans. These are commonly grown in most Asian countries including Cambodia, China, Indonesia, Lao PDR, Philippines, Thailand and Vietnam. Recent figures included in Table 1 below confirm their importance. While many vegetables are native to tropical Asia, a large number of them are temperate in origin. Hence, many vegetable crops perform best in cool temperatures found in production areas at higher altitudes. However, increasingly vegetable production is expanding into the lowlands with the release of new heat-tolerant varieties. The realization of optimal yields, particularly in the warm humid lowlands, is often constrained by pests and diseases. Poor horticultural practices further exacerbate these problems. Rather than ameliorate the situation, the application of pesticides most often causes resurgence of pest populations due to the destruction of their natural enemies (Shepard *et al.*, 2009).

Table 1. Recent data on area cultivated (ha) and production (tons/ha) of some vegetable crops in selected Southeast Asian countries

Country	Cabbages & other Brassica crops		Potatoes		Tomatoes		Beans (Dry + Green)	
	Ha	T/Ha	Ha.	T/Ha.	Ha.	T/Ha.	Ha. (d+g)	T/Ha.
Cambodia	n.a.	n.a.	n.a.	n.a.	n.a.	n.a.	65000	0/83 (d)
China	1768412	20.65	4436700	12.66	1453935	23.10	1163425	1.30 (d) 11.31 (g)
Indonesia	60711	21.22	62375	16.09	51523	12.33	451714	1.03 (d) 6.12 (g)
Thailand	24000	11.05	7981	15.75	12000	22.50	167298	0.78 (d) 4.00 (g)
Vietnam	40000	17.5	35000	10.57	n.a.	n.a.	205000	0.77 (d)
Lao PDR	n.a.	n.a.	5500	6.54	n.a.	n.a.	2450	1.08 (d)

FAO STAT 2009; n. a. not available; d- dry, g – green

Emerging Challenges Rresulting from Climate Change

Climate change induced changes are expected to facilitate the emergence and resurgence of invasive pests and plant disease vectors (e.g. whitefly, *Bemisia tabaci*; thrips (many species), aphids (several species) and plant hoppers). Imbalances in host-natural enemy population synchrony are some of the commonly observed impacts of climate change, which are upsetting the natural ecosystem balances across crops and geographical regions (Thomas *et al.*, 2009). It is expected that pests will have shorter life cycles, increased fecundity, increased herbivory due to poor quality food and in many cases movement towards new geographical areas and new crop hosts. That said, natural enemy populations are also expected to benefit from above processes and might well be able to keep pest populations in check (Kiritani, 2006). Currently, prevailing pest forecasting and management strategies are poorly equipped to deal with these rapid and dynamic changes. To deal with these ever increasing levels of biotic and abiotic stress, quantum changes in knowledge-attitude-practice of smallholder farmers will be of paramount importance as to enable them to continue to grow healthy and high-yielding crops with minimal inputs of agro-chemicals.

Problems Associated with Misuse and Overuse of Chemical Pesticides

The misuse and overuse of chemical pesticides in vegetable production in tropical Asia remain serious and acute (Shepard *et al.*, 2009). In the year 2005, 790,000 ton of various pesticide active ingredients, worth more than US$ 8.3 billion, were used in Asian countries (*see* Fig. 1 for detail). Pest problems in tropical vegetable production are frequent, often acute and difficult to manage. Yields are highly variable whereas market prices can vary considerably on a day-to-day and season-to-season basis. Vegetable production has become an ever riskier business providing a clear stimulus for farmers to minimize risk and maximize preventive pesticide application.

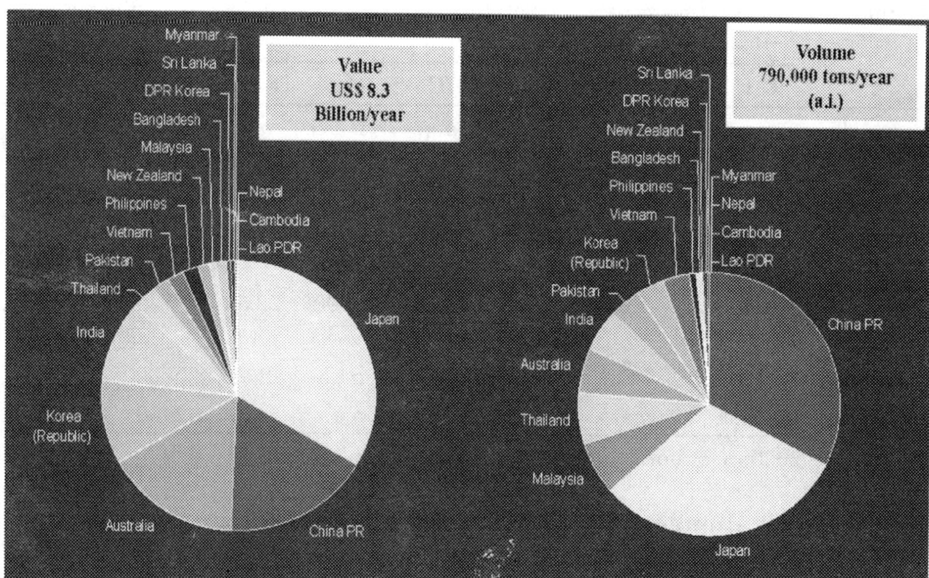

Figure 1. Status of pesticide consumption in the Asia and Pacific Region (Source: FAO Regional Workshop on Implementation of Code of Conduct on the Distribution and Use of Pesticides, 2005; supplemented with data from Agrow, various country reports and own estimates)

For some vegetable crops the average frequency of application of chemical pesticides is 10-20 times per season, with up to 80 applications per season for brinjal (eggplant) production in parts of South Asia (*e.g.*, Bangladesh). In such extreme cases in Bangladesh, cocktails of chemicals, mixing insecticides with fungicides and herbicides, are applied every other day during the six-month eggplant growing season. Often cost of pesticides alone constitutes over 50% of total cost of production (SUSVEG Asia, 2007; (*See* a recent review Srinivasan, 2008). These cocktails also include banned or restricted and often highly hazardous insecticides such as endosulfan, chlordane, sodium cyanide, methyl parathion and mono/dicrotophos. In Cambodia and Lao PDR, labeling of pesticide products is often insufficient or/and inappropriate as labels are usually in foreign languages (*e.g.*, Thai and Vietnamese). Original products are often re-packaged and contain no label at all (EJF, 2002). For example, a study conducted in Cambodia revealed that only 8 of 77 pesticide traders said they could read foreign labels on pesticides they sold, whilst 97.5% of the pesticides were labeled in a foreign language (CEDAC, 2000)[1]. The quality of imported pesticides in these GMS countries is often sub-standard. A recent quality check of 7 pesticides originating mostly from China and collected at random at markets in northern Lao PDR revealed that 5 out of 7 formulations did either not contain what was on the label or was sub-standard and/or contaminated with chemicals produced in the pesticide production process (Van der Wulp *et al.*, 2008). The extensive use of chemical pesticides results in a range of unsustainable production practices arising from undesirable externalities. The application of pesticides causes resurgence of pest populations due to the destruction of their natural enemies. Resistance of target pests against pesticides has become a serious problem for most important vegetable pest problems, *e.g.*, Diamond Back Moth (*Plutella xylostella*) on crucifers and Fruit and Shoot Borer (*Leucinodes orbonalis*) on eggplants. Widespread environmental pollution has an important impact on quality of natural resources such as soils, water and fish populations (*e.g.*, Settle & Whitten, 2000). The health hazards of chemical pesticide application to farmers and their families are serious as for example documented in various important health studies conducted among farmers who have been exposed to chemical pesticides (*e.g.*, Mancini *et al.*, 2009, Murphy *et al.*, 1999). Greater awareness among consumers in urban communities of the dangers of toxic residues on vegetables has created an ever growing demand for vegetables free of pesticide residues. In line with the FAO Codex Alimentarius (see http://www.codexalimentarius.net/web/index_en.jsp), governments in many Asian countries have established Maximum Residue Levels (MRLs) for pesticides on imported vegetables. During the start of this decade, the export of okra (ladyfinger) from the Philippines and Thailand to Japan and Korea was jeopardized because of unacceptable high levels of pesticide residues. Unacceptable levels of

---

1. For more graphic details on the problems associated with the –often illegitimate- trade in hazardous pesticides, the reader is advised to consult the following website: www.ToxicTrail.org

pesticide residues on Chinese export of fresh and processed vegetables and food products to Japan illustrate more recent food safety concerns. Clearly, the importance of negative international trade implications of the overuse of toxic pesticides on vegetable production can no longer be ignored. In addition to these problems associated with use of chemical pesticides, vegetable farmers are overusing chemical fertilizers. In short, vegetable production in tropical and subtropical Asia remains in a 'crisis phase' (Shepard *et al.,* 2009), requiring urgent attention to safeguard the production of healthy and safe food and producers' livelihoods Asia-wide.

Vegetable IPM: the Answer to the 'Crisis Phase' in Asian Vegetable Production

Integrated pest management (*see* box 1) has evolved in many developing countries from a mix of various technical components (Smith & Reynolds, 1966) to a farmer-led program (FAO, 2002). Successful IPM has always followed an ecological approach (Kogan, 1998). The decline of a chemical control paradigm came about because it could not successfully handle the problems of resistance, resurgence and secondary pest outbreaks, along with environmental and health hazards (Perkins, 1982). Alternatives to the chemical paradigm developed. The most successful of these was IPM. This approach encouraged the development of non-chemical methods that promote the growing of a healthy crop and at the same time protecting natural enemies, thus ensuring a low population of pest species. A participatory approach towards IPM demands that farmers need to be educated in the language of ecology to better appreciate the advances in an ecological pest management approach. In the last decade or so, advances in IPM in a number of Asian countries took the form of farmer education through Farmer Field Schools (FFSs) and follow-up activities.

More specific to the origins of *vegetable* IPM in Asia and FAO's involvement, an expert consultation on the status and management of major vegetable pests in the Asia-Pacific Region (with special emphasis on Integrated Pest Management) was held from 14-16 November 1988 (Lim & Di, 1989). This consultation marked the starting point for efforts by FAO to implement IPM in vegetables in Asia. The expert consultation found that there was overriding concern by most governments in the Asia-Pacific Region regarding pesticide misuse, pest resistance, environmental pollution, health hazards and toxic residues in vegetables, and that the region has entered the "crisis phase". Based on this consultation, it was strongly recommended that all efforts be exerted by national and international bodies to develop IPM programmes for vegetables in the Asia and Pacific Region. The consultation agreed that "in selected vegetable production systems adequate background information to develop IPM programmes are generally available". The consultation further agreed that "the proposed regional project should primarily concentrate on implementing practical vegetable IPM in the field." To reach the target beneficiaries with the technical information available, it was strongly felt that the "vegetable IPM programmes should focus on human resource development through training, communication, and promotions".

Box 1: Integrated Pest Management: Evolution of a concept

The label 'Integrated Pest Management' has enjoyed different definitions and meanings over the last four decades and has considerably evolved. FAO, through it's Rome based Crop Protection Service (AGPP) and its Panel of International IPM Experts, played a leadership role in the evolution of the concept between 1960 and 1980. Within FAO the concept of IPM has evolved from a strictly crop protection related concept to a much more holistic and ecological approach to crop production and protection. And the evolution of the concept hasn't stopped since. Through the more recent pioneering work of the FAO IPM Programmes in Asia and a range of international and local development organization partners, IPM has become synonymous with education and human resource development programs. These programs often had (and continue to have) farmer training in crop protection and reduction of chemical pesticides as entry points but have subsequently broadened to include overall crop production. The more recent term *"Integrated Production and Pest Management (IPPM)"*, widely used in FAO IPM training programs, especially on the African continent, more appropriately describes the curriculum content of IPM training programs. And, equally important, the training curriculum also addresses social issues and community development aspects. Thus, the 'IPM' label has evolved from a strictly technical to a much more holistic approach to crop production, human resource- and rural development. The nature of this new meaning can be gauged for example by reference to material posted on the website (www.communityipm.org) and the FAO publication titled "From Farmer Field Schools to Community IPM: Ten Years of IPM Training in Asia" (FAO, 2002). Recently, FAO has developed a website dedicated to information sharing of the innovative vegetable IPM programmes developed in the Asia region during the last decade, which can be accessed at: www.vegetableipmasia.org

THE FAO REGIONAL VEGETABLE IPM PROGRAMME IN ASIA

Scope and Rationale for the FAO Programme

The above described problems associated with misuse and overuse of pesticides in vegetable production in tropical Asia provide the rationale for FAO implementing a Regional Programme for the development and application of vegetable IPM in vegetable production systems in several Asian countries. The FAO programme operates currently in 10 member countries, e.g. Bangladesh, Bhutan (recently joined), Cambodia, China PR, Indonesia, Lao PDR, Nepal, Philippines, Thailand and Vietnam. The FAO programme works with National IPM programmes that assist farmers through training to better understand vegetable crop ecology and to minimize use of pesticides in their local crop production systems. The programme has, since 1996, carried out applied research, extension and farmer education activities to promote and support the development and application of vegetable IPM by Asian smallholder farmers. The programme recently completed its 2nd phase (2002-8). For more information on FAO's support for IPM and Pesticide Risk Reduction efforts in Asia, see website: www.vegetableipmasia.org.

Training Strategy and Methodology

The standard training approach employed in all FAO IPM Programmes is the so-called 'Farmers Field School (FFS)' approach. Farmers Field Schools are 'schools without walls' which are held in farmers' fields. Typically, a group of about 25 to 30 vegetable farmers meet on a weekly basis and grow and study together in a vegetable crop of local relevance and choice from sowing to harvest. Through the educational tool 'Agro-ecosystem Analysis' farmers monitor crops and record data on observations of the crop situation. Data are analyzed by farmers and discussed in small groups. Decisions are made and implemented with regards to the needed crop management. This Agro-EcoSystem Analysis (AESA) is done by farmers on a weekly basis and allows farmers to deepen their understanding of local crop ecology and needs for intervention. FFSs are facilitated by expert IPM trainers, who can either be Government or NGO extension workers or be FFS-Alumni farmers themselves. IPM Expert trainers have themselves undergone season long vegetable IPM training in so called 'Training of Trainers Courses'[1].

Working strategy of the training program is that farmers, who have undergone IPM FFS training, are better able to identify pest and disease problems and have acquired the necessary skills and confidence required to avoid the unnecessary use of chemical pesticides without yield losses. IPM farmers are then also in a much better position to make informed decisions about growing a healthy, sustainable and profitable vegetable crop. Farmers are set on a path to explore new innovative and more environment friendly options for pest and disease management (such as bio-pesticides) during follow up group learning activities as to further reduce dependency on inputs of chemical pesticides.

MAKING VEGETABLE IPM WORK: THE CASE FOR FARMERS AS IPM EXPERTS

One of the most important lessons of FAO's involvement in the development and application of IPM relates to the need for *farmers* to become IPM Experts. If vegetable IPM is to establish itself in a sustainable manner, it will need to depend on farmers themselves making vegetable IPM work on their farms. In the following sections, three case studies derived from the work of FAO in collaboration with various partner institutions are presented in order to illustrate this point.

(1) Brassica IPM in Highland Production Systems in Thailand

Diamond back moth (DBM, *Plutella xylostella*) is one of the major production constraints in crucifer production in most Asian countries including

---

1. For more details on the FFS approach to IPM training, various crops covered and an updated overview of # of farmers and trainers trained in each of the FAO Programme Member Countries, the reader is advised to consult the following website: www.vegetableipmasia.org.

Thailand. Despite many applications with a wide variety of pesticides, farmers still cannot control the DBM. On the other hand, the heavy reliance on chemical pesticides has continued to cause many undesirable problems, e.g. more DBM resurgence, human health problems, environmental pollution and reduced market access resulting from pesticide residues exceeding MRLs. Thus, an alternative approach for managing the DBM is considered crucial and is urgently needed.

The Royal Thai Government –through its National and Regional Biological Control Centers, coordinated by the Bangkok-based Pest Management Division at the Department of Agricultural Extension- has been on the forefront of mass rearing, introduction and promotion of various natural enemies (Table 2) among smallholder farmers in Thailand since many years.

Table 2. Commonly mass-reared natural enemies in Thailand (Lim G.S., 2008)

Bioagents	Category	Main target pest(s)	Main crop(s)
Anastatus	Egg parasitoid	Sucking bugs, including Nezara sp.	Tangerine and Longan
Trichogramma	Egg parasitoid	Borers	Sugarcane
Earwig	Predator	Borers	Sugarcane
Lacewing	Predator	Aphids	Vegetables and Tangerine
Sycanus	Predator	Caterpillars	Vegetables
Eocanthecona	Predator	Caterpillars	Vegetables
Beauveria bassiana	Microbial	Sucking insects (e.g. leaf- and planthoppers)	Rice
Metarhizium anisopliae	Microbial	Sugarcane beetle borer and flea beetle	Sugarcane and Vegetables
Trichoderma	Microbial	Diseases	Vegetables

Beginning 2005, the Thai Government, with technical and financial support from the FAO Regional Vegetable IPM Programme and a DANIDA-funded IPM project, engaged in the introduction, rearing and field releases of the parasitoid *Diadegma semiclausum.* This wasp was introduced for population regulation of DBM in highland brassica production systems in both Thap Boek (Lomkao district, Phetchabun province) and the organic farms of the Royal Project Foundation (RPF) in Doi Angkhang, Chiangmai.

During the early 1990s, AVRDC had introduced *D. semiclausum* from Taiwan and releases took place in Tap Boek, Phetchabun. The introduction failed, most likely as a result of farmers' continued pesticide sprays. In 2004/5, the DoAE conducted surveys to ascertain the status of DBM parasitoids prevalent in highland brassica production systems in Thailand (DoAE/FAO, 2005). The survey also served to assess farmer's knowledge on DBM management (Upanisakorn *et al.,* 2009; *see* Box 2 for key survey findings). The surveys confirmed the absence of *D. semiclausum* and the rationale for introduction of this parasitoid. *D. semiclausum* was subsequently introduced from the Cameron Highlands in Malaysia with CABI/MARDI assistance. Following IPPC ISPM-3 guidelines for the introduction of biological control agents, the parasitoids underwent quarantine screening and basic rearing at the Department of Agriculture prior to introductions in brassica crops on the organic farms operated by the Royal Project Foundation in Doi Angkhang, in Northern Thailand. A recent (January 2010) FAO survey confirmed that the parasitoid is established and provides good DBM population control with an average of 60% larval parasitisation observed in Doi Angkhang. Rearing efforts in RPF-Doi Angkhang were also successful (DoA/RPF, 2006). Parasitoid shipments were handed over to the Department of Agriculture Extension in Phetchabun Province for mass rearing and field releases in cabbage farms in conjunction with season-long IPM Farmer Field Schools (FFS). In a recent field survey carried out in 2008 (Upanisakorn *et al.*, 2009), it was clear that the *D. semiclausum* had been established in the farmer's field. Other benefits are the reduction in overall pesticide use, enhanced knowledge of farmers on pests and natural enemies and further augmentation of local natural enemies for effective natural pest management (Box 3 for details).

Box 2: Status of the DBM management in 2004-05 (Lim, G.S., 2008)

- Field observations and farmer feedback in the surveys confirmed that DBM is still a very important pest that is very difficult to manage.
- Although the time of DBM abundance (which normally peaks in mid-season) is generally governed by the crop stage, it often is affected by the pest management practices of farmers, e.g. very high levels of DBM infestations may continue late into the season when farmers apply pesticides intensively (up to 18-20 times).
- In general, the population levels of DBM in farmer practice fields were significantly higher than those in both organic and IPM fields.
- No egg parasitoid of DBM was found in the surveys.
- From DBM larvae, two kinds of parasitoids were recovered, i.e. a larval parasitoid *C. plutellae* and a larval-pupal parasitoid *Macromalon orientale*. Both parasitoids were present in all the study sites.
- The overall larval parasitism varied from 13.6% to 34.2%, averaging 23.2%. In general, the parasitism was higher in organic and IPM fields than in the farmer practice fields. Larval mortality from other causes was about 9.8%.
- From DBM pupae, only *Diadromus collaris* was recovered, with an average parasitism of 21.9% (range: 2.1% to 40.3%). It was fairly common in all the sites surveyed. Pupal mortality from other causes was 9.4%.
- Adult emergence from *C. plutellae* cocoons was particularly high, averaging 94.7%, with 3.9% hyper-parasitism and 1.4% other mortality.
- All farmer practice fields have comparatively lower parasitism than in the organic and IPM fields, suggesting that frequent pesticide applications have significant detrimental effects on the parasitoids. This

in turn has bearing on the DBM population, which was higher when parasitism was lower. In contrast, DBM population was less in fields with no chemical application. These relationships suggest that parasitoids are exerting an influential role on DBM and the extent of pesticide use can be critical in determining their impacts on DBM.

- In terms of parasitoid dominance, *C. plutellae* was usually higher in the early crop stage, while *M. orientale* and *D. collaris* during the later stages.
- Both temperature and relative humidity have no clear relationship with the occurrence or dominance of the parasitoids.
- On average, the overall combined parasitism from *C. plutellae, M. orientale* and *D. collaris* was 39.9%. Both larval and larval-pupal parasitoids contributed 20.3% and pupal parasitism 19.6%.
- Past experiences indicated that for effective DBM control, a combined parasitism of at least 70% sustained over an extended period is usually necessary. Normally, this can only occur in fields with no chemical pesticide or when only parasitoid-safe pesticides are used. Although such conditions occurred in the surveyed fields (organic and IPM fields), yet the overall parasitism was still below that needed to control DBM effectively, suggesting the present parasitoid complex in the highlands is still inadequate in controlling DBM.
- Although all three parasitoid species can be commonly found, farmers were not aware of their occurrences nor did they know of their potential roles. Farmers therefore continued to apply pesticides regularly and indiscriminately without due regard for the parasitoids.
- The present surveys failed to recover *D. semiclausum* in both the highlands of Chiangmai and Petchabun. Its absence provided clear evidence that the earlier attempt to introduce the parasitoid into Pechabun highlands was unsuccessful. This is unfortunate because *D. semiclausum* is the most sought among DBM parasitoids due to its known superior performance in controlling DBM in many parts of the world.

The parasitoid *D. semiclausum* is now established and provides effective biological control of DBM populations in Doi Angkhang, Chiangmai and Tap Boek, Petchabun. Despite of arguments to the contrary provided by Grzywacz *et al.,* 2009, this most recent successful case study of effective ecosystem services for biological control of DBM negates the need for introduction of Bt-brassica crops, at least in highland brassica production systems. As the technical capacity and rearing facilities are in place, the parasitoid will now also be distributed to other highland cabbage fields in Thailand by the pest management center staff. As a result of participation in Farmers Field School, the IPM farmers in Tap Boek have learned to appreciate and make biological control for pest management. Farmers have adopted IPM and reduced pesticide sprays, gaining an overall higher profit from their farm production. The higher quality brassica produce has also been Q-GAP certified and better market access is thus guaranteed. As for the Thai Government, lessons learned include the notion that successful biological control implementation needs to be done step by step as per inter-nationally agreed upon guidelines (IPPC-ISPM-3). An important lesson learned was also that parasitoid field introductions need to be undertaken simultaneously with farmer education efforts as to ensure that farmers appreciate and make optimal use of the introduced parasitoids for biological control of pest problems. A similar result was recently reported from Da lat, Vietnam, where FFS-based farmer training coupled with introduction of parasitoids supported by FAO in the early 1990s, has resulted in a decline of pesticide use and increased income of smallholder farmers (Nga and Kumar, 2008). Finally, the collaborative nature of this work supported by a wide variety of Thai government and international partner organizations also established that it is possible to work together on a successful biological control project that ultimately benefits smallholder farmers.

Box 3. Status of DBM management in 2008 (Upanisakorn *et al.*, 2009)

- The Phetchabun Provincial Agricultural Extension Office has set up a parasitoid rearing facility in Tub-berg Highlands as to mass produce the DBM parasitoid, *Diadegma semiclausum.*
- FAO and DoAE were geared towards technical and financial support to the Phetchabun Provincial Agricultural Office for rearing -and field releases- of the parasitoids.
- Monitoring and Evaluation activities were undertaken as to determine establishment of the parasitoid and evaluate their capacity to regulate DBM populations.
- Some 33,742 parasitoid (Ds) were reared and released to 10 IPM fields and after that, the 20 fields were monitored to evaluate the efficiency of Ds. It was found that the Ds can control DBM effectively.
- Farmer reduced pesticides use from 7-12 times/crops prior to parasitoid introduction to 3 times on average.
- Some farmers stopped spraying altogether since they joined the project. There are some other parasitoids (e.g. *Diadromus collaris* , *Macromalon sp. and Cotesia plutellae)* which work in tandem with Ds to regulate DBM populations.
- They also found that the Ds spread naturally to the un-released fields far away from the released fields.
- Hence it was concluded that Ds can establish well and spread naturally, especially in the non-sprayed fields.

(2) Potato IPM in Lowland Rice-based Production Systems in Vietnam

Potato is an important food crop in Vietnam, which is increasingly providing raw materials for the food-processing industry and stable income to the smallholder farmers in the Red-River Delta region. As evident from Table 3, there has been sluggish growth on account of area and total production during recent years in Vietnam. This could be primarily attributed to the lack of quality seeds and high labor costs. Generally, low level of mechanization leads to very high labor requirement for potato cultivation *i.e.,* soil preparation, intercultural operations and pest management.

With the multiple objectives including reducing labor costs, an innovative pilot project was set up under the auspices of the Vietnam National IPM Program/ Plant Protection Sub Department with the purpose to promote rice-potato farm system development. Whereas Vietnamese farmers generally burn rice stubbles after harvest, there is good potential for growing potatoes on these rice stubble crops as practiced elsewhere in the region (*e.g.,* Guangxi, China PR). A group of five IPM farmers (four women and one man) of Phat Loc Tien hamlet, Thai Giang Commune, Thai Thuy District, Thai Binh province carried out a field study on minimum tillage on potato production during the winter – spring season (2008-9). In this experiment the 'minimum tillage' (MT) (see box 4) was compared with conventional farmer's practices (FP) in potato production. Farmers gathered data on indicators like plant height, number of plant/hill; pests, natural enemies, yields and conducted an economic analysis as to determine differences among treatments.

Box 4. Planting method of minimum tillage potato in Vietnam: Planting holes are dig along the specified spacing (1,400 hill/Sao (360 m^2). These holes are filled with a basal dose of well-decomposed manure. The remaining depth is filled with soil cover. Then, all total amount of phosphate is mixed with one third of urea and applied around the hills. After completion of basic fertilization, one potato tuber seed are placed over each hole, and then covered with a shallow soil layer on the top. 7-10 cm thick rice stubbles follow that with some water application to facilitate sprouting. When the sprouts are about 7-10 cm height (rice straw starts decomposing), additional application of straw are carried out to cover the fertilizers and tubers from direct sunlight

Potato Growth, Insect-pests and Diseases

Potatoes grew faster in MT treatment compared to the FP and similarly more number of plants/m^2 developed in MT treatments. Since plant density is directly related to the size of the tubers, farmers had a choice either to produce seed by maintaining higher plant density or after thinning produce market size potato. The army worm (*Spodoptera* spp), which can be a major potato pest, appeared early in low population density in MT treatment but no serious losses were observed. Late blight (*Phytophthora infestans*), first observed at 30 DAS, spread rapidly and severity increased as the growth season progressed. The cumulated infection % stood at 59 and 63 in FP and MT treatments respectively at harvest time, but did not results into any significant overall yield loss. Similarly, no significant differences in natural enemy populations were found in these two treatments (*see* Table 4).

Yield and Economics

A total yield of 760 Kg/Sao (360m^2) was obtained in MT against 702 Kg in FP treatment (Table 4).

Table 4. Yield components

Indicator	Minimum tillage (MT)	Control Practice (FP)
# hill /360m2	1,400	1,400
# tubers /hill	13.5	12.3
tuber weight/hill (kg)	0.62	0.58
Actual yield/Sao (360 m2)	760	702

The economic comparison provided the most important results, where the MT treatment resulted in much higher net return compared to the farmer's practice (Table 5). Clearly, these profit gains are due to reduction in labor cost component of the economic analysis.

Table 5. Economic analysis

Indicator	Price per (kg)	Minimum tillage		Farmer Practice (FP)	
		Amount (in kg)	Cost /value (in VN Dong)	Amount (in kg)	Cost/value (in VN Dong)
Seed tuber (Dong10/kg)	7,000	40	280,000	40	280,000
Urea	7,000	7	49,000	7	49,000
Phosphate	3,000	10	30,000	10	30,000
Potassium	15,000	8	120,000	8	120,000
Manure	200	700	140,000	700	140,000
Soil preparation labor cost	30,000	1	30,000	3	90,000
Planting labor cost	30,000	3	90,000	3	90,000
Field maintenance labor cost	30,000	7	210,000	9	270,000
Harvesting labor cost	30,000	2	60,000	3	90,000
Total input costs			1,009,000		1,159,000
Total gross income	2,600	760	1,976,000	702	1,825,200
Economic benefit			967,000		666,200

This IPM farmer and trainer-led adaptation of the potato planting and production method is an excellent innovation for the given bio-physical and socio-economic conditions of agricultural production systems in the Red River Delta in northern Vietnam. This method of potato planting significantly increases the net returns from potato farming in an increasingly labor-scarce socio-economic production environment in North Vietnam without loosing productivity.

(3) Tomato IPM in Processing Tomato Production in the Philippines

Bemisia tabaci is currently the most serious pest of tomatoes (Kumar *et al.,* 2005; Kumar and Poehling, 2009), causing direct damage as well as transmitting tomato leaf curl virus (ToLCV) in processing tomato production in Ilocos Norte & Sur, Northern Philippines. ToLCV incidence is positively correlated both with the numbers of *B. tabaci* present and with yield loss

(Saikia & Muniyappa, 1989). When populations of *B. tabaci* are high early in the season, 90-100% of plants can become infected, resulting in a yield loss of 40-100% (Saikia & Muniyappa, 1989). With associated risks both to their health and the environment, farmers currently use large quantities of costly broad-spectrum insecticides in increasingly unsuccessful attempts to control *B. tabaci* and ToLCV.

ACTION RESEARCH AND IPPM TRAINING

Upon request of the semi-private sector Northern Food Corporation (NFC), who contracts tomato growers producing for its processing plant, and the Philippine National IPM Programme/Department of Agriculture, FAO provided technical assistance for better management of ToLCV in processing tomato production.

A series of IPPM training and action research activities were initiated with the aim to develop capacity of the trainers and farmers in understanding life-cycle, biology of the whitefly and identification of vector-virus like secondary hosts using rapid diagnostic kits. Whereas adherence to a mandatory three month host-free period, as for example practiced in processing tomato production in Mexico, appears difficult to promote and enforce among smallholder farmers in the Philippines, alternative management approaches were introduced. These action research and training activities led to the development of reliable and safe management options for this vector and virus. Based on technical advice from US-based Campbell Soup –and its Seed Research/IPM unit based in Davis, California in particular-, a two-tier strategy of vector and virus management was developed; first, to minimize the virus acquisition at seedling stage (see box 4) and second, to reduce vector immigration to the open tomato field crops by using reflective mulches and mineral oils.

Farmers quickly learn the importance of raising a healthy tomato crop using tunnel-grown and virus-free seedlings (see Box 5). Healthy seedlings for transplanted vegetable crops are crucial for raising a healthy crop, which is part of the first principle of IPM, "Growing a Healthy Crop". In absence of alternative practical virus management options for field-grown tomato crops, tunnel-raised seedlings delay the infection and allow for higher yield and net returns. UV-blocking plastic mulch and mineral oils are currently being tested as part of ongoing action research efforts for reduced immigration and development of vector populations, The introduced vector management strategies have allowed the economically-important processing tomato industry in northern Philippines to survive and flourish.

MAJOR LESSONS LEARNED FROM FAO SUPPORTED IPM WORK IN THE ASIA REGION

The FAO Programme, through its training and action research work over the last decade (1996-2008), has learned important lessons that provide direction for future work in the field of vegetable IPM training. Some of the most important lessons are summarized below.

Box 5: Minimizing virus acquisition in nursery (by introducing low-cost covered nursery)

Most farmers raise tomato seedlings in open nurseries resulting into early virus acquisition by tomato seedlings. Since ToLCV symptoms development takes 3-4 weeks after vector-transmission, farmers can not differentiate between virus infected and healthy seedlings. And, once such seedlings are transplanted, within a few weeks, full symptoms develop, which can result into 100% crop losses. To raise healthy seedlings, which is the first principle of IPM, low-cost nursery experiments were set-up, where closed nursery raised seedlings were compared with the open-air grown conventional seedlings. Pot-studies were set up with 3 replications, where each replication had 9 pots planted with seedlings from protected nursery and open-air. The farmers and trainers observed the symptoms development 4 weeks after transplanting using the standard symptoms guides. The potted tomato plants grown using protected nursery seedling had 22% virus at the end of the 4 weeks observation period, whereas this rate was at 44% (almost double; see Fig. 2) for open-air grown seedlings.

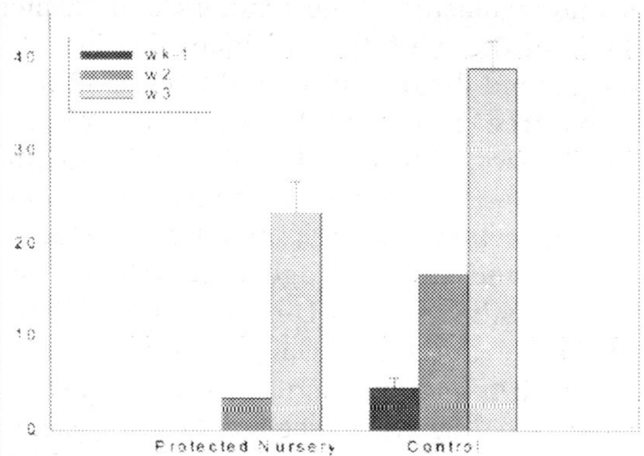

Fig. 2. Weekly cumulative Tomato Yellow Leaf Curl Infected tomato plants expressing disease symptoms in pot studies, where seedlings are either grown inside protected Nursery or in Open Field (control) (F = 11.35; df = 1.5; P = 0.0281)

Clearly, this work helped to disseminate the crucial message among farmers and trainers on the benefits of using protected nursery for raising seedling as to delay virus infections.

The Need for Vegetable Farmers as IPM Experts

A most crucial lesson learned from FAO work over the last two decades and re-enforced by experiences of the FAO Vegetable IPM Programme, relates to the need for vegetable farmers as IPM Experts. As is relevant for rice (Gallagher *et al.*, 2005), the ecological heterogeneity of local vegetable crop production systems is enormous and no standard extension message can adeptly address local production issues and constraints. Farmers need to thoroughly understand the local ecology in their crops and make informed decisions on crop management. There are numerous well documented case studies that confirm the notion that, rather than 'outsiders', farmers themselves need to become IPM experts. This important notion is reinforced, elaborated on and practically illustrated with case studies in this paper.

Farmer Field School of Crucial Importance for IPM Education of Vegetable Growers

Training results have shown that vegetable farmers appreciate the non-formal education- based training methodology applied in Farmers Field Schools. Learning is a consequence of experience and FFSs provide the necessary practical experience and enabling group learning environment to understand and practice vegetable IPM. 'Living' problems faced in local vegetable production provide momentum for a self-discovery learning process of problem identification, analysis and trying out potential solutions to resolve problems.

Vegetable Integrated Pest Management: The Need for 'Informed Intervention'

Most vegetables produced in Asia for local consumption and for market production are exotic in nature. Many popular cash crops such as tomatoes, crucifers, potatoes were introduced in Asia only a couple of hundred years back. Similarly, many important pest and disease problems are exotic, such as the Diamond Back Moth, which was introduced from Europe into Asia without its naturally existing complex of natural enemies. Crop protection problems in vegetable production are often severe and quite acute and, if not swiftly and adequately managed, can lead to total crop failure. Vegetable ecosystems are much less stable compared to, for example, rice (*Oryza sativa*) which is indigenous to Asia and for which ecosystems have evolved into extremely stable systems over a period of thousands of years. The IPM management strategy to be employed in Asian vegetable production, particularly with relevance to crop protection, should be based on 'informed (and pro-active or preventive) intervention' rather than 'informed non-intervention' which is the management strategy employed in tropical irrigated rice (Whitten & Ketelaar, 2002).

Lack of Knowledge on Vegetable Ecology and Good Agricultural Practices

In an increasingly globalized market, farmers will have to obtain GAP-certification as to ensure market access. This will require additional training efforts as to reach if not all a majority of smallholder farmers Asia-wide. In order to further reduce the use of chemical pesticides in conventional vegetable production, there is a need to better understand the ecology of vegetable production systems. The FAO Programme continues to support action research programmes that aim to investigate vegetable ecosystems in order to determine the roots of serious and often acute occurrence of pest and disease problems and to develop an ecosystem approach to prevent and solve problems. Active and functional links with the scientific community and the private sector are essential as to ensure that vegetable farmers have access to the most updated information, technologies and production inputs as to be able to compete in a globalized economy.

Need for Innovative Curriculum Development

The job of vegetable IPM trainers is quite demanding and requires the knowledge, experience and skills to be creative and innovative when it comes to curriculum development for IPM training. The vegetable IPM trainer must develop the skills and creativity needed to adapt the IPM training content and methodology to a wide range of circumstances and be able to respond pragmatically to complex problems that may emerge during training implementation. Curriculum development for a variety of different vegetable crops needs more focused attention, including the need for the inclusion of Good Agricultural Practice topics as crop nutrition needs, soil management, bio-control, post-harvest storage and processing, and marketing. Challenges to agricultural production and protection posed by climate change (*e.g.*, flood, drought, increased pest/disease pressure) will also require creative curriculum development, taxing the innovativeness of trainers to ensure adaptation of rural communities in the years and decades to come.

Need for Farmer Organization among Vegetable Producers

Vegetable farmers will be better able to protect, expand and sustain IPM-based production if they can organize themselves into groups and associations. They will then be better capable of taking a role in the facilitation of continued group learning and training, the production of agricultural inputs such as compost and bio-control agents and to establish and implement effective 'bargaining power' to achieve better prices for their vegetable produce. Increasingly, smallholder farmers are signed up as contract farmers in large-scale contract farming business schemes. This more recent development poses both opportunities and challenges with regards to up-scaling of IPM & GAP practices among smallholder vegetable farmers.

Need for Pesticide Policy Reform and Regulatory Action

IPM can only take root among smallholder farmers if pro-actively promoted and endorsed by government policy that favors IPM development and regulates pesticides available on the local market. The distribution and use of all WHO Class I pesticides should be terminated. China PR is taking the lead in Asia with regards to better pesticide regulatory action, as exemplified by the banning of production and use of 5 highly hazardous organophosphates and strengthening of pesticide labeling requirements in recent years. That said, much work remains to be done, particularly also in the Greater Mekong Subregion, where capacity for such regulatory action and enforcement remains weak.

CONCLUSION

The problems associated with misuse and overuse of pesticides in vegetable production in South and Southeast Asia remain serious and acute. With global warming now an established reality, it is likely that farmers will face a range of new pest and diseases. Without appropriate training in Integrated Production and Pest Management, farmers will continue to try to control these pests through higher inputs of agro-chemicals without –or too late- realizing the negative consequences. The key to on-farmer sustainable establishment of vegetable IPM, as argued above, is to facilitate farmers becoming vegetable IPM experts themselves.

The FAO Inter-Country Programme for Vegetable IPM in South and Southeast Asia is currently developing a third Phase (2010-15). With a particular focus on pesticide risk reduction and climate change risk mitigation and adaptation, the programme will work to enhance the capability and capacity of Governments, NGOs and farmer groups to effectively reach out to farmers Asia-wide. The FAO Programme will continue to support applied research, extension and farmer education activities to promote and support the development and implementation of Good Agricultural Practices–and use of Integrated Production and Pest Management as key component- among Asian smallholder farmers. These activities will be carried out in close collaboration with other regional, national and local IPM-related projects funded by Governments, donor agencies and NGOs. Close functional links with appropriate and action research-oriented research institutions will be pursued as to institute sustainable arrangements for the solution of on-farm technical problems. The FAO programme will also continue to engage with contract farming schemes and actively support the creation and strengthening of smallholder farmer groups and their associations so that they can take collective action in support ecologically-based vegetable production and marketing.

No doubt the job is daunting. Newly emerging pests and diseases, some of these brought about by climate change, will pose new crop production challenges. Yet, arguably the best-and as of now fairly untapped and under-developed- resource to make vegetable IPM work and prepare communities for climate change adaptation, is that of the human resource. The key to development and sustainable establishment of vegetable IPM is the ability of farmers to understand local ecosystems and implications of their management interventions on such ecosystems. Reduction of chemical pesticides in vegetable production will become a reality, only if farmers become 'ecosystem literate' and feel confident that they can manage vegetable production with lesser inputs of chemical pesticides. Developing farmers into Vegetable IPM Experts will remain a crucially important task if vegetable IPM is to establish itself in a sustainable manner and flourish Asia-wide.

ACKNOWLEDGEMENTS

The authors are grateful to all vegetable IPM farmers, trainers, researchers, colleagues within the FAO IPM programmes and their partner institutions, and all others, who have positively contributed to the practical development and application of vegetable IPM and from whose hard work this paper could be conceptualized.

REFERENCES

CEDAC., 2000. Pesticide pollution in the Tonle Sap catchment, Project Progress Report (Sept. 1999- Aug. 2000), CEDAC, Phnom Penh, Cambodia.

DoAE/FAO., 2005. Survey of Diamond Back Moth Parasitoids on Highland Crucifers in Northern and Northeastern Thailand, Report prepared under Letter of Agreement with FAO-RAP, unpublished, 26 pp.

DoA/RPF., 2006. Introduction of *Diadegma semiclausum* for the control of Diamond Back Moth (*Plutella xylostella*) in Thailand, Report prepared under Letter of Agreement with FAO-RAP, Unpublished, 23 pp.

EJF., 2002. *Death in Small Doses: Cambodia's Pesticides Problems and Solutions*, Environmental Justice Foundation, London, UK. 37 pp.

FAO, 2002. From Farmer Field School to Community IPM: Ten years of IPM training in Asia, FAO-RAP, Bangkok. 106 pp.

Gallagher, K., Ooi, P.A.C. Mew, T. Borromeo, E. Kenmore, P. and Ketelaar, J.W., 2005. Ecological basis for Low-toxicity Integrated Pest Management in Rice and Vegetables. In: Pretty, J. (Ed.), 2005, *The Pesticide Detox: Towards a more Sustainable Agriculture*. Earthscan: London, UK. 294 pp.

Grzywacz, D., Rossbach, A. Rauf, A. Russell, D.A. Srinivasan, R. and Shelton, A.M., 2009. Current control methods of Diamond Back Moth and other brassica insect pests and the prospects for improved management with lepidopteran-resistant Bt vegetable brassicas in Asia and Africa, Crop Protection (2009), doi: 10.1016/j.crppro.2009.08.009

Kiritani, Keizi., 2006. Predicting impacts of global warming on population dynamics and arthropod populations in Japan, *Population Ecology,* **48**: 5-12

Kogan, M., 1998. Integrated pest management: Historical perspectives and contemporary developments, *Annual Review of Entomology* **43**, 243-270.

Kumar, P., Poehling, H. M. and Borgemeister, C., 2005. Effects of different application methods of neem against sweetpotato whitefly *Bemisia tabaci* Gennadius (Homoptera: Aleyrodidae) on Tomato plants. *J. Appl. Entomol.*, **129**:489-497.

Kumar P. and Poehling, H. M., 2009. Novel strategies for management of tropical insect-pests of tomato. *In*: *Tomatoes: Agricultural Procedures, Pathogen Interactions and Health Effects*. Nova Science Publishers Inc. USA *(In Press)*.

Lim, G.S., 2008. Mission Report (11-15 February, 2008), FAO Regional Vegetable IPM Programme in Asia (GCP/RAS/209/NOR), FAO-RAP, Bangkok, Unpublished

Lim G. S. and Di Y. B. (eds), 1989. Status and management of major vegetable pests in the Asia-Pacific Region (With special focus towards Integrated Pest Management), RAPA, FAO, Bangkok, Thailand.

Mancini, F., Jiggins, J. and O'Malley, M. 2009. Reducing the incidence of pesticide poisoning by educating farmers on integrated pest management in South India, *Int. J. Occupational. Environ. Health,* **15**,:143-151

Murphy, H. H., Sanusi, A. Dilts, R. Djajadisastra, M. Hirshhorn, N. and Yuliatingsih, S., 1999. Health effects of pesticide use among Indonesian women farmers: Part 1. Exposure and acute health effects. *J. Agromedicine*, **6**: 61–85.

Nga, L.T. and Prabhat Kumar, 2008. Contributions of parasitoids and *Bacillus thuringiensis* to the management of Diamondback Moth in highland crucifer production in Da Lat, Vietnam, *J. Asia Pacific Entomol.,* **11**: 59–64.

Perkins, J.H., 1982. Insects, Experts, and the Insecticide Crisis. The quest for new pest management strategies, Plenum Press, New York. 304 pp.

Saikia AK and Muniyappa, V., 1989. Epidemiology and control of tomato leaf curl virus in southern India, *Tropical Agric.*, **66**(4), 350-354.

Settle, W. H. and Whitten, M. J., 2000. Plenary Lecture: "The Role of Small Scale Farmers in Strengthening the Link between Sustainable Agriculture and Biodiversity". **In**: *The XXIst Congress of Entomology*, Iguassu, Brazil (August 2000).

Shepard, B. M., Hammig, M.D. Carner, G.R. Ooi, P.A.C. Smith, J.P. Dilts, R. and Rauf, A., 2009. Implementing integrated pest management in developing and developed countries, **In**: *Integrated pest management: Dissemination and impact;* R. Reshn and A. K. Dhawan (Eds.), pp, 275-305, Springer Netherlands.

Smith, R.F. and Reynolds, H.T. 1966. Principles, definitions and scope of integrated pest control. **In**: *Proceedings of the FAO Symposium on Integrated Pest Control,* FAO, Rome. pp. 11-17.

Srinivasan, R., 2008. Integrated pest management for eggplant fruit and shoot borer (*Leucinodes orbonalis*) in south and Southeast Asia: past, present and future. *J. Biopesticides*, **1**(2):105 – 112.

SUSVEG-Asia, 2007. SUSVEG-Asia Brinjal integrated pest management (IPM). http://susveg-asia.nri.org/susvegasiabrinjalipm4.html (Accessed on 30 Oct.2009)

Thomson, L. J., Macfadyen, S. and Hoffmann, A. A., 2009. Predicting the effects of climate change on natural enemies of agricultural pests, Biological Control, doi:10.1016/j.biocontrol.2009.01.022

Upanisakorn, A., 2009. Final Report (LOA/RAP/2008/23); Crucifer IPM for the Northern and Northeastern Highlands of Thailand Under project GCP/RAS/ 209/ NOR., Department of Agricultural Extension, Pest Management Division, Bangkok, Thailand

Van der Wulp, H., Louanglath, K. and Taipangnavong, T., 2008. Back to Office Report on Mission to Northern Lao PDR, Unpublished. 19 pp.

Whitten, M.J. and Ketelaar, J.W., 2002. Farmer Field Schools: From Crop Protection to Crop Husbandry, *In:* K. Eveleens, J. Jiggins & G.S. Lim (eds.*), Farmers, FAO and Field Schools: Bringing IPM to the Grass Roots in Asia*, FAO, Rome: *(In publication).*

17. INTEGRATED BIO-INTENSIVE PEST MANAGEMENT IN HORTICULTURAL CROPS

P. Parvatha Reddy

Former Director, Indian Institute of Horticultural Research, Bangalore-560 089

ABSTRACT

Integrated Pest Management (IPM) is an important principle on which sustainable crop protection can be based. IPM allows farmers to manage pests in a cost effective, environmentally sound and socially acceptable way. According to FAO, IPM is defined as "A pest management system that in the context of the associated environment and the population dynamics of the pest species, utilizes all suitable techniques and methods, in a compatible manner as possible and maintains the pest populations at levels below those causing economic injury".

Biointensive Integrated Pest Management (BIPM) is a systems approach to pest management based on an understanding of pest ecology. It begins with steps to accurately diagnose the nature and source of pest problems, and then relies on a range of preventive tactics and biological controls to keep pest populations within acceptable limits. Reduced-risk pesticides are used if other tactics have not been adequately effective, as a last resort, and with care to minimize risks. An important difference between conventional and biointensive IPM is that the emphasis of the latter is on proactive measures to redesign the horticultural ecosystem to the disadvantage of a pest and to the advantage of its parasite and predator complex. While developing the BIPM technologies, the following aspects are to be considered:

- Economically viable. The input use efficiency should be given utmost importance.
- Bring down the pest population below economic injury level.
- Environmentally safe as enough damage has already been done to the environment by indiscriminate use of chemical pesticides.
- Conserve the biodiversity and natural enemies that are already present in the crop ecosystem. Hence, crop-mixes and variety-mixes should be adopted in the cropping system
- Safe to human beings, animals, birds, honey bees, fishes and beneficial soil organisms including earthworms.

- Not come in the way of export, since there is a growing demand for organic food in western countries. There should not be any pesticide residues in the farm produce.

The BIPM case studies on pest management in cabbage and cauliflower using mustard as a trap crop and spray of NSKE; tomato fruit borer management using African marigold as a trap crop and spray of *Ha*NPV; foot rot and nematode management in black pepper using soil solarization, organic amendments, AMF and *Trichoderma harzianum*; capsule rot management in cardamom by integrating *T. harzianum* with spray of potassium phosphonate; and root knot nematode management in tomato and brinjal using neem cake and *Paecilomyces lilacinus*/*Pochonia chlamydosporia*/*T. harzianum* have been discussed.

INTRODUCTION

Through 'Green Revolution' in late 1960's, India achieved self sufficiency in food production, which was hailed as a breakthrough on the farm front by international agricultural experts. But still the country has not achieved self sufficiency in production of horticultural crops. Most of the growth in food production during the green revolution period may be attributed to the use of improved crop varieties and higher levels of inputs of fertilizers and pesticides. The modern agricultural techniques such as use of synthetic fertilizers and pesticides are continuing to destroy stable traditional ecosystems and the use of high yielding varieties of crops has resulted in the elimination of thousands of traditional varieties with the concurrent loss of genetic resources. The introduction of high yielding varieties changed the agricultural environment leading to numerous pest problems of economic importance. In the process of intensive farming, the environment has been treated in an unfriendly manner.

Prof. Swaminathan emphasized the need for 'Ever Green Revolution' keeping in view the increase in population. The increase in population and diminishing per capita availability of land demands rise in productivity per unit area. In India, annual crop losses due to pests, diseases and weeds have been estimated to be about Rs. 60,000 crores in 2005. Increasing yields from existing land requires effective crop protection to prevent losses before and after harvest. The challenge before the plant protection scientist is to do this without harming the environment and resource base.

INTEGRATED PEST MANAGEMENT (IPM)

Integrated pest management is an important principle on which sustainable crop protection can be based. IPM allows farmers to manage pests in a cost effective, environmentally sound and socially acceptable way. According to FAO, IPM is defined as "A pest management system that in the context of the associated environment and the population dynamics of the pest species, utilizes all suitable techniques and methods, in a compatible manner as possible and maintains the pest populations at levels below those causing economic injury".

Biointensive Integrated Pest Management (BIPM)

Biointensive integrated pest management is a systems approach to pest management based on an understanding of pest ecology. It begins with steps to accurately diagnose the nature and source of pest problems, and then relies on a range of preventive tactics and biological controls to keep pest populations within acceptable limits. Reduced-risk pesticides are used if other tactics have not been adequately effective, as a last resort, and with care to minimize risks.

An important difference between conventional and biointensive IPM is that the emphasis of the latter is on proactive measures to redesign the agricultural ecosystem to the disadvantage of a nematode and to the advantage of its parasite and predator complex.

BINM options may be considered as proactive or reactive.

Proactive options: Proactive options, such as crop rotations and creation of habitat for beneficial organisms, permanently lower the carrying capacity of the farm for the pest. The carrying capacity is determined by the factors like food, shelter, natural enemy complex and weather, which affect the reproduction and survival of a pest species. Cultural control practices are generally considered to be proactive strategies. Proactive practices include crop rotation, resistant crop cultivars including transgenic plants, disease-free seed and plants, crop sanitation, spacing of plants, altering planting dates, mulches, etc.

The proactive strategies (cultural controls) include:

- Healthy, biologically active soils (increasing below-ground diversity).
- Habitat for beneficial organisms (increasing above-ground diversity).
- Appropriate plant cultivars.

Reactive options: The reactive options mean that the grower responds to a situation, such as an economically damaging population of pests, with some type of short-term suppressive action. Reactive methods generally include inundative releases of biological control agents, mechanical and physical controls, botanical pesticides and chemical controls.

While developing the BIPM technologies, the following aspects have to be kept in mind:

- The technologies developed should be economically viable. About 70% of our people depend on agriculture for their livelihood and more than 80% of them are small and marginal farmers. Hence, there is a need to develop low input crop protection technologies. The inputs should be efficiently used. The input use efficiency should be given utmost importance.
- The technologies developed should bring down the pest population below economic injury level.
- The technologies developed should be environmentally safe as enough damage has already been done to the environment by indiscriminate use of chemical pesticides. It is heartening to note

that the pesticide consumption in India has come down from 75,033 tonnes in 1990-91 to 47,020 tonnes during 2001-02.

- The technologies developed should conserve the natural enemies that are already present in the crop ecosystem. Hence, only safer and eco-friendly pesticides should be recommended. In this context, Avermectins, Strobilurin fungicides and pheromones are safe to natural enemies.
- The technologies developed should conserve the biodiversity. Hence, crop-mixes and variety-mixes should be adopted in the cropping system. In this context, use of trap crops like Indian mustard and African marigold in developing IPM technologies in cabbage and tomato, respectively are relevant.
- The technologies developed should not pose risk to human beings, animals, birds, honey bees, fishes and beneficial soil organisms including earthworms.
- The technologies developed should not come in the way of export, since there is a growing demand for organic food in western countries. There should not be any pesticide residues in the farm produce.

Some of the recent advances in plant protection which can lead to enhanced crop productivity are summarized below:

MANIPULATION OF AGRO-ECOSYSTEM

The manipulation of agro-ecosystem should be to make it less favourable to the pests and more favourable to their natural enemies with a view to prohibit, reduce and delay pest population increase. Pest avoidance should receive greater attention in IPM.

Biodiversity

Agricultural production is more sustainable in areas where proper crop-mixes and variety-mixes are adopted in a given crop season. Monocultures and overlapping crop seasons are more prone to severe outbreaks of pests and diseases. Diversity of natural enemy complexes attacking various stages of pests prevalent in poly-crop and inter-cropping systems also tend to prevent severe pest outbreaks. A poly-crop ecosystem of pulses, oil seeds, vegetables and cotton kept the populations of the polyphagous pest, the gram pod borer, *Helicoverpa armigera* and the leaf hopper, *Amrasca devastans* well below ETL for considerable time during the crop season compared to the pure crop ecosystem.

Pest-resistant Cultivars

Perhaps the greatest single technological achievement is the advancement in breeding crops for resistance to pests. Cultivation of resistant varieties is the cheapest and best method of controlling pests. One of the important components of IPM is the use of resistant cultivars to key pests. Under All India Coordinated Research Projects of Indian Council of Agricultural Research, New Delhi, large number of highly/moderately resistant varieties are released to the farmers.

Transgenic crop varieties in horticultural crops (tomato, potato, brinjal, beans, cabbage, cauliflower, musk melon, banana, coffee) have been developed by cloning *Bt* endotoxin genes which are cultivated in large areas. In 2006, India was the 5th largest GM crops growing country (3.8 million ha) in the world only next to USA (54.6 million ha), Argentina (18 million ha), Brazil and Canada.

Semiochemicals and Behavioural Pest Control

Semiochemicals are more selective and biologically active ingredients which are generally target specific and used at very low application rates. Pheromones are a class of semiochemicals, which are used commercially for indirect (monitoring, quarantine and time of pest control) and direct (mass trapping) pest control.

Botanical Pest Control

Neem products such as cake, oil, Neem Seed Kernel Extract (NSKE), Neem Seed Powder Extract (NSPE), Pulverized NSPE and soaps are being used extensively to manage horticultural crop pests (Beanfly, *Ophiomyia phaseoli*; serpentine leaf miner, *Liriomyza trifolii* on several crops; cucurbit fruitfly, *Bactrocera cucurbitae*; tomato fruit borer, *Helicoverpa armigera*; brinjal fruit and shoot borer, *Leucinodes orbonalis*; water melon and chilli thrips, *Thrips* spp.; chilli yellow mite, *Polyphagotarsonemus latus* and okra leaf hopper, *Amrasca biguttula biguttula*) (Krishna Murthy and Krishna Kumar, 2002).

Biological Control

Of late, biological suppression of pests has become an intensive area of research because of environmental concerns. About 60% of the natural control of insect pests are by the natural enemies of pests such as parasitoids, predators and pathogens. The Australian lady bird beetle, *Cryptolaemus motrouzieri* has been found very effective against mealy bugs infesting grapes, guava, citrus, mango, pomegranate, ber and custard apple. The encyrtid parasite, *Leptomastix dactylopi* is effective against mealybug, *Planococcus citri* on guava, citrus, pomegranate, ber and custard apple (Mani, 2001). *Bacillus thuringiensis* (*Bt*) is effective against tomato fruit borer, okra fruit borer and diamondback moth on cabbage and cauliflower.

Several methods of enrichment and conservation of natural enemies include providing nesting boxes for wasps and predatory birds; retaining pollen and nectar bearing flowering plants like Euphorbia, Wild Clover on bunds to provide supplementary food for natural enemies; placing bundles of paddy straw in fields for attracting predatory spiders. In addition, erecting perching sites, water pans, retaining bushes (Acalypha, Hibiscus, Crotons) helps in retention of predatory birds.

The last decade has witnessed a tremendous break through in biological control of diseases and nematodes like *Rhizoctonia, Pythium, Fusarium, Macrophomina, Ralstonia* and *Meloidogyne* in banana, tomato, egg plant, pea, grapes, cucumber, black pepper, cardamom, ginger and turmeric, especially by using species of *Trichoderma, Pochonia, Pseudomonas* and *Bacillus.*

PLANT GROWTH PROMOTING RHIZOBACTERIA (PGPR)

PGPR are root colonizing bacteria with beneficial effects including plant growth promotion and induction of systemic resistance in plants against wilt and damping-off diseases. *Pseudomonas fluorescens, P. putida, Bacillus subtilis* and *B. pumilis* as seed treatment and soil application have been found effective against anthracnose, *Fusarium* and bacterial wilt, CMV and angular leaf spot on cucumber; halo blight of beans; damping off, CMV and root-knot nematode on tomato and *Fusarium* Wilt, *Alternaria* and *Pseudomonas* leaf spots on radish.

Foliar Spray Theraphy

Spray application of K_2HPO_4 or KH_2PO_4 at 3.5 g/litre of water has been reported to control powdery mildew in rose and carnation. Similarly, the above treatment was also found effective for the management of powdery mildew on mango, grapes and cucurbits.

Non Pathogenic Strains (NPS)

A novel strategy of using non-pathogenic strains of fungi is gaining importance in the management of plant diseases.

NPS of *Fusarium* (Fusaclean Fo 47) is being used widely to control *Fusarium* wilt in tomato, melons and carnation over thousands of hectares in Europe. In Japan, a NPS of *Fusarium* is successfully used for the management of *Fusarium* wilt in sweet potato. In Italy, NPS of *Fusarium* (251/2) is being successfully used for the management of *Fusarium* wilt in carnation and other horticultural crops.

NPS are known to compete with pathogens for nutrients in the rhizosphere, compete for infection sites on the rhizoplane and induce systemically acquired resistance in the host.

Avermectins

Avermectins are the secondary metabolites of an Actinomycete, *Streptomyces avermitilis,* which are highly effective against mite pests and plant parasitic nematodes at a very low concentration. Scientists at the Indian Institute of Horticultural Research, Bangalore, for the first time in India, have isolated 6 strains of *S. avermitilis* and showed their effectiveness for the management of root-knot nematodes infecting tomato, egg plant, chilli, carnation and gerbera (Parvatha Reddy and Nagesh, 2002).

Strobilurin Fungicides

Strobilurin fungicides are also called Qo inhibitors as they act on cytochrome Qo of the fungi. The Basidiomycetous fungus, *Strobilurus tenacellus* produces antibiotics to ward off competition from other fungi. Based on this principle, several fungicides have been developed namely Azoxystrobin, Kresoxy methyl, Metominostrobin, Trifloxystrobin, Picoxystrobin, Pyraclostrobin, Famoxadone and Fenomidone during 1996-2001. Within 4 years, sale of these fungicides totaled $620 million, accounting for 10% of total fungicide market in the world. This success is unparalleled in the history of fungicides.

Strobilurin fungicides are naturally occurring compounds and hence eco-friendly, highly systemic, have unique mode of multisite action, hence development of resistance is remote. They have broad spectrum of activity on all groups of fungi and registered in 72 countries on 84 crops representing over 400 crop/ disease systems.

CASE STUDIES OF BIOINTENSIVE INTEGRATED PEST MANAGEMENT

Use of African marigold (*Tagetes erecta*) as a trap crop for the management of fruit borer on tomato involves planting one row of 45 day-old marigold seedlings after every 16 rows of 25 day-old tomato seedlings and spraying of *Ha* NPV at 250 LE/ha or 4% NSKE or 4% pulverized NSPE, 28 and 45 DAP coinciding with peak flowering (Srinivasan *et al.*, 1994).

IPM using Indian mustard as a trap crop involves planting of paired rows of mustard after every 25 rows of cabbage/cauliflower and spraying of 4% NSKE at primordial formation. Two more sprays of 4% NSKE may be given at 10-15 day interval after the first spray. The IPM gave 152% more returns than pure cabbage crop. IPM controls diamondback moth (*Plutella xylostella*); leaf webber (*Crocidolomia binotalis*); stem borer (*Hellula undalis*); aphids (*Brevicorne brassicae, Hyadaphis erysimi*) and bug (*Bagrada cruciferarum*) (Srinivasan and Krishna Moorthy, 1991).

Integrated management of foot rot (*Phytophthora capsicii*) and nematodes (*Meloidogyne incognita* and *Radopholus similis*) on black pepper was achieved by

i. mixing VAM and *Trichoderma harzianum* in solarised nursery mixture to raise healthy and robust seedlings,
ii. application of *T. harzianum* and FYM in planting pit,
iii. field application of neem cake at 1 kg/vine mixed with 50 g of *T. harzianum* during August (Sarma, 2003).

Management of capsule rot (*Phytophthora spp.*) of cardamom was achieved by two applications of *T. harzianum* at 1 kg/plant (grown on decomposed coffee pulp and FYM in 1:1 ratio) during May and July integrated with foliar spray of Akomin (potassium phosphonate) (Anandraj and Eapen, 2003).

The root-knot nematodes (*Meloidogyne incognita*) on tomato and brinjal were effectively managed by integration of neem cake with *Paecilomyces lilacinus*/*T. harzianum*/*Pochonia chlamydosporia*. The above treatment also increased the fruit yield, root colonization, parasitization of eggs and females with the bioagents (Parvatha Reddy and Nagesh, 1998).

TRANSFER OF TECHNOLOGY

The constraint for wider adoption of eco-friendly crop protection technologies is transfer of technology. Crop protection technologies developed have not reached the small and marginal farmers. Unless these technologies are assessed in farmer's fields and refined to suit local conditions, the fruits of research will not benefit farmers. Researchers and extension personnel should work hand in hand for successful transfer of crop protection technologies. Communication media such as radio, TV, audio and video cassettes, agri portals, farmer's field schools and KVK's should be used for effective transfer of technologies.

CONCLUSIONS

Globalization driven by WTO is opening up fantastic opportunities to export horticultural products and processed food from India. It is a revolution, which is taking place, and our farmer's will miss this golden opportunity if they are not equipped with the right crop protection technologies to produce horticultural products of international standards without pesticide residues. The challenge before the plant protection scientists is to prevent crop losses due to pests before and after harvest without harming the environment. We have to develop low input and eco-friendly crop protection technologies so as to be very competitive in the international market.

Padma Vibhushan Prof. M.S. Swaminathan, an eminent agricultural scientist of international repute, has said that "The ever-green revolution will be triggered by farming systems that can help produce more from the available land, water and labour resources without either ecological or social harm." Let us rededicate ourselves to achieve the dream of "Ever-Green Revolution" of Prof. Swaminathan.

REFERENCES

Anandraj, M. and Eapen, S.J. 2003. Achievements in biological control of diseases of spice crops with antagonistic organisms at Indian Institute of Spices Research, Calicut. In: *Current Status of Biological Control of Plant Diseases using Antagonistic Organisms in India* (B. Ramanujam and R.J. Rabindra, eds.), pp. 189-215.

Krishna Moorthy, P.N. and Krishna Kumar, P.N. 2002. Advances in the use of botanicals for the IPM of major vegetable pests. *Proc. of the Intn. Conf. on Vegetables*, Bangalore. Dr. Prem Nath Agri. Sci. Foundn., Bangalore, pp. 262-272.

Mani, M. 2001. Biological control of fruit crop pests. In *Integrated Pest Management in Horticultural Ecosystems* (P. Parvatha Reddy, A. Verghese and N.K. Krishna Kumar, eds.), pp. 93-107. Capital Publishing Co., New Delhi.

Parvatha Reddy, P. and Nagesh, M. 1998. Integrated nematode management in horticultural crops: Eco-friendly approaches for the future. In: *Advances in IPM forHorticultural Crops* (P. Parvatha Reddy, N.K. Krishna Kumar and A. Verghese, eds.), pp. 275-300.

Parvatha Reddy, P. and Nagesh, M. 2002. *Avermectins: Isolation, Fermentation, Preliminary Characterisation and Screening for Nematicidal Activity*. Tech. Bull. – 17, Indian Inst. of Hort. Res. (IIHR), Bangalore, 28 pp.

Sarma, Y.R. 2003. Recent trends in the use of antagonistic organisms for the disease management in spice crops. In: *Current Status of Biological Control of Plant Diseases using Antagonistic Organisms in India* (B. Ramanujam and R.J. Rabindra, eds.), pp. 49-73.

Srinivasan, K. and Krishna Moorthy, P.N. 1991. Indian mustard as a trap crop for management of major lepidopterous pests on cabbage. *Top. Pest Mangmt.*, **37**: 26-32.

Srinivasan, K., Krishna Moorthy, P.N. and Raviprasad, T.N. 1994. African marigold as a trap crop for the management of the fruit borer *Helicoverpa armigera* on tomato. *Int. J. Pest Mangmt.*, **40**: 56-63.

18. PESTICIDES AND URBAN VEGETABLE PRODUCTION PATTERN IN SOUTH - WESTERN NIGERIA

O. Olubunmi, Fadina, and O. A. Toriola

Department of Crop Protection and Environmental Biology, University of Ibadan, Ibadan, Nigeria

ABSTRACT

African economies are heavily dependent on agriculture which even under the most environment friendly approach for pest control, the use of pesticides is deemed necessary. In Nigeria, urban agricultural production has become an important supplement to household income. The urban farmer irrigates his farm with contaminated water that carries industrial effluents, domestic wastes and pesticides from drifts. The research was carried out in LASU – OJO peri-urban vegetable growing area in Lagos State, Nigeria. The study carried out to find the major pest management practices adopted by farmers to protect their vegetable crops and the effects of the management practices on food safety.

Attempts were also made to identify viable alternative methods of pest control that can reduce the dependence of pesticide usage. Nine major insect pests were identified on the farms and high incidence of nematode induced galls was also observed. The study reported a high level of unsafe practices in the use of pesticides among the vegetable farmers while a viable alternative method was not found.

The findings of this study was used in proposing recommendations for the development of government policy to monitor the quality of agricultural produce in the country.

INTRODUCTION

The important roles of vegetables in promoting good health have well been documented and the urban and peri-urban market gardens are important sources of fresh vegetables for the teeming population in the urban areas (Williams, 1985).

Urban vegetable production also provides employment opportunities for city dwellers, helps in making use of some urban waste thereby contributing to the general cleanliness and purification of the city environment (UNDP, 1996).

Nigeria is capable of producing myriads of vegetables in enough quantities for domestic and export purposes but the country is yet to achieve the 5% total caloric intake of non-starchy vegetables as recommended by the FAO (Adedoyin, *et al.,* 1996).

The major problem facing the farmers is the lack of adequate storage facilities and the menace of pests and diseases. Intensive production of vegetables is therefore often accompanied with frequent spraying of pesticides to improve cosmetic values thereby increasing profits (Deang, 1991). However, sustainable agricultural production must not only allow farmers to earn adequate profit, but must also provide the consumers with wholesome safe food and minimize adverse impacts on the farmers and the environment.

The study was thus carried out to highlight some pesticide use patterns and make suggestions for improved crop protection practices in urban vegetable production.

MATERIAL AND METHODS

LASU – OJO peri-urban vegetable growing area was chosen for the research because it is one of the four zones of intensive vegetable production in Ojo Local Government area of Lagos State (Adedoyin *et al.,* 1996). Ojo Local Government is the most important source of fresh leafy vegetables for the teeming population in Lagos State.

Investigation was done to determine the major types of crops grown, agronomic practices and the type of crop protection practices used by the farmers.

Interviews were carried out with the use of a structured questionnaire schedule for data collection purposes. The open-ended and close ended questions were designed to determine:

i. Farmers' compliance with the recommended practices in the use of pesticides by the Agrochemical Association of Nigeria.
ii. The occurrence of symptoms of pesticide poisoning among the vegetable farmers.
iii. The levels of vegetable production and the channels of distribution of the vegetable to the markets
iv. Any other alternative to pesticide use
v. Signs of pesticide poisoning

Administration of questionnaires was carried out by trained enumerators and the results obtained were analysed using descriptive statistics and presented in Tables.

The area was sampled for pests on different types of vegetables. The foliar pests were caught and killed with a kliner jar, identification was done with the help of a field handbook.

Those that could not be identified on the field were identified at the insect museum of the Department of Crop Protection and Environmental Biology, University of Ibadan. The area was also sampled for the root knot nematode by taking count of the number of samples with galls.

RESULTS AND DISCUSSION

Different types of crops ranging from fruit trees to vegetables were the most prominent. Three types of leafy vegetables were the most prominent and these are *Amaranthus hybridus, Celosia argentea* and *Corchorus olitorus.*

Also, pesticides regularly used by farmers are lambdachyalothrin, cypermethrin and gammalin (Lindane-based WHO Class 1a pesticide). Results in Table 1 showed that 75% of the farmers spray their vegetable farms as many times as possible without following manufacturers instructions. In a previous report, Fadina and Adeoti (2000) highlighted the lack of technical advice on the entire farming systems in the country and this may further explain why gammalin formulated for use on tree crops is being sprayed on vegetable plants.

Table 1. Frequency distribution of number of spraying of vegetable beds

No. of sprayings	Absolute frequency	Relative frequency	Cumulative frequency
1 – 2	3	15	15
3 – 4	15	75	90
5 – 6	1	5	95
7 – 8	1	5	100
Total	20	100	

Source: Survey data 2007.

Of all the respondents, only 10% had received any formal training in pesticide application; 85% were thought by acquaintances or by observing others (Table 2).

Table 2. Frequency distribution of form of training in pesticide application

Type of training Received	Absolute frequency	Relative frequency	Cummulative frequency
Formal Training	2	10	10
Informal Training	17	85	95
No Training	1	5	100
Total	20	100	

Source: Survey data 2007.

Out of all the respondent farmers, only 25% of them use protective clothing; 10% use only nose mask and 15% only on overall (Table 3). Majority of the respondents carry out spraying operations simply clad in their normal clothes.

Table 3. Frequency distribution on the use of protective clothing

Type Used	Absolute Frequency	Relative Frequency	Cummulative Frequency
Nose mask	2	10	10
Overall	3	15	25
None	15	75	200
Total	20	100	

Source: Survey data 2007.

Based on the various routes of exposures, all the interviewed farmers often complain of one or more of the symptoms as indicated in Table 4; there is therefore an urgent need for a thorough medical examination to be carried out on all farmers.

Table 4. Frequency distribution of the occurrence of symptoms of poisoning among farmers.

Symptoms	Absolute frequency	Relative frequency	Cummulative frequency
Weakness and fatigue	1	5	5
Headache (alone)	0	0	0
Skin irritation	7	35	40
Breathing difficulty	0	0	0
Skin irritation + Burning eyes	1	5	35
Skin irritation + Headache	4	20	65
Skin irritation + headache + burning eyes	2	10	75
All the above symptoms	1	5	80
No symptoms	4	20	100
Total	20	100	

Source: Survey data 2007.

CONCLUSION

Urban and peri-urban vegetable production is gaining prominence in many developing countries and it has become an integral part of the city life. It is therefore important that clear-cut policies are formulated by the government to monitor the quality of urban produce. Also, the agricultural development programmes extension workers in the various states of the federation could mount campaigns on the judicious and safe application of pesticides. These campaigns should be linked with arrangements to assist farmers in the purchase of pesticides and regular training of farmers on safe application techniques.

Furthermore, IPM must be encouraged among the farmers while research studies need to be addressed to meet the farmer's circumstances. Donor countries of food aids should also support the production of indigenous biocontrol formulations and other non-chemical crop protection methods.

The FAO code on pesticides and the Prior Informed Consent Protocols should be fully adopted by the people and government of the federation.

All these will go a long way to increase farmer's productivity and ensure the production of "healthy" vegetables for consumers.

REFERENCES

Adedoyin, S.F., Dixon, O.T. and Ogunbanwo, A.S. 1996. Technologies for leafy vegetable production. A case study of urban farmers in Ojo Area of Lagos State, In: *Proc. of the 14th HORTSON Conference*, Ago-Iwoye. 1st–4th April, 1996. pp. 94-97.

Deang, R.T. 1991. Health considerations in crop protection for resource poor farmers, In: *Proc. of a seminar on crop protection for resource poor farmers*, CTA-NRI, 4th – 8th Nov, 1991, pp. 131-135.

Fadina, O.O. and Adeoti, B. 2000. Survey of downly mildew disease of maize (*P. sorghi)* for appropriate control strategies, *Nigerian Agric. Development Studies,* **1**(2): 62 – 69.

FAO, 1990. *International code of conduct on the distribution and use of pesticides* (amended version), Rome, Italy, Food and Agricultural Organization of the United Nations.

UNDP, 1996. *Agriculture, Food, Jobs and Sustainable Cities.* United Nations Development Programme. New York, USA. pp. 16.

Williams, S.R. 1985. *Nutrition and Diet Therapy*, 5th Ed. Times Mirro/Mosby College Publishing Co. St Louis Toronto, Santa Clara, USA. pp. 25.

19. UNDERUTILISED AND UNDEREXPLOITED TROPICAL LEAFY VEGETABLES A SOURCE OF FIBRE AND ANTI-OXIDANTS

P.G. Sadhan Kumar and S. Nirmala Devi

Professors, Kerala Agricultural University, KAU P.O. 68065, Thrissur, Kerala

ABSTRACT

Leafy vegetables are a rich source of fiber and antioxidants. They help in meeting the nutrient requirements of the poor who cannot afford high value food. Dietary fibers move food through the digestive system, absorbing water and easing defecation. Fiber has many advantages like assisting the body with healthy digestion, lowering cholesterol, reducing the risk of cancer of colon, reducing constipation and maintaining blood sugar level. They also contain a good amount of antioxidants like carotenoids, vitamin C, vitamin E, bioflavanoids, Selinium and Zinc. Antioxidants help in preventing oxidation, cardiovascular diseases, aging, certain cancers and cateracts. They are involved in the prevention of cellular damage. Main function of antioxidants is blocking the action of free radicals.

In addition to the commonly consumed leaf vegetables in India, great variety of less familiar greens are also used locally in different parts of the country. Most common among them are amaranthus, drumstick, chekkurmanis, waterleaf, basella, horse purslane, water convolvulus, ponnanganni greens, fenugreek, Indian pennywort, bathua, coriander, mint, agathi, drumstick, pisonia, celosia, roselle, thyme, common purslane blacknight shade and curry leaf. Leaves of a few tuber crops and other vegetables like taro, sweet potato, cocoyam, ash gourd, pumpkin, bottle gourd, mustard, winged bean, cowpea and radish are also consumed to a limited extent.

INTRODUCTION

Leafy vegetables are rich source of fibre and antioxidants. But most of them are not utilized to their potential. These are well adapted to the local environmental conditions where they thrive best with least care. They help in meeting the nutrient requirements of the poor who cannot afford high value vegetables, fish, meat and animal products.

Dietary fibers are the indigestible portion of plant foods that helps to move food through the digestive system, absorbing water and easing defecation. Dietary fibre consists of non-starch polysaccharides such as cellulose and many other plant components like dextrins, inulin, lignin, waxes, chitins, pectins, beta-glucans and oligosaccharides. Fibre has many advantages like assisting the body with healthy digestion, lowering cholesterol, reducing the risk of cancer of colon, reducing constipation and maintaining blood sugar level. Leaf vegetables are a good source of dietary fibre, particularly soluble fibre. They also contain a good amount of antioxidants like carotenoids, vitamin C, vitamin E,bioflavanoids, selinium and zinc. Carotenoids are a group of fat soluble pigments. They include beta carotene, lutein and lycopene. Antioxidants help in preventing oxidation, cardiovascular diseases, aging, certain cancers and cateract. They are involved in the prevention of cellular damage. Main function of antioxidants is blocking the action of free radicals.

In addition to the commonly consumed leaf vegetables in India, great variety of less familiar greens are also used locally in different parts of the country. Kerala, located in the warm humid tropics enjoy a wide range of climatic conditions suitable for growing variety of indigenous leaf vegetables. Most common among them are amaranth, drumstick, chekkurmanis, water-leaf, horse purslane, water convolvulus, ponnanganni greens, fenugreek, Indian pennywort, basella, agathi and pisonia.

AMARANTHUS

Amaranthus is a leafy vegetable which can be grown through out year. *A. tricolor (A. tristis, A gangeticus), A. dubius, A. hypochondriacus (A. flavus), A. cruentus (A. paniculatus)* and *A. caudatus (A. edulis)* are the important cultivated species. *A. spinosus* and *A. viridis* are two weed species whose leaves are also consumed by local people. A *blitum (A. lividus)* includes both weedy and cultivated species. *Amaranthus* sp is a rich source of fibre (1g/100 g) and antioxidants like beta carotene (5520 μg/100 g) and vitamin C (99 mg/100 g). It also contains 178 mg/100 of zinc.

In amaranth both red and green varieties are available. Arun *(A. tricolor),* Pusa Lal Chauli *(A. tricolor),* Arka Arunima *(A.tricolor)* and Kannara Local *(A. tricolor)* are red leaved varieties. CO1 *(A. dubius),* CO 2 *(A. tricolor)*, CO3 *(A. tristis),* CO4 *(A.hypochondriacus),* CO5 *(A. tricolor),* Mohini *(A. tricolor),* Renushree *(A .tricolor),* Arka Suguna *(A. tricolor),* Chloti chaulai *(A. blitum),* Badi chaulai *(A. tricolor)* Pusa Kirti *(A. tricolor)* and Pusa Kiran *(A. tricolor)* are high yielding green leaved varieties. Red leaved varieties are nutritionally better for their pigment contents.

Presence of antinutrient factors like oxalate and nitrate is a limiting factor in the large scale consumption of amaranthus. It is reported that about 40% of oxalates found in amaranth is in free state and affect the calcium metabolism. Another anti nutrient factor present in amaranth is free nitrates. The oxalate and nitrate content of various *Amaranthus* sp. were reported by Devadas and Mallika (1991) (Table 1).

Table 1. Oxalate and nitrate content of *Amaranthus* sp.

Species	Oxalate (%)	Nitrate (%)
A tricolor	1.30	1.37
A lividus	4.3	0.33
A viridis	4.27	0.63
A spinosus	0.77	0.69
A dubius	3.77	0.20
A hypochondriacus	2.03	2.47
A cruentus	3.94	0.53
A caudatus	3.6	0.48

According to Krishnakumary (2000) accumulation of oxalates was higher in leaves than in stem and during summer than in rainy season.

CHEKKURMANIS

The pioneer institutes working on this crop are Tamil Nadu Agricultural University, Coimbatore, Indian Agricultural Research Institute, New Delhi, Indian Institute of Horticultural Research, Bangalore and Kerala Agricultural University, Vellanikkara, Thrissur.

Chekkurmanis (*Sauropus androgynous* Merr.) (Syn. *S. albicans Blume*, *S. gardnerianes* wight., *S. sumatrans* Mig.) is a perennial leafy vegetable and called as multivitamin and multi mineral packed leaf vegetable. It has a unique position among leaf vegetables because of its high nutritive value and multifarious uses. Its leaves are very rich in protein, minerals and vitamins A, B and *C. Chekkurmanis* contains 1.4 g fibre, 5706 μg beta carotene and 247 mg vitamin C per 100g edible portion (Indira and Peter, 1988). The crop is being grown in Southern India, Indonesia and Singapore. The leaves can be cooked like other greens. The leaves are used to give a light green colour to pastry and to fermented rice in the Dutch East Indies. Leaves are used for preparation of soup in Java. The plant is also useful for growing as a hedge around home gardens. In Java, it is often planted in live fences and in the midst of garden beds to provide light shade for the other vegetables planted in the beds. Besides other uses, the leaves are used as cattle and poultry feed in certain parts of the country and the plants are planted as a soil binder to prevent soil erosion.

Chekkurmanis has several medicinal properties. The juice of leaves pounded with roots of pomegranate (*Punica granatum*) and leaves of jasmine (*Jassminum sambac*) is used against eye troubles. A decoction of its roots is often recommended for fever in rural areas. Pounded roots and leaves are used as poultice for ulcers in the nose.

Chekkurmanis is usually grown as a live fence for kitchen garden and is propagated by semi hard wood cuttings. The first harvest is done 3-4 months after planting. The annual productivity is 30-50 t/ha with per plant yield of 1-3 kg/ year. The tender shoots and leaves are used for culinary purposes. No released variety is available in this crop.

WATERLEAF

Water leaf (*Talinum triangulare* (Jacq.) Willd.) belonging to family Portulacaceae is a soft mucilaginous shade loving vegetable. It belongs to the genus Talinum which has nearly 50 other species of annual or perennial herbs, more or less succulent, sometimes woody at base. The related species in which leaves are eaten as vegetable are *T. arnotti, T. caffrum, T. patens, T. portulecefolium, T. crassifolium, T. cuneifolium* and *T. indicum*.

It contains 0.7 g fibre and 102 mg vitamin C per 100 g edible portion. Leaves and tender shoots are used as vegetables in soups and stews. The plant is a herbaceous perennial erect at base with succulent purple stem and triangular light green shining fleshy leaves. It is propagated through seeds and cuttings. Tender shoots can be harvested to 6 to 8 weeks after planting. The yield is about 4 to 6t/ ha. greens. At present it is being cultivated as a minor leafy vegetable in India, Malaysia, Indonesia, Arabian countries, USA and West Indies.

There are no released variety in this crop. The high oxalate content of the leaves limits its consumption. Therefore, breeding programme should be directed towards evolving varieties or types with low oxalate content.

BASELLA

Basella (Syn. *Malabar spinach*, Indian spinach, Vine spinach, Poi, Chinese spinach, Malabar climbing spinach and East Indian spinach) is a popular leaf vegetable grown in almost all parts of India. Fresh tender leaves and stems are consumed as leaf vegetable after cooking. The mucilaginous qualities of the plant make it an excellent thickening agent in soups, stews etc.

The leaves contain 1.55 g fibre, 9571.7 μg β carotene and 112.9 mg vitamin C per 100 g edible portion. The average nutritive value of basella is 16.62. The mean oxalate content is 0.23 g and nitrate content is 0.11 mg per 100 g edible portion.

Basella originated in Asia and more particularly in India. Two species *B. alba* and *B. rubra* are available. It is widely grown in North East and South India. It is also cultivated in Tropical Asia and Africa. The leaf juice is a demulcent, diuretic, febrifuge and laxative. A paste of the leaves is applied externally to treat boils. Basella belonging to the family Basellaceae is a fleshy annual and biennial, twining, much branched herb with alternate, broadly ovate leaves which are pointed at the apex. Fresh tender stem and leaves are consumed after cooking. Colouring matter present in red cultivar is used for making dye in China. Deep violet colouring matter is used for colouring food. Crop yields 14.0-19.5 tonnes/ha.

HORSE PURSLANE

Horse purslane (*Boerrhavia diffusa*) belonging to family Nyctaginaceae. It is a diffusely branched pubescent or glabrous prostrate herb. The leaves contain 2.4 g fibre and 77.9 mg vitamin C per 100g edible portion. There are two types one with red flowers and another with white flowers. The red flowered type is used in curries and soups. It is given internally as blood purifier and to relieve muscular pain (Mini and Krishna kumary, 2004)

PONNANGANNI GREENS

Ponnanganni greens (*Alternanthera sessilis*) is a small annual or perennial herb belonging to family Amaranthaceae. Alternanthera is grown throughout the hotter parts of India. The greens contain 2.8 g fibre, 1926 μg carotene and 17 mg vitamin C per 100g edible portion. The leaves are variegated green and yellow or bronze and green or red and pinkish brown. The leaves are 2.5 to 7.5 cm long and are fleshy when grown in watery places. The leaves and tender shoots are clipped and used like spinach and in soups. It is propagated by cuttings and first clipping of tender leaves and stems can be done 45 to 60 days after planting.

WATER CONVOLVULUS (Kang Kong)

Water convolvulus (*Ipomoea aquatica*) belongs to the family Convolvulaceae. It is an aquatic trailing or floating herb with hollow stem, rooting at the nodes. It is ideally suited to areas where water stagnation is a problem. It is commonly seen in paddy field and hence known as swamp cabbage.It is tolerant to saline-alkaline conditions. Kang Kong contains 1.5 g fibre, 22147μg ß carotene and 76.3 mg vitamin C per 100g edible portion. The leaves are a good source of minerals and vitamins especially carotene. The carotenoids include ß carotene, xanthophylls and traces of taraxanthin. The young terminal shoots and leaves are used as vegetable and salads. The leaves and stem are said to be cooling. The first harvest is done 40 days after sowing and thereafter leaves can be cut at intervals of 40days for 4 months.

FENUGREEK (Methi)

Fenugreek/Methi (*Trigonella foenum graecum*) is a herbaceous annual belonging to family Fabaceae. It is cultivated for green leaves that are used as vegetable and fully matured grains that are used as condiment. Leaves contain1.1g fibre, 2340 μg carotene, 52.0 mg vitamin.C and 358 mg Zn per100 g edible portion.(Gopalan *et al.,*1995). Flavanoids like kaemferol are reported in fenugreek leaves. Carotenoids like lutein which has got antioxidant property is present in its leaves.The fresh tender leaves, pods and stems are consumed as curried vegetable either alone or in combination with potato. Pusa Early Bunching, Lam Selection, Methi No. 47, Methi No. 14, Pusa Kasuri, Rajendra Kranti, RMT-1 and Co.1 are important varieties.Fenugreek is propagated by seeds.The tip of young shoots are nipped off 10-30 days after sowing.at a height of 4-5 cm above ground level.The subsequent leaf cuttings are possible at 15 days interval.Common methi yields 7-8 tonnes of leaves/ha while kasuri methi gives a yield of 9-10 tonnes of leaves/ha.

INDIAN PENNY WORT

The Indian penny wort (Vallarai greens) known as *Centella asiatica* belong to the family Apiaceae. It is known as "Kudangal" or "Buddhi cheera" in Malayalam since the greens are found to improve memory power. It is a prostrate, faintly aromatic, stoloniferous perennial with glabrous stem rooting at the nodes and fleshy leaves. The edible leaves contain 4.08 g fibre, 4007μg carotene, 127mg vit. C and also antinutrients oxalate (0.69 g) and nitrate (0.32g) per100 g edible portion (Maya, 2000). Tender leaves and shoots are used as vegetable in several delicious preparations.

PIGWEED /BATHUA

Pigweed (*Chenopodium album*) grows wild in India. It belongs to the family chenopodiaceae. Leaves are dull green with pale pink centre, waxy coating and wavy margins. However, it is not grown commercially as a leafy vegetable. 100 g leaves contain 0.8 g fibre, 1.74 mg vitamin A and 35 mg vitamin C. Improved variety Pusa Bathua 1 developed at IARI New Delhi is 2.25 m tall having purplish green leaves and stems. Tender leaves are harvested 45 days after planting and yield is 30 t/ha. Pusa Bathua from IARI, New Delhi and Ooty-1 from TNAU,Coimbatore are two released varieties.

CORIANDER

Coriander (*Coriandrum sativum*) belonging to the family apiaceae is an important leafy vegetable for flavouring curry and soups. It is cultivated in Rajasthan, Andhra Pradesh, Uttar Pradesh, Bihar, Maharashtra and Gujarat. It is considered as a tropical and subtropical crop and grows well in moderately cool climate and warm weather. It is susceptible to frost and high temperature. 100 g coriander leaves contain 1.2 g fibre, 6918 microgram carotein and 135 mg vitamin C.

Only a few varieties are suitable for greens. They are Co-1, Co-2 and Co-3 (TNAU, Coimbatore), Sadhana and CS-7 of ANGRAU, Hyderabad, Gujarat-1 and Gujarat-2, Hisar Anand (HAU, Hisar) and IIHR-13 (IIHR, Bangalore). Harvesting begins 25-30 days after sowing. Yield varies from 6-8 t/ha.

MINT/PUDINA

Mint (*Mentha arvensis*) is grown mostly in western Himalayas, Punjab, Jammu and Kashmir, UP, Uttaranchal and parts of Maharashtra.It belongs to the family labiatae. Leaves having strong aroma are used for flavouring foods and preparation of chutney. 100 g leaves contain 2 g fibre, 5.48 mg vitamin A, 27 mg vitamin C and 438 mg of zinc.

AGATHI

Agathi (*Sesbania grandiflora* Pers) is a perennial tropical, quick growing and soft wooded tree belonging to the family Fabaceae. It has ornamental, food and fodder values. The flowers are also eaten as vegetable. Flowers are good for making "bajji". The tree can be used as a standard for pepper and betelvine. In Tamil Nadu, agathi is grown around banana as a wind-break and coconut seedlings as a shade plant. The bark yields good fibre and a gum and the juice of the flowers is said to improve the sight, when squeezed into the eyes. Leaves and flowers of agathi has nutritional and medicinal properties.

Leaves are rich in fibre (2.2 g), vitamin A (9000 IU) and vitamin C (169 mg) per 100 edible portion. Agathi is one of the few vegetables which provides iodine (2.3 mg/100 g of the edible portion). Hence regular consumption is good for goiter patients (Indira and Peter, 1988).

It is a native of Malaysia. It is not grown on large scale for vegetable purpose. It is grown in parts of Punjab, Delhi, Bihar, Orissa, Assam, Bengal, Tamil Nadu and Kerala. There are two forms of agathi, one with red flowers and the other with white flowers. The white flowered agathi is suitable for kitchen garden. Agathi grows to a height of 12 m with straight stem and spreading branches. The roots are copiously covered with large root tubercles which make them a good green manure crop. Agathi is an under exploited crop and no breeding work is reported in this crop.

DRUMSTICK

Drumstick (*Moringa oleifera*) is a vitamin rich mineral packed nutritious vegetable belonging to family Moringaceae. *M. oleifera* is characterized by its trip innate leaves with 12-18 mm long leaflets having yellow or white petioles. Even though it is mainly grown for pods, it is also considered as a leaf and flower vegetable. The leaves and flowers are also edible and nutritionally rich. (Indira and Peter, 1988). Jaffna is a perennial type drumstick which bears long pods with soft tasty flesh. Other types are Chavakacheri murunga, Chemmurunga, Kattumurunga, Palmurunga, Chemmurunga flowers throughout the year. The improved varieties developed by TNAU are PKM 1, PKM2 and KM1. There are annual types and perennial types. Annual drumstick is propagated by seeds and perennial types by limb cuttings. The leaves of both types are edible. The leaves and flowers rich in minerals and nutrients are used in different preparations and curries. Moringa leaves contain 0.9 g fibre, 11300 IU vit A 220 mg ascorbic acid, 7.4 mg tocopherol per 100 g and 163 mg of zinc. As little as 20 g of leaves would provide a child with all the vitamin a and vitamin C (Natarajan and Beaulah (2009).

CURRY LEAF

Curry leaf (*Murraya koenigii*) is a large shrub or small tree native of south east Asia belonging to the family rutaceae .It is a tropical crop suited to warm humid climate and tolerate a temperature upto 37°C . One hundred grams of curry leaf contain 6.4 g fibre, 7.56 microgram carotene, 4 mg vitamin C and 196 mg of zinc.

The breeding work done in curry leaf has resulted in development of two varieties at UAS, Dharwad (DWD-1 and DWD-2). In parts of Tamil Nadu, a local cultivar "Senkaampu" is popular.

Besides these, there are many more leaf vegetables consumed in different parts of the country depending on local availability. The important ones are listed below:

Crop	Scientific name	Family
Pisonia	(*Pisonia grandis* Syn. *P.alba*)	Nyctaginaceae
Celosia	*Celosia argentea*	Amaranthaceae
Roselle	*Hibiscus subdariffa*	Malvaceae
Thyme	*Thymus vulgaris*	Labiatae
Common purslane/Kulfa	*Portulaca oleraceae*	Portulacaceae
Black nightshade	*Solanum nigrum*	Solanaceae

Leaves of a few tuber crops and other vegetables are also consumed to a limited extent. Some of them are listed below.

Crops	Scientific name	Family
Taro/Arvi	*Colocasia esculenta*	Araceae
Sweet potato	*Ipomoea batatas*	Convolvulaceae
Cocoyam	*Xanthosoma sagittifolius*	Araceae
Onion, Garlic, Leek, Shallot	*Allium* sp	Alliaceae
Ash gourd	*Benincasa hispida*	Cucurbitaceae
Pumpkin	*Cucurbita moshata*	Cucurbitaceae
Bottle gourd	*Lagenaria siceraria*	Cucurbitaceae
Mustard	*Brassica juncea*	Brassicaceae
Winged bean	*Psophocarpus tetragonolobus*	Fabaceae
Cowpea	*Vigna unguiculata*	Fabaceae
Radish	*Raphanus sativus*	Cruciferae

The leafy vegetables are rich in carotenoids, vitamin C, vitamin E and other phytochemicals that act as powerful antioxiadants. They are also rich in fibre. At the same time they also contain antinutrient factors like oxalates and nitrates. The nutraceutical properties of the leaf vegetables should be validated before attempting improvement of the crop. The breeding objectives will therefore vary with the crop. Since India is attempting to achieve Nutritional Food Security in this millennium, efforts should be made to utilize these underexploited underutilized leafy vegetables in our diet.

REFERENCES

Devadas, V.S. and Mallika V.K., 1991. *Review of research on vegetables and tuber crops - Amaranthus*. Kerala Agricultural University, Vellanikkara. p 59.

Gopalan,V., Ramasasthri, B.V. and Balasubramanian, S.C., *Nutritive value of Indian foods*. National Institute of Nutrition, Indian Council of Medical Research, p 155.

Haines, N.H. 1910. *Forest Flora of Chota Nagpur, Part I and II*, M/S Bishen Singh Mahendra Pal Singh, New Connaught Place, Dehra Dun, and M/S Periodical Experts, 42-D Vivek Vihar, Delhi.

Indira, P. and Peter, K.V., 1988. *Under exploited tropical vegetables.* Kerala Agricultural University, Vellanikkara.p 68.

Krishnakumary, K., 2000. *Genotypic and seasonal influence of leaf spot disease in amaranth.* Ph.D. thesis, Kerala Agricultural University, Vellanikkara.p 199.

Mini, C. and Krishnakumary, K., 2004. *Leafy vegetables*, Agrotech Publishing Academy, Udaipur. p 294.

Natarajan, S. and Beaulah, A., 2009.Nutritive values and importance of leafy vegetables in human diet. *Leafy vegetable research in India*, Tamilnadu Agricultural University. p 132.

20. CROP DIVERSIFICATION WITH HORTICULTURAL CROPS FOR ENHANCING INCOMES AND IMPROVING LIVELIHOODS OF POOR FARMERS IN DRYLAND AREAS

C L Laxmipathi Gowda[1], D Pasternak[2], S Kumar[3], A Nikiema[2], and L Woltering[2]

[1]*International Crops Research Institute for the Semi-Arid Tropics (ICRISAT), Patancheru, Andhra Pradesh, India*
[2]*ICRISAT, Niamey, Niger*
[3]*AVRDC- The World Vegetable Center-ICRISAT Project, Niamey, Niger*

ABSTRACT

The International Crops Research Institute for the Semi Arid Tropics (ICRISAT) focuses in the improvement and promotion of five mandate crops (sorghum, millet, groundnuts, chickpeas and pigeon peas) that are the lifeline of the Semi Arid Tropics (SAT) of Africa and South Asia. However, ICRISAT and its partners have initiated crops diversification using high value fruit and vegetables in West and Central Africa and South Asia. These crops bring much higher and additional income per unit land area than staple crops, thus helping in the alleviation of poverty. Furthermore, vegetables and fruit are a rich source of vitamins, anti oxidants, and micronutrients needed to combat the "hidden hunger" that prevail in most of the SAT regions of both Africa and Asia. Much of the crop diversification work is in collaboration with AVRDC-the World Vegetables Center. Research is conducted on both irrigated and rain fed indigenous as well as introduced horticultural crops.

Vegetable research focuses on selection and breeding of important exotic and indigenous vegetable crops. Most varieties of the exotic vegetables in Africa's SAT are imported from Europe and the US, and are not well adapted to the tropical conditions that prevail during the rainy season and to the predominantly poor soils. So far several heat tolerant open pollinated varieties better adapted to tropical conditions and produce higher yields and quality have been identified.

Among the major constraints in fruiticulture in the SAT of West Africa is the narrow range of varieties and the insufficient information on suitable rootstock for the production zones and ICRISAT work focuses on those issues.

So far 140 fruit trees accessions were introduced and planted in a Mother Plantation at Sadore (Niger) for initial selection and distribution to research and development partners. Fruit trees research includes: identification of best rootstocks, clonal selection, methods of vegetative propagation, and adaptation to West African soils and climate.

Three production systems have been targeted to allow optimal production and productivity of the vegetables and fruits by the small farmers in West Africa: (i) The African Market Garden (AMG) is an innovative production system for vegetables incorporating low-pressure drip irrigation combined with a comprehensive management package. (ii) The Dryland Eco Farm (DEF) is a rainfed system in which drought tolerant fruit trees (notably Zazyphus or Ber) are intercropped with high value annuals such as watermelons and cowpeas; and (iii) Bioreclamation of Degraded Lands (BDL) is a system for reclamation of degraded fine textured and crusted soils. Trees such as Ber and Moringa stenopetala are planted in micro-catchments. The trees are intercropped with annual hardy vegetables such as okra, Senna obtusifolia and roselle planted in special planting pits called zai.

The paper summarizes research and development efforts on vegetable crops, fruit trees; while highlighting the best available crops and genotypes, the economics of the various production systems, and the role of fruits and vegetables in improving the nutrition and livelihoods of farmers in the drylands.

INTRODUCTION

Despite droughts, famine, utter poverty and hunger, people in Africa's drylands want a better future, as they are willing to cultivate lands in more productive and sustainable ways. Drought is common in these areas, but if we look outside the box of current farming systems, we can find ways to mitigate drought risks, raise incomes and improve livelihoods (Dar 2007). The International Crops Research Institute for the Semi Arid Tropics (ICRISAT) was established in 1972, with its Headquarters in Patancheru, near Hyderabad, India, to conduct research in the semi-arid tropics (SAT) of Asia and sub-Saharan Africa, where agriculture is the backbone of economies. ICRISAT's research efforts cover three major regions globally: Asia, Eastern and Southern Africa (ESA), and West and Central Africa (WCA). Nairobi, Kenya is the hub for ESA region, with country programs in Malawi, Zimbabwe and Mozambique. Niamey, Niger is the regional hub for WCA, with country office in Mali. Encompassing 55 countries, the drylands of SAT are home to the poorest of the poor, with very fragile ecosystem and extreme climatic variations. ICRISAT addresses these critical challenges by developing technologies through impact-driven research to solve the problems faced by the rural communities. The challenges of persistent poverty, food security, malnutrition and health are addressed by working with diverse partners for enhancing productivity and production of five mandate crops (sorghum, pearl millet, groundnut, chickpea and pigeonpea) that are the lifeline of the Semi Arid Tropics (SAT) of Africa and South Asia. As a part of its eco-regional mandate, ICRISAT also works on agroecosystems development and management, crop diversification, farmer-to-market linkages, seed systems, and policy advocacy for generating on farm impacts.

For the last eight years, ICRISAT in collaboration with AVRDC-The World Vegetable Center and its partners has been involved in diversifying production systems, especially involving horticultural crops (vegetables and fruits). The Sudano-Sahelian zone of West Africa with one of the most fragile ecosystems for agriculture and poor infrastructure in the world (Sivakumar, 1986), has been a major target zone of these efforts. Although little recognized, horticulture has always played a very important role in the diet and income of the poor farmers in the Sudano-Sahel. Farmers (mostly women) collect, process and preserve wild leafy vegetables that are important in the daily diet, as they are used as sauces for porridges produced from millet, sorghum or corn meal. These indigenous vegetables are also sold in local markets bringing revenues to women (Abasse *et al.,* 2007). Women farmers also cultivate rain-fed vegetables such as okra and roselle. However, for commercial purpose, increased emphasis has been towards cultivation of exotic vegetables such as tomato, cabbage, onion, hot and sweet peppers, lettuce etc. These are produced in peri-urban gardens, in small village gardens and in areas specializing in production for regional exports.

Common fruit trees found on rain-fed farmland in the semi arid zones are *Vitellaria paradoxa* (shea tree), *Parkia biglobosa* (néré), *Tamarindus indica* (tamarind), *Ziziphus mauritiana* (jujube or ber), *Mangifera indica* (mango), *Saba senegalensis* etc. These trees grow mixed with annual crops forming agro-forestry parklands (Nikiema, 2005). People in the rural areas rely strongly on parkland tree products for their daily life. Tree crops of the parklands including fruits contribute 20 to 30% of household income in the rural areas (Boffa, 2000). Most of these trees have a high socio-economic value at the local and/or regional level, with important contribution to income generation for rural farmers, especially for women. In the past, many countries in the region have attempted to develop commercial fruit production (mostly mango and citrus) with limited success mostly because of lack of quality and adapted varieties, insufficient and inadequate propagation facilities, inadequate pest and disease management capacities, poor irrigation management, lack of post harvest handling, and processing facilities and poor market access (Rey *et al.,* 2004). ICRISAT has developed integrated fruit-vegetables production systems to enhance the incorporation of the new vegetables and fruit trees into the production systems of the Sudano-Sahel. Most of the research activities are carried out at the ICRISAT Station at Sadore, situated 45 km south-east of the Niamey, the capital of Niger in West Africa.

INTEGRATED HORTICULTURE SYSTEMS

Traditional horticulture production systems are inefficient resulting in low yields, poor product quality and low profits. The traditional parkland system is not conducive to introduction of more intensive rain-fed fruit trees production. Hence, there is a need to find means to increase rain-fed production. To solve these constraints, ICRISAT has developed following three integrated production systems that predominantly produce high value fruits and vegetables:

- African market garden (AMG) that uses a low pressure drip irrigated system,
- Sahelian eco farm (SEF), for non-degraded lands in rain-fed systems, and
- Bioreclamation of degraded lands (BDL) for degraded lands in rain-fed systems.

The African Market Garden (AMG)

The African market garden (AMG) is based on low-pressure drip irrigation. It integrates and optimizes all aspects of production, socio-economics and marketing in one system aimed to enhance household nutrition and income of smallholder vegetable producers in dry areas of Sub-Saharan Africa. Drip irrigation improves growing conditions of crops while at the same time saves water, labour and energy (Sivanappan and Padmakumari, 1980). The standard (commercial) AMG model is comprised of a reservoir and a 500 m^2 plot with drip kit (Pasternak and Bustan, 2003). The reservoir allows mixing of soluble fertilizer in the irrigation water regulates the amount of water and provides pressure for the drip kit. The drip kit requires only 1 meter of pressure to have uniform distribution of water. Over the last seven years, ICRISAT developed four AMG models to suit a range of scenarios (Woltering *et al.*, 2009b). The cluster and communal AMG systems were developed for producers to benefit from collective use of water, land and energy resources reduce installation and operational costs (Table 1). The payback period for these systems is only six months against 14 months for the traditional or commercial system. Studies carried out by ICRISAT (Oumarou, 2009, Woltering et al., 2009a, Mahamadou, 2005) show that the profitability of the AMG is double that of traditional irrigated vegetable gardens, and the returns to labour are six times and to water three times higher (Table 2). Improved varieties of fruit and vegetables are produced throughout the year allowing for continuous cash flow and continuous supply of fresh products.

Table 1. Comparison of three African market Garden (AMG) models

	African Market Garden models		
	Commercial	Cluster	Communal
Description	A concrete reservoir serves one or two 500 m2 garden units	One central water source provides water to multiple kits of 500 m2	5,000 m2 field is subdivided to individual producers
Target vegetable producers	Farmers with access to land and water year-round	Group of independent individuals sharing common water source (mostly rural men)	Association of producers (mostly rural women groups)
Advantages	- Option to connect two 500 m2 kits - Easy incorporation	- Lower set-up costs per farmer than Commercial AMG	- Community provision of basic services for producers (water,

	of fruit trees	- Compatible with existing infrastructure (dams, water towers, artesian wells) - Mutual learning, joint marketing	fertilizer, spraying, maintenance) - Cost effective -Compatible with solar powered pumps
Disadvantages	- Requires capital investment in garden (reservoir) - Need for motor pump and clean water source	- Requires engineering for design and installation of water distribution network - Requires well-organized community to realise advantages	- Requires capital investment in garden (reservoir) - Requires cooperative group organization

Table 2. Returns to resources from lettuce production under irrigation systems

Return to	Unit	AMG- drip	Watering can	Farmers practice
Land	(\$/m^2)	1.8	1.0	0.8
Labor	(\$/mhr)	6.0	1.5	1.0
Water	(\$/m^3)	8.4	4.7	2.7

Research for development of the AMG by ICRISAT culminated in the development of a sustainable and effective method for horticulture production by small producers. The pilot demonstration plots established in six Sahelian countries (Niger, Benin, Burkina Faso, Ghana, Mali and Senegal) prompted governments, development agencies and NGOs to mass-introduce the AMG to the region. It is expected that by the end of 2010 there will be 1,500 hectares of AMG in West Africa, benefiting more than 30,000 farmers. The profit from a 500 m^2 AMG unit is as high as \$1,000 per year.

The Sahelian Eco-Farm (SEF)

The Sahelian eco-farm is an integrated system for production of rain-fed high value crops along with millets. The components of the SEF as developed for the Sahel are: Pomme du Sahel (the domesticated ber or *Z. mauritiana*) trees planted inside micro-catchments at a spacing of 8 m x 8 m or 10 m x 10 m. The Pomme du Sahel trees are planted in *Zai* holes and intercropped with dual purpose cowpea and watermelon (*Citrulus lanatus*) (Fatondji *et al.*, 2008). The field is surrounded by a living hedge. Research on the SEF is continuing. We are in the beginning of the learning curve. Forty pilot sites established in Niger should provide valuable information as to the strategies needed for up-scaling of the SEF. Ex-ante analysis of the SEF with high value fruit tree showed that profits of this system are 4-6 times higher than profits from a traditional millet-cowpea system over time duration of 5 to 15 years.

Bioreclamation of Degraded Lands (BDL)

The BDL is a system that combines a series of water harvesting technologies with hardy fruit trees and traditional vegetables to reclaim degraded lateritic soils in the Sudano-Sahel (Pasternak *et al.,* 2009). The system involves their empowerment of women groups. The BDL has the potential to bring to the poorest and most remote villages highly nutritious vegetables and fruits and thus become an important tool in the fight against malnutrition, what is now called "the hidden hunge". The water harvesting technologies are micro-catchments, planting pits and trenches (Fatondji *et al.,* 2008). Trees are planted in the micro-catchments, vegetables in the planting pits and the trenches collect surplus runoff water for the vegetables. Two trees species were selected; Pomme du Sahel and *Moringa stenopetala* (originating from Ethiopia). Okra is the main vegetable species. In addition, two traditional leafy vegetables, viz., *Hibiscus sabdariffa* and *Senna obtusifolia* are also cultivated. The BDL is in the research and pilot study phase and the information on yields from fruit tree is yet to be recorded. However, two successive years of a test with a group of 20 women each having 1000 sq m plot showed that rain-fed okra yield can be more than 5-6 t/ha. The estimated revenue from one hectare BDL can be as high as $1,500--ten times higher than the revenue from millet-cowpea fields.

VEGETABLES RESEARCH

The Sudano-Sahel is typified by a dry climate, with a relatively cool winter that is conducive to vegetable production. Production is mostly restricted to the October-March periods, when nights are relatively cool. Lack of improved genotypes that are well adapted to high temperatures has been a major constraint for increased productivity, year-round production, and income generation through regional exports. Most of the vegetable seeds are imported from Europe and the US, and these varieties are not well adapted to the tropical conditions that prevail during the rainy season and to the predominantly poor soils. Since 2002, ICRISAT and partners have been selecting and promoting vegetable genotypes that are better adapted to the hot climate, and with higher yields and quality (Pasternak *et al.,* 2006). To further strengthen work on vegetable research, AVRDC and ICRSAT jointly appointed a vegetable breeder for the Sudano-Sahelian region in 2007. Local and regional market surveys revealed that onions, hot peppers, tomatoes and potatoes are the main exotic vegetables traded in the region (Mohamed, 2006). There are very big fluctuations in vegetables prices due to climatic constraints. In the rainy season (June-October) high humidity and high night temperatures are a constraint and in the hot dry season (March-June) the very high temperatures (up to 45°C) are a limiting factor. Lack of appropriate storage facilities results in large fluctuations in onions and potatoes prices. The vegetables diversification program has taken an integrated approach and aims at:

- Selection of suitable new vegetable varieties (Table 3)
- Maintenance breeding of local varieties that have deteriorated (e.g., onion)
- Germplasm regeneration and breeding of new okra varieties that are suitable for less input conditions and resistant to major pests and diseases

- Training of small seed producers on vegetable seed multiplication and storage

Table 3. List of vegetable genotypes available and promoted (updated from Pasternak *et al.,* 2006)

Crop	Name of line	Remarks
Tomato	ICRI-Xina	Selected from Xina bred at ISRA, Senegal and very popular in the region for rainy season cultivation
Okra	Konni	Day natural line purified from local population with short plant height, early maturing, pod quality needs improvement
???	Terra	A photoperiod sensitive line purified from local population, late maturing but with preferred pod quality because of high mucilage content
Hot pepper	Safi, Hot Habenaro	Both belong to Capsicum chinense. Safi was bred at ISRA, Senegal
Lettuce	Maya	Introduction from Israel, light green leaf, heat and bolting tolerant
Onion	#13 and #28 Violet de Galmi	Local populations from Niger, #13 has high yield while #28 has longer shelf life
Melon	Ein Dor	Introduction from Israel, heat tolerant plants with high TSS fruits
Watermelon	Malali	Drought tolerant and has ability to produce fruit on residual soil moisture
Cucumber	Beit Alpha	Introduction from Israel, heat tolerant
Carrot	Uberlandia, Brasilia	Heat tolerant populations currently evaluated for bolting without vernalization and its consequences on root development
Moringa	PKM-1	Introduction from India, better leaf yield and preferred taste over locally grown lines (how about early bearing?)

Dissemination of Vegetables Varieties

The ultimate objective of the vegetables improvement program is to provide small vegetable producers of the region with superior genotypes to address the many constraints of the small-scale horticulture industry. International commercial seed companies are not producing these seeds, with the exception of a small operation of Technichem in Senegal. ICRISAT-AVRDC therefore developed a strategy for dissemination of the selected varieties. Local seed producers of the region are provided with foundation seeds and are trained in seed multiplication. So far some 140 seed producers were trained and we are now forming a network of small vegetable seed producers. These farmers will receive further training, be updated on the new varieties, and will be supplied with foundation seeds. The efforts are underway for the establishment of a foundation seeds production program in the region.

FRUIT TREES RESEARCH

The fruit trees diversification activity includes:

- Introduction and evaluation of new fruit trees species and varieties
- Domestication of selected fruit trees
- Evaluation of suitable rootstocks for the potential trees
- Training of nurserymen on fruit trees production

Germplasm Collection

Since 2001, a total of 136 accessions of 21 fruit tree species have been collected, maintained and are being evaluated. Most of these species are those which can fit in the existing rainfed or irrigated cropping systems. The introduced fruit tree varieties are subjected to on-station adaptation trails before its promotion in the region.

Domestication

Many indigenous fruit trees are highly valued and farmers have developed local management practices to stimulate production. The efforts made by local farmers to domesticate indigenous fruit trees is usually limited to natural regeneration, and selection based on desired traits such as fruit taste and size (Boffa, 2000; Maranz and Wiesman, 2003). Traditional tree management systems cover only those indigenous trees considered to be compatible with the local farming system. Semi-domesticated parkland trees as well as important wild fruit plants should therefore undergo domestication process in order to improve the quality of the fruits and increased yields while developing appropriate agronomic and propagation techniques (Nikiema, 2005).

Currently the domestication work is centered on *Saba senegalensis*, belonging to the Apocinaceae family and occurs along seasonal rivers and woodlands in the Sudan savanna of Africa. Because of its climbing habit, it grows very often next to other trees. Saba fruits are appreciated for its tasty sweet and sour and yellow pulp. It is widely consumed as a fruit, used as a food additive, and more recently for production of a sweet and sour juice. Saba is ranked by many authors (Bonkoungou *et al.,* 2002; Nikiema, 2005) to be among the highest valued indigenous food plants of semi-arid West Africa. Fruits are available only from June till August. Natural populations of Saba are declining and subject to severe threat because of lack of regeneration in the agricultural lands (Nikiema, 2005). In order to improve the supply of fruit while ensuring the regeneration of the resource, a domestication program on *Saba senegalensis* was started in 2004 with a short-term objective to evaluate variability in fruiting and fruit quality between and within provenances, select high yielding Saba plants; and develop propagation techniques. The preliminary results show that irrigated plants start giving fruits from second year after planting, and fruits can be harvested over a span of eight months in a year.

Ziziphus mauritiana (ber) occurs naturally in the tropical semi arid zones of Africa and Asia. Five improved Indian and Israeli varieties were introduced in 1997 in four countries of the Sahel. The fruits of the improved *Ziziphus* varieties are about 10 times bigger, much more tasty and juicy than the local genotypes. On station trials indicate that promising lines gave 10 kg under rainfed and 22 kg under irrigated conditions after 3 years from planting. The use of appropriate rootstock helps overcome constraints such as water and nutrient deficiency, pests and diseases. More information on fruit trees research can be found in Nikiema *et al.* (2008).

Dissemination of Fruit Trees

Two strategies were developed for fruit trees dissemination

- *Regional nursery.* This nursery produces each year about 100,000 grafted fruit trees that are sold to private sector and small farmers or supplied to various projects
- *Establishment of village nurseries.* About 300 nursery-men from region have been trained so far on nurseries methods. ICRISAT is supporting the construction and operation of scores of nurseries in the region.

FUTURE OUTLOOK AND CONCLUSIONS

The joint AVRDC-ICRISAT research on vegetables is bearing fruits. Newly introduced and selected improved varieties of tomato, okra, onion and leafy vegetables are encouraging and being adopted by local farmers. We are encouraging the private sector to take over the seed production and distribution of improved vegetable varieties in Sub-Saharan Africa, and hope to share results with Asian farmers also. Countries such as Burkina Faso, Mali and Senegal are showing increased interest in enhancing fruit production. We need to increase the fruit tree nurseries in these countries so that they can provide seedlings to farmers, including tamarind, ziziphus, saba, and date palm.

The integrated horticultural systems, especially the Africa market garden needs to be up-scaled to benefit poor farmers in these countries, as there is demand for fruits and vegetables for local consumption and export.

The multifaceted approach taken by ICRISAT and its partners for promotion of the horticulture industry in the Sudano Sahel is starting to bear fruits. New and better vegetables and fruit trees varieties are being introduced and new more effective horticulture production systems are being adapted. This unique effort is likely to increase income of a large number of the poor smallholders of the region, thus achieving the objectives of crop diversification, increasing incomes and enhancing livelihoods of people in the drylands of Sub-Saharan Africa.

REFERENCES

Abasse, T.A., Gouzayé, A., Woltering, L. and Pasternak, D., 2007. The role of indigenous leafy vegetables on daily diet and rural and urban economy in Niger, *Acta Hort.,* **752**: 35-40.

Boffa, J. M., 2000. Agroforestry parklands in sub-Saharan Africa, *FAO Conservation Guide,* **34**: 1-230.

Bonkoungou, G.E., Djimde, M., Ayuk, E.T., Zoungrana, I., Tchoundjeu, Z., Niang, A., N'Diaye, S., Malaki, A., Ouédraogo, J.S. and Yossi, H., 2002. L'agroforesterie, un outil performant pour la gestion des ressources naturelles et la lutte contre la désertification au Sahel: Bilan de dix années d'expérience en recherche-développement et perspectives. CILSS-INSAH; ICRAF-SALWA. Les monographies SAHELIENNES No 11. 88 p.

Dar DW., 2007. *Hope from the High-Value Crops*, International Crops Research Institute for the Semi-Arid Tropics, Patencheru, Andhra Paradesh, India.

FAOSTAT. 2006., http:/faostat.fao.org/site/567/default.aspx

Fatondji, D., Pasternak, D. and Woltering L., 2008. Watermelon production on stored rainwater in Sahelian sandy soils. *African J. Plant Sci.*, **2**: 151-160.

Mahamadou, O.A. 2005. Diffusion des systèmes d'irrigation goutte-à-goutte dans la zone péri-urbaine de Niamey et dans la région de Dosso au Niger, Mémoire de Fin d'études l'Institut de formation et de Recherche appliquée IPR/IFRA de Katibougou, Mali.

Maranz, S. and Wiesman, Z, 2003. Evidence for indigenous selection and distribution of the shea tree, *Vitellaria paradoxa*, and its potential significance to prevailing parkland savanna tree patterns in sub-Saharan Africa north of the equator, *J. Biogeography*, **30**: 1505-1516.

Mohamed, S., 2006. *Potential regional markets for the Burkina Faso horticulture industry*. Report submitted to USAID-WA, ICRISAT, Niamey, Niger.

Nikiema, A., 2005. Agroforestry Parkland Species Diversity: Uses and Management in Semi-arid West Africa (Burkina Faso). PhD Thesis, Wageningen University, Netherlands.

Nikiema, A., Pasternak, D., Fatondji, D., Senbeto, D., Ndjeunga, J., Woltering, L. and Abdoussalam, S., 2008. Fruit trees for the Sudano-Sahel region of West Africa. *Chronica Hort.,* **48**: 24-29.

Oumarou, S., 2009. Étude comparative de l'irrigation goutte à goutte à basse pression JPA et de l'arrosage manuel sur la production de la laitue en zone sahélo soudanienne du Niger, Mémoire de Fin d'études à l'Institut Polytechnique Rural de Formation et de Recherche Appliquée IPR/IFRA de Katibougou, Mali.

Pasternak, D., and Bustan, A., 2003. The African Market Garden. In: *Encyclopaedia of Water Science*. B.A. Stewart and T. Howell (eds).. Marcel Dekker Inc., NY, USA, pp. 9-15.

Pasternak, D., Nikiema, A., Senbeto, D., Dougbedji, F. and Woltering, L., 2006. Intensification and improvement of market gardening in the Sudano-Sahel region of Africa, *Chronica Hort.*, **46**: 24-28.

Pasternak, D., Senbeto, D., Nikiema, A., Kumar, S., Fatondji, D., Woltering, L., Ratnadass, A. and Ndjeunga, J., 2009. Bioreclamation of degraded African lands with women empowerment. *Chronica Hort.,* 49: 24-27.

Rey, J.Y., Diallo, T.M., Vannére, H., Didier, C., Kéita, S. and Sangaré, M., 2004. La mangue en Afrique de l'Ouest francophone : syntese historique. Fruits, **59**: 121-129.

Sivakumar, M.V.K., 1986. Soil climatic zone for West African semiarid tropics: implications for millet improvement. *Regional Pearl Millet Workshop,* 6-12 Sept., Niamey, Niger.

Sivanappan, R. K. and Padmakumari, O., 1980. *Drip Irrigation.* College of Agricultural Engineering. Tamil Nadu Agricultural University. Coimbatore, Tamil Nadu, India.

Woltering, L., Ndjeunga, J. and Pasternak, D., 2009a. *The Economics of the African Market Garden: results from a case study in Niger,* International Crops Research Institute for the Semi-Arid Tropics; Working paper series, Socioeconomics and Policy, Patancheru, India (in press).

Woltering, L., Ndjeunga, J. and Pasternak, D., 2009b. The African Market Garden: Lessons learned for up-scaling of an integrated horticulture production system for Africa's smallholders, *J. Appl. Irrigation Sci.*, (communicated).

21. SUCCESSFUL MODELS OF CROP DIVERSIFICATION

Usha R. Palaniswamy

Chair, Division of Natural Sciences and Mathematics, Excelsior College, Albany, NY 12203

ABSTRACT

Merits and demerits of crop diversification are varied with the crops of choice and sites of implementation. Reported benefits of crop diversification range widely from an overall increased income to specific benefits such as mitigation of drought and increased water use efficiency and overall yield of the farming systems in the upland areas. Studies have documented crop diversification as significantly improving technical efficiency on farms in Papua New Guinea, but as reducing technical efficiency in Indonesian farms and economic efficiency in Ethiopian farms. Despite such contradicting reports, crop diversification could be an effective strategy for food and nutrition security and poverty alleviation as well as natural resource management for sustainable agricultural development. Crop diversification is also an important instrument for economic growth, because it is intended to give a wider choice in production of a variety of crops in a given area so as to expand production related activities on various crops and also to decrease the risk of no returns due to unexpected events and/or natural disasters. Remarkable progress has been made in crop diversification in certain countries while in others the progress is very limited. This is because the ability for a country to diversify successfully will depend on opportunities, farmers' responsiveness to diversification initiatives, governmental policies and the demand for crops and crop products introduced due to crop diversification. Movement toward crop diversification include shift from low value to high value crops, crops with a high production cost to those with low production costs, high maintenance to low-maintenance crops, mono-cropping systems to multiple/mixed cropping systems, from the practice of only cropping to the practice of combinations of crop-livestock-fish-apiculture, from simple agricultural production to include processing and value addition to generate multiple household income. Due to the accumulating evidence on the health benefits of horticultural crops- especially vegetables and fruits- there is a growing demand for these products in the market. The demand for fruits and vegetables coupled with the low cultivation/maintenance costs offer a promise of high returns to the farmers. Thus many crop diversification models have come to include fruits and vegetables. In this paper, selected models of crop diversification will be presented and compared for the features that contributed to their success.

INTRODUCTION

The number of hungry people has increased by more than 80 million since the 1990–92 base period and the number of undernourished has increased by 75 million between 2003-2005 and 2007 (Food and Agriculture Organization, 2008). With the recent increase in the number of hungry people in the world to 923 million, the vision of a world free of hunger and poverty has been long-standing and continues to be a priority goal to be achieved. Hunger and malnourishment are caused by poverty, a circumstance resulting from peoples' livelihood income. The term livelihood refers to peoples' ability to support themselves and those depending on them for their basic needs or subsistence. A person's livelihood comprises of the capabilities, assets and activities required for means of support. A livelihood will be "sustainable when it can cope with, and recover from, stresses and shocks and maintain or enhance its capabilities and assets both at the present time and in the future, while not undermining the natural resource base" (Chambers and Conway 1992).

Most poor people live in rural areas of developing countries and are dependent on agriculture for their livelihood. Dependency on farming for livelihood is a major cause of poverty, and intensified farming for poverty alleviation causes natural resources depletion and degradation (Thapa, 2001). Population growth continues to put a pressure on land that is available for farming. Populations who engage in crop farming as their only occupation for livelihood are faced with the uncertainty of basic livelihood income due to the numerous variables involved in crop production. Insecurity of livelihood income by farming will also urge the rural farming community to look for alternate occupations and migration to the urban areas to earn income leading to the reduction in the size of populations engaged in farming. However, the current world food crisis due to inadequate production to feed the burgeoning population and high food prices associated with economic decline, offer new opportunities to the farmers; on the other hand the need for sustainable growth puts a pressure on the farmers to make choices that are ethical and sustainable.

A farmer faces risk and uncertainty about the economic returns that can be expected from the farm. Most often, the farmer is also uncertain of the environmental consequences caused to the farms by the farming practices due to limited knowledge and ability to predict variables such as the weather and biological responses to different farming practices (*Pannell et al., 2000*). In attempts to reduce the risks the farmers take into consideration the national and international policies as well as the scientific and technological advances and the local community trends. The uncertainty of income coupled with decreasing natural resources and biodiversity of species sends a clear message to the farmers to turn to practicing farming in an environmentally sound, ethical, economically viable and socially acceptable manner. Clearly, a farmer does not decide independently on how to react to a policy, a potential technological innovation or any new information:the farmer is influenced by the social milieu (*Anderson et al., 1985*). For most farmers, livelihood strategies include agricultural intensification, diversification, and migration (Hussein and Nelson, 1998). In embracing such measures, farmers are constantly seeking out strategies for income growth, poverty alleviation, employment generation, sensible use of land, water and other resources, sustainable agricultural development, and environmental and ecological management/ improvement practices. In the long term, the agriculture sector's potential to respond to opportunities of higher prices and sustainable farming must be strengthened by sound knowledge and information to make the right decisions through support from the scientists as well as the governments and policy makers in developing productivity-enhancing measures, including increased access to land and water, technology, markets, infrastructure and credit.

Crop diversification is a strategy in agricultural diversification to design poly-culture or multiple-cropping systems, by which the farmers move away from mono-culture or single crop systems to growing different plant species. This can refer to cultivation of more than one plant species in various combinations either in crop rotation or multiple cropping systems as a means to reduce the risk of no/low farm income resulting from unpredictable variables affecting a specific crop such as weather, price change or sudden attacks of pests and diseases. Where there is more than one crop in the field, when one crop performs poorly, because of drought or pest epidemic, for example, the other crop(s) in the field can compensate, using the space and resources made available by structural diversity increasing the yield stability both agronomically and economically.

METHODOLOOGY

Described benefits of crop diversification range from a simple increase in yield in terms of greater mass produced per unit of production area to specific benefits such as reduced pest incidences and increased soil nutrients and quality. Crop diversification could be an effective strategy in this direction and a means towards poverty reduction and food and nutrition security, two major goals of the international communities, and listed as two of the major United Nations Millennium Development Goals: (1) end poverty and hunger, and (2) ensure environmental sustainability. Reducing the number of undernourished and eradication of poverty are international commitments (Food and Agriculture Organization, 2008).

Diverse cropping systems that can be designed include the cultivation of multiple plant species or multiple cultivars of a single crop in a given production area which should also focus on attributes such as conservation of natural resources, improvement of environment quality and the human health. Several models of crop diversification have been designed and implemented specific to the region, culture and the prevailing agro-climatic conditions. This paper focuses on a "biodiversity conservation model" in crop diversification that describes mixed cropping practiced to include traditional varieties and modern hybrid varieties in the same field and the suitability and benefits of vegetable crops in diversification practices.

Biodiversity Conservation Model

Loss of biodiversity due to selective crop production is a major concern and a majority of the programs which are in place to collect, conserve, characterize, and ensure the continued availability and exchange of genetic resources for planting and plant breeding are mostly of *ex-situ* in nature as samples are maintained outside their natural habitat. They are also extremely expensive and not effective in the preservation of the actual ecosystems they come from. Advances in science and technology has ushered in new improved varieties and in modern agriculture by which many traditional varieties have been replaced by a few improved "high-yielding" varieties and intensive cultivation practices are adopted on a massive scale to "produce more in less time." In such adaptations, more resources are used by way of fertilizers and the damages they cause to the natural resources are huge. This has resulted in serious genetic erosion by which traditional varieties have been lost from the agoecosystems (Porceddu *et al.*, 1988; Singh, 1999).

Since time immemorial, farming communities have engaged in the practice of growing crops/varieties that are of interest to the locale. Often, unknown to the farmers, their choice to cultivate a single crop/cultivar in their farms have had serious consequences on the ultimate survival or extinction of particular species/cultivar. Plant breeders are constantly in search of gene pool for better traits which are frequently present in traditional varieties or land races of the plant species they are working to improve for a specific trait. Not all traditional varieties and land races are maintained in the *ex-situ* systems due to the high-cost involved and they soon become extinct due to non-cultivation and consumption, as farmers move to high-yielding hybrids and shift their diets to consume what they grow or what is readily available in the market for a low price. There is a need to design cropping patterns as integral part of farming that can serve as *in-situ* conservation plots in which farmers can cultivate the traditional varieties with the modern varieties in the same field to conserve the biodiversity in a community without a reduction in the farm income.

Example 1: A case study in China

Indeed, farmers play an important role in conserving the traditional varieties, by making a choice about what to grow in their farms. In Yunnan Province, China, the traditional rice varieties that produced lower yields and were more susceptible to the blast disease were replaced by high-yielding hybrid varieties. Although, the local community had a preference for the traditional varieties in their diets, they shifted to growing the improved varieties for higher yields and related economic benefits. A crop diversification program was taken up to conserve traditional varieties of rice by cultivating local traditional varieties with the high-yielding rice varieties in various combinations in the same field. In this region, such an approach of *in situ* conservation by cultivation also conveyed a clear message to the farmers that their decisions ultimately determine whether these traditional varieties are maintained or abandoned. This model emphasized the choice of right combinations of traditional and selected varieties and the cultivation model to be adopted (Zhu *et al.,* 2003).

In this model, the selection for mixed planting was based on genetic background, differences in resistance to rice blast, and agro-economic characteristics. The traditional and hybrid varieties were genetically divergent with a genetic dissimilarity coefficient greater than 70 percent, as measured by the resistant gene analogue (RGA) fingerprinting technique and had different patterns of resistance to rice blast. For mixed cropping the traditional varieties with good market demand, high quality, and palatability were chosen. Traditional variety plants were taller (~30 cm) than the hybrid varieties. The hybrid varieties were high yielding, rice blast resistant and short, and the difference in time to maturity between the two varieties was less than 10 days. In this diversification model in Yunnan Province, the main hybrid varieties were Shanyou-63 and Shanyou-22, and the most common traditional varieties for mixed planting included Huangkenuo, Zinuo, Xianggu, Huangbansuo, Milexiangmi, and Jiugu. This kind of mixed planting had great success in controlling disease and increasing yield (Zhu *et al.,* 2000), and reduced the need for pesticides. Additionally, the farmers received greater price for the traditional varieties as the traditional glutinous varieties fetched higher prices than modern hybrid rice. In mixed planted fields, the disease occurrence was reduced and fetched a greater price to the farmers due to the preference for traditional varieties in the communities. Due to the benefits experienced by the farmers, this became very popular and in this process some traditional rice varieties that were formerly locally extinct, such as Ersigu and Dajiujia, were restored to agricultural commun-ities, and some rare and endangered varieties also became more popular in Yunnan Province. As demonstrated in this region, conservation of traditional crop varieties can be well integrated into modern agroecosystems through diversity management where appropriate economic incentives are created. This model also proves the value of indigenous knowledge of the farmers that was helpful in developing such a crop diversification cropping system, in the first place. It is evident that scientifically based but gener-alized and simplified messages from the indigenous knowledge can then be returned to the local community to guide farming practices. By focusing on how mixed cultivation can reduce disease problems in rice, Zhu *et al.* (2000) were able to recommend simple techniques that are effective on a large scale and solve the issues related to the traditional varieties, which were still the preferred varieties in the local diets, and design a cropping technique to grow them without having to forego the benefits of high yields from growing high yielding varieties. As a result, the area of mixed cultivation of hybrid and traditional rice varieties increased to 167,000 hectares in 2001 played a great role in rice production and conservation in this region. This model for *in-situ* conservation of traditional varieties through crop diversity management is beneficial for the agroecosystems, preserving the germplasm and promising for future breeding efforts utilizing traditional varieties. This kind of diversity management method has long-term potential for overcoming the commonly perceived constraints of on-farm/*in situ* conservation.

Example 2: A case study in India

The M.S. Swaminathan Research Foundation (MSSRF), a non-profit private Trust founded and led by Dr. M.S. Swaminathan, recipient of the first World Food prize, currently has several active projects that encourage biodiversity conservation model of crop diversification in Jeypore (Orissa), Wayanad (Kerela) and the Kolli Hills, Namakkal (Tamil Nadu) for developing models for conservation of agro-biodiversity for food security, reduction of poverty and management of natural resources by increasing and diversifying the food basket. The MSSRF strongly believes that anything that is used will be conserved and hence is developing conservation models that practice cultivation, consumption and commercialization. In its developmental activities for poverty eradication, and food and nutrition security, the MSSRF takes a pro-poor, pro-nature and pro-women approach and is engaged in harnessing science and technology for environmentally sustainable and socially equitable development. Its pathways include conservation and enhancement of natural resources, and promotion of sustainable livelihoods; its activities take a multi-pronged bottom-up participatory approach for building community trust and partnership and empowerment in multiple ways (eg. education, communication, training, capacity building). The MSSRF recognizes that the development of on-farm conservation by crop diversification will require collaborative efforts and action between government, non-governmental, community, and peoples' organizations, and an interdisciplinary approach to technical and scientific questions by plant-breeders, biologists, nutritionists, socio-economists, NGOs and farmers.

An example of MSSRF's biodiversity conservation by crop diversification is its project in Kolli Hills, in the Eastern Ghats of southern India which is an important center of diversity for millets. The Malayalis (tribes living in this region) use nine landraces of little millets (Samai), eight landraces of Italian millets (Thinai), four landraces of Kodo millets (Varagu). In growing these crops, they evolved a variety of cropping practices that includes intercropping, crop rotation, mixed cropping and mixed farming fitting to the local agro climatic conditions to support their food and nutritional security. Traditionally, they were engaged in mixed cropping with coriander, greens, and pulses in a rotation on their farms which preserved the soil nutrients and organic matter and adding diversity to their diets. However, in the last two decades, promotion of commercial crops like tapioca, coffee, and pepper had resulted in reducing the area under their traditional food crops, increasing the use of chemical fertilizers and loss of their unique and healthy cultural values and practices leading to a food insecure future for the region. A decision to grow crops fetching greater price and adopting practices that are not sustainable, without insights into local tradition and culture can lead to loss of traditional value, destruction of traditional values, and unsustainable development of the region. This can also result in loss or lack of knowledge about the importance and nutritive values of millets, especially among the younger generation, at a time when the rest of the world is "discovering" the exceptionally high nutrient content of millets and associated health benefits of "traditional" diets!

Although there is a consumption preference for the traditional minor millets based diets, reasons given by the Malayalis for not growing and consuming millets are that "hand-pounding millets is drudgery," minor millets do not have any "marketing value," Government policies related to crop loan, crop subsidies that favor "commercial" agriculture than "traditional" agriculture. They were also influenced to make the easy choice to consume food items such as rice and wheat that are distributed by the Indian Public Distribution System at a reasonable cost, alleviating the need to laboriously hand-pound millets for daily consumption.

The MSSRF recognized that all these conditions have had a cumulative effect in bringing down the cultivation of millets and assisted the community in getting government loans, bank credits to procure de-husking mills, and also educated and trained the community to prepare a variety of processed food from the millets to increase their income, by developing minor millets value addition program and creating marketing linkages. In its community based agro-biodiversity conservation and management approach, MSSRF increased the marketing potential of these minor millet crops by adding value to them and helped the farming communities to maintain the existing diversity among the minor millet crops by providing economic incentives for their conservation efforts. The Malayali farmers made a choice to return to their traditional cropping systems and practices of diversification with pulses and vegetables as the crop diversification model based on cultivation, commercialization and consumption practices offered sustainable livelihood, increased their income significantly and also maintained their traditions (King, 2009).

CROP DIVERSIFICATION WITH VEGETABLES

The high yielding potential of vegetable crops per unit area of production, high nutritional value of vegetables, scientific evidence of health benefits associated with increased vegetable consumption, call for diverse diets and increasing the choices of foods for consumption justifies the inclusion of vegetable crops in crop diversification efforts. Vegetable consumption is also linked to income and the significant income growth observed in the economically developing nations (*eg*., Brazil, Russia, India, and China) is bound to put pressure on vegetable demand. Micronutrient deficiencies have become a serious global problem, especially in areas where the diet lacks variety (Welch and Graham, 1999 and Kennedy *et al.,* 2003). Thus the demand for vegetables is higher in terms of quantity as well as diversity.

The concerns for sustainability in the continuous cereal-cereal rotations (Cassman and Pingali, 1993) as well as the need for diversification of risk, income, and food, and enhanced efficiency of farm resources have called the attention of farmers to the benefits of crop diversification. The move away from monoculture of rice and other cereals is also due to the declining rice prices (Pingali 1992), shortages of water due to deteriorating irrigation infrastructure, reduced profitability of irrigation investment, and/or competing water demand for domestic use. Vegetables utilize water most efficiently in terms of both production and economic efficiency (Ali 1999), and vegetable production engages more labor of vulnerable population groups, such as women and children (Braun *et al.* 1989; Wann *et al.* 2000). Vegetables in the cropping system can help break the pathogen cycle in cereal-cereal rotations, and the integration of leguminous vegetables, such as mungbean, can improve the productivity and profitability of the cereal-cereal system (Ali 1998; Ali *et al.* 1997). Intercropping with legumes under very low soil N conditions can contribute up to 25% of N requirement of the main crop (Eaglesham *et al.,* 1981). The practice of intercropping with leguminous vegetables can enhance crop productivity compared to the growth of sole crops through more effective use of water, nutrients and solar energy (Midmore, 1993), had a positive effect on soil conservation and improvement of soil fertility (Jarenyama *et al.*, 2000), and produced more stable yields (Horwith, 1985), and reduced pest and disease attacks (Theunnissen, 1997).

Benefits associated with multi-species cultivation include the long term benefits due to the non-harvested plant parts which play a key role in the functioning of the agroecosystems, its sustainability and long-term productivity, and its function to other parts of the society at large, such as the non-harvested parts of total biomass production that are used to enhance and maintain the soil nutrients and organic matter (Vandermeer *et al.,* 1989) when returned to the land in the form of manure and organic matter. The mixing of crops to include species with different rooting patterns and N fixing capabilities can enhance soil nutrients. Selective inclusion of legumes in the cropping systems has been practiced widely for benefits associated with N fixation (Wetselaar, 1967; Ladd *et al.,* 1983; Suwanarit *et al.,* 1986; Dakora *et al.,* 1987; Peoples and Herridge, 1990; Herridge *et al.,* 1995). Studies by Dakora and Keya (1997) found that rotational cropping of legumes with cereals to be a more sustainable system for increasing food production than monoculture or intercropping as both monoculture and intercropping as yields of sole cropped and intercropped maize were only 35% and 38%, respectively of yields from crop rotation (Dakora and Keya, 1997). This benefit is mostly due to the increase in soil N and organic matter.

The choice of crops also depends on the length of growing period available to the farmer to farm, while planning a cropping system: when it is less than 75 days the farmer goes for perennial vegetation and monocropping of short duration pulses; with 75 to 140 days of growing period mono-cropping is most suitable with some inter cropping with short term crops such as short duration vegetables; with 140 to 180 days farmers practice intercropping and when the growing period is higher than 180 days they go for double cropping. Choice of base and intercrops can be decided based on rainfall distribution. In regions with uni-modal rainfall pattern and shallow soils, base crop should be of shorter duration and the companion of longer duration. In case of medium to deep soils, base crop should be of longer duration while the companion crop can be of shorter duration. In bi-model distribution, choice of the crops should be such that the peak growth period of the base and companion crops

Mixed cropping of more than one vegetable crop in the farm could lead to better land use efficiency as an important component of sustainable farming (Guvenc and Yildirim, 1999).

Cultivars chosen for intercropping should enhance the complementary effects between species (Baumann *et al.*, 2001). Advantages of intercropping with legumes have been demonstrated in numerous studies; tomato or okra with cowpea (Olasantan, 1991), cabbage with bean (Poniedzialek *et al.*, 1989), watermelon with soybean (Sharaiha and Hattar, 1993), and chilli with bean (Costa and Perera, 1998). These studies have indicated that intercropping was more productive than sole cropping because of the complementary effects of intercrops. Furthermore, many authors showed favorable aspects of intercropping with non-legume vegetables under field conditions (Brown *et al.*, 1985; Baumann *et al.*, 2001). Intercropping can result in an increase in the productivity of vegetables per unit area, and improve net income. Cauliflower intercropped with other vegetables like cos lettuce, leaf lettuce, bean or onion would be a remunerative cropping system which produced higher yields and economic returns when compared to a monocrop of cauliflower (Yildirim and Guvenc, 2005). Intercropping based on cauliflower could not only use limited areas for crop production more efficiently but also increase income for small farmers with limited resources.

There is increasing consensus that wild foods could significantly contribute to alleviating hunger and malnutrition (Burlingame, 2000). Many studies confirm the importance of wild vegetables as sources of micronutrients (Booth *et al.* (1992), Freyre *et al.* (2000), Ogle *et al.* (2001), Sundriyal and Sundriyal (2004) and Gupta *et al.* (2005). Integrating wild vegetables into diets has been promoted as the most practical and sustainable way to achieve nutrition security (Chadha and Oluoch, 2003), since such vegetables are efficient sources of several important micronutrients, both with respect to unit cost of production and per unit area of land (Ali and Tsou, 1997). In view of this, neglected crops, non-commercial foods, and wild foods are receiving renewed attention, with the recognition that they could become useful parents in breeding programs, convenient sources of income, and vehicles for improved nutrition and increased food supply (Burlingame, 2000; Grivetti and Ogle, 2000; Padulosi *et al.*, 2002; Olorode, 2004). Also, there is increasing evidence to suggest that nutritional success is linked to diversification of the food base (Altieriet *et al.*, 1987), which could be achieved by modifying agricultural practices and increasing collection of wild plant germplasm (Grivetti, 1979).

Higher returns under intercropping systems have been reported in cucumber, pepper and tomato:lettuce (Erdogan and Karatas, 2000), garlic: bean (Abidin *et al.,* 1989) in cabbage:bean, Quayyum and Akanda, 1990) and in okra:bean (Prabhakar and Shukla, 1990, 1991) intercrop systems.

Ecological theory and experimental evidence suggest that taxonomic diversity plays an important role in the functioning of ecosystems (Hooper *et al.*, 2005). Increasing cropping system diversity has been advocated as a potential means of decreasing the need for intensive chemical inputs for weed control (Liebman and Gallandt, 1997).

Agricultural production is the "most weather-dependent of all human activities" (Oram, 1989) and its vulnerability due not only to the year-to-year variability of the climate per se, but also of its unpredictability every year. Crop diversification offers a viable strategy to minimize the risk imposed in production and is driven by the demands from consumers who now show a greater awareness of the advantages of diversifying a cereal-based diet with vegetables and fruits. It is an important instrument for economic growth and the ability to diversify will depend on the opportunities for diversification, the farmers' responsiveness as well as the governmental policies. In designing a suitable crop diversification model and making changes to increase the farm income and reduce poverty, the farmers not only consider their economic background and their own needs but also the social parameters like attitudes, values, traditions, peer group pressure and the local culture (Dent, 1990).

REFERENCES

Abidin, Z., Subhan, R., Basuki, R.S. 1989. Experiments on multiple cropping of garlic with red bean and red pepper, *Hort. Abst.*, **59**(3), 1968.

Ali, M. 1998. Green manure evaluation in tropical lowland rice systems, *Field Crops Res.*, **61**:61-78.

Ali, M. 1999. *New paradigm of vegetable cultivation in Asia,* Asian Vegetable Research and Development Center, Shanhua, Tainan, Taiwan.

Ali, M. and Tsou, C.S. 1997. Combating micronutrient deficiencies through vegetables — a neglected food frontier in Asia, *Food Policy*, **22**:17–38.

Ali, M., Malik, LA., Sabir, H.M. and Ahmad, B. 1997. *The mungbean green revolution in Pakistan*, Shanhua, Taiwan, Asian Vegetable Research and Development Center, Technical Bulletin No. 24, 66 p.

Altieri, M.A., Merrick, L.C. and Anderson, M.K. 1987. Peasant agriculture and the conservation of crop andwild plant resources, *Conservation Biol.*, **1**:49–58.

Anderson, J., Dillon,J. and Hardaker, J. 1985. Socio-economic modelling of farming systems. In: J.B. Hardaker, Editor, *Agricultural systems research for developing countries: Proceedings of an International workshop*. ACIAR Proceedings Series, Australian Centre for International Agricultural Research, Hawkesbury Agricultural College, Richmond, NSW, Australia (1985), pp. 77–88 2–15 May 1985.

Baumann, D.T., Bastiaans, L. and Kropff, M.J. 2001. Competition and crop performance in a leek-celery intercropping system, *Crop Sci.*, **41**:764-774.

Booth, S., Bressani, R. and Johns, T. 1992. Nutrient content of selected indigenous leafy vegetables consumed by the Kekchi people of Alta Verapaz, Guatemala. *J. Food Composition and Analysis*, **5**, 25–34.

Braun, J.V., Hotchkiss, D. and Immink, M. 1989. *Nontraditional export in Guatemala: effects on production, income, and nutrition, Research Report* 73. International Food Policy Research Institute (IFPRI) and Institute of Nutrition of Central America and Panama (INCAP), Washington, DC, USA.

Baumann, D.T., Bastiaans, L. and Kropff, M.J. 2001. Competition and crop performance in a leek-celery intercropping system, *Crop Sci.*, **41**:764–774.

Brown, J.E., Splittstoesser, W.E. and Gerber, J.M. 1985. Production and economic returns of vegetable intercropping systems, *J. Am. Soc. Hort. Sci.,* **110**(3), 350–353.

Burlingame, B. 2000.Wild nutrition, *J. Food Composition and Analysis*, **13**: 99–100.

Cassman, K.G. and Pingali, P.L. 1993. Extrapolating trends from long-term experiments to farmers field: the case of irrigated rice systems in Asia, *Paper presented at the working conference on measuring sustainability using long-term experiments*, Rothamsted Experimental Station, 28-30 April 1993.

Chadha, M.L. and Oluoch, M.O. 2003. Home-based vegetable gardens and other strategies to overcome micronutrient malnutrition in developing countries, *Food, Nutr. Agric. Series*, **32**, FAO, Rome, Italy.

Chambers, R. and Conway, G. 1992. Sustainable rural livelihoods: practical concepts for the 21st Century, IDS DP296, Brighton

Costa, W.A.J.M. and Perera, M.K.K.W. 1998. Effects of bean population and row arrangement on the productivity of chilli/dwarf bean intercropping in Sri Lanka, *J. Agron. Crop Sci.,* **180** (1), 53–58.

Dakora F. D., Aboyinga R. A., Mahama Y. and Apaseku J. 1987. Assessment of N2 fixation in groundnut (*Arachis hypogea* L.) and cowpea (*Vigna unguiculata* L. Walp.) and their relative N contribution to a succeeding maize crop in Northern Ghana. MIRCEN, *J. Appl. Microbiol. Biotechnol.*, **3**:389-399.

Dakora, F.D. and Keya, S. 1997. Contribution of legume nitrogen fixation to sustainable agriculture in sub-saharan Africa, *Soil Biol. Biochem.*, **29**: 809-817,

Dent, J.B. 1990. Optimising the mixture of enterprises in a farming system. **In**: J. Jones and P. Street, Editors, *Systems theory applied to the agriculture and the food chain*, Elsevier Applied Science, London (1990), pp. 113–130.

Eaglesham, A.R.J., Ayanaba, A., Ranga Rao, V. and Eskew, D.L. (1981). Improving the nitrogen nutrition of maize by intercropping with cowpea, *Soil Biol.Biochem*, **13**:169- 171.

Erdogan, H. and Karatas, A., 2000. An investigation on intercropping production of tomato, cucumber and pepper together with crisphead and cos lettuce in heated glasshouse by geothermal energy, **In**: *Proceedings of the vegetable symposium*, 11–13 September, Isparta, Turkey, pp. 296–302.

Food and Agriculture Organization 2008. The state of food insecurity in the world 2008, ftp://ftp.fao.org/docrep/fao/011/i0291e/i0291e00a.pdf

Freyre, M.R., Baigorria, C.M., Rozycki, V.R., Bernardi, C.M. and Charpentier, M. 2000. Suitability of wild underexploited vegetables from the Argentine Chaco as a food resource, *Archivos Latinoamericanos de Nutricion,* **50**:394–399.

Grivetti, L.E. 1979. Kalahari agro-pastoral-hunter-gatherers, The Tswana example, *Ecol. Food Nutr.*, **7**:235–256.

Grivetti, L.E. and Ogle, B.M. 2000. Value of traditional foods in meeting macro and micronutrient needs: the wild plant connection, *Nutr. Res. Rev.*, **13**:31–46.

Gupta, S., Lakshmi, A.J., Manjunath, M.N. and Prakash, J. 2005. Analysis of nutrient and antinutrient content of underutilized green leafy vegetables. LWT—*Food Sci. and Technol.*, **38**:339–345.

Guvenc, I. and Yildirim, E. 1999. Multiple cropping systems in vegetable production. **In:** *Proceedings of the organic agriculture symposium*, 21–23 June, Izmir, Turkey, pp. 288–296.

Herridge D.F., Marcellos, H., Felton, W.L., Turner, G.L. and Peoples M.B. 1995. Chickpea increases soil- N fertility in cereal systems through nitrate sparing and Na fixation, *Soil Biol. Biochem.*, **21**:5245-5551.

Hooper, D.U., Chapin, F.S., Ewel, J.J., Hector, A., Inchausti, P., Lavorel, S., Lawton, J.H., Lodge, D.M., Loreau, M., Naeem, S., Schmid, B., Setala, H., Symstad, A.J., Vandermeer, J. and Wardle, D.A. 2005. Effects of biodiversity on ecosystem functioning: a consensus of current knowledge. *Ecological Monographs*, **75**:3–35.

Horwith, B. 1985. A role for intercropping in modern agriculture, *Bioscience*, **35**:5.

Hussein, K., Nelson, J. 1998. *Sustainable livelihoods and livelihood diversification.* IDS working paper 69, Brighton.

Jarenyama, P., Hesterman, O.B., Waddington, S.R. and Harwood, R.R. 2000. Relay-intercropping of sunnhemp and cowpea into a smallholder maize system in Zimbabwe, *Agron. J.*, **92**: 239–244.

Kennedy, G., Nantel, G. and Shetty, P. 2003. The scourge of "hidden hunger": global dimensions of micronutrient deficiencies. *Food, Nutr. Agric. Series*, **32**. Food and Agriculture Organisation, Rome, Italy.

King, I.O. E.D. 2009. Personal communication by the Kolli Hills Program Director Dr. King during a site visit to the Malayali community.

Ladd. N., Amato M., Jackson R. B. and Butler J. H.A. 1983 Utilization by wheat crops of N from legume residues decomposing in soils in the field, *Soil Biol. Biochem.*, **15**:231-238.

Liebman, M. and Gallandt, E.R. 1997. Ecological management of crop–weed interactions, *Ecology in Agriculture* (ed. L.E. Jackson), pp. 291–343. Academic Press, San Diego, CA.

Midmore, D.J. 1993. Agronomic modification of resource use and intercrop productivity, *Field Crops Res.*, **34**:357–380.

Ogle, B.M., Johansson, M., Tuyet, H.T. and Johannesson, L. 2001. Evaluation of the significance of dietary folate from wild vegetables in Vietnam. *Asia Pacific J. Clinical Nutr.*, **10**:216–221.

Olasantan, F.O. 1991. Response of tomato and okra to nitrogen fertilizer in sole cropping and intercropping with cowpea. *Hort. Sci.*, **66**:191–199.

Olorode, O. 2004. Conservation of plant genetic resources. *African J. Traditional, Complementary and Alternative Medicines,* **1**:4–14.

Oram, P.A. 1989. Sensitivity of agricultural production to climatic change, an update, **In**: *Climate and food security.* IRRI, Manila, The Philippines, pp. 25–44.

Padulosi, S., Hodgkin, T., Williams, J.T. and Haq, N. 2002. Underutilized crops: trends, challenges and opportunities in the 21st Century, **In**: Engels, J.M.M., Ramanatla, R., Brown, A.H.D., Jackson, M.T. (Eds.), *Managing genetic diversity.* International Plant Genetic Resources Institute, Rome, pp. 323–338.

Pannell D.J., Malcolm B. and Kingwell R.S. 2000. Are we risking too much? Perspectives on risk in farm modelling, *Agric. Econ.*, 23:69–78.

Peoples M. B. and Herridge D. F., 1990. Nitrogen fixation by legumes in tropical and subtropical agriculture, *Adv. Agron.*, **44**:155-223.

Pingali, P.L., 1992. Diversifying Asian rice-farming systems: A deterministic paradigm, **In**: Barghouti, S., Garbux, L., and Umali, D. (ed.) *Trends in agricultural diversification: regional perspectives*, World Bank Technical Paper no. 180, The World Bank, Washington, D.C. pp: 107- 126.

Poniedzialek, M., Zacharias, A., Kunucki, E., Suchodolska, R., 1989. Effect of cabbage, French bean and snap bean intercropping on the level and quality of yield, *Folia Hortic.*, **1**(2): 37–51.

Porceddu, E., Ceoloni, C., Lafiandra, D., Tanzarella, O., Scarascia, A. and Mugnoza, G.T. 1988. Genetic resources and plant breeding: Problems and prospects. Pages 7-21 in Miller TE, Koebner RMD, eds. Proceedings of the 7th international wheat genetics symposium. Cambridge (United Kingdom): Institute of Plant Science Research.

Prabhakar, B.S. and Shukla, V. 1990. Crop land use efficiency in sequential intercropping systems with vegetables, *Indian J. Hort.*, **47** (4): 427–430.

Prabhakar, B.S. and Shukla, V. 1991. Crop land use efficiency sequential intercropping systems with vegetables. *Hort. Abst.* 61 (9), 7856.

Quayyum, M.A. and Akanda, M.E.M. 1990. Productivity and profitability of cabbage intercropped vegetables. *Hort. Abst.,* **60** (9):7201.

Sharaiha, R.K. and Hattar, B. 1993. Intercropping and poultry manure effects on yields of corn, watermelon and soybean grown in a calcareous soil in the Jordan Valley, *J. Agron. Crop Sci.*, **171**:260–267.

Singh, R.K. 1999. Genetic resource and the role of international collaboration in rice breeding. Genome, **42**: 635-641.

Sundriyal, M. and Sundriyal, R.C. 2004.Wild edible plants of the Sikkim Himalaya: nutritive values of selected species, *Econ. Bot.,* **58**:286–299.

Suwanarit A., Suwannarat C. and Chotechaungmanirat S. 1986. Quantities of fixed N and effects of grain legumes on following maize, and N and P status of soil as indicated by isotopes, *Plant and Soil*, **93**:249-258.

Thapa, G.B. 2001. Changing approaches to mountain watersheds management in mainland south and Southeast Asia, *Environ. Management J.*, **27** (5):667–679.

Theunnissen, J. 1997. Intercropping in field vegetables as an approach to sustainable horticulture. *Outlook Agric.,* **26** (2): 95–99

Vandermeer, J. 1989. *The ecology of intercropping*. Cambridge University Press, Cambridge, UK, p. 237

Wann, J.W., Peng, T. K., and Wu, M.H. 2000. Taiwan. **In**: Ali, M. (ed.) *Dynamics of vegetable production, distribution, and consumption in Asia*. Shanhua, Taiwan, Asian Vegetable Research and Development Center, AVRDC Publication No. 00-498,470 p.

Welch, R.M. and Graham, R.D. 1999. A new paradigm for world agriculture: human needs productive, sustainable, nutritious, *Field Crops Res.*, **60**:1–10.

Wetselaar, R. 1967. Estimation of Nz fixation by four legumes in dry monsoonal area of northwestem Australia. *Australian J. Exp. Agric. and Animal Husbandry*, **7**: 518-522.

Yildirim, E. and Guvenc, I. 2005. Intercropping based on cauliflower: more productive, profitable and highly sustainable, Europ. *J. Agron.*, **22**:11–18.

Zhu, Y., Wang, Y., Chen, H., and Lu, Bao-Rong 2003. Conserving Traditional rice varieties through management for crop diversity. *Bioscience*, **53**(2):158- 162.

Zhu, Y., Chen, H., Fan,J., Wang,Y., Li, Y., Chen, J., Fan, J., Yang, S., Hu, L., Leung, H., Mew, T.W., Teng, P.S., Wang, Z., and Mundt, C.C. *2000. Genetic diversity and disease control in rice.* Nature, ***406***:718-722.

22. TAMIL NADU PRECISION FARMING PROJECT: A PRELUDE TO NEXT GREEN REVOLUTION

E. Vadivel[1], N. Sriram and I. Muthuvel

Director of Extension Education, Tamil Nadu Agricultural University, Coimbatore – 3, Tamil Nadu

ABSTRACT

The Tamil Nadu Precision Farming Project was implemented at Tamil Nadu to promote market led horticulture and technically, economically and socially empower farmers and cluster level precision farmers, registered associations and to provide hands on training on hi-tech horticulture in built with precision elements. Assessing impacts of a project is essentially management requirement. Accordingly this study was conducted with an aim to examine the impact of the Precision Farming Project on different stakeholders within and outside the state through farmers' participatory operational research mode. The study was conducted at nine focus districts of Tamil Nadu through Farmers' Participatory Operational research mode. The project was assessed in terms of inputs delivered, activities carried out, outputs obtained (qualitative and quantitative outputs) and impacts created due to the Tamil Nadu Precision Farming Project. It was also found that, this project has travelled across all the districts of Tamil Nadu and other States of India. The study also revealed that the precision farming technology traveled across this globally only because of the strength of the precision technologies and process of implementation. The strong feedback was also communicated to the scientific community for refining R&D programmes. The study concluded that the overall impact of the Tamil Nadu precision farming project is excellent on the basis of productivity increase, quality improvement in the produce, income increase, marketing efficiency, negotiating and bargaining power, cost of cultivation, retention of soil fertility, water saving, energy conservation, farmers confidence level, farmers' proficiency in cultivational practices, seeds and seedlings usage, duration of crops and market orientation. It could also be concluded that the precision farming technologies have been accepted by all the stakeholders and have started cultivating with their own investments. Productivity and production have been increased tremendously through precision protocol but needs to realize the profit through proper marketing strategies. Almost all the farmers of precision farming have felt that need to establish their own mega market to realize higher margin and profitability.

INTRODUCTION

Tamil Nadu Precision Farming Project is a State sponsored 'Turn key Project' executed by Tamil Nadu Agricultural University in Dharmpuri and Krishnagiri districts in an area of 400 ha with participation of 400 farmers @ one ha per farmer during 2005-07. It was a state level demonstration to provide hands on training to farmers, extension officers, agri-horti entrepreneurs and students with objectives *viz.*, (i). promotion of market led horticulture, (ii). technically, economically and socially empower the farmers and cluster level precision farmers registered associations, and (iii). provide hands on training on hi-tech horticulture in built with precision elements.

Based on the unprecedented level of acceptance and adoption by the farmers, the precision farming programme was extended subsequently to all the districts of the Tamil Nadu in about 50,000 ha in the next three years. Assessing impacts of a project is essentially a management requirement and provides the basis for improved project screening. Accordingly this study was conducted with an aim to examine the impact of the Precision Farming Project on different stakeholders within and outside the state through farmers' participatory operational research mode.

METHODOLOGY

The impact study on precision farming project was carried out by involving the farmers as key partners. Farmers' observations, remarks and experiences have taken into account as key indicators of this study. The researchers facilitated the farmers to analyse the impact in terms of technical, economical and social aspects due to intervention of the precision farming project. The study also explored the diffusion effects assessed through SWOC analysis and analyzed the operational blues while implementation of the TNPF which are when such a project was scaled up and policy implications are proposed to Government. The study was conducted at districts of Tamil Nadu *viz.*, Dharmapuri, Krishnagiri, Salem, Namakkal, Villupuram, Ramnathapuram, Perambalur, Coimbatore and Dindigul which has been identified exclusively for implementation of the NADP programmes during 2007-08 by the Government of Tamil Nadu. Thirty beneficiary farmers from each focus districts were selected using participatory tool viz., snow ball technique for this study. Accordingly the 30 beneficiary farmers per focus district were selected proportionately from Department of Agriculture, Department of Horticulture and TNAU KVKs. Participatory data collection tools which included (a) participatory listening and observation; (b) visual tools such as maps, daily activity diagrams, institutional diagrams and venn diagrams, flow diagrams and livelihood analysis; (c) semi-structured interviews; and (d) focus group discussions (d) brainstorming sessions (e) farmers' interactive meetings *etc.*, were employed for this study. Data has also been collected through e-mail, newspapers, magazine, letters, phone calls, fax, mobile SMS and websites.

A. IMPACT OF THE PRECISION FARMING PROJECT ON BENEFICIARY FARMERS

Results and Discussion

a. Impact of the Precision Farming Project on Beneficiary Farmers

The impacts of the precision farming project on beneficiary farmers analyzed through farmers participatory mode and the critical observation are presented in Table 1.

Table 1. Impact of precision farming

Project **INPUTS** Human, material and other resources required to initiate project activities	→	Technical team Drip & Fertigation Units Community Nursery Seed Materials Water Soluble Fertilizers Plant Protection Chemicals, Packing crates *etc.,* Proactive, active and indifferent farmers
↓		↓
Project **ACTIVITIES**	→	Organizing cluster level associations Assessment of quality of water and fertility status of soil Field visit for beneficiary farmers Facilitating nursery growing at community nursery centres Farm Records Hands on training on fertigation Plant protection Post harvest management Exposure visit to markets
↓		↓
Project **OUTPUTS** a. Qualitative elements	→	Cluster level associations have grown as an institutional mechanism to sustain the project after project period. Farmers, farm women and labourers were technically empowered to handle fertigation, portray nursery and IPM. Scientists vs Farmer and Farmer Vs Farmer mode of TOT ensured 100% adoption level Farmers transformed themselves as innovators and introduced new methods,

		procedures and systems. Sensitivity to market demand and consumer preference was enhanced. Added value to the technologies, scientists and institutions. Fostered healthy competition among the farmers to excel in yield. Incorporation of producer company under company Act and planning for establishment of mega market. Farmers becoming resource persons to organise training and consultants to fellow farmers.
↓		↓
b. Quantitative elements	→	100 % higher yield in 45 crops 90 % plus first grade marketable produces 30% premium price in the market 5-6 days more shelf life Less labour dependence 30-40% Water Economy and electricity power economy Extended Crop Harvest
↓		↓
Project **Impact** The ultimate objective of taking up the Project	→	Scaled upto 50,000 ha across the state within 3 years of completion of the project in 400 ha. Economic empowerment viz., minimum net profit Rs.1.00 lakh / ha and maximum Rs.8.5 lakhs / ha and exceptional Rs.45.00 lakh / ha / annum. Releived the farmers from debt burden, enabled them to expand the area under precision farming on their own to the rest of the land. Capability to raise the crop throughout the year with limited water resource. Buy back tie ups for gherkin, carrot, brinjal, tapioca and banana. Market driven production system gains momentum. Triggered interest of business sector to venture large scale precision farming on their owned estates.

		Self esteem to be farmer and involved the younger generation in Agriculture. Healthy relationship between scientists and farmers besides fostering healthy relationship among farmer themselves. Viability of agriculture as a profitable venture was proved beyond any doubt.

DIFFUSION EFFECTS (WITHIN THE STATE)

1. Dutch rose

Dutch rose cultivation was only in green houses with an investment @ Rs. 1000/ cm^2 at locations of mild climates like Hosur. For the first time, Dutch rose was cultivated at Saragapalli, Hosur in the open in 0.5 ha. The yield was 1,00,000 bunches (20 stems) and marketed at the rate of Rs. 16/ bunch in two years. Though the blooms were not of export quality, it was enough for the retail markets around temples and decorative purpose during social functions which last not more than 12 hours as reported by Mr. Raveendra Reddy of Soragapalli cluster.

There was a spurt among the project and non project farmers to undertake Dutch rose cultivation in the open in an area of 300 ha within a span of three years. The notable issue here is the buy back tie-up for all the flowers from 300 ha. This has happened purely by farmers initiatives and with their own investment.

2. High starch sago

Cassava cultivation under precision system not only ensured 60% yield increase but also phenomenally increased the starch content beyond 30 points. When conventionally grown tubers record 22-25% starch level, the precision grown tubers recorded 30 points, the maximum unit available in the measuring scale. The sago factories offered Rs. 50 extra premium prize / bag for the tubers from the precision fields. Now sago factories extend support in collaboration with drip companies to extend the area of cassava under precision system.

3. Farmers with drip engineering skills

When drip installation was being done, the firms used to unduly delay the process as the 'fitters' were inadequate and most of them were erratic in attending to the work. Many a times, missing the season and over ageing of seedlings happened. That too when the installation was for a group of farmers in a location, extreme delays caused a concern. The precision farmers who underwent a training at Jalgaon, undertook the assignment by themselves and have become trained and skillfull fitters. Introduced new innovative ideas *viz.*, extending the inlet pipe in fertigation tank upto the bottom to ensure uniform flow of fertilizers, fixing 2nd pressure gauge before filter to understand. The level of dust, laying main, sub and gatevalves underground instead of looping above ground. Above all, advising fellow farmers to use urea and white potash as source of water soluble fertilizer to cut the cost of nutrition signifies their competency to handle the system.

4. Precision farmers producer company

Dharmapuri Precision Farmers Agro Services Limited was incorporated under company's act and owned by 166 precision farmers as stakeholders. 'Quality input' concept was propagated through the firm which ultimately reduced the residual toxicity load in the produce. Similarly Erode Precision Farmers Producer Company was incorporated in Erode district. Salem, Trichy, Madurai and Tirunelveli farmers are preparing themselves for incorporation. The firms not only intend to sell inputs but plan for establishing a megha market and a network of 'Farmers own' retail outlet on four lane roads in Tamil Nadu.

5. Supply chain management

Farmers association conducts buyer seller meet regularly involving all the stake holders in the value chain. Visits to markets, participating in trade fairs and interacting with marketing agencies have helped the farmers to have the better understanding of the market trend. TNAUs support in providing on-line mobile phone market information service (www.tnau.ac.in) on whole sale and retail price of the market is fully availed by the farmers for the better advantage.

6. Social transformations

When summer starts, the irrigation water source dwindles and brothers/ neighbours used to pick up quarrel for water. A beaming farmer at Tanjore said 'All three of my brothers have adopted the precision protocol and the feud over irrigation water persisting over generations vanished as they have plenty of irrigation water during summer'. Now there is cordiality and harmony in our family.

'My well is the deepest in my village, when I draw for irrigation, the well goes dry during summer and this deprives the rest of the farmers in my vicinity as their wells also dry up. When I transformed the entire area under precision system, now I have water in my well even during summer and all the neighbouring farmers wells also have water for irrigation. This ensures social harmony in the village.

7. Better livelihood system

Many a farmer's had plans to migrate from the village as the income dwindled and debt mounted. After the harvest of two or three crops, now they have rooted there it self and impart training to rest of the farmers who visit from other districts. In fact many of the farmers have become 'technical consultants' to farmers of other districts. Of course the TOT is through mobile phone from seed to harvest. 'Conducted tours' are also being organized with 'guides' to visit the project site.

8. Growing together concept

The link between TNAU and precision farmers has crossed business and official boundaries and has become emotional one at this stage. From the farmers stand point, the intervention by TNAU through precision farming project has not only revitalized their farming but regained the status of being a farmer. The farmers value the way the scientists lived with the farmers and growing with the crop.

9. Red alert signal to scientists

The process of implementation and values realised through the precision project have sent few red alerts to the scientific community also. There should be a multi disciplinary approach and farmers participatory mode of research management system to investigate any issue. More than the issue, the 'surrounds' of the issue should be taken into account while generating the technologies. The 'surround' for each location is specific with macro climate, micro climate, edapic, socio, politico, community, racial, flora and fauna variations.

10. Agri-business sector

The seed firms have experienced 100% expression of the genetic potential of their variety in precision fields, eg. 500 MT brinjal yield in (ii) a duration of 450 days under (iii) staked conditions. Many of the firms have hiked the price of seed and some renamed the same variety and sold at a highest cost. Firms like Thrinethra, Reliance *etc.*, were not accepted by the precision farmers in as much as farmers wanted a forward contract while the firms are ready for local market price plus Re. 1.00 or spot price. Water soluble fertilizer industry was given a fillip and the products were made known throughout the state. The employees of information technology sector, big business houses have availed the consultancy and undertook precision farming in a big way.

11. Press and media

There was a competition among press and media to highlight the achievement, publish interviews. This includes BBC, London also. In fact having published a write up on precision farming system, reporter himself ventured to undertake cultivation in seven acres.

DIFFUSION OUTSIDE THE STATE

1. Farmers from Pondicherry visited the project site on their own and reported back to their government. The Lt. Governor and Minister

for Agriculture made visits to project site and sanctioned 1000 ha programme for Government of Pondicherry.

2. The public and policy makers of Kerala had seen the write up in Indian Express and the Government had sent a delegation to project site and planning to introduce the programme in Kerala.
3. As in Pondicherry, the farmers from Karnataka had visited the project site and on their return to Karnataka demanded the same project through Agricultural University and the Karnataka Government has sanctioned Rs. 7.50 crores for the project.
4. Two of the farmers from precision site made presentation to Secretary (Agriculture), Government of India and Agricultural Production Commissioner of 32 states and the State of Maharastra, Sikkim, Thripura and Nagaland had sent delegations to study the project on site.

CONCLUSION

The overall impact of TNPFP is excellent on the basis of: productivity increase; quality improvement in the produce; income increase; marketing efficiency; negotiating and bargaining power; cost of cultivation; retention of soil fertility; water saving; energy conservation; farmers confidence level, farmers' proficiency in cultivation practices and related activities; seeds and seedlings usage; duration of crops and market orientation. It could be concluded that the precision farming technologies have been accepted by all the stakeholders and started cultivating the same with their own investments. Productivity and production have been increased tremendously through precision protocol but needs to realize the profit through proper marketing strategies. Almost all the precision farmers have felt that need to establish precision farmers owned mega market to realize higher margin and profitability. This could be possible through public private partnership mode, for that the Government should synergy all the schemes to promote market led agriculture for accelerating the livelihood status of the farming community of country.

REFERENCES

Arockiam Justin W. 2007, Agripreneurship under Tamil Nadu precision farming project -A case study. Unpub. Thesis. Department of Agricultural Extension and Rural Sociology, TNAU, Coimbatore.

Boopathy. C. 2009, Banana, brinjal and watermelon cultivation under precision farming, Morappur Cluster, Dharrnapuri District

Chidmbaran'.G 2008. Taniil Nadu precision farming project: An assessment, State Planning Commission, Government of Tamil Nadu

Chinnasamy, P.M 2008, Tomato and brinjal cultivation under precision farming technology, Somanahal Ii Cluster, Dharmapuri District.

Howlett Peter and Aasish Velkar, 2008. Tamil Nadu precision farming project: An evaluation, London School of Economics, London

Ponnuvel. K 2008, Chillis cultivation under precision farming technology, Agaram Village, Dharmapuri District

Ramareddy. P 2008, Paprika, tornata cultivation under precision farming. Soragapally cluster, Hosur, Krishnagiri Dt.

Ravindra Reddy. N 2008, Dutch rose under precision farming, Soragapally cluster. Hosur. Krishnagiri Dt.

www. thehinduhusmesslinc corn, 2006 (10 17± pages) Precision farming paying dividends 05 'w'av kc ac uk collcctions cconomicl-jistory pdf I AC! SPI)! 708VlkaiHosslctt pdf (44 pages)

23. IMPACT OF IMPROVED VEGETABLE CULTIVARS IN OVERCOMING FOOD INSECURITY

João Carlos da Silva Dias

Technical University of Lisbon, Instituto Superior de Agronomia, Tapada da Ajuda, 1349-017 LISBOA, Portugal

ABSTRACT

Vegetables make up a major portion of the diet of humans in many parts of the world and play a significant role in human nutrition and health. "Hidden hunger" or micronutrient deficiency, affects the health of between 2 and 3.5 billion people worldwide in the developing world. Vegetables are grown worldwide in almost 200 countries, on large and small farms, on good and marginal land, and by large commercial growers and small subsistence farmers. Asia produces 74.7% of the world's vegetables (671 million t) and currently China produces over 50% of the world's vegetables and 313 Kg per capita. China is the largest consumer of vegetables in the world with a per capita consumption of 154 kg.

Breeding is probably nearly as old as agriculture itself. Early vegetable breeders developed landrace cultivars by selection of favorable variations in horticultural traits, yield and resistance to diseases and other problems. Later new breeding methods were developed, including hybridization techniques, culminating with the use of recently developed molecular tools, all leading to modern improved vegetable cultivars. In modern vegetable production the seed business is usualy conducted as a distinct industry. Vegetable seed companies do not only produce and distribute seeds but also play a major role in developing new improved seed cultivars mainly F1 hybrids. Nowadays commercial vegetable seed market is concentrated in the hands of limited number of players. Great emphasis on protection of cultivars by seed companies, including development of F1 hybrids, plant cultivar protection and patenting have been done. There were 392 vegetable crops cultivated worldwide but only slightly over one half of the total number of them have attracted commercial breeding attention.

Many vegetables are consumed near where they are produced. The rapid transformation of the vegetable market with the spread of super and hypermarkets worldwide, market and trade globalization, the improvement of packing and display facilities, the large offer systems and commodities for the production of vegetables all the year round have forced the majority of vegetable growers to use commercial improved or hybrid vegetable cultivars to rise the quality standards and uniformity demanded in those developed markets. There is a general concern about the ability of small growers, mainly from the developing world, to meet exacting quality and safety standards of commercial markets. The production of vegetables with improved or hybrid cultivars offers opportunities for poverty reduction and for overcoming food insecurity by providing more productivity, more marketable produce, more employment, and more income than the traditional landraces.

The decision of grower to purchase hybrid seeds depends on factors such as lack of capital, confidence in the produce market and availability and capital to buy inputs such as fertilizers and pesticides. Almost half of the world vegetable growers are poor and can't afford to buy hybrid seed every growing season. Their economics and logistics make them difficult to buy expensive improved or hybrid cultivars. These cultivars often require an intensive input regime and need large labor inputs for planting and harvest that cannot be met by them, and tend to be riskier than landraces, since the higher costs associated with production impose a greater income risk due to yield and market fluctuations. The lack of capital by subsistence growers denies them the opportunity to invest in vegetable production inputs. If credit facilities and other input facilities are offered, improved or hybrid cultivars can have also a great impact on subsistence vegetable growers overcoming their poverty and food insecurity.

Genetically modified (GM) vegetable cultivars enable breeders to bring favorable genes, often previously inaccessible, into already elite cultivars, improving their value considerably and offer unique opportunities for controlling insects and pathogens. The best opportunities for GM vegetable cultivars are in developing countries where more than 80% of the world's 6.8 billion population live, the majority of vegetables are produced and pest problems are most acute. The high deregulation costs are the main obstacles for the developing of these GM cultivars and so they need to be developed through partnerships between the private and public sector.

In last 50 years vegetable genetic resources are being lost, on a global scale at the rate of 1-2% per year. The multinational seed companies' concentration in huge corporations have merged or canceled some vegetable breeding programs to reduce costs. Then there will be fewer vegetable breeders in the future and the growers will be dependent on a narrower genetic background, that could contribute in a near future, for food insecurity. There is a need of investment in research breeding and cultivar development in traditionally open-pollinated cultivars and in the minor and so-called "forgotten" vegetables. More investments in this area will mean cheaper cultivars for growers to choose from and more preservation of vegetable biodiversity.

In recent years, private plant breeding programs have increased in size and number. Financial investment also increased, as well as interest in intellectual property protection. Protective measures, especially patenting, must be moderated to eliminate coverage so broad that it stifles innovation. The intellectual property protection laws for plants must be made less restrictive to encourage research and free flow of materials and information. Public sector breeding must remain vigorous, especially in areas where the private sector does not function. This will often require benevolent public/private partnerships as well as government support. Active and positive connections between the private and public breeding sectors and large-scale gene banks are required to avoid a possible conflict involving breeders' rights, gene preservation and erosion.

We must ensure that society will continue to benefit from the vital contribution that plant breeding offers, using both conventional and biotechnology tools, because improved and hybrid vegetable cultivars are, and will continue to be, the most effective, environmentally safe, and sustainable way to ensure global food security in the future.

INTRODUCTION

Plant breeders play a key role in determining what we eat, since the cultivars they develop begin the dietary food chain. Every vegetable product we see in the market has benefited from plant breeding in one way or another. Breeding started in the early days of farming when vegetable crops were first domesticated. Without understanding the science behind it, early farmers saved the seed from the best portion of their crop each season. Over the years they selected the traits that they liked the best, transforming and domesticating the vegetable crops they grew.

In the 18th and 19th centuries the Vilmorin-Andrieux family, owner of the first commercial seed company, played an important role in a number of theoretical and technical advances in commercial vegetable breeding, such as producing the first vegetable seed catalog for horticulturists, developing the principles of genealogical breeding programs, improving seed quality through cross-breeding initiatives, and creating disease-resistant and hybrid cultivars of vegetables (Gayon & Zallen, 1998). In 1856 Louis Vilmorin published his "Note on the creation of a new race of beetroot and considerations on heredity in plants", establishing the theoretical groundwork for the modern vegetable seed-breeding industry. The first suggestion to exploit hybrid vigor or heterosis in vegetables was made by Hayes & Jones (1916) for cucumber. Commercial hybridization of vegetable species began in the United States in the middle 1920's with sweet corn, followed by onions in the 1940's. Since that time, private breeding companies are placing more and more emphasis on the development of vegetable hybrids, and many species of vegetables have been changed into hybrid cultivars for the marketplace. Besides heterosis, hybrids also allow breeders to combine the best horticultural traits and multiple disease and stress resistances. Furthermore, if the parents are homozygous, the hybrids will be uniform, an increasingly important trait in commercial vegetable market production.

In the 1970's breeders' rights protection has been provided through UPOV (International Union for the Protection of New Cultivars of Plants), which coordinates an international common legal regime for plant cultivar protection. Protection was granted for those who develop or discover cultivars that are new, distinct, uniform, and stable. Cultivars may be either sexually or asexually propagated. Coverage is for 20 years for herbaceous species. Protective ownership was extended by UPOV in 1991 to include essentially derived cultivars. At the same time, the farmer's exemption (that permitted farmers to save seed for their own use) was restricted, giving member states the option to allow farmer's to save seed. In addition, after 2001, in the

United States plant breeding companies can take advantage of the United States utility patent law to protect not only the cultivar itself but all of the plant's parts (pollen, seeds), the progeny of the cultivar, the genes or genetic sequences involved, and the method by which the cultivar was developed. Protection is for 20 years. The seed can be used only for research that does not include development of a commercial product i.e., another cultivar, unless licensed by the patent older. The United States utility patent is considered the ultimate protective device alowing neither a farmer's exemption nor a breeder's exemption (that permitted that the protected cultivar could be used by others in further breeding to create new cultivars). The creation of vegetable hybrid cultivars requires homozygous inbred parental lines, which provide a natural protection of plant breeders' rights without legal recourse and ensure a market for seed companies. Research and development (R&D) for improved seed development is expensive. Such product protection has presented a business incentive to corporations to invest in the seed industry, which supported an enormous increase in private (R&D) leading to strong competition in the marketplace between the major seed companies. The majority of current vegetable cultivars sold nowadays are proprietary products developed by private R&D. A significant consequence of this increase in R&D has been a reduction in public breeding programs. As a result, the cost for R&D to develop new cultivars is shifting from the publicly supported research programs to the customers of the major seed companies.

One of the main factors that determine success in vegetable production is genetic capacity. No practical breeding program can succeed without large numbers of lines (genotypes) to evaluate, select, recombine and inbreed (fix genetically). This effort must be organized, so valid conclusions can be reached and decisions made. Scientists, breeders, support people and facilities, budgets, and good management are requirements to assure success in the vegetable seed business. Science must be state-of-the-art to maximize success in a competitive business environment. Since the continued need for fundamental breeding research is critical to support development of new technology and expansion of the knowledge base that supports cultivar development, competition between proprietary cultivars results in owner-companies striving to do the best possible research to develop their own products and to compete on genetic and physiological quality of vegetable seed in the marketplace. Reasonable profit margins are necessary to pay back the R&D costs to the owner and to fund future research on developing even better vegetable cultivars to stay competitive. There is considerable genetic variation within the various vegetable species, which can be exploited in the development of superior proprietary cultivars. The consequences of this dynamic situation will mean relatively short-lived cultivars replaced by either the owner of the cultivar or a competitor seed company. This intense competition means constantly improved and more sophisticated cultivars for the vegetable industry. Seed companies are in the business of manipulating genes to improve plant cultivar performance for a profit. The success of the research is judged by the success of the product in making a reasonable profit. The research must improve economic performance starting with the seed production costs and include the grower-shipper/processor and the end user. If any link in this sequence of events is weak or broken, the new cultivar will likely fail.

Biotechnology is a new, and potentially powerful, tool that has been added by all the major seed corporations to their vegetable breeding research programs, and is part of ongoing public research for developing transgenic vegetable projects. It can augment and/or accelerate conventional cultivar development programs through time saved, better products, and more genetic uniformity, or achieve results not possible by conventional breeding. The special techniques of seed production and the combination of breeding improvement programs such as development of hybrids and the incorporation of biotechnology, cannot be efficiently carried by individual growers. As a consequence a high tech seed industry is a key part of modern vegetable production that combines seed production, genetic improvement, storage, and distribution.

The first commercially grown genetically modified (GM) crop was Flavr Savr™ tomato which was released by Calgene in 1994. This tomato contains an antisense version of the *poligalacturonase* (*PG*) gene that was the result of many years of research on several genes involved in fruit development and tomato ripening. They were identified, cloned and characterized for the developing of genetically modified tomato cultivars (Kramer & Redenbaugh, 1994). This tomato was deemed to have a long shelf-life. However, this GM tomato cultivar failed in the market since the cultivar was considered inferior by growers, and was rapidly withdrawn from the market. An important lesson was learned by plant genetic engineers: the importance of cooperation with breeders. New vegetable cultivars must be tested for performance in all markets before sales. Engineered tomato cultivars with delayed ripening and a higher lycopene content are now on the horizon.

In 2008 the global vegetable seed market was estimated at US$3.5 billion with the following share of vegetables: solanaceous (30%), cucurbits (21%), roots and bulbs (16%), brassicas (13%), large seed (13%), leafy and other (7%) (Monsanto, 2009). In the last 8 years global commercial vegetable seed sales had an annual growth rate of 5.8%. With the increase in world population and consumption, and the advent of a high degree of added value through biotechnology, the global market of vegetable seeds is expected to expand in future years.

Food availability and population growth are related, when increases in population exceed increases in food supply, poverty and starvation rapidly ensues. There are now over 6.8 billion human beings inhabiting this planet, and it has been projected that worldwide population growth may exceed 70 million annually over the next 40 years. It is expected to reach approximately 9.5 billion by 2050, when approximately 90% of the global population will reside in Asia, Africa and Latin America. Today 840 million people in the developing countries suffer from malnutrition and 1.3 billion are affected by poverty and food insecurity. Food security is not only about solving the urgency in the short-term; it is also addressing the long-term issue of poverty alleviation and economic growth. Greater investment in agriculture, including breeding and cultivar development, more effective development aid, and reforms to trade and domestic policies are all part of the solution. As worldwide health awareness increases and household income grows, an increasing global demand for vegetables is expected. At the same time, available arable land and a suitable water supply is lessening, so energies should be directed to enhance vegetable productivity and quality to face these real challenges in the society. Also biodiversity is considered essential for food security and nutrition.

In recent times there have been new challenges in the marketing, patenting and in the vegetable breeding domain. The objective of this paper is to discuss these challenges and to highlight the importance of vegetable production, marketing and trading of vegetables, as well as the social aspects of vegetable production and commercial vegetable breeding and their impacts in diversity and world food security.

WORLD IMPORTANCE, MARKETING AND TRADING OF VEGETABLES

World Importance

Vegetables are considered essential for well-balanced diets since they supply many of the vitamins, minerals, dietary fiber and phytochemicals not found in many of the field crops. "Hidden hunger" or micronutrient deficiency, is a pernicious problem that is caused by a lack of vitamins and minerals such as vitamin A, iodine and iron in the human diet and affects the health of between 2 and 3.5 billion people worldwide in the developing world (Peiffer & McClafferty, 2007). Nutrition is both a quantity and quality issue, and vegetables in all their many forms can bring a much-needed measure of balance back to diets. Additionally, there is compelling evidence that a diet rich in vegetables can improve the gastrointestinal health and vision, and reduce the risk for some forms of cancer, heart disease, stroke, diabetes and other chronic diseases (Prior & Cao, 2000; Southon, 2000; Goldberg, 2003). Women in particular benefit from good vegetable nutrition, which is beneficial during later pregnancy, and lactation. The interplay of the different micronutrients and antioxidants found in vegetables has important health impacts, explaining for instance the higher birth weight of children in India, when mothers consumed higher rates of green leafy vegetables and fruits during pregnancy (Rao *et al.*, 2001). Because of their nutritional importance vegetables have been recognized within the nutrition and medical communities and there is a general sense that the promotion of a greater consumption of vegetables will improve health. There is also an increasing awareness among the general public of the advantages of diets rich in vegetables to ensure an adequate intake of most micronutrients, dietary fibers, and phytochemicals that promote health.

Vegetables are grown worldwide in almost 200 countries, on large and small farms, on good and marginal land, and by large commercial growers and small subsistence farmers. According to FAO statistics, the production of vegetables in the world in 2007 was almost 900 million tons (FAO, 2009). Asia produced 74.7% of the world's vegetables (671 million t) on 72.8% of the world's vegetable production area (52.7 million ha). China has always been a large contributor to world vegetable production, and currently this country produces over 50% of the world's vegetables. This percentage translates to 313 kg per capita. In other developing countries of southern Asia, with a mid-subtropical climate favorable for vegetable production, the per capita production is only 80 kg. India is the second largest producer of vegetables in the world but almost a six-fold lower than the production of China. In contrast, the production of vegetables in Africa is lagging far behind the worldwide average. Annually only about 50 kg of vegetables per capita are produced in Africa, which is less than half of the average production on the other continents of the world. Vegetable production during the period 1965-2005 grew at an average annual rate of 5.6% in China and 3.5% in other developing countries in Asia. Worldwide the area of arable land devoted to vegetables is expanding at 2.8% annually, higher than fruits (1.75%), oil crops (1.47%), root crops (0.44%) and pulses (0.39%), and at the expense of cereals (-0.45%) and fiber crops (-1.82%) (FAO, 2009). For example in recent years large areas in Shandong province (China) have been converted from grain to vegetable production to supply inexpensive vegetables for consumers in nearby Japan. The share of cultivated areas allocated to vegetables in 2007 was high in east and southeast Asia (around 10%) and in China (around 20%) but low in south Asia (5%). Although this cultivated area is largest in China, growers in other regions of Asia and in the Pacific have also found it profitable to expand production of vegetables.

The worldwide consumption and importance of vegetables in the diet is difficult to estimate owing to scant production statistics. China is the largest consumer of vegetables in the world. Vegetables make up about 35% of per capita food consumption in China, a much higher share than the world average (Gale, 2002). Besides India, other southern Asia countries such as Bangladesh, Cambodia, Vietnam, Laos and Philippines are also high producers and great consumer of vegetables. For example, vegetables comprise 40% of the Bangladeshi diet (Rich, 2008). Many vegetables are consumed near where they are produced, especially in China, India and other Asian countries. The per capita consumption of vegetables in Asia has increased from 41 kg to 137 kg between 1975 and 2003 (FAO, 2009), particularly in China where the per capita consumption has increased from 43 kg (1975) to 154 kg (2003). In contrast in Africa, the per capita consumption of vegetables lags far behind the worldwide average with less that the WHO/FAO recommended minimum uptake of 200 g of vegetables/day (73 kg/year). In the developed countries there was a remarkable consumption of vegetables but since the 1970s the global amount consumed is increasing at a slow pace. Desire for year-round availability and increased diversity, as well as growing health awareness, have also been important reasons for increased consumption of vegetables in developed countries. For example, the dietary benefit of fresh produce is the major reason for the 25% increase in fresh vegetable consumption in the United States during the 1977–1999 period (Regmi & Gehlar, 2001). Rapid growth in mean per capita incomes in developed countries during the 1990s enabled consumers to purchase a broader range of relatively expensive vegetable commodities such as off-season produce, relatively new or renewed vegetables, and organic produce. Higher incomes of consumers in developed countries have also raised the demand for other attributes such as better quality vegetables and more variation in their daily menu. In developing countries consumption and domestic vegetable markets are also expanding because of an emerging educated middle class with increasing incomes. China, India and Indonesia, countries with the largest populations, have large emerging middle class, which impact the demand for high-value vegetables (Senauer & Goetz, 2003). A great and growing number of urban consumers are demanding safe vegetables in the markets of Hanoi (Figuié, 2004). Increasingly, consumers in developed countries are also concerned about quality and safety of their food, as well as the social and the environmental conditions under which it is produced

A world vegetable survey indicated 392 vegetable crops cultivated worldwide, representing 70 families and 225 genera (Kays & Dias 1995, 1996); non-cultivated species, lower organisms (e.g. fungi), most trees and woody shrubs, and plants grown in or gathered from salt water were excluded. Vegetable crops in which the leaves or young leafy shoots are consumed made up the most common group of vegetables utilized (53% of the total), followed by vegetable fruits (15%). Below ground crop vegetable organs ranked as follows in frequency of use: roots>tubers>rhizomes>corms>stolons, and together comprised 17% of the total number of vegetables consumed. Many vegetable crops have more than one part used. Most of the vegetables are marketed fresh with only a small portion processed. Of these market vegetables, only 67 (17%) have attracted great breeding attention by international seed companies due to their large area of production and consumption, 52 (13%) were considered minor and 85 (22%) were considered rare. In Guangxi province (China), one of the largest production areas of vegetables in the world, only 33 of the 229 total cultivated vegetable species of 56 families were considered major.

Vegetable Marketing

Vegetables are typically perishable products that are of specific high value and that are usually sold through specialized markets. Currently more than 60% of the vegetables produced in the world are sold by vegetable growers to wholesale dealers or huge supermarket chains, and relatively few growers sell their product at retail prices to consumers in farm markets. Globally growers only receive 30% or less of the retail price. This situation is a serious problem for growers.

Domestic and international markets for vegetables are changing rapidly all over the world, partially fuelled by the spread of supermarkets. Consumers increasingly purchase their vegetables and other foods in large convenience stores such as super- and hypermarkets which creates both challenges and opportunities for vegetable producers. Indeed supermarkets may contribute to a higher demand for horticultural products and increase expectations for quality, safety, and presentation, while simultaneously excluding small growers from participating in procurements and contracts. The importance of supermarkets is growing especially fast in the southeast Asia region. Supermarkets are expanding in China 30-40% per year, 2-3 times faster than other developing countries and, according to USDA, China could support at least 9100 hypermarkets (Bi *et al.,* 2004). In Indonesia, Malaysia, and Thailand, supermarkets account for 33% of the market while in Taiwan and Philippines the figure is closer to 63% (Weinberger & Lumpkin 2005). In Indonesia the local retail supermarket chain Ramayana is growing at 25% a year (Manalili & Tumlos, 2004). Otherwise in Africa, supermarket sales in Kenya are growing at 18% a year and they will become the dominant urban food retailers by 2011 (Neven *et al.,* 2006). This is already having a profound effect on the way that vegetables are produced in these and other developing countries and some supermarket chains have implemented programs of assistance and consulting for growers and suppliers to help them in upgrading the quality, marketability, and competitiveness of their products.

The growing importance of convenience supermarket outlets has implications of its own, regarding their methods of procurement and their quality standards. Supermarkets in the cities are bringing quality in vegetables with well packet and presented, providing scope for premium quality as well as novelty items. The difficulties that growers can experience is reflected in fairly rapid declines in the numbers of growers involved, as companies tend to de-list suppliers who do not meet expectations in terms of volume, quality, and timely delivery. In Malaysia, for example, the Giant supermarket chain had 200 vegetable suppliers in 2001, but by 2003, this was down to just 30. In Thailand, similar changes have been seen following the introduction of a distribution center for the TOPS supermarket chain (Shepherd, 2005). Small-scale growers can find it difficult to produce and deliver according to the supermarket chain's standards (Reardon *et al.,* 2003; Boselie *et al.,* 2003). In Brazil the Wal-Mart supermarket chain has about 1500 associated growers producing vegetables from which 90% are large producers and only 10% are small producers. The standards of quality, safety and presentation make it difficult for the small producers to compete. In Portugal and in other European countries the situation is similar. The market is getting more refined in terms of quality and yield expectations, and there is a clear demand for excellent hybrid vegetable cultivars. Success for vegetable growers will depend on their ability to access diverse markets and respond promptly to changes in market conditions. Growers grow vegetables for immediate marketing and their produce is subjected to competition to decide the prices. Hence they cannot compromise the quality of the seeds they use for fear of rejection of their produce. They buy the best seeds, mainly hybrids, and frequently try new products to remain successful. The seed companies in turn get instant response and success if they develop new promising hybrids. Price of seeds is a more critical factor in marginal vegetable areas, where capital and spending input regimes are low, but less important where high yields can be obtained and grower produce can be sold profitably. Providing that the benefits of the hybrid seed are understood, rather than price of seeds the decision of grower to purchase hybrid seeds are in other factors such as lack of capital, confidence in the produce market and availability and capital to buy inputs such as fertilizers and pesticides.

Worldwide the vegetable industry is now in a transition period that is changing from expanding quantity to increasing quality and efficiency. This is being achieved by the introduction of innovative production technology and improvement of market competitiveness. This involves improving the grower's vegetable crop production knowledge, adapting more convenient management techniques and cultural practices such as irrigation and fertigation, protected cultivation, and transplant technology as well as the use of promoting better hybrid vegetable cultivars.

It is expected that the assurance of "safe" vegetable products will become increasingly important. Food safety legislation in the European Union and in the United States is introducing increasingly stricter standards. In general, vegetable production requires more plant protection products per hectare. Because of their cultivation intensity, they suffer from many biological stresses including pests, diseases, and weeds. Because of the diversity within and between plant families, their pest loads are far more varied and complex compared to field crops. Considerably fewer resources have been directed at improving their production and pest and disease management options compared to field crops such as rice, wheat and maize (Lumpkin et al., 2005). Since vegetables are high value commodities with high "cosmetic" standards, the main method for control of pests, diseases and weeds has been the use of pesticides. Vegetables are often consumed in fresh form, so pesticide residue and biological contamination in them is a serious issue. Vegetables account for the major share of the global pesticide market. Almost 25 kg per ha of active pesticide substances are used on average in vegetable production in the European Union (OECD, 1997). Vegetable production accounts for less than 1% of the United States crop area, but 14% of its total pesticide use (Osteen, 2003). In the case of insecticides, nearly 20% of the worldwide US$8.1 billion annual insecticide products are applied to vegetables (Krattiger, 1997). Insecticides are regularly applied to control a complex of insects that cause damage by feeding directly on the plant or by transmitting pathogens that harm plants. Studies on pesticide levels in vegetables, in some developing and developed countries, found that the pesticide residues of vegetables sometimes exceeded tolerance limits (Mansour, 2004; Ferreira, 2009). For example, in India a survey of pesticide residues in vegetable crops taken at the farm gate and in markets from 1999 to 2003 confirmed that of the 3,043 samples, two-thirds had pesticide residues, which were within accepted tolerances, whereas 9% contained residues above the minimum recommended levels (Choudhary & Gaur, 2009). The increasing amount of insecticide residues in vegetables is a major concern to consumers and there are some concerns about the ability of small growers, mainly from the developing world, to meet exacting quality and safety standards of commercial sector.

Pesticide residues are often attributed to the failure of growers to restrain application before harvesting and use of prohibited pesticides. Aside from the effects on growers and consumer health and the environment, the presence of pesticide residues has significant trade implications as well. In South Asia, pest and disease vectors of eggplant, tomato and legumes, notably eggplant fruit and shoot borer, cotton bollworm, root knot nematode, white fly, and legume pod-borer have been identified as the major targets of pesticide use and abuse with frequent and excessive applications of pesticide. For the eggplant fruit and shoot borer pheromone traps and nethouses developed by the World Vegetable Center (known by its acronym AVRDC, Taiwan) have helped reduce pesticide application significantly. The promotion of integrated and biological pest control is expanding worldwide. The use of vegetable improved cultivars with resistance or tolerance to diseases and pests can also contribute to the reduction of pesticide applications and pesticide residues.

Transgenic, or genetically modified (GM), vegetables also offer unique opportunities for controlling insects and the pathogens they transmit. They have provided some successes. Growers in the United States have benefit from having GM virus resistant squash cultivars resistant to *Zucchini yellow mosaic virus* (ZYMV), *Watermelon mosaic virus* (WMV), and *Cucumber mosaic virus* (CMV). These were deregulated and commercialized in 1996 (Tricoli *et al.,* 1995; Acord, 1996). *Bt* sweet corn has proven effective for control of some lepidopterous species and continues to be accepted in the fresh market in the United States where *Bt* fresh-market hybrids are released each year. Biotechnology products will be successful if clear advantages and safety are demonstrated to consumers. Countries vary in their market standards of acceptance of GM products.

Consumers want also more vegetable diversification and continuous supply. Vegetables are purchased on eye appeal, which means that the development of desire to consume increases market demand. Diversification also tends to increase consumption. Product differentiation, including new or renewed product introductions, is still a key strategy for expanding sales in vegetable markets. For example, the fresh tomato category has been differentiated to more than 10 offerings (beefsteak tomato, Roma type tomato, vine-ripe tomato, cocktail tomatoes on vine, tiny-plum tomatoes, mini-plum tomatoes, red cherry tomatoes on vine, attractive yellow and orange cherry tomatoes, mini San Marzano-type tomatoes, tear drop or pear shaped tomatoes, super or premium taste tomatoes). The introduction of specialty fresh baby leaf vegetable salads and fresh-cut products has opened new opportunities for domestic producers. The recent world increased production of baby leaf vegetables is intended to increase desire of elite consumers and is also an excellent way of supplying micronutrients. To exploit such opportunities, it is important to continue research and to disseminate information regarding the benefits of vegetables, develop new improved vegetable cultivars and processed products, evaluate the economic opportunities and the market scope of these new products, and identify marketing trends and alternatives.. Increasingly consumers also demand processed, ready-to-eat vegetables and this is true not only for developed but also for developing countries. The growth rates in retail sales of processed vegetables in several Asian countries over the past years has been tremendous. For example the total percentage of retail sales of processed vegetables during the 2001-2005 period in Vietnam, China, Indonesia, South Korea, Thailand, India and Philippines was 11.2, 9.6, 7.8, 6.8, 6.1, 6.0, and 5.1% respectively (FAO, 2007). The food processing industry has also improved dramatically over the past 30 years in transportation technology, sanitation packaging, storage, and product and packaging appeal. Taken together, these changes have sharply increased the value-added share of vegetable processing and distribution, opening up new marketing opportunities.

Vegetable International Trading

As a result of the increased frequency of world travel, there is a substantially greater variety of vegetable crops available worldwide in many local markets. Vegetables, in particular, are attractive candidates for introduction into a new environment, as most vegetables tend to be fairly short season crops and, as a consequence, lend themselves for cultivation in many diverse areas. Superior vegetable crops and cultivars are often assimilated into the cuisine of the original inhabitants.

Globalization has also now become increasingly important among vegetable producers as indicated by the recent massive increases in international trade of fresh and processed horticultural crops. Worldwide the total volume traded in vegetables has increased fivefold from 1965 to 2005 (FAO, 2007). The European Union (EU), followed by North America and Japan, are the world's most important fresh vegetables import regions. A large share of this comes from Africa, Latin America and the Caribbean (LAC), and from China, each contributing roughly one-tenth to overall vegetable exports. While more than half of global vegetable production takes place in Asia and in the Pacific, only about 12% of the total value of exported horticultural products is generated in Asia and in the Pacific (FAO, 2007). Because of the large share that Asia and the Pacific has in global population, domestic markets and their requirements are likely to be more important in the near future than the attraction of export markets. In Asia, China, Thailand and the Philippines are the largest net exporters of vegetables, while Singapore, the Republic of Korea and Malaysia are the largest net importers of vegetables (FAO, 2007). In sub-Saharan Africa, Kenya and Côte d'Ivoire are two of the most important vegetable producers for export.

While many other agricultural commodities face stagnation and declining world prices, vegetables exports grow dramatically. In many African countries, export of fresh vegetables has become a bright spot in an otherwise dim agrarian landscape (Dolan & Humphrey, 2000). The increase in total volume of vegetables traded worldwide has been dramatic. Still, compared to overall exports of agricultural products, the importance of vegetable exports remains minor, comprising less than 10% of the total value. However, in recent years the share has been rising and is projected to continue to rise faster than other agricultural products. International supermarket chains and large processors are becoming the main buyers of exported fresh vegetables (Reardon *et al.,* 2003) and small-scale growers worldwide need to be trained and organized to meet the challenge of supplying these international players. The major constraints against the participation of small-scale growers in international vegetable exports are the increasing attention that food quality and safety are receiving in food trade and an expansion in the number of non-tariff measures that developed countries apply to horticultural products (Henson & Loader, 2001; Dinham, 2003; Henson *et al.,* 2005). Traceability, phytosanitary, infrastructure, and productivity issues will continue to be a barrier for participation in the vegetable trade for most of the developing world. Application of agricultural chemicals is often poorly regulated and industrial pollutants are common hazards in the soil, water, and air of developing countries. In the future, the inability of these countries to meet increasingly strict phytosanitary and traceability requirements for food products will constrict exports to developed countries. Small-scale growers and processors in developing countries will thus have to learn to supply safe products with traceability labels if their participation in global trade is to continue and expand. Technologies for safe and environmentally friendly vegetable production as well as capacity building should therefore gain particular attention for training to enable small-scale growers to participate in the vegetable production for international markets.

Social Importance of Vegetable Production

Commercial vegetable production is a high input and labor-intensive activity from cultivation to processing. Furthermore, many vegetable crops require careful monitoring of plant health, and careful attention to weed control, irrigation, fertilization, harvest timing, and handling. So the production of vegetables offers opportunities for poverty reduction and for overcoming food insecurity by creating and providing employment, because it is more labor intensive than the production of field crops. Often, vegetable production requires twice as much, sometimes up to four times as much, labor as the production of cereal crops. In Kenya, the production of snow peas and French beans, the two most widely grown horticultural export vegetables, require 600 and 500 labuor days per ha, respectively (Dolan, 2002). In Mexico, the vegetable sector accounts for more than 20% of the total labuor days within the agricultural sector. Shifting from cereal production to vegetable production generates additional employment. Greater employment opportunities result in greater incomes for poor households. Diversification of vegetable production can affect both the structure and the level of employment. But where labour is scarce, availability of hired labour may actually be a limiting factor to vegetable production. Labour demands also arise in the post harvest sector, since transports, packing, sorting, grading, and cleaning are all labour-intensive activities (Weinberger & Genova, 2005). Generally, the vegetable export industry also generates substantial employment. Many work tasks, such as chopping, washing, labeling, and bar coding, are increasingly being transferred to developing countries and are generating many new jobs, particularly among the unskilled segment of the workforce. Usually employment is generated on farms owned by the major exporters and on independent large farms producing for these exporters under contract (Maertens, 2006).

Production of vegetables is becoming attractive for many poor farmers worldwide because it is profitable. Growers engaged in the production of vegetables often earn higher net farm incomes than farmers that are engaged in the production of cereal crops, even though there is greater risk, and requires better and more innovative management. A study in Kenya that sampled small-holder vegetable growers that produce for export found that net farm incomes were five times higher per family member compared to small holders that did not grow vegetable products (McCulloch & Ota, 2002). Lumpkin *et al.* (2005) show that also the per capita farm income of vegetable growers is up to five fold higher than for cereal producers in Asia. Maertens (2006) estimates that households participating in export vegetable production in Senegal earn incomes that are 120% higher than the average household income in the region. The relative profitability of vegetable crops compared to cereals has been shown to be a determining factor for crop diversification into vegetable production in India (Joshi *et al.,* 2003).

The production of vegetable crops is generally more knowledge and capital intense than cereal crops and then is often riskier, because these crops are much more costly to produce per hectare than traditional crops, and because yields and prices are more variable than for field crops (Key & Runsten, 1999). Small-scale and poor growers must thus be supported by an enabling institutional environment, such as access to credit and capital and must be provided with access to market price information.

Vegetable production activity is booming in many world regions and countries with mild climates favorable to vegetable production. In these regions and countries vegetable production is the most important industry in rural economy and also serves as an important social function against crisis. The development of rural markets is considered to be a prerequisite for poor agrarian regions. Growth in vegetable production can contribute to the development of supporting and secondary industries through several means.

Since vegetable producers are usually better integrated into markets than field crop producers, the production of vegetable crops contributes to commercialization of the rural economy, which is characterized by increased trade and marketing. Studies show that commercialization of vegetables stimulates and benefits the rural economy and contributes to the growth and development process through generation of employment and increasing agricultural productivity (Pingali & Rosegrant, 1995; Braun, 1995).

VEGETABLE BREEDING AND FOOD INSECURITY

The impact of vegetable breeding on vegetable production is dependent upon the complex relationships involving the growers, the cultivars available to them, and the developers of those cultivars. Vegetable growers consist of commercial producers with varying size land holdings ranging from moderately small farms to very large ones, and poor growers many of them subsistence farmers with small farms often on marginal lands. Several types of cultivars are available. The least sophisticated in terms of the method of development are landraces. Modern cultivars consist of those developed by crossing and selection alone, those developed by crossing and selection but with specific important improvements often obtained from crosses with wild species or by transgenic methods, and F1 hybrids between desirable inbred lines. The developers of landraces are usually the farmers themselves, and are obtained by repeated simple selection procedures generation after generation. Improved cultivars and hybrids are created either by public sector breeders or seed companies.

Growers in some cases can plant and save their own vegetable seeds, but there are real problems in this system in commercial production, where typically many different species may be grown. In grower-grown seeds, viability may be low, due to poor seed storage environment, pollination is often uncontrolled, genetic improvement is lacking and seed born diseases including virus may be a problem. Thus in modern vegetable production the seed business is most efficiently conducted as a distinct industry dominated by international private seed companies.

The consequences of these relationships may be quite profound for the growers at each level, the seed producers, the availability of food worldwide, and the future of crop diversity and sustainability. Therefore, it is worthwhile to examine vegetable breeding and its connections to assess our future expectations.

Objectives and Techniques of Vegetable Breeding

Vegetable breeding has to address and satisfy the needs of both the consumer and the grower. The general objectives for growers are good yield, disease and pest resistance, uniformity and abiotic stress resistance. Objectives for consumers are quality, appearance, shelf life, taste, and nutritional value. Quality in vegetable crops, in contrast to field crops, is often more important than yield. For growers to survive, cultivars must be accepted by the market. Thus, color, appearance, taste, shape, are usually more important than productivity.

Private breeding companies are placing more and more emphasis on the development of hybrids to exploit heterosis, and to combine multiple disease and stress resistance, but also for economic purposes to ensure growers must purchase seed for each planting. Control of the parents prevents other seed companies from reproducing the hybrid.

Since vegetables are rich in vitamins, minerals and other micronutrients, and therefore vital for health, breeding objectives should include improving the nutritional value. Nutritional content would add value for poor, malnourish populations. Breeding for provitamin A carotenoids, iron, and zinc is of keen interest as a strategy to alleviate nutrient deficiencies in developing countries (Khush, 2002; Carvalho *et al.,* 2006; Graham *et al.,*

2007). An example is the "golden tomato" which contains three to six times more provitamin A carotenoids than standard tomatoes. Developed at AVRDC with conventional breeding techniques, these improved nutritionally-rich tomato lines could help prevent many children of developing countries going blind since vegetarians and populations with limited access to animal products depend on provitamin A carotenoids for vitamin A. One "golden tomato" can provide a person's full daily vitamin A requirements. Similarly, yellow sweet-potatoes are much more nutritious than white ones since they are high in provitamin A carotenoids. Unfortunately, popularity of white-fleshed sweet-potato cultivars in many tropical regions may complicate the acceptance of more nutritious orange ones. Recent studies across a range of Andean potato cultivars show wide variation in calcium, iron and zinc content (Andre *et al.,* 2007) as well as anthocyanins (Brown *et al.,* 2005) due to the existence of red-, blue- and purple-fleshed potatoes. Although a vegetable in order to have impact for nutrients, must be appealing to consumers. Sensory appeal, including color, is an attribute important to consumers when selecting many vegetables.

Growers pay all the breeding work and seed marketing costs when purchasing. improved or hybrid vegetable seed. International seed companies are mainly interested in the breeding and production of vegetable seeds with a high commercial value. Traditional vegetable landraces have largely been neglected by seed companies, policymakers, and researchers. But while their production often takes place under low-input conditions, they contribute substantially to household food and livelihood security, particularly for small resource-poor farmers (Cavendish, 2000; Weinberger & Msuya, 2004). For example, in Africa traditional landraces constitute an important source of micronutrients, contributing between 30% and 50% of iron and vitamin A consumed, respectively, in poor households (Gockowski *et al.,* 2003; Weinberger & Msuya, 2004).

Breeding companies strived more and more to bring hybrid seeds onto the market. Worldwide the share of hybrid seed is increasing at a fast pace of 8 to 10% annually in most of the vegetables. More than two-thirds of the 5000 non-hybrid vegetable cultivars available in 1984 seed catalogs from North America were dropped during last 25 years. A hybrid is produced by crossing two carefully selected parent lines to produce seed combining the best characteristics of each. Hybrids often exhibit higher productivity and vigor than open pollinated cultivars due to heterosis. The superior characters of hybrid plants, unlike that of open pollinated cultivars, cannot be maintained by growers by saving their seeds for growing in the next growing season since the uniformity, vigor and overall performance of the hybrid is lost during seed multiplication. Therefore growers are obliged to buy seeds from the seed companies every growing season if they want to compete in the marketplace. Many growers have been skceptical about the cost of hybrid seeds but found that they gave excellent returns. The main reasons which influenced the growers' decision to adopt improved or hybrid vegetable seeds are the ability of the product to meet the market demand, for high productivity, uniformity, resistance or tolerance to diseases or pests, better response to costly inputs, high quality, and storability. Hybrid vegetable cultivars are used increasingly in large-scale intensive production because they provide increased marketable product for commercial growers and thus add to commercial incentives for the seed companies.

Open-pollinated cultivars are derived by repeated selection of superior plants from within the same line and are genetically uniform for appearance traits. In naturally self-pollinated plants such as tomato, lettuce, and legumes, inbreeding leads to homozygosity. It has generally been found that self-pollinating vegetables such as tomato, after they have become stabilized, do not substantially change their genetic constitution (Kerr, 1969). Thus it is easy to maintain the purity of self-pollinated vegetable crops by removing occasional variants. In the case of cross-pollinated plants such as cucurbits, cole crops, sweet corn, beet, and spinach, uniformity can be achieved by rigorous selection and confining natural crossing within the selected population. Isolation distances of at least 1,500 m between different cultivars of the same species are recommended for all cross-pollinated crops that are to be harvested for seed.

Hybrid vegetable technology has made significant impact on most vegetable crops in developed countries, but a major limitation to vegetable production in many developing countries is the unavailability of high quality seeds. Hybrid seed production is a high level technology and cost intensive venture. Only well organized seed companies with good scientific manpower and well-equipped research facilities can afford hybrid seed production. The public sector in developing countries frequently does not have sufficient capacity to supply adequate quantities of good quality vegetable seed to poor growers and at present, there are few private sector seed companies adapting cultivars to local environments, especially in the poorer countries (Rohrbach *et al.*, 2003). Growers themselves often produce seeds of locally preferred or traditional landraces, as the individual markets are too small and private companies have little interest in producing open pollinated cultivars (Weinberger & Msuya, 2004). Without proper seed production, processing technology, quality assurance, and management supervision, locally produced seeds are often contaminated by seed transmitted viruses and other disease organisms, and are genetically diverse. Lack of proper storage facilities and an effective monitoring mechanism often leads to low or uncertain seed viability and vigor. Moreover, low capital resources and poor market information discourage the development of seed-related agribusinesses. Seed quality and treatment are keys to product quality, and there is a need for upgrading quality control laboratories to meet international standards.

By those reasons in modern vegetable production the seed business is usually conducted as a distinct industry. The global seed trade is now dominated by international corporations whose vast economic power has effectively marginalized the roles of public sector plant breeding and local, small scale seed companies. Thirty years ago there were thousands of seed companies in the world, most of which were small and family owned. Today, the top six global seed companies control almost 50% of the commercial seed trade. Some of these companies belong to worldwide corporations that are also involved with pesticides and biotechnology.

The vegetables attracting the most breeding attention vary considerably between small and huge seed companies/corporations. Small seed companies have a tendency to specialize in a few vegetable crops. In large international companies the breeding activity is more diverse, but is concentrated on the more economically important crops. In these companies, marker assisted selection (MAS) has become an integral component of many commercial vegetable breeding programs (Dias, 1989). The aim of molecular breeding is to supplement conventional methods with faster and more efficient breeding through MAS and/or marker-assisted backcrossing (MAB). Molecular markers that are closely linked to the trait of interest may be identified and applied in gene pyramiding, facilitating introgression of desirable traits into cultivars, early selection, etc. For more complex traits conferred by polygenes, quantitative trait loci (QTL) analysis is carried out. The decision to implement molecular markers as a selection tool is based on the fact that the cost of molecular marker analyses has steadily decreased due to the development of new, less expensive marker technologies, and by the implementation of automation for marker detection. Thus, for phenotypes that are simply inherited, screening with markers is often less expensive, and may be more precise, than screening for the phenotype in the field. QTL analysis is more complex and therefore may be more expensive. Markers bring additional value when they can be used to accelerate the development of new improved vegetable cultivars. This may occur when phenotypes can be assessed only in specific seasons (since markers can be done anytime), or by improving the efficiency of selecting a recurrent parent for backcrossing or for selecting multi-locus genotypes. For example, in southern India, molecular-based tools were used to identify strains of the tomato leaf curl virus (ToLCV) and to select genes of resistance from wild tomatoes that were then bred into cultivated lines. ToLCV-resistant lines were developed that produce twice the yield of the most popular cultivars in the region after only a few years (Hanson *et al.,* 2000). Tomato resistant to geminivirus were also developed with the help of the molecular markers. SCAR and CAPS markers linked to the beta gene in tomato have been developed for MAS (Zhang & Stommel, 2001). MAS for higher provitamin A carotenoids content is under way in muskmelons. Recently developed markers for watermelon lycopene β-cyclase allows selection differentiating lutein yellow carotenoid plants from lycopene red fruited plants early in development (Bang *et al.,* 2007).

In vitro haploidization techniques like anther and microspore culture and *in vitro* gynogenesis, are also used in many breeding seed companies to accelerate the production of homozygous inbred parental lines and new hybrid vegetable cultivars. For example successful *in vitro* microspore culture of broccoli and an improvement in the existing cole protocols, to make this technique available for the purpose of routine breeding, was described by Dias (2001, 2003) in 10 different broccoli genotypes. This protocol is now used by several seed companies worldwide to produce doubled haploid lines for pure inbred progenitors to obtain and ensure uniformity in generated hybrid cultivars. Developmental cycles of commercial hybrid cultivars range between 5 and 12 years. Besides breeding lines necessary for creating hybrid seed need to be refined for specific markets. All these high significant R&D investments make commercial vegetable breeding expensive.

Genetically modified (GM) vegetables have recently been introduced in the market and more are expected to be licensed for sale. Pest and virus-resistant transgenic plants are particularly valuable if no genetic sources of resistance have been identified or if host resistance is difficult to transfer due to genetic incompatibility or links to undesired traits. In such cases, engineered resistance may be the only viable option to develop pest and virus-resistant cultivars. However, vegetables are considered minor crops and traditionally have had fewer resources channeled to them compared to field crops. While it is becoming less expensive to create GM crops for pest management, developing a marketable product and a regulatory package remains costly. Development and regulatory costs can be more readily recouped if the product is grown on an extensive area, as would be done with field crops, but which is not generally the case for individual vegetable crops. For example, the large agricultural biotechnology companies have for the most part abandoned the development of GM vegetable crops because of the high costs associated with product development and deregulation. For vegetables, there are many cultivars of the same crop and the expected life of a particular cultivar can be quite limited. Introducing a GM trait into a breeding program can be complicated and cost prohibitive, especially in crops where backcrossing is difficult or impossible (e.g. potatoes). In most countries, deregulation of a GM trait is event specific. For many vegetable crops, it is not possible to develop a single GM event that can be converted into many different cultivars of a single or closely related group of vegetable species via conventional breeding. For example, *Brassica* contains about 40 closely related commercialized crops, including cabbage, cauliflower, broccoli, Brussels sprouts, turnip, Chinese cabbage, pak-choy, various mustards, swede, vegetable rapes, etc. (Kays & Dias, 1995,1996). No single parent exists that can be used to backcross the transgene into the many different types of *Brassica* botanical varieties. Individual events would have to be developed for many of the crop types and deregulation of more than one event for a single protein is problematic for most business models. For the few transgenic vegetable crops that are being developed, novel or unconventional strategies have been employed to bring the crops to market based generally on private and public partnerships, in which the private sector would focus on selling hybrids to higher end producers while the public sector would focus on low resource growers. Consumers concerns on GM vegetables, in particular in Europe and Japan, is also important. Globally it is widely accepted that the introduction of GM cultivars of vegetables is some way off, but it remains a future possibility mainly in developing countries where more than 80% of the world's 6.8 billion population lives, the majority of the vegetables are produced and pest problems are most acute.

A greater desire for year-round availability of vegetables has had a significant impact on seed companies, requiring full year-round production and consequently a global presence. Unlike for instance agrochemicals "where one size fits all", seed cultivars need to be adapted and differentiated to suit the agronomic needs of the respective region where the vegetable is grown. Active international trade and overseas vegetable seed production by contract is common in many countries. Each multinational company vies to provide better vegetable seeds to compete with domestic seed producers. In China, whose seed market is estimated to be valued at more than US$1.4 billion, the increased recognition of new and high-yielding hybrid cultivars has encouraged the local development of a large number of vegetable seed producers and distributors. Four types of vegetable seed producers were established: public seed companies, research institutes, foreign seed companies, and local seed companies. Private seed companies have been expanding rapidly in recent years and there are now thousands of small firms. Some companies have started to breed their own cultivars and establish marketing networks. They play a strong role in the Chinese vegetable seed industry. About 60 foreign seed companies have opened branch companies or stations in China. Most of them not only sell their vegetable seeds but also have established breeding stations. In other southeastern Asian countries such as India, Indonesia, Vietnam and Malaysia, the percentage of hybrid vegetable cultivars is lower than in China, and so a large expansion of seed companies has not yet occurred. However in India, vegetable seed import policies were liberalized in 1988 and the "New Policy on Seed Development" was instituted to augment productivity and output quality. These events stimulated major growth in the industry and attracted more investment in the seed business from domestic and international seed companies. Consequently the estimated area under vegetable hybrids has gone up from 192,100 ha in 1993-94 to 416,013 ha in 1999-2000. The seed replacement by hybrids during the 2001-2006 period in the main vegetables was: tomato (99.3%), eggplant (63.4%), chili (83.7%), peppers (95%), cauliflower (86.4%), cabbage (100%), radish (96.5%), onion (87.3%), okra (92.4%), muskmelon (71%), watermelon (89.2%), cucumber (72.6%) and gourds (73.5%) (ICRP, 2007). The share of vegetable hybrids have the following distribution: private (64%), imported (33%) and public (3%). Today the hybrid vegetable seed industry in India is estimated to be worth US$40 million and the open-pollinated industry to be around US$118 million. (MAGV, 2008). While there has been rapid growth in the seed markets of developing countries due to a shift away from farm saved seed, the seed markets in developed countries, particularly those of Europe and Japan are stagnant. In Europe and the United States, the seed industry has been concentrated and is largely in the hands of large corporations and many small firms are closing.

Vegetable breeding strategy and targets are dependent on market trends. Successful breeders anticipate changes in the market by developing new cultivars that are ready to be released to the growers when their demand increases. Therefore it will be interesting to see how breeding companies react to changes in vegetable consumption and to evaluate the potential influence that the vegetable market and growing systems may have on breeding targets and priorities.

Prospects for Poor Vegetable Growers

Nearly half of the world's vegetable growers are poor and can't afford to buy hybrid seed every growing season. What are the prospects for these growers since they produce 15-20% of the world's vegetables and they directly feed almost one billion people in Asia, Latin America, and Africa. Capital and risk factors are the key constraints that limit the adoption of improved vegetable cultivars by small-scale and poor growers, because these vegetables generally are much more costly to produce per hectare than traditional landrace cultivars (Key & Runsten, 1999; Ali & Hau, 2001), and most growers require credit to finance their production. While landraces are usually cultivated using a level of input intensity appropriate to the financial resources available within a household, improved vegetable cultivars often require an intensive input regime, including large labour inputs for planting and harvest that cannot be met with family labour alone (Weinberger & Genova, 2005). For small-scale and poor growers, improved vegetable cultivars also tend to be riskier than landraces, since the higher costs associated with seeds and production impose a greater income risk. Small-scale growers may have lower production costs with landraces because they achieve adequate yields with fewer inputs. In addition, the profits from improved cultivars or hybrids tend to be more variable. Yields are often higher but prices are more variable. From another perspective, prices for improved or hybrid vegetable cultivars are more variable because the variability in yields increases the variability in market supply (Key & Runsten, 1999).

Similarly, lack of education, resources, skill training, and support prevent these growers from using improved cultivars and then to generate a stable income from their production. In addition, governments usually do not regulate the price of vegetable crops or even provide market information, unlike for field crops. Improving market information systems for vegetable crops and facilitating farmers' access to credit are then essential components of a strategy to enable poor growers to grow improved vegetable cultivars and to overcome the insecurity of their food supplies. The problem of food insecurity in this situation, like that of poverty, is thus frequently traceable to macroeconomic conditions and market failures due to actions of exploitative intermediaries, including land owners, money lenders, and traders. One of the most severe problems in vegetable production is the shortage of affordable credit. In some cases vegetable growers must pay high interest rates of 15 to 25 percent per 100 days. Desperate for cash, subsistence growers are forced to sell their crops immediately after the harvest to middlemen or their creditors at unfavorable prices. Credit facilities and other inputs must be part of these vegetable production systems, so that the use of improved vegetable cultivars can help subsistence vegetable growers overcome their poverty and food insecurity.

The benefits from the use of improved cultivars are shown by a project supported by the United States Agency for International Development (USAID) and conducted by AVRDC from 1991–2000 at Bangladesh, in two districts (Jessore and Savar), with the aim of overcoming constraints in vegetable production (Weinberger & Genova, 2005). The two districts were selected because they have large vegetable production areas. In Savar three-quarters of all agricultural land is in vegetable production, while in Jessore, the share is 50%. Possible technological interventions included germplasm evaluation and cultivar development for many vegetables, off-season production technologies, and grafting technologies for tomato and watermelon to control soil-borne diseases. Between 1996 and 2000, after cultivar development and the implementation of some facilities, rural infrastructures and extension services, vegetable production grew at an average annual rate of 5.4%. In a survey conducted with 300 growers, the adoption of improved cultivars (42%) and hybrid seed (30%) were the most responsible for this increase in vegetable production of all the technologies used. Among 27 vegetable crops propagated by seed there were 5 (19%) where the only change in vegetable production was the adoption of improved cultivars or hybrid cultivar seeds; in 14 (52%) there was the additional adoption of simple cultural practices like row sowing and fertilization. Tomato grafting and other more sophisticated practices were not implemented. Of the hybrid cultivars used, 92% were of cross-pollinated species in which hybrids are important for uniformity. Vegetatively propagated vegetables did not receive selection or breeding inputs. Eager to increase their production, the majority of the growers (91%), regardless of farmer type, invested in some new vegetable technology over the last five years of the project. The average proportion of growers who adopted an improved technology was 43%, and the average adoption rate among all technologies was 31%. Improved or hybrid vegetable cultivars were 72% responsible for the increase of vegetable production, since it was easier and cheaper for the growers to buy improved cultivar seeds than adopting other technologies. This fact per se highlights the importance of improved cultivars and the need to invest in varietal improvement research, since it requires fewer behavioral changes compared to adopting new crop management practices. In terms of growers' receptiveness to these improved technologies, small-scale growers, particularly small land owners, tend to be late adopters due to skepticism about the cost of improved and hybrid seeds and capital and risk constraints compared with larger-scale growers with large cultivated areas (Collins 1995).

Increased vegetable production has also resulted in important employment benefits for the community such as: new employment opportunities, substitution of family labour for hired labour, and increased wages. The average hired labour man-days in the cultivation of vegetables were 170 per ha (excluding labor from contracted companies), half of the total labour requirements. Likewise, the total value added in wages was approximately US$400 per ha, 7.5 times higher than valued added through employment in rice. In particular, small-scale growers benefit from additional employment opportunities. In small-land owning households cultivating small areas, more than half of men and 16% of women seek employment outside their own farm. One-third of these employment opportunities are vegetable-related. These off-farm employment activities were mainly at the production and post-harvest levels since the vegetable processing sector is not yet fully developed in

Bangladesh. Local support industries have also benefited from the expansion of vegetable cultivation both on the input and output side. A higher degree of input commercialization was observed for vegetables as compared to cereals and included all inputs such as seed, inorganic fertilizers, pesticides, farm manure, plastic, mesh netting and bamboo poles. In general, a higher share of vegetable output was sold on markets as compared to the production of cereals. Vegetable growers were highly integrated into markets, selling a large share of their products and retaining a small portion for home consumption. This was true for both small-scale and large-scale growers. Since supermarkets continue to play a minor role in Bangladesh, most of the vegetable produce was sold either in the local markets or to wholesalers.

In general, the survey project found that vegetable production has contributed to widespread welfare effects and poverty alleviation in Bangladesh. While nearly all communities agreed that they were benefiting from increased vegetable production (either in terms of enhanced consumption, enhanced investment, saving opportunities, or increased welfare), the grower level data also showed that larger-scale growers have been able to capitalize more. On average, 90.3% of households experienced an improvement in their lives over the past five years, but large-scale growers reported greater increases in well-being as compared to smaller-scale growers. The study has also shown that more impact can still be expected, particularly if agro-technology industries develop further. However, the availability of cheap, high quality vegetable seed may be restricted and a major impediment to progress.

Similar projects were implemented by AVRDC in other Asia-Pacific countries and in Africa and the results were similar showing that farmers receive more income from vegetables per hectare than grain crops, and that efficient vegetable production using cheap improved seeds contributes to poverty reduction and less food insecurity. These projects show that improved vegetable cultivars could benefit poor growers and landless laborers by increasing both production and employment. It could benefit the rural and urban poor through growth in the rural and urban non-farm economy and by making food available that is high in nutrients.

Improvement through Genetic Modification

Another means for improving vegetable cultivars is through GM technology. Two successful examples are GM squash and eggplant cultivars.

Pumpkin, squash and gourds are grown throughout the temperate, subtropical, and tropical regions of the world. They collectively rank among the top 10 vegetable crops produced worldwide with nearly 1.6 million ha devoted to production (FAO 2009). In the United States squash yield losses due to viruses often range from 20% to 80% in summer squash (*Cucurbita pepo*) with a reported US$2.6 million economic loss alone in the state of Georgia in 1997 (Gianessi *et al.,* 2002). Three of the most important viruses affecting squash production are *Zucchini yellow mosaic virus* (ZYMV), *Watermelon mosaic virus* (WMV), and *Cucumber mosaic virus* (CMV) (Zitter *et al.,* 1996). No summer squash cultivars with satisfactory resistance to

CMV, ZYMV and WMV has yet been developed by conventional breeding (Gaba *et al.,* 2004). Control of squash viruses has focused on cultural practices, including delayed transplanting relative to aphid flights, use of reflective film mulch to repel aphids, and application of stylet oil to reduce virus transmission, in combination with insecticides to reduce aphid vector populations (Perring *et al.,* 1999). In the state of Georgia, it is estimated that ten applications of stylet oil and insecticide sprays are made routinely to control aphids thereby, limiting virus incidence and transmission (Gianessi *et al.,* 2002). Two lines of squash expressing the coat protein (CP) gene of ZYMV, WMV and CMV were deregulated and commercialized in 1996 (Medley, 1994). Subsequently, many squash types and cultivars have been developed by crosses and backcrosses with the two initially deregulated lines. This material is highly resistant to infection by one, two or all three of the target viruses. (Ochoa *et al.,* 1995; Tricoli *et al.,* 1995; Fuchs et al., 1998; Schultheis & Walters, 1998). Virus-resistant transgenic squash limits virus infection rates by restricting challenge viruses, reducing their titers, or inhibiting their replication or cell-to-cell or systemic movement. Therefore, lower virus levels reduce the frequency of acquisition by vectors and subsequent transmission within and between fields. Consequently, virus epidemics are substantially limited. The adoption of virus resistant squash cultivars has steadily increased in the United States since 1996. In 2005, the adoption rate was estimated at 12% (approximately 3,100 ha) across the country with the highest rates in New Jersey (25%), Florida (22%), Georgia (20%), South Carolina (20%) and Tennessee (20%) (Shankula, 2006). Virus-resistant transgenic squash has allowed growers to achieve yields comparable to those obtained in the absence of viruses with a net benefit of US$22 million in 2005 (Shankula, 2006) that contribute to improvement of their lives. Engineered resistance was the only approach to develop cultivars with multiple sources of resistance to CMV, ZYMV and WMV in those markets where GM squash is allowed.

Another example is GM eggplant in India. Eggplant is a popular vegetable crop grown in many countries throughout the subtropics, tropics and Mediterranean area (since it requires a relatively long season of warm weather to give good yields) on a total of 2 million ha in 2007 (Choudhary & Gaur, 2009) for a total production of 32 million tons. Asia contributed 91.5% of the world production (FAO, 2009). India produces 8 to 9 million tons, one quarter of the global production, making this country the second largest producer in the world after China. It is commonly known as brinjal in India (550,000 ha) and Bangladesh (64,208 ha) and is the most popular vegetable grown in the Philippines (20,000 ha). The crop is often considered a "poor man's vegetable" and is mainly cultivated on small family farms. A total of 1.4 million small marginal and resource-poor growers grow eggplant in all eight vegetable growing zones of India. It is an important source of nutrition and cash income for many resource-poor growers, since they transplant it from nurseries at different times of the year to produce two or three crops, each of 150 to 180 days' duration. Growers start harvesting fruits at about 60 days after planting and continue the harvest for 90 to 120 days, thereby providing a steady supply of food for the family and a stable income. Eggplant was one of the first vegetable crops adopted by growers in India as hybrids. Hybrids occupy more than 50% of the eggplant planted area, the balance being planted with open-pollinated cultivars.

Eggplant is an annual plant. It is attacked by a number of insects (including thrips, cotton leafhopper, jassids and aphids); however, the most damaging is the eggplant fruit and shoot borer (FSB, *Leucinodes orbonalis*), which is considered the most serious and destructive pest of eggplant in Asia and Africa (Palada et al. 2006). Infestation is caused by adults migrating from neighboring fields, from eggplant seedlings, or from previously grown eggplants in the same planting area. Damage from *L. orbonalis* starts at the nursery stage and continues after crop transplanting until harvest. Losses have been estimated to be between 54 and 70% in India and Bangladesh and up to 50% in the Philippines (Choudhary & Gaur, 2009). Recommended insect pest management practices include the prompt manual removal of wilted shoots, trapping male moths using pheromones to prevent mating, ensuring regular crop rotation, and using nylon net barriers. These methods, however, are not widely adopted by growers because of time and resource constraints or lack of awareness. There are no known eggplant cultivars resistant to the FSB, so the use of insecticide sprays continues to be the most common control method used by growers. The borer is vulnerable to sprays only for a few hours before they bore into the plant. Therefore, growers in India spray insecticides as many as 40 to 80 times over a 7-month cropping season (AVRDC, 2001; Choudhary & Gaur, 2009). Growers may even spray every other day, particularly during the fruiting stage, which contributes to the presence of pesticide residues. But despite the application of many insecticides, the eggplant fruits sold in the Indian market are still of inferior quality, infested with larva of FSB (Choudhary & Gaur, 2009). A survey of pesticide use in Central Luzon in the Philippines indicates also that growers there spray up to 56 times with insecticides during a crop season to protect their eggplant crops against the borer (Palada et al., 2006).

The decision of growers to spray is influenced more by subjective assessment of visual presence of the insect rather than guided by the more objective science-based methodology of threshold levels. This reliance on visual assessment leads to gross over-spraying with insecticides, higher insecticide residues, and unnecessary increase in the growers' exposure to insecticides. On average 4.6 kg of active ingredient of insecticide per hectare per season is applied on eggplant (Choudhary & Gaur, 2009). Such pesticide use, besides being detrimental to the environment and human health, also increases the cost of production making this humble vegetable expensive for poor consumers. In Asia, chemical spraying for this insect accounts for 24% of the total cost of production (Choudary & Gaur, 2009). Intensive use of insecticides raises serious concerns for environmental and human health if not properly applied. A study conducted in the Jessore district of Bangladesh found that "98% of farmers felt sickness and more than 3% were hospitalized due to various complexities related to pesticide use" (AVRDC, 2003). FSB-resistant *Bt* eggplant was genetically engineered by Mahyco under a collaborative agreement with Monsanto® and the first *Bt* transgenic eggplant with resistance to FSB was produced in 2000. This GM eggplant incorporates the *cry1Ac* gene expressing insecticidal protein to confer resistance against FSB. This *Bt*-eggplant was effective against FSB, with 98% insect mortality in *Bt*-eggplant shoots and 100% in fruits compared to less than 30% mortality in non-*Bt* counterparts (ISAAA, 2008). Similar to *Bt*-sweet corn it is expected that *Bt*-eggplant cultivars –soon to likely released for growing in India– will reduce pesticide applications, contribute to poverty reduction and to overcome food insecurity. Krishna & Qaim (2008) indicated that simulations using farm-survey data suggest the aggregate economic surplus gains of *Bt*-eggplant hybrids could be around US$108 million per year in India. Eggplant consumers will capture a large share of these gains, but farmers and the seed company will also benefit. By sharing this technology with the public sector, *Bt* open-pollinated eggplant cultivars may become available with a certain time lag (Krishna & Qaim 2007; Kolady & Lesser 2008), thereby making this technology more accessible for resource-poor farmers.

Besides this research work in FSB *Bt* eggplant the genetic engineering approach seems to be very promising for the improvement of eggplant as transgenic lines with interesting traits, particularly resistance or tolerance to other insects, nematodes, diseases, and abiotic stress, as well as parthenocarpic fruits.

Future of Breeding and Diversity

About one half (52%) of the total number of vegetables cultivated in the world get commercial breeding attention by seed companies and, of those, only 17% are in large scale breeding programs, fostering a need for serious attention to maintenance of vegetable crop biodiversity. There has been a severe decline in the vegetable cultivar genetic base, as evidenced by the significant reduction, especially within the last 50 years, in the number and range of vegetable cultivars grown. During this period vegetable genetic diversity has been eroding all over the world and vegetable genetic resources are disappearing, on a global scale, at an unprecedented rate of 1.5–2% per annum. Widespread adoption of simplified vegetable systems with low genetic diversity carries a variety of risks including food insecurity. In the short term, such systems risk potential crop failure. In the longer term, they encourage the reduction of the broad genetic base that contributes to high yields, quality traits, disease and pest resistance, etc. and thus compromise the future genetic health of vegetables. Especially prominent among the "enemies" of genetic diversity are the commercial markets and economic social pressures that have influenced breeding practices that promote uniformity, encouraging extensive cultivation of preferred improved and hybrid vegetable cultivars with insufficient diversity. In addition, globalization has stimulated the consolidation of vegetable seed companies into huge corporations and the decline of small seed companies that serve local and regional markets. In consequence some vegetable breeding programs have been merged or eliminated to reduce costs. Thus fewer and fewer companies/

corporations are making critical decisions about the vegetable research agenda, and the future of vegetables worldwide. Inevitably, two things will happen. There will be fewer vegetable breeders in the future and growers will be dependent on a narrower genetic background that could contribute in the near future to food insecurity for poor growers and consumers. Also, with the advent of genetic engineering, these huge seed corporations are also assuming ownership of a vast array of living organisms and biological processes. Of equal concern are expanded uses of legal mechanisms, such as patents and plant breeder's rights that are removing vegetable plant germplasm from general public use (Ryder, 2005). Intelectual property rights for plants was intended as a defensive mechanism to prevent the loss of invented cultivars to competitors. However, with the more stringent enforcement of plant breeding rights, and particularly with the application of the utility patent law in the United States to protect all forms of an innovation, this has become an offensive weapon to stifle competition and inhibit the flow of germplasm and information. This can have serious implications for the future conservation of vegetable genetic resources and for world food security.

Some landraces and old open-pollinated cultivars of vegetables have existed for long periods outside the commercial and professional plant breeding circles because they have been kept alive within communities by succeeding generations of seed savers. Unfortunately, active seed savers tend to be more and more a minority among the millions of vegetable growers. Due to the demand of commercial markets and the professionalization of the sector, many growers are no longer saving seeds. This is an additional threat to genetic diversity. So continued survival of landraces and open-pollinated cultivars of vegetables depends largely on popular interest and initiative as well as preservation in gene banks. We should be alerted and concerned about the loss of biodiversity in vegetables and about this impact on food security.

Vegetable growers have an important role in conserving and using vegetable biodiversity. The future of world food security depends not just on stored vegetable genes, but also on the people who use and maintain crop genetic diversity on a daily basis. In the long run, the conservation of plant genetic diversity depends not only on a small number of institutional plant breeders and seed banks, but also on the vast number of growers who select, improve, and use vegetable diversity, especially in marginal farming environments. That is why we should be also alerted and particularly alarmed by the current trend to use improved and hybrid vegetable cultivars exclusively. Growers do not just save seeds, they are plant breeders who are constantly adapting their vegetable crops to specific farming conditions and needs. For over 300 vegetable generations, vegetable growers have been selecting seeds and adapting their plants for local use. This genetic biodiversity is the key to maintaining and improving the world's food security and nutrition. No plant breeder or genetic engineer starts from scratch when developing a new cultivar of tomato, pepper, cabbage or lettuce. They are building on the accumulated success of generations of growers, who have selected and improved vegetable seeds for thousands of years. If poor small-scale growers in marginal areas stop saving seeds, we will lose genetic diversity. Growers will lose the means to select and adapt vegetable crops to their unique farming conditions, which are characterized by low external inputs. Hybrid seed technology is designed to prevent growers from saving seed from their harvest, thus forcing them to return to the commercial seed market every year. Hybrid vegetable seeds alone, and used globally, can be a dead-end to biodiversity. If growers abandon completely their traditional vegetable landraces in the process of adopting only hybrids, crop genetic diversity achieved over centuries will be lost forever. Many agronomic benefits will be lost to worldwide growers and thus to consumers.

The exclusive adoption of hybrid cultivars in marginal areas may restrict the vegetable producing capacity of growers, it will destroy biodiversity, and it may contribute in the long-term to food insecurity. A study by Daunay *et al.* (1997) points out that the release of F1 hybrids (in Europe and some Asian countries like China and Japan) displaying higher productivity, but with poor phenotypic variability, has contributed to the losses of eggplant landraces, thus inevitably leading to genetic erosion of *S. melongena.* Moreover, some African cultivated eggplants have been lost following social, economic, and political changes (Lester *et al.*, 1990). Therefore, the cultivated eggplant has been considered a priority vegetable species for the preservation of genetic resources since 1977. Several studies have been carried out in Asia and Africa (Lester *et al.*, 1990; Gousset *et al.*, 2005), and collections built up, particularly in China (Mao *et al.*, 2008).

Fortunately, in some developed countries new independent seed companies, offering unique collections of regionally adapted landrace vegetable cultivars, have recently emerged. Furthermore vegetable hobbyist groups, mainly from organic horticulture, are thriving and maintaining old vegetable landraces, in organizations known as "Seed Savers". In this way traditional landraces are being restored to native growers and urban and peri-urban growers. Some of these traditional landraces display combinations of traits that make them especially responsive to local or regional conditions, or are well-suited to particular growing methods such as those used in organic horticulture or low-external-input systems, or are tolerant to local pests and diseases, or other stresses and constraints. Organic growers who seek to grow "full-cycle" or seed-to-seed, are also working to ensure the continued availability of organically grown seeds. There are also considerable ongoing efforts by national governments and international organizations to preserve plant vegetable germplasm in gene banks. This is a valuable but static approach, as further evolutionary changes and improvements will not occur until the seeds are planted. It is also an activity that relies heavily on continued political stability and support, including sustained governmental funding. Active and positive connections between the private breeding sector and large-scale gene banks are required to avoid possible conflict involving breeders' rights and gene preservation.

The diversity of crop species will be promoted by the maintenance of crop gene banks by governments and non-governmental organizations, the continued use of diverse sources by plant breeders, especially in the public sector, and by the use of local cultivars and landraces by farmers.

CONCLUSION

Domestic and international vegetable markets are changing rapidly and a variety of factors such as supermarkets and improvements in transportation and refrigeration have largely contributed to this development. Direct investment from global retailers has also had an impact on the structure of retailing. Trade liberalization has impacted on the increasing importance of exports, which are increasing for high value vegetable crops and processed foods alike. Increasing urbanization, with increasing incomes mainly of growing middle classes in most parts of the world, requires large quantities of vegetables. These may be produced locally or at great distances from where they are consumed, with effects on vegetable post-harvest processing and value-added activities.

The standards for participation in high value vegetable markets have increased, both in developed and developing countries and supply chains are increasingly complex, undergoing rapid changes, and often based on strong vertical integration. Informed policies and a conducive regulatory environment will increase the incentives for agents in the supply chain to accept the produce of small-scale growers as inputs, and improve the growers' capacity to meet the product attributes required in a rapidly modernizing agricultural marketplace. The participation of these small-scale growers in dynamic vegetable markets for higher value vegetables is a major challenge. Participation requires, particularly in developing countries, a set of institutional changes, training, and credit facilities to allow them to compete in increasingly competitive global markets that demand safe, uniform and high-quality produce.

Among the obstacles to participation in modern markets is that yield improvements in vegetables have been lower than in cereals. Concerted actions are required to raise the average yields of vegetable crops. Yields will increase by focusing on inherent productivity and on maintaining stability of yields. More emphasis will have to be placed on the development of modern cultivars, using the natural vigor of hybrids to fight stresses of disease, heat, and drought. Molecular tools will be useful to select genes of resistance, and to increase quality, nutritional value, and yields. These traits plus food safety are important aspects of future breeding efforts. Overall, there is great genetic and phenotypic diversity for types and amounts of micronutrients in the different vegetables. Consequently there is a good potential to increase micronutrient content and thus enrich the diet of the average consumer. More research is needed towards these goals to provide benefit to poor and malnourished populations.

Creation of vegetable hybrids is a key means towards the development of cultivars for modern vegetable production. Hybrid seed production is high technology and a cost intensive venture. Only well organized seed companies with good scientific manpower and well equipped research facilities can afford seed production. Due to globalization most vegetable breeding research and cultivar development in the world is presently conducted and funded in the private sector, mainly by huge multinational seed companies. Public vegetable breeders and cultivar development activities and research are disappearing worldwide. This means in general that there are fewer decision-making centers for vegetable breeding and cultivar development. This has also resulted in the focus on relatively few major vegetables produced world-wide, to the detriment of all the other cultivated vegetables. It is imperative that national governments and policymakers, as part of a social duty, invest in breeding research and cultivar development of traditional open-pollinated cultivars and in the minor and so-called "forgotten" vegetables. More investments in this area will mean less expensive seed for growers to choose from, and an increased preservation of vegetable biodiversity. The accompli-shment of this goal may require new approaches to vegetable breeding research and development by both the public and private sector.

Until recently vegetable breeding research and development that targets small-scale and poor vegetable growers has largely been undertaken by public sector institutions and national agricultural research institutes. However, the capacity to undertake the work was mainly dependent on national or international funding and expertise. The work has been limited by the capacity of these institutions to pay for it. As a result horticultural advancement has varied enormously among countries and even within regions in developed and developing countries. In the area of plant breeding the process of producing improved cultivars is slow, and it requires long-term sustained commitment that may not fit the continuing changes in the national and international politics of research funding. The application of biotechnology promises acceleration of some aspects of plant breeding, but the adoption of more advanced technologies significantly raises the cost of research at a time when investment funding has diminished.

Public plant breeding remains a key component of vegetable research systems worldwide, especially in developing countries. However, the incre-asing presence of private sector breeding and a decrease in national and international support makes it difficult for the public sector to continue operating in the traditional manner. Declining funding for public vegetable breeding coupled with the rapid increase of vegetable production and consumption and an urbanizing population, has created a difficult situation. More public sector vegetable breeders are needed worldwide to select and to produce non-hybrid cultivars of the minor and "forgotten" vegetables. This will benefit small-scale growers, and will safeguard biodiversity and food security in developing countries. A good example is China where there are four types of vegetable breeders and seed producers: public seed companies, research institutes, foreign seed companies, and local seed companies.

While the maintenance of vigorous public sector breeding programs in areas where private companies are not interested in providing low cost cultivars is highly desirable, an additional approach to maximize vegetable and horticultural research input would be the development of global programs with public–private partnerships. The public sector may support portions of vegetable and horticultural R&D that are not attractive to the private sector, and feed improved breeding lines and systems to the private sector for exploitation in regions where the private sector is active, and nurture private sector development in regions where it is lacking. Many in the public and private sectors support such a complementary approach to overcome poverty and malnutrition in developing countries.

In recent years, private plant breeding programs have increased in size and number. Financial investment also increased, as well as interest in intellectual property protection. The spirit of the original attempts to protect plant breeders' rights was that the granting of a certificate of protection should not inhibit the flow of information and products through continued research by the entire plant breeding community. In the classic sense, the patent is a defensive tool to prevent competitors from reaping some of the benefits rightfully belonging to the inventor. In the modern context, it is an offensive weapon, to stifle competition, prevent further innovation by others and maximize income. The United States utility patent is a means to slow the flow of progress of plant breeding research, except within the company holding the patent. While obviously benefiting that company, it is a big step backwards for the plant breeding community and by extension, for agriculture itself. The intellectual property protection must encourage research and free flow of materials and information. Protection should be for the cultivar only. There should be no constraint against other breeders using that cultivar in further research, including further breeding. Another breeder should be free to use the protected cultivar in a cross, followed by further development through pedigree breeding. Another breeder should also be free to transfer genes controlling economic traits into the protected cultivar by the backcross method or by genetic transformation procedures.

Genetically modified (GM) vegetable cultivars enable breeders to bring favorable genes, often previously inaccessible, into already elite cultivars, improving their value considerably and offer unique opportunities for controlling insects and pathogens. The best opportunities for GM vegetable cultivars are in developing countries where more than 80% of the world's 6.8 billion population live, the majority of vegetables are produced and pest problems are most acute. The GM eggplant in India is an excellent working example of a model pro-poor philanthropic-public-private sector partnership. *Bt* eggplant technology has been generously donated by its private sector developer, Mahyco to public sector institutes in India, Bangladesh and the Philippines for incorporation in open-pollinated cultivars of eggplant for the use of small resource-poor farmers. The development and regulation of Bt eggplant in India, will also serve the common regulatory needs in Bangladesh and the Philippines and generally will facilitate the regional harmonization of biotech vegetable crop use in Asia. It can also serve as a model to facilitate regional harmonization in other developing countries in Asia, the Pacific, sub-Sahara Africa or Latin America and the Caribbean, where the need for simplified, responsible and appropriate regulation is great. It is the key barrier that denies growers in the developing countries timely access to the significant benefits that biotech vegetable crops offer. This sharing of knowledge and of experience with the regulation process could greatly simplify and lighten the regulatory burden by eliminating duplication of the significant effort already expended by India, thereby contributing to the important goal of harmonizing regulations among countries.

Molecular tools will be useful for selecting resistance genes, and increasing quality, nutritional value, and yields. These value-added traits plus food safety will be important aspects of future GM vegetable breeding efforts. Biotechnology provides the ability to produce a broad array of insect-resistant and pathogen-resistant cultivars that also express a variety of other value-added traits such as nutritional and post-harvest traits. As the number of value-added, GM traits increases, the number of potential combinations of traits that could be stacked within individual cultivars increases geometrically, as to the cost associated with maintaining inventories of geographically adapted cultivars expressing different combinations of traits. Consequently, we can expect that commercially available, GM vegetable cultivars of the future will express multiple, unrelated, transgenic traits.

Many current vegetable breeding efforts remain underfunded and disorganized. There is a great need for a more focused, coordinated approach to efficiently utilize funding, share expertise, and continue progress in horticultural technologies and programs. Vegetable production can help the poor escape poverty and malnutrition in the 21st century, but only if enough investments are made to improve and sustain the breeding and productivity of vegetable crops. Policymakers and investors (including international aid and philantropy) have to turn their attention to enhanced funding for the vegetable and horticultural sector, allowing growers to compete with their products on a world market increasingly determined by market quality standards and phytosanitary concerns and regulations. Only then will the silent vegetable and horticultural revolution currently underway benefit a significant portion of the world's poor nations, growers, and landless laborers and enable us to overcome poverty, malnutrition and food insecurity.

In summary, we must ensure that society will continue to benefit from the vital contribution that plant breeding offers, using both conventional and biotechnological tools, because improved and hybrid vegetable cultivars are, and will continue to be, the most effective, environmentally safe, and sustainable way to ensure global food security in the future.

REFERENCES

Acord, B.D. 1996. Availability of determination of non-regulated status for a squash line genetically engineered for virus resistance. *Federal register* **61**:33484-33485.

Ali, M. & V.T.B. Hau 2001. Vegetables in Bangladesh. *Technical Bull.* **25**. AVRDC, Shanhua, Tainan.

Andre, C.M., M. Ghislain, P. Berlin, M. Oufir, M.R. Herrera, L. Hoffman, J.F. Hausman, Y. Larondelle & D. Evers 2007. Andean potato cultivars (*Solanum tuberosum* L.) as a source of antioxidant and mineral micronutrients. *J. Agr. Food Chem.* **55**:366-378.

AVRDC (Asian Vegetable Research and Development Center), 2001. Fact Sheet: Eggplant Fruit and Shoot Borer. AVRDC, Shanhua, Tainan.

AVRDC, 2003. Socio-economic Parameters of Eggplant Pest Control in Jessore District of Bangladesh. *Technical Bull.* **29**. AVRDC, Shanhua, Tainan.

Bang, H.J., S.G. Kim, D. Leskovar & S. King, 2007. Development of a codominant CAPS marker for allelic selection between canary yellow and red watermelon based on SNP in lycopene beta-cyclase (LCYB) gene. *Mol. Breed.* **20**:63-72.

Barrett, D.M., E. Garcia & J.E. Wayne, 1998. Textural modification of processing tomatoes. *Crit. Rev. Food Sci. Nutr.* **38**: 173–258.

Bi, X., X. Dong, J. Huang, D. Hu & S. Rozelle, 2004. Securing small producer participation in restructured national and regional agri-food systems - *China Country Report. Regoverning Markets* - Center for Chinese Agricultural Policy. Chinese Academy of Sciences, Beijing.

Boselie, D., S. Henson & D. Weatherspoon, 2003. Supermarket procurement practices in developing countries: Redefining the roles of the public and private sectors. *Am. J. Agr. Economics* **85**: 1155–1161.

Braun, J. von, 1995. Agricultural commercialization: Impacts on income and nutrition and implications for policy. *Food Policy* **20**: 187–202.

Brown, C.R., D. Culley, C.P. Yang, R. Durst & R. Wrolstad, 2005. Variation of anthocyanin and carotenoid contents and associated antioxidant values in potato breeding lines. *J. Am. Soc. Hort. Sci.* **130**:174-180.

Carvalho, P.G.B., C.M.M. Machado, C.L. Moretti & M.E.N. Fonseca, 2006. Vegetable crops as functional food. *Hort. Brasileira* **24**:367-404.

Cavendish, W., 2000. Empirical regularities in the poverty–environment relationship of poor households: Evidence from Zimbabwe. *World Development* **28**: 1979–2003.

Choudhary, B. & K. Gaur 2009. The development and regulation of Bt Brinjal in India. International Service for the Acquisition of Agri-Biotech Aplications (ISAAA). ISAAA Brief 38.

Collins, J., 1995. Farm size and non traditional exports: Determinants of participation in world markets. *World Development* **23**: 1103-1114.

Daunay, M.C., R.N. Lester & G. Ano, 1997. Les Aubergines. **In**: Charrier, A., M. Jacquot, S. Hamon & D. Nicolas, (eds.) L'amélioration des Plantes Tropicales. Repères. CIRAD-ORSTOM, Paris. pp. 83–107.

Dias, J.S., 1989. The use of molecular markers in selection of vegetables. SECH, *Actas Hort.* **3**:175-181.

Dias, J.S., 2001 – Effect of incubation temperature regimes and culture medium on broccoli microspore culture embryogenesis. *Euphytica* **119**:1-6.

Dias, J.S., 2003 – Protocol for broccoli microspore culture. In: Maluszynski, M., K.J. Kasha, B.P. Forster & I. Szarejko (eds.). Doubled haploid production in crop plants. A manual. Kluwer Academic Publishers, Dordrecht. pp. 195-204.

Dinham, B., 2003. Growing vegetables in developing countries for local urban populations and export markets: Problems confronting small-scale producers. *Pest Manag. Sci.* **59**: 575–582.

Dolan, C. S., 2002. Gender and witchcraft in agrarian transition: The case of Kenyan horticulture. Develop. Change 33: 659–681.

Dolan, C. S. & J. Humphrey, 2000. Governance and trade in fresh vegetables: The impact of UK supermarkets on the African horticulture industry. *J. Develop. Studies* **37**: 147–176.

FAO, 2007. FAOSTAT data. http://www.fao.org.

FAO, 2009. FAOSTAT data. http://www.fao.org.

Ferreira, J. (ed.), 2009. As bases da Agricultura Biológica. Tomo I – Produção Vegetal. EDIBIO, Castelo de Paiva.

Figuié, M., 2004. Consumer perception of vegetable (tomatoes and water morning glory) quality in Hanoi. Sustainable Development of Peri-urban Agriculture in South-East Asia Project (Kingdom of Cambodia, Lao PDR, Vietnam RS) Project Report. CIRAD, Hanoi.

Fuchs, M., D.M. Tricoli, J.R. McMaster, K.J. Carney, M. Schesser, J.R. McFerson & D. Gonsalves, 1998. Comparative virus resistance and fruit yield of transgenic squash with single and multiple coat protein genes. *Plant Disease* **82**: 1350–1356.

Gaba, V., A. Zelcer & A. Gal-On, 2004. Cucurbit biotechnology - the importance of virus resistance. *In vitro Cell. & Develop. Biol. - Plant* **40**: 346–358.

Gale, F., 2002. China's food and agriculture: Issues for the 21st century. Market and Trade Economics Division, Economic Research Service. USDA, Washington DC.

Gayon, J. & D.T. Zallen, 1998. The role of the Vilmorin Company in the promotion and diffusion of the experimental science of heredity in France, 1840-1920. *J. History Biol.* **31**: 241-262.

Gianessi, L.P., C.S. Silvers, S. Sankula & J.E. Carpenter, 2002. Virus resistant squash. **In**: Plant Biotechnology: Current and Potential Impact for Improving Pest Management in U.S. Agriculture. An Analysis of 40 Case Studies. National Center for Food and Agricultural Policy, Washington, DC, USA.

Gockowski, J., J. Mbazo'o, G. Mbah & T. Moulende, 2003. African traditional leafy vegetables and the urban and peri-urban poor. *Food Policy* **28**: 221–235.

Golberg, G. (ed.), 2003. Plants: Diet and Health. The report of a British Nutrition Foundation Task Force. Blackwell Science, Oxford.

Gousset, C., C. Collonnier, K. Mulya, I. Mariska, G.L. Rotino, P. Besse, A. Servaes & D. Sihachakr, 2005. *Solanum torvum*, as a useful source of resistance against bacterial and fungal diseases for improvement of eggplant (*S. melongena* L.). *Plant Science* **168**: 319–327.

Hanson, P. M., D. Bernacchi, S.K. Green, S.D. Tanksley, V. Muniyappa & A. Padmaja, 2000. Mapping a wild tomato introgression associated with tomato leaf curl virus resistance in a cultivated tomato line. *J. Am. Soc. Hort. Sci.* **125**: 15–20.

Hayes, H.K. & D.F. Jones, D.F. 1916. First generation crosses in cucumber. *Republic Conf. Agr. Exp. Stat.* **5**:319-322.

Henson, S. & R. Loader, 2001. Barriers to agricultural exports from developing countries: The role of sanitary and phytosanitary requirements. *World Devel.* **29**: 85–102.

Henson, S., O. Masakure & D. Boselie, 2005. Private food safety and quality standards for fresh produce exporters: The case of Hortico Agrisystems, Zimbabwe. *Food Policy* **30**: 371–384.

ICRP (India Coordinate Research Program) 2007. Vegetable seed industry prospect and retrospect. India Coordinate Research Programme. Project Directoreat of Vegetable Research. New Dheli.

ISAAA, 2008. *Bt* brinjal in India. Pocket K 35. International Service for Acquisition of Agri-Biotech Applications, Ithaca, New York.

Joshi, P.K., A. Gulati, P.S. Birthal & L. Tewari., 2003. Agriculture diversification in South Asia: patterns, determinants, and policy implications. MSSD Discussion Paper 57. IFPRI, Washington DC.

Kays, S.J. & J.S. Dias, 1995. Common names of commercially cultivated vegetables of the world in 15 languages. *Economic Botany* **49**:115-152.

Kays, S.J. & J.S. Dias, 1996. Cultivated Vegetables of the World. Latin Binomial, Common Names in 15 Languages, Edible Part, and Method of Preparation. Exon Press, Athens, Georgia.

Kerr, E.A., 1969. Do tomato cultivars deteriorate after they are introduced? Reports of the Horticultural Research Institute of Ontario, Vineland Station, Ontario, Canada. p. 75–80.

Key, N. & D. Runsten, D. 1999. Contract farming, small-holders, and rural development in Latin America: The organization of agroprocessing firms and the scale of outgrower production. *World Devel.* **27**: 381–401.

Khush, G.S., 2002. The promise of biotechnology in addressing current nutritional problems in developing countries. *Food Nutr. Bul.* **23**:354-357.

Kolady, D. & W. Lesser, 2008. Can owners afford humanitarian donations in agbiotech - The case of genetically engineered eggplant in India. Elect. J. Biotech. 11, http://www.ejbiotechnology.info/content/vol11/issue2/full/5/

Kramer M.G. & K. Redenbaugh, 1994. Commercialization of a tomato with an antisense polygalacturonase gene: the FLAVR SAVR™ tomato story. *Euphytica* **79:** 293–297.

Krattiger, A.F. 1997. Insect Resistance in Crops: A case study of *Bacillus thuringiensis* (*Bt*) and its transfer to developing countries. International Service for the Acquisition of Agri-Biotech Aplications (ISAAA). ISAAA Brief 2.

Krishna, V.V. & M. Qaim, 2007. Estimating the adoption of *Bt* eggplant in India: Who benefits from public-private partnership? *Food Policy* **32**: 523-543.

Krishna, V.V. & M. Qaim, 2008. Potential impacts of *Bt* eggplant on farmers' health in India. *Agricultural Economics* **38**:167-180.

Lester, R.N., P.M.L. Jaeger, B.H.B. Spierings, H.P.O. Bleijendaal & H.L.O. Hooloway, 1990. African eggplant: a review of collecting in West Africa. *Plant Gen. Resources Newsl.* 81–82: 17–26.

Lumpkin, T., K. Weinberger & S. Moore, 2005. Increasing income through fruit and vegetable production opportunities and challenges. Consultive group on International agricultural research science Forum. CGIAR Priorities: Science for the Poor. CGIAR, Marrakech, Morocco.

MAGV (Ministry of Agriculture Government of India), 2008. Indian Horticulture Database. National Horticulture Board. Ministry of Agriculture Government of India. New Dheli.

Maertens, M., 2006. High value supply chains, food standards and poor farmers in developing countries: the case of vegetable exports from Senegal. American Agricultural Economics Association Annual Meeting. Long Beach, California.

Manalili, N. & L.A.C. Tumlos, 2004. Regoverning market:securing small producer participation in restructured national and regional agri-food system - Southeast Asia Overview. Regoverning Markets. SEARCA, Los Banos.

Mansour, S. A., 2004. Pesticide exposure – *Egyptian scene. Toxicology* **198**: 91–115.

Mao, W., J. Yi & D. Sihachakr, 2008. Development of core subset for the collection of Chinese cultivated eggplant using morphological-basal passport data. *Plant Gen. Res.* **6**: 33–40.

McCulloch, N. & M. Ota, 2002. Export horticulture and poverty in Kenya. IDS Working Paper 174. IDS, Brighton, Sussex.

Monsanto, 2009. Supplemental toolkit for investors. Updated June 2009. Monsanto.

Neven, D., T. Reardon, J. Chege & H. Wang, 2006. Supermarkets and consumers in Africa: The case of Nairobi, Kenya. *J. Int. Food Agrib. Marketing* **18**: 103–123.

Ochoa, J.P.A., F. Dainello, L.M. Pike & D. Drews, 1995. Field performance comparison of two transgenic summer squash hybrids to their parental hybrid lineage. *Hort. Science* **30**: 492–493.

OECD, 1997. Impacts of fruit and vegetable production on the environment and policy responses: European Union community instruments relating to the environmental aspects of the fresh fruit and vegetable sector. OECD, Paris.

Osteen, C., 2003. Agricultural resources and environmental indicators: Pest management practices. In: Heimlich, R. (ed.). Agricultural resources and environmental indicators, 2003. Agriculture Handbook AH722, Ch. 4.3. Washington DC.

Palada, M.C., J-F. Wang, R. Srinivasan & C. Ma, 2006. AVRDC-The world Vegetable Center's Present and Future. Approaches to Good agriculture Practices. AVRDC, Shanhua, Tainan.

Perring, T.M., N.M. Gruenhagen & C.A. Farrar, 1999. Management of plant viral diseases through chemical control of insect vectors. *An. Rev. Entom.* **44**: 457–481.

Pfeiffer, W. & B. McClafferty, 2007. HarvestPlus: breeding crops for better nutrition. *Crop Sci.* **57**:s88-s105.

Pingali, P. & M. Rosegrant, 1995. Agricultural commercialization and diversification: Processes and policies. *Food Policy* **20**: 171–185.

Prior, R.L. & G. Cao, 2000. Antioxidant phytochemicals in fruit and vegetables, diet and health implications. *Hort Science* **35**: 588-592.

Rao, S., C.S. Yajnik, A. Kanade, C.H.D. Fall, B.M. Margetts & A.A. Jackson, 2001. Intake of micronutrient-rich foods in rural Indian mothers is associated with the size of their babies at birth: Pune maternal nutrition study. *J. Nutr.* **131**: 1217–1224.

Reardon, T., C.P. Timmer, C.B. Barrett & J. Berdegue, 2003. The rise of supermarkets in Africa, Asia, and Latin America. *Am. J. Agri. Economics* **85**: 1140–1146.

Regmi, A. & M. Gehlar, 2001. Consumer preferences and concerns shape global food trade. *Food Rev.: Global Food Trade* **24**: 2–8.

Rohrbach, D. D., I.J. Minde & J. Howard, 2003. Looking beyond national boundaries: Regional harmonization of seed policies, laws and regulations. *Food Policy*, **28**: 317–333.

Ryder, E.J., 2005. Intellectual property rights for plants. The case for a new law. *Chronica Horticulturae* **45(2)**:5-11.

Schultheis, J.R. & S.A. Walters, 1998. Yield and virus resistance of summer squash cultivars and breeding lines in North Carolina. *Hort. Science* **8:** 31–39.

Senauer, B. & L. Goetz, 2003. The growing middle class in developing countries and the market for high-value food products. *Economic Research Service*. USDA, Washington DC.

Shankula, S., 2006. Quantification of the Impacts on US Agriculture of Biotechnology-Derived Crops Planted in 2005. http://www.ncfap.org/

Shepherd, A., 2005. The implications of supermarket development for horticultural farmers and traditional marketing systems in Asia. FAO, Rome.

Southon, S., 2000. Increased fruit and vegetable consumption within the EU: Potential health benefits. *Food Res. Int.* **33**: 211–217.

Stevens, M.A., 1986. Inheritance of tomato fruit quality components. *Plant Breed. Rev.* **4:**273-311.

Tricoli, D.M., K.J. Carney, P.F. Russell, J.R. McMaster, D.W. Groff, K.C. Hadden, P.T. Himmel, J.P. Hubbard, M.L. Boeshore & H.D. Quemada, 1995. Field evaluation of transgenic squash containing single or multiple virus coat protein gene constructs for resistance to cucumber mosaic virus, watermelon mosaic virus 2, and zucchini yellow mosaic virus. *BioTechnology* **13**: 1458–1465.

Zhang, Y.P. & J.R. Stommel, 2001. Development of SCAR and CAPS markers linked to beta gene in tomato. *Crop Sci.* **41**:1602-1608.

Zitter, T.A., D.L. Hopkins & C.E. Thomas, 1996. Compendium of Cucurbit Diseases. APS Press, St. Paul, MN, USA.

Weinberger, K. & C. Genova, 2005. Vegetable production in Bangladesh: Commercialization and rural livelihoods. *Technical Bull. 33*. AVRDC, Shanhua, Tainan.

Weinberger, K. & T.A. Lumpkin, 2005. High value agricultural products in Asia and the Pacific for smallholder farmers: Trends, opportunities and research priorities. GFAR, Rome.

Weinberger, K. & J. Msuya, 2004. Indigenous vegetables in Tanzania: Significance and prospects. *Technical Bull. 31*. AVRDC, Shanhua, Tainan.

24. VEGETABLE CROPS FOR FOOD, NUTRITION SECURITY AND IMPROVING LIVELIHOOD

A.S. Sidhu and Sabina Islam*

Director, Indian Institute of Horticultural Research,
Bangalore-560 085, Karnataka, India
**Scientist, Division of Vegetable Science, IARI, New Delhi, India*

ABSTRACT

Vegetable production has established its prime place in mitigating the global food crisis, which is afflicting the whole world - developed, developing and underdeveloped. Vegetables with their, high yield potential, diverse crop varieties, nutritional superiority, labour intensive method of cultivation, high market price and prospects of different processed products play an important role in food and nutrition security, crop diversification, employment generation, and improving livelihood. The productivity of major vegetables is at least 3-4 times more than the major cereal crops and they are rich in nutrient content also. The productivity of rice, wheat, and maize are 22.5, 38.95 and 26.98 tons/ha, respectively, while the productivity of potato, tomato, sweet potato are 198, 156 and 83 tons/ha, respectively. For proper well being, an adult of moderate work style requires 2500 kcal and consumption of 300g vegetables comprising 125 g leafy vegetables, 100g root and tuber vegetables and 75g other vegetables has been recommended by ICMR and NIN, Hyderabad. Besides food security, nutritional security is a major concern. A global survey report estimates that more than 200 million children in developing countries suffer from malnutrition. Leafy vegetables are rich reservoir of vitamins and minerals, root and tuberous vegetables are good source of carbohydrate and leguminous vegetables are good source of protein. Vegetables being labour intensive, generate lot of employment opportunity. The profit from vegetables by way of fresh sale, processed product sale, export and seed business is quite higher than other crops. Besides this, vegetables contain a large number of non nutritive pytochemicals, which help in prevention of many diseases. The prominent protective vegetables are bitter gourd, bottle gourd, garlic, onion, tomato etc. So in the era of rapid global climate change, increased industrialization and urbanization and decreased input use efficiency, vegetable crops are the safe option for diversification of our cereal based cropping system to ensure food, nutrition and livelihood security.

INTRODUCTION

Hunger and malnutrition are the two major global challenges afflicting the population as a whole particularly those of developed and developing world. Globally, 800 million people are going to sleep hungry every night. In the present year only, due to global fuel, food and financial crisis an additional 100 million people would be bogged in hunger and poverty. Vegetables with their high yield potential, diverse crop varieties, nutritional superiority, labour intensive method of cultivation, high market price and prospects of different processed product play an important role in food and nutrition security, crop diversification, employment generation, and improving livelihood. It has been estimated that yield potential of already low yielding cereal crops is bound to decrease by 10-40% by 2100. In this context, vegetable production has emerged as suitable alternatives to cereal crops and established its prime place in generating sustained and increased income to farmers and in providing employment both in the rural and urban areas. Different group of vegetables cater to the different dietary requirement, for example, potato and other underground tuberous vegetables are good source of carbohydrates, peas, beans and legumes are good source of proteins, leafy vegetables are good source of minerals and almost all the vegetables are endowed with different phytochemicals having potent anti-oxidant property.

VEGETABLES FOR FOOD SECURITY

Global food crisis is inflicting one and all be the developed, developing or under developed countries. The recent food mayhem has been attributed to inadequate supply of food grains on the part of developing countries. But in reality, the global cereal stocks have been found inadequate to feed ever-increasing population of the world. The subsequent fuel crisis and financial turmoil globally have worsened the situation. There is an urgent need to increase food production manifold in a shorter span of time to get relief from the world food crisis. The productivity of major cereals in India, highest productivity in the world has been compared in Table 1. When we compare the productivity of major cereals and vegetables, vegetables stand far ahead to meet the global food challenge.

Table 1. Average and highest productivity of major cereals in India and world

Crop	Average productivity in India (tons/ha)	Highest productivity in World (tons/ha)
Rice	2.90	3.96
Wheat	2.71	2.87
Maize	1.18	3.38

Facts of Vegetable Production in India

India is the second largest producer of vegetables in the world after China occupying an area of 7.16 million hectares accounting only 2.8% of the cropped area with a production of 109.05 million tonnes consisting of 15% of the total world production (690 million tonnes). The production of vegetables in the year 1960-61 was at the level of 23.4 million tonnes. In 2005-06, country has witnessed more than 5% increase in area under vegetables with more than 50% increase in production. The present average productivity is 15.2 t/ha. The wheat, rice and cotton occupy 91.83% (42.95, 32.92 and 5.96%, separately) of the total cropped area. The potato shares the maximum production followed by brinjal and tomato crop in total vegetable production. Although our productivity is low as compared to some developed countries but in some vegetables like cabbage, pea and potato our productivity is higher than that of the world. India ranks first in production of cauliflower, second in onion, third in cabbage and potato, although there is wide range of productivity ranging from one t/ha (Lakshadweep) to 28 t/ha (Tamil Nadu) followed by Uttar Pradesh, Karnataka and Punjab. After independence, the production of vegetables in India has increased by six times. The availability of vegetables is 210 g per capita per day, which is less than dietary requirement of 300 g per capita per day. Further, our population is growing at the rate of 1.8% per year. At this rate of population growth, we will have population of over 1.6 billion by the year 2020 and our demand will be around 200 million tonnes.

Scope of Vegetable Production

I. Increasing area

The vegetable production in India can be boosted by increasing area under vegetable cultivation and by development and adoption of high yielding, disease and pest resistant varieties/hybrids.

Increasing area under vegetables

Mixed cultivation with cereal crops: It is quite difficult nay impossible to convince cereal grower farmers to switch to vegetable cultivation, but the area can be increased by following intensive cultivation of mixed or less duration vegetable crops in relay cropping in cereal based cropping system like cauliflower, radish in sugarcane crops.

Kitchen gardening: Kitchen gardening is one of the most important alternatives for increasing production of vegetable production. Considering the fact that there are almost 5 lakh 95 thousand inhabited villages, if at least four kitchen gardens of 500m^2 size are developed in each village, the supply of vegetables can be increased by 2 million tonnes.

The urban and peri-urban vegetable cultivation (in box and terrace cultivation): It can support to increasing production of vegetable in towns and cities, where open space is a limiting factor.

Moreover, riverbed cultivation also can be a positive support to vegetable production of the country.

II. Increasing Yield per Unit Area

Although India ranks second in global vegetable production map, the productivity is quite low when compared to the global best performers. There is an urgent need to improve the productivity levels of some of the major vegetables. For this genetic improvement as well improved cultivation practices need to be followed.

Development of high yielding, disease pest resistant varieties/ hybrids: India ranks second in global vegetable production, but there is a huge gap in productivity of major vegetables in India and other developed countries. The gap can be minimized by exploiting the genetic resources and developing high yielding varieties/hybrids and popularizing them. The yield potential of the varieties/hybrids already in cultivation get reduced due to occurrence of wide array of pests and diseases like downy mildew and several viral diseases in cucurbits, early and late blight in potato, fruit and shoot borer in brinjal, leaf curl in tomatoes and chillies, YVMV in okra and many others. Therefore, development of resistant /tolerant varieties along with increased yield potential is must to increase the productivity level. Several varieties and hybrids have been developed by the public and private sectors, but still lots of unidentified varieties are being used in India.

III. Other Techniques for Increasing Production

Off-season and protected cultivation: To meet the demand of vegetable crops in India the off season, green and net house technologies are being adopted which are somewhat increasing the productivity and vegetable availability during the off season period (Table 2). But still there is urgent need to adopt these technologies in large scale to fulfill the required demands.

Table 2. Yield and availability of high value vegetables in low cost poly-house

Crop	Estimated yield (q/ha)	Period of availability (days)
Cucumber	1150	3rd week of Dec. to 2nd week of March
Summer squash	900	3rd week of Dec. to 2nd week of March
Capsicum	520	1st week of Feb. to 1st week of April
Tomato	1100	3rd week of January to 1st week of April

Improving land use efficiency through inter-cropping, relay cropping and mixed cropping

Intensive vegetable crop rotation are;

Radish (Jan.-Feb.)	Cowpea (March-June)	Early cauliflower (July-Oct.)	Turnip (Nov.-Dec.)
Palak (Jan.-Feb.)	Okra (March-June)	Cucumber (July-Oct.)	Knol khol (Nov.-Dec.)
Turnip (Jan.-Feb.)	Amaranth (March-May)	Round gourd (June-Sept.)	Early potato (Oct.-Dec.)

Inter-cropping

- Radish, turnip, carrot and beet root in the bunds of potato, cauliflower, cabbage, fenugreek etc.
- Snap melon and cabbage in kharif sweet corn
- Palak in pumpkin during summer
- Radish in watermelon during summer
- Amaranth in ash gourd during summer

Mixed cropping

- Radish + turnip
- Chinese cabbage + vegetable mustard
- Cauliflower + Cabbage
- Cauliflower + Broccoli
- Carrot + Beet root

Relay cropping

- Planting of early cucurbits (end of Jan.-Feb.) in late potato crop
- Cowpea and cluster bean (June-July) in spring-summer okra crop

IV. Grafting in Vegetables

Grafting has been proved to be the important technology for the production of vegetable crops like watermelon tomato and cucumber. The grafting like cabbage on radish, Chinese cabbage on radish, tomato on potato and eggplants on potato has been successful. Compare to other countries Indian is still behind in adopting this technique but it is being adopted in other counties like Japan and Korea (Table 3). This technique is good for stress tolerance in high yielding varieties or hybrids. By using a simple grafting machine up to 1200 tomato seedlings can be grafted in one hour.

Table 3. Cultivation area of some vegetable and percentage of grafted plants in Japan and Korea

Crop	Japan				Korea			
	Field + Tunnel		Greenhouse		Field + Tunnel		Greenhouse	
	Total area (ha)	Grafted (%)	Total area (ha)	Grafted (%)	Total area (ha)	Grafted (%)	Total area (ha)	Grafted (%)
Cucumber	10,160	55	5,440	96	1,728	42	5,964	95
Egg plant	11,815	43	1,785	94	650	0	413	2
Melons	6,142	0	8,258	42	1,047	83	9,365	95
Pepper	2,684	- (3)	1,468	-	75,574	0	5,085	5
Tomato	6,459	8	7,141	48	258	0	4,752	5
Watermelon	14,017	92	3,683	98	13,200	90	21,299	98

V. Improving Water use Efficiency

Water is a scarce resource and availability of good quality irrigation water is decreasing day by day. Therefore, to minimize water usage improve yield and quality drip irrigation system have been developed. The percentage yield increase and water saving following drip method is presented in Table 4.

Table 4. Yield increase and water saving in vegetable through drip irrigation

Vegetables	Yield increase (%)	Water saving (%)
Tomato	50	39
Watermelon	88	46
Okra	16	40
Brinjal	14	53
Bitter gourd	39	53
Ridge gourd	17	59
Cabbage	02	60
Radish	02	77
Beet root	07	79
Chilli	44	62

VI. Mechanized Cultivation

Due to urbanization and industrialization, the man power in field related operations are bound to decrease. Vegetables being labour intensive venture, need suitable alternatives. Numerous instruments have been developed world-wide for mechanized cultivation and the prominent ones are;

- Vegetable transplanter
- Direct seeding of onion
- Vegetable seed extracting machine
- Garlic planter

VII. Production of Healthy Seedlings and Good Quality of Vegetable Seeds

Seed is the most delicate input in successful crop production. In India the approximate annual seed production of vegetable crops is around 32,000 tones while the demand is 70,000 t. National Seed Corporation, different ICAR institutes mainly IARI, IIVR, IIHR and different SAUs are doing commendable work to supply the quality seed of improved varieties of vegetables to growers at a nominal cost. Besides a large number of seed companies and NGOs are involved in supplying good quality planting materials of vegetables. In order to reduce losses, plug tray nursery raising technology of seedlings have been standardized in different vegetable crops, where individual seeds are sown in single hole and each seed is expected to produce healthy seedling.

VII. Other Factors Contributing Towards Increasing Production

Pre and post harvest management of vegetable crops:

- Timely hoeing, weeding, irrigation
- Effective plant protection measures
- Control of physiological disorders
- Control of nutritional deficiencies
- Proper care to check mechanical damages while performing inter-cultural operations
- Harvesting of the crop at proper stage of maturity
- Proper curing, sorting, grading and storage
- Natural air-ventilated onion storage structure and zero energy cool chamber

VEGETABLES FOR NUTRITIONAL SECURITY

Vegetables are rich in vitamins, minerals which are essential for proper development of human body mind and health. ICMR, New Delhi and NIN, Hyderabad recommends intake of 300 g vegetables/ head/day consisting of 125 g leafy vegetables, 100 g fruit and other vegetables and 75 g root vegetables. Different vegetables supply wide variety of nutrients like potato, sweet potato, tapioca, yam are rich in carbohydrate, leguminous vegetables are good source of proteins while leafy vegetables abound in vitamins and minerals. Besides, numerous phyto-chemicals have been reported from different vegetables which act as antioxidant and prevents many deadly diseases. The prominent ones are cheratin in bitter gourd, diphenylamine in onion against diabetes. The effectiveness of different vegetables in prevention of many lifestyle diseases is presented in Table 5.

Table 5. Vegetables suited for prevention of diseases

Vegetables	Diseases
Onion	Diabetes, blood purifier
Bitter gourd	Diabetes, blood purifier, hyper tension, dysentery
Garlic	Joint pain, Hypertension, Gout, Cholesterol
Radish	Jaundice, piles, gastric, diarrhea
Broccoli	Cancer
Chilli	Diarrhoea, joint pain
Cabbage	Gout, diarrhea
Tomato	Blood purifier, Dyspepsia
Brinjal	Cancer, diabetes

Leafy vegetables are quite rich in antioxidant and antioxidant activity of different leafy vegetables. Mint has been found to be richest in antioxidant activity. The top ten healthy and nutritious vegetables in Indian diet are presented in Table 6.

Table 6. Top ten healthy and nutritious vegetables in Indian diet

Vegetables	Nutrients content
Asparagus	Phenolics, flavonoids
Fenugreek	Phenolics, flavonoids
Red capsicum	Capsanthin, Ascorbic acid, phenolics
Spinach	Flavonoids
Tomato	Lycopene
Brinjal	Phenolics, flavonoids
Red cabbage	Anthocyanins
Broccoli	Sulphuron, ascorbic acid
Onion	Quercetin
Cabbage	ß- carotene, phenolics

Consumption of Pesticide Residue Free Vegetables

India is one of the lowest pesticide user countries in the world, but their injudicious application make vegetable consumption unhealthy and health hazardous. It is common practice among the Indian vegetable growers to dip the freshly harvested vegetables to dip in pesticide solution before marketing or spraying in the morning and harvesting in the same day evening or the next day. Therefore, judicious use of pesticides and adoption of integrated pest and disease management strategy is must to supply safe and quality vegetable. The maximum residue limits of pesticides in different vegetables are presented in Table 7. Excessive residues can be prevented by adopting integrated pest management techniques, creating awareness among farmers regarding potential hazard, restricting the use of more persistent insecticides, developing pesticides with greater efficacy and less persistent nature and developing efficient method of pesticide application.

Table 7. Maximum residue limit of pesticides in vegetables

Vegetables	Name of pesticides	Maximum residue limit (mg/kg)
Potato	Aldicarb	0.50
Okra	Carbaryl	10.00
Leafy Vegetables		10.00
Potato	Chlorpyriphos	0.01
Onion		0.01
Cauliflower		0.01
Cabbage		0.01
Other vegetables		0.20
Brinjal	Cypermethrin	0.20
Cabbage		2.00
Okra		0.20
Cauliflower	Diazinon	0.50
Tomato	Dithiocarbamates	3.00
Cucumber (Squash)	Ethion	0.50

Onion	Fenthion	0.10
Beans		0.10
Pea		0.50
Tomato		0.50
Muskmelon		2.00
Cauliflower	Fenvalerate	2.00
Brinjal		2.00
Okra		2.00
Carrot	Monocrotophos	0.05
Turnip		0.05
Potato		0.05
Onion		0.10
Pea		0.10
Tomato	Phorate	0.10
Other Vegetables		0.05

VEGETABLES FOR IMPROVING LIVELIHOOD

Employment generation through vegetable cultivation is more than any other crop cultivation. Apart from cultivation huge employment is generated in the process of hybrid seed production, processing and marketing of vegetables. The employment and income generation through hybrid seed production of different vegetables are presented in Table 8.

Table 8. Employment and income generation through hybrid seed production of different vegetables

Crop	Total man days employed (million)	Net income of each farm family/unit area (Thousand Rs.)
Chilli	0.67500	19.0-64.0
Capsicum	0.02990	121-151
Tomato	0.75984	20.6-55.6
Brinjal	0.06900	19.5-34.5
Okra	0.41994	19.3-49.3
Cucumber	0.08100	14.6-29.6
Watermelon	0.17870	36.4-38.4
Muskmelon	0.07000	21.2-35.2

VEGETABLES IN NATIONAL ECONOMY

Vegetables are important source of farm income and provide employment to the poorest people. The horticulture crops contribute 25% to agricultural GDP from less than 8.5% of total agricultural area. Approximately, 994 tonnes of hybrid seed is being produced in vegetables from an area of 7,959 acres leading to employment generation of 3.28 million-man days/year with additional income of Rs.130.28 million.

Income Generation through Export of Fresh and Processed Vegetable

Export of fresh and processed vegetable from India during 2005-06 was of Rs. 1803.00 crores, which helps in increasing the national economy. India's export of processed vegetable is only 2 % to the world export market, while, Malaysia (83%), South Africa (80%), Philippines (78%), Brazil (70%) and USA (65%) are the main exporters of processed vegetables. India exported Potato, onion, garlic, green chillies and red chillies in bulk to Pakistan recently during 2006.

CONCLUSION

There is a long way to go to meet the United Nations first millennium promise. Apart from reducing hunger and poverty concerted attention is needed to alleviate malnutrition and nutrition deficiency. Vegetables with their high yield potential, richness in nutritional compounds and huge potential for employment generation is ideal for fulfilling this objective.

25. BIOTECHNOLOGICAL INTERVENTIONS IN HORTICULTURE AND SUSTAINABLE LIVELIHOOD

C. Aswath and Vageeshbabu S. Hanur

Division of Biotechnology, Indian Institute of Horticultural Research, Hesaraghatta Lake Post, Bangalore-560 089

ABSTRACT

The ever increasing demand for food, feed, nutrition and other basic needs like jobs and health, as compounded by the burgeoning population has put a lot of pressure on the agricultural and horticultural sectors in terms of addressing the accompanied challenges associated with increasing crop productivity while maintaining environmental sustainability. At the same time, we are witnessing the potential benefits of the revolutions propelled by the science of biotechnology. The realistic utilization of biotechnology in horticulture has not only brought about profound changes in the crop productivity and quality, but has also opened up newer and hitherto unforeseen potential vistas including engineering novel traits, successful commercial micropropagation, manipulating plant architecture and molecular farming. Challenges like biotic and abiotic stresses, gradually reducing crop productivity and environmental safety, demand for newer products in terms of phytoneutraceuticals, exotic plant types, healthy planting materials, post harvest losses, export quality plant products and designer horticultural crops have been constantly addressed successfully using modern horticultural biotechnology. A major challenge of sustainable livelihood in the developing nations can be met with the judicious and rational application of horticultural biotechnology along with other equally powerful crop production systems. How this challenge can be realistically addressed by exploiting the opportunities provided by biotechnology-will be discussed in this paper.

INTRODUCTION

Biotechnology offers immense opportunities to manipulate traits of agronomic importance for the improvement of agricultural and horticultural crops. Utilizing the burgeoning knowledge and tools in the field of modern molecular biology, it is now possible to rationally address the various challenges of crop improvement and productivity and bring about affirmative changes in any crop cultivar by the power of genetic engineering. The last decade has witnessed considerable advances in the development and successful commercialization of transgenics in a number of crops world wide. Transgenic plants are those whose genome has been altered by incorporating genes from other species using recombinant DNA methods. Although conventional plant breeding techniques have made considerable progress in the development of improved disease and pest resistant varieties, they have not been able to keep pace with the increasing demand for vegetables and fruits in the developing countries. Therefore an immediate need is felt to integrate biotechnology to speed up the crop improvement programs. Biotechnological tools have revolutionized the entire crop improvement programs by providing new strains of plants, supply of planting material, more efficient and selective pesticides and improved fertilizers. Crop varieties have been generated by breeders in the past by forcing hybridization between unrelated species or even genera in the so-called wide crosses by resorting to techniques such as protoplast fusion, embryo rescue, mutagenesis, etc. Transgenic approach is just an extension of the breeder's method with a further refinement consisting of transferring only a small segment of DNA responsible for the trait of interest rather than chunks of uncharacterized chromosomes. Transgenic plants are being developed for a variety of reasons such as insect resistance, disease resistance, drought and salinity resistance and extending shelf life.

World wide transgenic plants of over 20 plant species that are resistant to more than 30 different viral diseases have been produced using pathogen derived strategies. Similarly insect resistant plant varieties have been produced in several important cultivated crops varieties. Transgenic approach to increased shelf life in crops like papaya and tomato can benefit the farmer in developing countries where inadequate storage facilities, high cost of transportation coupled with cycles of glut often result in poor returns. One of the ways to increase production would be to expand cultivation to poor and marginal lands with little assured irrigation. This is possible through varieties that are resistant to drought and salinity.

TISSUE CULTURE AND COMMERCIAL MICROPROPAGATION

Tissue culture and transformation protocols are now readily available in most crops. Using these, techniques like *in vitro* germplasm conservation and exchange, rapid multiplication of elite cultivars using micropropagation, synseeds using somatic embryogenesis, somaclonal variations, somatic hybridization, and double haploid production- are being attempted in a number of horticultural crops, apart from developing transgenics. Commercial production of banana in terms of millions of plants per month has become a reality, due to improved micropropagation technologies reducing cost, input use and somaclonal variations. Even this rate is insufficient to cater to the burgeoning demand for banana plantlets. More than a dozen major companies are involved in the production, sale and export of tissue culture banana presently in India. Even in rose, orchids, pachouli and other important horticultural crops, micropropagation based commercial production of disease-free, uniform and elite varieties are underway.

Fermenter and bioreactor based production of planting material, secondary metabolites and important compounds are now order of the day in the commercial sector. Medicinal, aromatic, ornamental, spice and other forest tree species have been used to produce important secondary metabolites, phytoceuticals and pharmacologically useful compounds *in vitro* in bioreactors on a commercial basis.

TRANSGENICS AND TRAIT IMPROVEMENT

Transgenics have been developed in a number of crops of horticultural importance in respect of myriad of traits. Resistance to biotic stresses (insects, nematodes, bacterial, fungal and viral pathogens), resistance to abiotic stresses (herbicide, drought, salinity, heat, flooding etc), plant architecture (flowering, harvest index) post harvest quality traits (shelf life, ripening, color, aroma etc) and novel traits (expression of theraupeutics, secondary metabolites and drugs)- have been some of the traits that have so far been exploited in the development of transgenic horticultural crops. Herbicide tolerant and insect pest resistant transgenic crops have been the major products of plant biotechnology. For example, Bt transgenic technology has revolutionized the production and productivity of cotton globally in the last 15 years, although Bt cotton source has been primarily from the private sector.

This success story has also been demonstratably repeated in India in the last five years when Bt cotton was introduced. The area, production, yield and quality of cotton in India have substantially increased in a short span of time. Resistance to bacterial, fungal and viral pathogens has been addressed using genes isolated from other microbe and plant sources. Pathogen derived resistance has also been exploited in introducing virus resistance using coat protein, replicase and other viral genes. Chitinases, glucanases, antimicrobial peptides, ribosome-inactivating proteins and such similar sources have been used in fungal and bacterial transgenic resistance. Even though traits like abiotic stress are difficult to address using single or few genes, it is now possible to attempt to develop drought tolerant plants using transcription factors like DREB1A, which regulate a number of downstream genes involved in plant cell water balance and water use efficiency. Efflux and antiporter genes of sodium have also been used to develop salinity tolerant crops. Irrespective of the genetic information, newer techniques like RNA interference based gene silencing can be utilized to inactivate host or pathogenic gene expression and desirable phenotypes can be obtained. Already RNAi technology has been in use to develop viral resistant transgenics in a number of crops.

In developing transgenic crops, it is extremely important to carry out phenotyping of the transgenics in respect of the trait under study. Most traits are complex to manipulate, express and accommodate in the 'alien' environment of the genome of the recipient transgenic crops. The introduced trait may undergo known or unknown modifications in its expression, due to different levels of cellular tissue and organ level interactions.

MOLECULAR MARKER ASSISTED BREEDING

Besides the transgenic technology, biotechnology has made major contributions in the field of crop breeding in enhancing the efficiency of all steps of breeding activities. Molecular breeding, integrating conventional wisdom of breeding and powerful modern molecular biology, has resulted in the unparalleled improvement in terms of trait improvement, using the numerous forms of molecular markers. Marker assisted selection and marker assisted breeding have become an important integral part in the breeder's kitty. With the complete genome sequences available in major model plants like Arabidopsis and crops like rice, wheat and corn etc., exploiting synteny, molecular breeding has become more focused, targeted and informative in terms of identifying important genetic determinants of selected traits. Crop evolution, germplasm assessment, identification of parents, analysis of segregating populations, rational selection and other crop improvement methods have been substantially benefited by the exponentially increasing genetic data through genomics, proteomics, metabolomics, transcriptomics, interactomics and high throughput analysis and bioinformatics. Utilization of various molecular markers like simple sequence repeats, microsatellites, AFLP, SCARs, SNPs and high throughput methods like next generation sequencing, has enabled rational breeding and crop improvement.

Horticultural crops have also been the target of developing designer transgenic crops for use in pharmaceutical and other intensive industrial applications. Plantibodies, antibodies expressed in plants, against a number of human diseases have been developed by expressing in crops like tomato and potato. Edible vaccines incorporating engineered vaccine components have been expressed in crops like tomato and banana. Therefore, 'molecular pharming' has become a routine but impressive technology to address pharma and human health care industries.

BIOINFORMATICS

Bioinformatics is another important area where tools of information technology have been extensively exploited in biotechnology. Bioinformatics has been used in genomics, proteomics and other omics analysis, database and software driven molecular analysis, cellular interactions, drug designing, molecular modeling, genome analysis, maintenance and management of germplasm, breeding material and exchange etc. Novel genes and targets, cell metabolism, cellular interactions, signal transduction, pathway dissections, receptor-ligand binding and other routine molecular activities are facilitated by the use of bioinformatics. Most of the bioinformatics analysis softwares are either freely downloadable from major molecular database warehousing and mining sites or available for nominal prices for use in biotechnological analyses.

BIOTECHNOLOGICAL RESEARCH AT IIHR, BANGALORE

Research at IIHR, Bangalore, in the field of biotechnology has been multi-dimensional. Work in various fields including tissue culture, micropropagation, endophytic research, development of transgenic crops, molecular markers and marker assisted breeding, molecular entomology, is underway. Micropropagation of selected pre-release superior hybrids of hippeastrum has been achieved using bulb scale explants and hardened plantlets ready for transfer to field have been developed. Using hybrid embryo rescue in intractable crosses where conventional breeding does not succeed, seedless clone "Thompson Seedless" grape variety was used as the female parent and various pollen donor parents variously resistant to downey mildew fungal pathogen were used in breeding program for hybrid embryo rescue. In crosses effected so far, over 1500 hybrids have been generated *in vitro*, more than half of which have entered field testing. This is for the first time IIHR has generated hybrid material with "Thompson Seedless" as the maternal parent. Micropropagation of banana in Rasthali and Neypoovan varieties has been achieved using somatic embryogenesis approach. Endophytic microbes both cause havoc and beneficial effects in the commercial tissue culture industry. At IIHR, research on the management of endophytic microbes in tissue cultures specifically of important crops like banana is actively underway and important leads in terms of tissue indexing, identification and management of endophytes have been discovered.

Transgenic plants have been developed in important vegetable crops including tomato, brinjal, chillies, water melon and banana, incorporating insect pest resistance, disease resistance and drought resistance traits. Bt transgenic brinjal (cultivar Arka Keshav, purple long variety) and tomato (cultivar Arka Vikas) have been developed using *Cry1Ab*, *Cry1Aa3* and *Cry2A* synthetic Bt gene constructs. These genes have been developed in house and do not therefore carry any burden of IPR issues. The transgenics have been tested for the expression of Bt protein and functional resistance using insect bioassays in advanced generations. RNA interference based gene silencing has also been initiated to address the problem of Helicoverpa armigera as a model pest in tomato. Transgenic tomato and pepper plants carrying antifungal genes like chitinase and glucanase, antibacterial genes like antimicrobial peptide have been developed and are in advanced stages of trials. Transgenic banana with these genes for resistance to fungal diseases have also been produced. Transgenic tomato plants resistant to viral diseases like tomato leaf curl virus and tomato spotted wilt virus and transgenic watermelon resistant to peanut bud necrosis virus have been developed using pathogen derived resistance and are in different stages of advanced trials. Drought tolerant transgenic tomato and pepper plants have been developed and successfully tested using transcription factors like *DREB1A*, *DREB2A*, trehalose phosphorylase genes. Transgenic pomegranate resistant to the bacterial nodal blight caused by *Xanthomonas axonopodis* pv. *punicii* is under initial stages of development and testing.

In the area of molecular markers and marker assisted breeding programs at IIHR, various research projects including germplasm evaluation, development of crop specific molecular markers like STMS, SCAR, SSR and ISSR, analysis of segregating breeding lines and crosses using molecular markers, diversity assessment and marker assisted breeding are underway in fruit crops (mango, banana, papaya, guava, pomegranate) and vegetable crops (tomato, chilli pepper, okra, beans and peas etc) with active collaborations from various crop breeders of the Institute and other Institutes. Molecular entomology has also been intensively used to study pest vectors of plant pathogens and transmission dynamics in thrips, aphids and white flies.

Thus it is evident from the discussions that biotechnology has played and will continue to play prominent roles in improving yield and quality of crops, with a special reference to horticultural and seed spice crops. While on the one hand, transgenic crops with genetically engineered novel traits can be developed to meet specific and tailor-made demand and applications, on the other hand, molecular markers and other newer technologies including bioinformatics are becoming handy tools in significantly enhancing the efficiency of plant breeding using targeted approaches. Newer technologies like RNAi, systems biology, next generation sequencing and 'omics' could herald an entirely new way of improving quality and yield of crops in near future.

TRANSGENICS FOR TOLERANCE TO BIOTIC AND ABIOTIC STRESSES

The marketable yield and quality of many horticultural crops are severely affected by a number of pests and diseases. A summary of important biotic and abiotic stresses of selected horticultural crops is presented in Table 1, along with genes that have been utilized.

Table 1. Biotechnological interventions of mitigating stresses in horticultural crops

Crop	Stress	Gene(s) utilized
Tomato	TLCV	Pathogen derived genes, RNAi, Ty1-5 markers
	TOSPO	RNAi
	Fruit borer, *Helicoverpa armigera*	Bt
	Nematode	Mi gene
	Wilt	-
	Drought	DREB1A transcription factor
Eggplant	Shoot and fruit borer, *Leucinodes orbonalis*	Bt
	Wilt	-
Chillies pepper	*Alternaria*	Antifungal
Watermelon	Peanut bud necrosis virus	RNAi
Bhindi	Yellow vein mosaic	-
Potato	Late blight & early blight	Antimicrobial
	Nematode	-
Onion	Purple blotch	Antifungal
Cabbage,	Diamond back moth,	Bt

cauliflower	*Plutella xylostella*	
Banana	Panama	Antifungal
	BBTV, Bract mosaic	Pathogen derived, RNAi?
	Nematode	-
Papaya	Ringspot	Coat protein, N
Mango	Spongy tissue	???
Citrus	Tristeza and greening	-
Pomegranate	Bacterial nodal blight caused by *Xanthomonas axonopodis* pv. *punicii*	-
Grapes	Downey mildew	Antifungal?

5% and 1% level respectively. P- Phenotypic correlation coefficient, G- Genotypic correlation coefficient

PUBLIC PERCEPTION OF GMOs: CHALLENGES AND ACTION PLAN

Even though the revolution brought about by GM agriculture has already started yielding perceptible and positive, fruitful results worldwide, there is an equally challenging task crouching the biotechnological world, that is staunch opposition by NGOs and other self-help/green activists in regard to the safety hazards of the GM crops on human health, environment, monopolistic market control by multinationals and perceived hidden threats including horizontal gene transfer and genetic erosion. On the other hand, many farmers still need to be convinced of the benefits of the GM technologies.

The time has arrived for the scientific community involved in the GM crop development to engage policy makers, farming communities, academicians, NGOs and all concerned with agricultural biotechnologies in a participatory mode to sensitize about the true immense benefits of GM crops accrued by the present day agriculture and horticulture. Common misconceptions need to be removed and certain real issues can be thoroughly addressed by the researchers so that there is no scope for a trace of doubt regarding biosafety of GM crops on all fronts. Thorough and strict adherence of all stipulated biosafety guidelines need to be practiced and demonstrated to all the concerned so that critics of the GM technologies shall appreciate the comprehensive safety of cultivation and use of GM crops in agriculture and horticulture.

THE TAKE-HOME MESSAGE

Biotechnology offers immense opportunities in the present day agriculture and horticulture in improving yield, quality and value addition apart from sustainably addressing several threats including biotic and abiotic stresses, post harvest losses and environmental issues. A large gene pool exists across kingdoms which can be ingenuously and rationally utilized for targeted transgene insertions and development of directed specific GM crops. Conventional breeding has been a major benefactor which utilizes molecular marker assisted selections in precisely handling breeding populations, making valid selection criterion and hastening breeding programs. This has been further assisted by the voluminous knowledge accumulated by functional genomics and proteomics databases across crop species. A tremendous challenge is faced by the researchers in addressing a number of public concerns on the issues of biosafety, environmental sustainability and commercialization of GM crops. Efforts like PPP and network-based research systems can improve the overall efficiencies of the research systems. It's time we prepare ourselves and look forward to the opportunities provided by modern horticultural biotechnology and realize its fruits in a sustainable and judicious manner.

IV
Delivery of Technology and Products

26. HORTICULTURAL REVOLUTION THROUGH RESEARCH-EXTENSION-FARMERS LINKAGES IN MAHARASHTRA

K.D. Kokate

Deputy Director General (Agriculture Extension), Indian Council of Agricultural Research, New Delhi

ABSTRACT

Maharashtra state occupies a unique position in India by contributing 18% share in production of horticultural crops. The state is a leading producer of grape (76.9%), pomegranate (69.5%), orange (55.2%), onion (32.9%) and sapota (23.8%) in national production. 448 food-processing units are started by different sectors in the state. The ambitious programme 'Employment Guarantee Scheme linked Horticulture Development' launched by the Government of Maharashtra along with strong research-extension-farmer linkage led to horticultural revolution in Maharashtra state. This combined effort made a significant impact in increasing the area under horticultural crops from 2.42 lakh ha in 1990-91 to 14.96 lakh ha in 2006-2007.The production also increased from 9.45 lakh MT to 103.24 lakh MT.

The horticulture development in Maharashtra has been achieved through a strong and wide network among research institutes, state extension system and horticulture growers. The National Research Centers of ICAR and SAUs have done commendable research in horticulture. The ICAR-SAU-KVK based frontline extension system along with state extension system played a critical role in educating the farmers about improved horticultural technology. Under ATMA the Farmers Interest Groups (FIGs), Commodity Interest Groups (CIGs) and Self-Help Groups (SHGs) have been formed which has resulted in wider farmer participation and bottom-up programme planning. There is a strong co-ordination between SAUs and State Department of Agriculture through the Joint Agricultural Research and Development Co-ordination Committee, Zonal Research and Extension Advisory Committee, Monthly District Workshops, Diagnostic team visits, field visits etc. Farmers Field Schools (FFS) to promote organic farming, Integrated Pest Management are organized for different horticultural crops. The Maharashtra State Agricultural Marketing Board (MSAMB) has provided access to global market by promoting farmer's co-operatives for horticultural produce in Maharashtra. Maharashtra state has commodity based growers organizations, which led to expansion and sustainable horticultural development in the state. This institutional paradigm of research-extension-farmers linkage has brought revolutionary impact on development of horticulture sector in Maharashtra.

Innovative approaches like integrated cropping system, varietal improvement, quality planting material, Public-Private Partnership based mass dissemination, ICT based information system, print and electronic media and farmers-led extension concept of Farmers-Scientists Forum, decentralized, participatory and market-led extension strategies and human resource development programmes played a crucial role in strengthening the Research-Extension-Farmers linkages for horticultural development. In view of the global competitiveness and challenges before horticulture sector, these experiences of Maharashtra state are to be replicated in other states. This would lead to increased productivity of horticultural crops, which would sustain the country's economic growth, and ultimately the livelihood security of farmers.

INTRODUCTION

India is bestowed with diverse agro-climatic conditions and soil types, which are ideal for production of wide range of high value horticultural crops including fruits, vegetables, flowers, spices and condiments. The global horticultural scenario indicates that India achieved second position in fruit crops next for China by contributing large area (5.77 million ha) under fruit cultivation and 63.5 million tonne of total production. But, the average productivity of fruit crops (11 t/ha) is quite low as compared to other countries. The leading fruit crops grown in India are mango, papaya and banana contributing 41%, 30% and 29% share respectively in its total production in the world. Similarly, India ranks second in vegetable crops in the world with an area of 78.08 lakh ha and 12.58 million tonne of total production with low average productivity of 16 t/ha. India contributed 18% share in onion production and is a leading onion producer after China in the world.

A centrally sponsored scheme of National Horticulture Mission (NHM) has been launched in the country since 2005 for the development of horticulture duly ensuring an end to end approach having backward and forward linkages covering research, production, post-harvest management, processing and marketing. Similarly, National Horticulture Board (NHB) is involved in the development of high quality horticulture farms in identified belts and make such areas vibrant with horticultural activity which in turn will act as hub for developing commercial horticulture, development of post-harvest infrastructure, strengthening market intelligence system and horticulture data base, assisting research and development programmes to develop products suited for specific varieties with its improved production techniques, and capacity building of farmers through appropriate extension interventions. Due to such types of policies and institutional reforms, India has made significant progress in horticultural development during last two decades. The area and production of horticultural crops in the year 1990-91 was 12.8 million ha and 96.6 million tonne, respectively, which steadily increased up to 20.1 million ha area and production of 207.0 million tonnes in 2007-08. The average productivity of horticultural crops also increased by 4 t/ha during the same period. India exported 27, 54,181 tonne of horticultural produces and earned Rs. 5,05,659 lakh from foreign trade in 2006-07 (APEDA). This overwhelming progress is possible due to India's strength in different areas like;

- Different agro-ecological zones permit cultivation of high value horticultural crops
- Strong and wide network of National Agricultural Research System (NARS) e.g. ICAR institutes (IIHR), NRCs and SAUs
- Effective and efficient ICAR-SAUs based front-line extension system
- Increased role of private agencies in research, extension, post-harvest management and value addition
- Seasonal advantage for export to temperate countries
- Large Non-Governmental Organizations (NGOs) in horticultural sector development

STATUS OF HORTICULTURE SECTOR IN MAHARASHTRA

Maharashtra state occupies a unique position in the country with respect to wide range of horticultural crops. The state has revolutionized the cultivation of different fruits and vegetables. Maharashtra is famous for Alphonso mango, Nagpur *santra*, basarai banana, excellent quality of seedless grapes, soft seeded pomegranate and Poona fig. Other important fruit crops grown are orange, sapota, sweet orange (*Mosambi*), citrus and guava. Onion is a predominantly grown vegetable crop in the major parts of the state. Maharashtra is a leading state in fruit production in the country by contributing 18% share (National Horticulture Board, 2008). The area, production and productivity of major horticultural crops grown in Maharashtra state are presented in Table 1.

Table 1. Area, production and productivity of major horticultural crops in Maharashtra (2007-08)

Horticultural crop	Area (000 ha)	Production (000 t)	Productivity (t/ha)
Orange	723.7	796.1	1.1
Sweet orange (*Mosambi*)	617.0	678.7	1.1
Mango	455.8	710.9	1.6
Citrus	261.3	1627.7	6.2
Onion	254.5	4003.1	15.7
Kagzhi Lime	139.1	153.0	1.1
Pomegranate	90.5	596.2	8.2
Banana	80.0	4962.9	62.0
Sapota	65.3	294.1	4.5
Grape	45.6	1290.0	28.3
Guava	32.0	227.8	7.8

(Source: NHB, 2008)

Maharashtra contributed a major share in horticultural crops *viz.*, grape (76.9%), pomegranate (69.5%), orange (55.2%), onion (32.9%) and sapota (23.8%) in national production and occupies first position in the country. The other crops like sweet orange, citrus, banana and guava ranked second in its total production in the country (Table 2).

Table 2. Share of Maharashtra in different horticultural crop production (2007-08)

Horticultural crop	Share of Maharashtra in all India production (%)	All India Rank
Grape	76.9	I
Pomegranate	69.5	I
Orange	55.2	I
Onion	32.9	I
Sapota	23.8	I
Sweet orange (*Mosambi)*	23.0	II
Citrus	21.5	II
Banana	21.4	II
Guava	12.7	II
Kagzhi Lime	6.3	V
Mango	5.2	VII

(Source: NHB, 2008)

Maharashtra is a leading state in export of fresh fruits and vegetables in the country. The state accounts for a significant share in export of grapes (98.00%), mango (65.00%), onion (90.00%) and vegetable (68.00%) in total national export.

The major horticultural crops growing belts/districts of Maharashtra are:

- Banana: Jalgaon, Ahmednagar, Dhule, Nanded, Parbhani
- Citrus: Amravati, Nagpur, Akola, Aurangabad
- Kagzhi lime: Pune, Satara, Wardha, Solapur, Akola, Nagpur, Beed, Aurangabad, Ahmednagar
- Sweet orange: Ahmednagar, Aurangabad, Jalgaon, Amravati, Pune
- Orange: Nagpur, Akola, Amravati, Wardha
- Grape: Sangli, Satara, Nasik, Solapur, Pune, Ahmednagar
- Pomegranate: Solapur, Sangli, Nashik, Ahmednagar, Pune, Dhule, Aurangabad, Satara, Osmanabad and Latur
- Guava: Satara, Beed, Pune, Ahmednagar, Aurangabad, Amravati
- Mango: Ratnagiri, Sindhudurg, Raigad, Aurangabad, Parbhani, Jalana
- Onion: Nasik, Ahmednagar, Pune, Satara, Solapur, Dhule, Jalgaon

PROCESSING INDUSTRY IN MAHARASHTRA

The processing industry has opened new avenues for the horticulture farmers of Maharashtra. Efforts are being made to establish processing units particularly of fruits. This will increase value addition of horticultural crops, besides providing employment in rural areas. A total of 448 registered processing units have been started by different sectors in the state. The details of private, co-operative and public food processing units are given in Table 3.

Table 3. Registered food processing units in Maharashtra

Sector	Home scale	Cottage scale	Small scale A	Small scale B	Large scale	Total
Private	48	34	31	111	183	427
Co-operative	3	0	4	3	3	13
Public	5	0	1	1	1	8
Total	56	34	36	115	187	448

The state also tops in establishment of grape wineries with 51 wineries (95% of the country) with total annual wine production of 13.2 million litres with the investment of Rs. 2477.1 million. Thus, Maharashtra has achieved significant progress in sustainable horticulture development. The present paper deals with policy initiatives, linkages between research institutes, extension system and client's organizations, innovative approaches employed for strengthening Research-Extension-Farmers linkages, market support provided to horticultural produce and SWOT analysis of horticultural revolution in Maharashtra state.

EMPLOYMENT GUARANTEE SCHEME (EGS) LINKED HORTICULTURE DEVELOPMENT

The Government of Maharashtra (GoM) established a separate Department of Horticulture in 1981 to provide horticultural extension service. The diverse agro-climatic conditions and varying soil types of the state are congenial for cultivating the high value horticultural crops. Considering this strength, the Government of Maharashtra started an ambitious programme "Employment Guarantee Scheme (EGS) linked Horticulture Development" in 1990-91. This EGS programme along with strong research-extension-farmers linkage in real sense led to horticultural revolution in Maharashtra state.

Objectives of EGS

a. Utilize 29 lakh ha cultivable waste areas

b. Diversification of crops from low value to high value

c. Creation of employment opportunities in rural areas

d. Natural resource management i.e. soil, water etc.

e. Socio-economic improvement of the farmers

Salient features of the EGS

- 25 principle fruit crops covered under the scheme
- Each farmer can avail benefits upto 4 ha; However, upto 10 ha in Konkan region
- Cent percent subsidy on wages and inputs to small and marginal farmers and those belonging to SC, ST, VJ, NT and Nav Buddhist categories. However, other beneficiaries avail 100% assistance on wages and 75% assistance on inputs
- Subsidy amount ranges from Rs.17,281/- to Rs. 49,189/- per ha as per the crop
- Subsidy provided in terms of cash for wages and in kind like planting material, fertilizer, plant protection measures
- Subsidy amount is directly deposited by Demand Draft to beneficiaries' on bank account

Achievement under EGS

- Area covered during 1990-91 to 2006-2007 was 12.54 lakh ha, thus, total area under fruit crops in the state has gone up to 14.96 lakh ha. On an average, one lakh ha area per annum was covered under the scheme (Figure 1)
- The production of horticultural crops has increased from 9.45 lakh tonne (1990-90) to 103.24 lakh tonne (2006-07) under EGS programme (Figure 2)
- More than 2.76 lakh ha waste and fallow area (29%) has come under cultivation
- Numbers of beneficiaries are 15.51 lakh
- 7% SC and 11% of ST farmers benefited
- Employment generation is 23.33 crore mandays
- 35,525 villages covered
- Rs. 803.50 crore has been distributed as subsidy

RESEARCH-EXTENSION-FARMERS LINKAGES IN MAHARASHTRA

Strong and demand driven linkages between various stakeholders and farmers are important for the significant progress of any sector. In Maharashtra, the horticulture development has been achieved through a strong and wide network between various research institutes, state extension system and horticultural growers. The existing structure of Research-Extension-Farmers Linkages in Maharashtra is depicted in Figure 3.

Research and Front Line Extension System

- *ICAR*: Indian Council of Agricultural Research
- *MCAER*: Maharashtra Council of Agricultural Education and Research

- *SAUs*: State Agricultural Universities
- *NRCs*: National Research Centres
- *ATICs*: Agricultural Technology Information Centres
- *KVKs*: Krishi Vigyan Kendra's
- *NARP*: National Agricultural Research Project

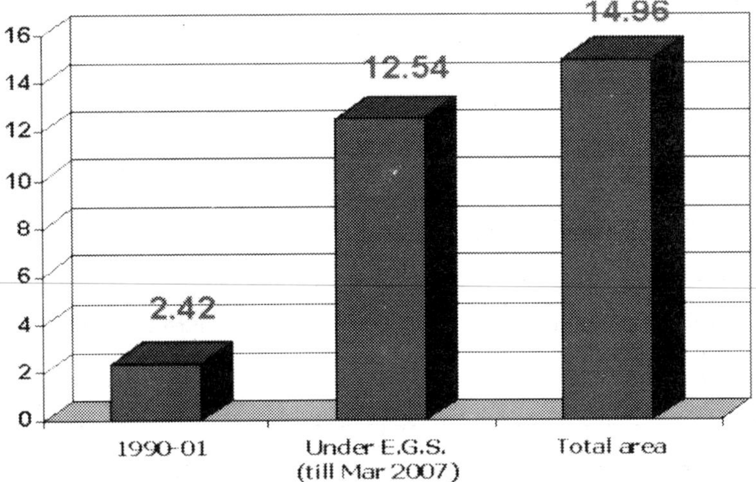

Figure 1: Area in horticulture

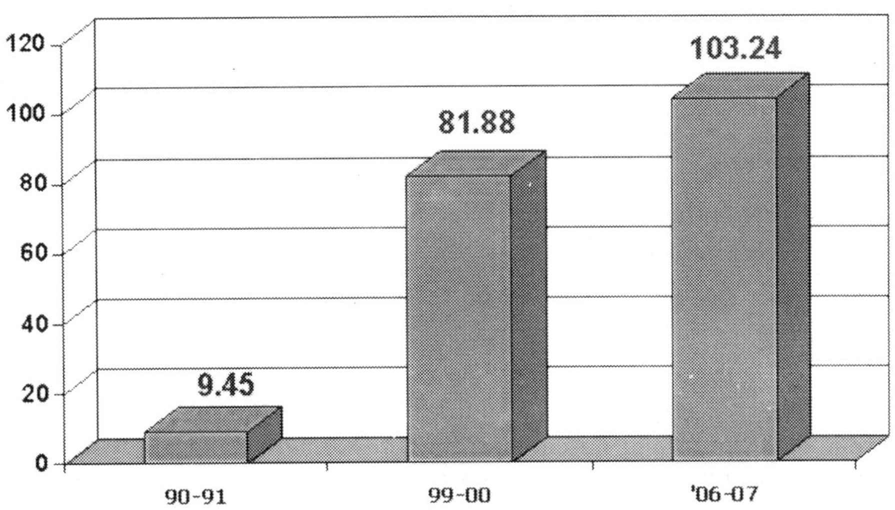

Figure 2: Production in horticulture (Production in lakh MT)

State Extension System

- *VANAMATI*: Vasantrao Naik Agriculture Management and Training Institute

- *SAMETI*: State Agriculture Management Extension Training Institute
- *RAMETIs*: Regional Agriculture Management Extension Training Institutes
- *ATMA*: Agricultural Technology Management Agency
- *DES*: District Extension System

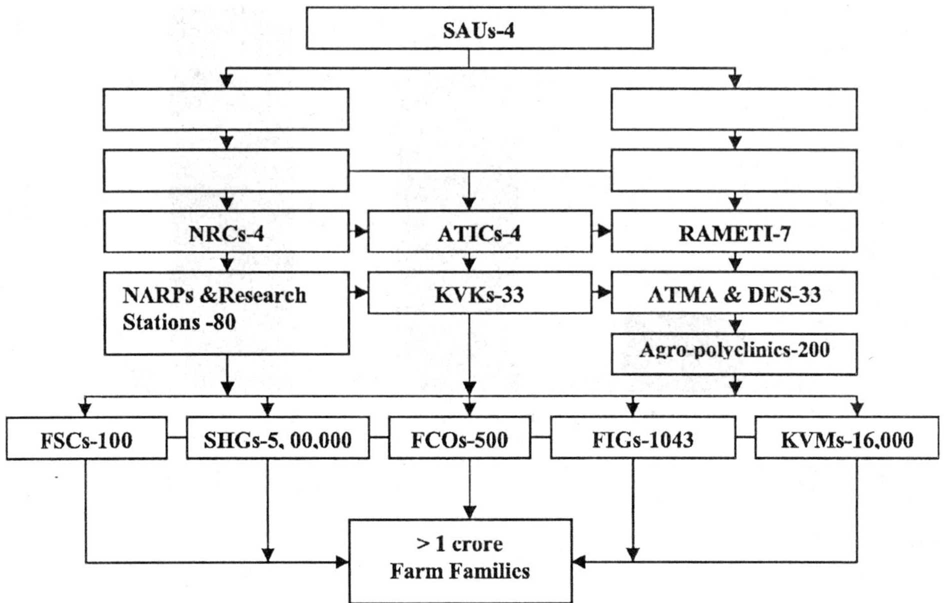

Figure 3. Institutional paradigm of Research-Extension-Farmers linkages in Maharashtra

Client System

- *FSCs*: Farmers-Scientist Clubs
- *SHGs*: Self-Help Groups
- *FCOs*: Farmers Cooperative Organizations
- *FIGs*: Farmers Interest Groups
- *KVMs*: Krishi Vigyan Mandals

As discussed above, the horticulture revolution in Maharashtra can be attributed to the strong Research-Extension-Farmers linkages. ICAR's four National Research Centers in the state *viz*., NRC on Grapes-Pune, NRC on Onion and Garlic-Rajgurunagar, NRC on Citrus-Nagpur and NRC on Pomegranate-Solapur along with four SAUs of Maharashtra *viz*., Dr. BSKKV, Dapoli, MPKV, Rahuri, MAU, Parbhani and Dr. PDKV, Akola have done commendable research on various fruits, vegetables and flowers crops and have developed varieties, designed equipments, processing technologies etc. of horticulture crops. SAUs have developed and standardized different grafting techniques for horticulture crops. These include layering in pomegranate and fig, stone and softwood grafting in mango, cashew, tongue layering in guava, saddle and softwood grafting in sapota. These propagation techniques are the best for proper quality planting material. The major research outputs include development of true-to-type mother plant, virus free planting material for lime fruit crops, early spot free pomegranate planting material and solar/chemical sterilization for nematode and fungal free planting material. The SAUs have a strong network of National Agriculture Research Projects (NARP) and Agricultural Research Stations (ARS) to promote location specific research.

The State Department of Agriculture, the main agency that is responsible for extension services in the state, disseminates these technologies to the farmers. In 1998, the State Government has merged all line departments under single umbrella in the name of single window system to provide all extension services i.e. agriculture, horticulture, animal husbandry, fisheries and soil and water conservation to farmers. The state has Vasantrao Naik Agriculture Management and Training Institute (VANAMATI) at Nagpur, State Agriculture Management Extension Training Institute (SAMETI) at state level, Regional Agriculture Management Extension Training Institutes (RAMETIs) at regional level for capacity building of extension functionaries. The district extension system consists of district superintendent agricultural officers at district level, sub-division agricultural officers at division level, taluk agricultural officers at taluk level, block agricultural officers, agricultural supervisors, and agricultural assistant at grassroots level. Two hundred Agro-polyclinics are also functioning at taluk level. The Agriculture Technology Management Agency (ATMA) was pilot tested in four districts of the state viz., Ratnagiri, Ahmednagar, Aurangabad and Amravati under the National Agriculture Technology Project (NATP) during 1999-2003 and the results were encouraging. Now this project has been scaled-up in the state by preparing Strategic Research and Extension Plans (SREPs) for each district. Under ATMA, the Farmers Interest Groups (FIGs), Commodity Interest Groups (CIGs) and Self-Help Groups (SHGs) have been formed which have resulted in wider farmers participation and bottom-up programme planning with help of Block Technology Team (BTT). SHGs are facilitating entrepreneurship development in horticulture sector. Similarly, 16,000 *Krishi Vigyan Mandals* are also engaged in participatory extension work. The SAUs on their part also carry the mandate of extension education through ATICs, KVKs, Regional Extension Centres, District Extension Centres, Agriculture Information Centres, College Development Blocks, and Communication Centre. The concept of Farmers-Scientists Forum has achieved remarkable success for dissemination of technologies in the state. Apart from this, the SAUs use mobile crop dispensaries equipped with soil-water testing kit, publications etc. for mass dissemination. Capacity building of farmers, extension personnel, officers of various line departments, KVKs, NGOs, private companies is regularly executed by the SAUs. Farmer's participatory approaches like front line demonstrations, on-farm trials, and experiments are conducted by SAUs and KVKs on large scale. All these efforts have resulted in mass dissemination of horticulture varieties and technologies. The

SAUs and State Department of Agriculture have developed good coordination. The Joint Agricultural Research and Development Co-ordination Committee (AGRESCO) meeting is regularly organized every year in which there is participation of scientists of SAUs, NRCs and higher level state extension functionaries. The new varieties and technologies developed by the research institutes are presented; thoroughly discussed and finally released. After the Joint AGRESCO, the Zonal Research and Extension Advisory Committee (ZREAC) meetings are organized in each SAU before *kharif* and *rabi* seasons for transferring the recently evolved varieties and technologies to the extension personnel for rapid and timely dissemination to the farming community. There is regular interaction between the scientists and extension personnel through the Monthly District Workshops, Diagnostic team visits, field visits etc. The Maharashtra Government has undertaken the Maharashtra Agricultural Productivity Improvement Campaign (MAHAPIC) for increasing the productivity of agricultural and horticultural crops. Besides this, the *Krishi Dindi* programme was implemented on massive scale during May 2009 with active participation of scientists, extension personnel, farmers and other stakeholders to create awareness about improved technology. Farmers Field Schools (FFS) on different horticultural crops are organized to promote organic farming, Integrated Pest Management etc. Under centrally sponsored NHM, the state is implementing a special campaign for management of pomegranate oily spot and technological mission on orange in suicide prone zone of Maharashtra. The National Research Centre for Citrus, Nagpur is acting as a nodal agency for this programme.

The innovative horticulture farmers of the state have organized themselves into commoditywise co-operative groups. Maharashtra state has commodity-based strong growers' organizations, which has led to expansion of horticulture area and sustainable horticultural development in the state. Some major commodity-wise growers' organizations are Grape Growers Association, Mango Growers Association, Pomegranate Growers Association, Banana Growers Association, Sapota Growers Association, Orange Growers Association, Sweet Orange Growers Association, Kagzhi Lime Growers Association, Guava Growers Association, Anola, Fig, Custard apple Growers Association, Vegetable Growers Association, Flower Growers Association, Western India Floriculturist Association (WIFA), Medicinal & Aromatic Plants Growers Association and Nursery Owners Association which are working effectively and efficiently. This institutional paradigm of research-extension-farmers linkage has brought revolutionary impact on development of horticulture sector in Maharashtra.

APPROACHES FOR STRENGTHENING RESEARCH-EXTENSION-FARMERS-MARKET LINKAGES

Varietal Improvement

The State Agricultural Universities, ICAR research institutes, private agencies and NGOs involved in research and development programmes have made significant contribution in development of location-specific varieties of horticultural crops by using hybridization techniques and biotechnology, which blended with Indigenous Technological Knowledge (ITKs) for genetic conservation. These varieties have high production capacity, insect-pest resistance, drought tolerance, long storage capacity and high export potential. These varieties have demand in domestic as well as international markets. The commodity-wise improved varieties predominantly grown in the state are given below:

Orange: Nagpur mandarin, Kinnow mandarin (Exotic introduction)

Mango: Alphonso, Kesar, Pairi, Ratna, Sindhu, Vanaraj, Totapuri, Sai-Sugandh

Pomegranate: Phule Bhagawa, Phule Arkta, Mridula, Ganesh, G-137

Banana: Grand Naine, Srimanti, Dwarf Cavendish, Basrai, Lal Velchi, Safed Velchi

Grapes: Thomson Seedless, Sonaka, Flame Seedless, Tas-e-Ganesh, Sharad Seedless, Manik Chaman

Guava: Lucknow 49, Nagpur Seedless

Kagzhi Lime: Sai Sharabati, Phule Sharabati

Sapota: Kalipatti, Cricket Ball

Onion: Baswant 780, Phule Samarth, N 2-4-1, OFLR

Supply of Quality Planting Material

Timely supply of good quality planting material like seedling, saplings and rootstocks are crucial for higher production of horticultural crops. In Maharashtra state, the public institutes, private agencies and NGOs have established large numbers of certified nurseries. Presently 1478 certified nurseries in the state comprise of public sector (136), SAUs (42) and 1300 nurseries established by private agencies. These nurseries provide quality planting-material at optimum rate to the farmers. The SAUs play a vital role by providing quality planting- material of the improved varieties developed by them. MPKV, Rahuri produced about 40,000 kg and Dr. BSKKV, Dapoli produced 5,000 kg of vegetable seeds. MPKV supplied 8.5 lakh planting material to the farmers during 2007-08. The KVKs of Maharashtra also produced more than 5 lakh of seedlings (ICAR report, 2008).

Integrated Cropping System Approach

Dr. BSKKV, Dapoli has developed coconut and spices mixed integrated cropping system module named *Lakhi Bag*. Konkan region of Maharashtra comprises of more than 30,000 ha of area under coconut and arecanut. In order to achieve sustainable production and higher economic returns, intercropping of spices and condiments in coconut based cropping system is evolved through *Lakhi Bag*. In this system, total 70 coconut trees are planted in one acre area along with intercropping of black pepper (140), cinnamon (123) and nutmeg (54). The system is popularized through different extension strategies in the region and farmers have realized more than 1 lakh net return in Konkan region.

Public-Private Partnership (PPP) Based Mass Communication

Agro-Technology Week-2005

Mass communication is an essential phenomenon for agriculture development. In view of this MPKV organized Agro-Technological Week at university head quarter in the year 2005, wherein agricultural exhibition of 250 stalls consisting of critical inputs and different improved technologies developed by the SAUs, ICAR institutes and private agencies. MPKV organized live demonstrations of different crops on 1700 acre area. More than three lakh farmers from Maharashtra and other states were benefited.

International Hi-Tech Agro Week-2007

The MPKV, Rahuri introduced the concept of PPP based mass dissemination of technology by organizing International Hi-Tech Agro Week at Pune in 2007. The programme was organized in partnership with Agrowon, a reputed vernacular agricultural news paper in the state. Nobel laureate Prof. M. Yunus inaugurated the event. Many international, national research institutes, SAUs from entire country, developmental agencies from public and private sector, leading multinational private companies, NGOs and SHGs participated in the mega exhibition programme. Live demonstrations of 17 different crops with 56 varieties, protected floriculture, exotic vegetables, cropping systems, Integrated Nutrients Management (INM) and fruit orchard management were organized on 135 acre area. In all 31 experts guided farmers on different topics like HRD, agro-processing and marketing, sustainable IFS model and ICT in agriculture. More than 5 lakh farmers, farm women, agri-entrepreneurs, SHGs members, students and youth across the country visited the event.

ICT-Based Information System

Access to information and knowledge has become a key element for agricultural competitiveness at household, regional, and international level. Thus, the role of ICT is important. In this connection, the Agricultural Technology Information Centres (ATIC) established at SAUs under NATP in Maharashtra play a vital role in providing on door information and critical inputs to farmers. ATIC is providing information through ICT based tools like telephone help line, photo gallery, electronic mails, mobile SMS, website, video-conferencing, e-dialogue activity and mobile dispensary. MPKV started the e-dialogue activity connecting 12 villages under its jurisdiction. The video-conferencing facility is created in the ATIC to act as central hub. The centre also consists of seed sale counter for providing better quality seed of improved varieties. Many KVKs in the state are providing weather and market based information through mobile SMS to the farmers. The information is also disseminated through FAQs, e-aqua portals, websites etc.

Use of Print and Electronic Media

The SAUs of Maharashtra regularly publish annual and seasonal publications like *Krishi Darshani, Krishi Sanvadini, Shree Sugi*, University News Letters, booklets, folders, posters etc. to provide location-specific information to the farmers. More than one lakh of these publications are sold in the state during every year. Agricultural scientists are regularly disseminating information through vernacular newspapers like Agrowon and monthly agricultural magazines. The university scientists also deliver talks on radio and TV. Community radio and farm schools disseminate agro-based information to the farmers.

Farmers-Led Extension Strategies

MPKV, Rahuri has implemented the novel concept of Farmers-Scientists Forum (FSF) as farmer-led extension approach since October, 2005. The FSF established at central campus, comprises of 40 Farmers-Scientist Clubs (FSCs) of 1500 member farmers in the jurisdiction of university. Each FSC has a coordinator and these FSCs are attached to the Agricultural Colleges, Agril. Research Stations and Agricultural Technology Schools of the university. These FSCs function as Commodity Interest Groups (CIGs) and started to educate the farmers about the innovative technologies generated by university in order to promote farmer to farmer extension. MPKV has developed model farms of these members in their villages and these farmers serve as farm leaders to disseminate the technologies in their social system. FSF promotes participatory research- extension approach. Regular monthly meetings and review meetings are organized, wherein interaction between FSC members and university scientists take place. The university has developed database of soil, water, cropping pattern, various enterprises of member farmers etc. and accordingly, and suggest location specific interventions for increasing their income through agro-based enterprises. Some case studies of members of FSC are presented below:

Market-Led Production of Exotic Vegetables

A woman member farmer of FSC Mrs. Vanita Gunjal from Kandali village of Pune district produces exotic vegetables (e.g. Red cabbage, Brocoli, Chinese cabbage, Pakchoy) for five-star hotels such as Taj and Oberai in metropolitan cities (Pune and Mumbai). She owns a meager land of 1.87 ha But, she has reaped rich benefits due to demand driven marketing under the guidance of university scientists. MPKV is also providing guidance on vegetable production techniques and post harvest management aspects. After grading and packaging, vegetables are dispatched to Mumbai by road transport. Marketing of these exotic vegetables is done under the trade name of *Green Pallavi Farms and Sales*. She is getting Rs. 25000/ month monetary returns and more prices to vegetables in off-season also, because she is cultivating vegetables as per demand of the hotels. Mrs. Gunjal is the winner of N.G. Ranga Farmer Award for Diversified Agriculture (2005-06) of the ICAR, New Delhi.

Bumper Pomegranate Production in Scarcity Zone

Another member farmer of FSC Mr. Hanumant Gajare from Aran village of Solapur district has bumper pomegranate production in scarcity zone. He has cultivated *Phule Bhagawa* pomegranate variety released by MPKV, Rahuri on 1.20 ha area. He has received technical information through participation in activities of FSC, discussion with the scientists of MPKV and NRC for Pomegranate. The pomegranate farm of this farmer is certified by EUROGAP under the guidance of University scientists. He produced an average 48 tonnes of pomegranate per ha and realized a total income of Rs. 17.50 lakh (2006), Rs. 31.90 lakh (2007) and Rs. 37.20 lakh (2008) through pomegranate export to European countries. He received market rate of Rs. 70/kg, whereas, he incurred an expenditure of only Rs. 3.0 lakh.

Decentralized, Participatory and Market-Led Extension Strategies

Low productivity of crops and less remunerative market prices are the major causes for declining accelerated growth of farming. Farmers need to transform themselves from mere producer to producer cum seller in a wider market to realize good returns on their investments, risks and efforts. In order to achieve this goal, farmers need to know answers to questions like what to produce, when to produce, how much to produce, when and where to sell, at what price and in what form to sell their produce. Hence, decentralized and participatory Agricultural Technology Management Agency (ATMA) established under National Agricultural Technological Project (NATP) funded by World Bank played crucial role in promoting and nourishing market-led production of horticultural crops in Konkan region of Maharashtra. Bottom-up and need-based Strategic Research and Extension Plan (SREP) were prepared by ATMA and Dr. Balasaheb Sawant Konkan Krishi Vidyapeeth, Dapoli for sustainable horticultural development in the region. Alphanso mango is a predominant horticultural crop in Konkan. About 99% of resource poor mango growers sell their produce at 1/4th price through intermediaries. ATMA of Ratnagiri district promoted 13 Farmers Interest Groups (FIGs) of mango growers. Dr. BSKKV, Dapoli and Krishi Vigyan Kendra, Shirgaon made capacity building of FIGs through training for intervention in different critical areas like quality mango production, plant protection measures, post harvest management etc. More than 2000 mango growers were trained to enhance their knowledge and management skill. ATMA, Ratnagiri, Dr. BSKKV, Dapoli, KVK, Shirgaon and other private agencies organized Mega Mango Festival in metropolitan city Mumbai on public-private partnership mode to create direct producers-consumers linkages. Each FIG established stall in this Mega Mango Festival. The farmers got four times more returns as compared to return from previous system and consumers got better quality mangos at cheaper rate than the regular market.

Human Resource Development Programmes

Agricultural universities in the state regularly organize human resource development programmes for capacity building and enhancing resource use efficiency of farmers. The cashew is principle fruit crop grown in the Konkan region, but 85% of cashew produced is processed out of Maharashtra. Dr. BSKKV, Dapoli identified key cashew producers and organized large-scale training programmes on post harvest management and value addition. More than 2500 cashew producers empowered in knowledge, skill and management for entrepreneurship development. Majority of the cashew growers were motivated and established 1500 small-scale cashew-processing units in the region. The processing capacity of each unit is 5 tonne and total investment accounting Rs. 36.0 crore. These processing units enhanced livelihood options in rural areas.

ROLE OF MAHARASHTRA STATE AGRICULTURAL MARKETING BOARD (MSAMB) IN HORTICULTURE

The Maharashtra State Agricultural Marketing Board (MSAMB), Pune established in 1984 has done pioneer work in agricultural marketing and achieved success in various areas. MSAMB has played an important role in developing and coordinating efficient marketing system in Maharashtra. Some major roles of MSAMB are to co-ordinate the functioning of Market Committees for development of markets and market related areas, advice, guide and supervise the Market Committees in the state, to prepare state plans for marketing of agriculture produces, maintain and administer the agricultural marketing development fund, capacity building of employees of Marketing Committees through seminar, workshop and training, and to provide market intelligence via different print and mass media.

Promotion of Farmers Partnership Cooperatives for Marketing

In the context of liberalization, privatization, commercialization and globalization, MSAMB played critical role in promoting and nourishing the Farmers Partnership Cooperatives for global marketing of horticultural produce. MSAMB provides technical guidance and financial support to these cooperatives to face the global competitiveness. MSAMB has made quality certification of different farms of horticultural crops *viz.*, 2000 grape, 200 mango and 50 pomegranate farms are certified by EUROGAP. The details of farmer's co-operatives in Maharashtra are given below:

1. MAHAGRAPES

"MAHAGRAPES" a co-operative partnership firm has been established at Pune in 1991. Total 16 Grape Grower Co-operative Societies are the members of Mahagrapes. Main objective of Mahagrapes is to boost the export of grapes for which facilities like pre-cooling and cold storage has been erected at each grape grower co-operative society. Due to its effective functioning, the brand of Maharashtra Grapes is penetrated in the international market. Mahagrapes established the brands particularly in European Union and Middle East market during last decade. Mahagrapes exported 40,000 tonne of grapes to European Union.

2. MAHAMANGO

"MAHAMANGO" is mango producers organization promoted by MSAMB established at Kudal, District Sindhudurg in 1991. Facilities like pre-cooling, cold-storage, pack house, grading packing line etc. are created with financial assistance from APEDA, New Delhi and MSAMB, Pune to boost up the export and domestic marketing. The facility centre is certified by the China delegation (AQSIQ department) and agreed to import Alphanso and *Kesar* mangos. Total 35,328 t of mangoes were processed and exported which accounts for 65% of total mango exports.

3. MAHABANANA

India is a leading producer of banana and Maharashtra contributed 21.4% share in national production. But there is hardly any export of bananas due to improper post harvest practices, lack of storage facilities and inefficient transport system. The produce is being wasted upto 25 to 40%. Taking all these facts into consideration, MSAMB initiated the efforts for co-ordination and established "MAHBANANA" organization located at Jalgaon in 2002. Considering the area, production and wide scope for banana export, MSAMB has proposed to establish two modern packhouses in banana growing belt *i.e.,* Jalgaon and Hingoli. Modern Packhouse includes automatic banana conveyor system, pre-cooling, cold storage and ripening chamber. This organization carries pivotal role in planning a strategy for banana export, supplying good quality tissue culture plants, organizing seminars / workshops/training programmes to boost production having export standards, and domestic and international marketing of produce. Total 4165 t of banana were processed and exported which accounts for 25% of total banana exports of the country.

4. MAHAORANGE

In order to promote export of orange in the international market and strengthening the domestic marketing system of orange, an apex body named "MAHAORANGE" has been established on 10th March, 2008. At present, there are five orange growers co-operative societies are the member societies of MAHAORANGE. The organization is established with aim to strengthen the domestic marketing system of orange as well as sweet lime, promote quality production to meet international requirement, establish direct linkages between producers to consumers, provide required infrastructural facility and erect orange processing as well as orange fruit juice parlors. In order to give maximum returns to the orange growers, this organization has taken initiative to arrange events like Orange festivals in metropolitan cities with the help of MSAMB.

5. MAHAANAR

Keeping in view the importance and widely growing trade of pomegranate at international level, an apex farmer's cooperative organization named "MAHAANAR" has been established on 1st August, 2005. The main purpose is to provide technical information about quality production of pomegranate that meets international standards and provide necessary infrastructure like pre-cooling, cold storage, handling line to the growers. At present, 3,500 individual pomegranate growers are the members of MAHAANAR.

Supply Chain Support System

In order to promote export of horticultural produce and earn foreign exchange, the MSAMB has taken concrete efforts for effective supply chain management in the state. The state is well equipped with adequate infrastructure and efficient public organizations for demand driven supply chain management in Maharashtra are furnished in Table 4.

Table 4. Public organizations for demand driven supply chain management

Particulars	Location
Inland Container Depots (ICD)	Mumbai, Chinchwad, Nagpur, Bhusawal, Aurangabad, Miraj
Regional Plant Quarantine Centre	Shiwdi (Mumbai)
Airport-Air Cargo Facilities	Mumbai, Pune, Aurangabad, Nagpur
Sea Port	JNPT Mumbai
Vapour Heat Treatment Plant	Washi (Mumbai)
Irradiation Plant	Lasalgaon (Nashik)
Phytosanitary Certificate facility for export	Pune, Nasik, Sangli, Solapur, Amravati, Ratnagiri, Sindhudurgh
Post Entry Quarantine (PEQ) facilities for import inspection	4 SAUs

Agri Export Zones

Maharashtra added more than 12 lakh ha area under horticulture crops during the last ten years. Substantial production from this newly added area is expected in forthcoming years. This means wide potential for marketing of horticulture produce from the state with possibility of increased export. Maharashtra is the leading state in export of fresh and processed fruits and vegetables from the country today. Keeping this in mind, the state has identified eight potential horticultural crops of commercial importance and created Agri Export Zones (AEZ), which will give required momentum for steady and systematic growth of export of horticultural produce from the State. The commodity wise AEZ established in the state are furnished in Table 5.

The specific interventions and agencies are identified for AEZ implementation. Funds are made available from public, private and co-operative sector on subsidies/grants basis, and loan availed from bank and financial institutions. Responsibility for coordination lies with the nodal agency, *i.e.,* Maharashtra State Agricultural Marketing Board. The nodal agency with support and cooperation from the Department of Horticulture, Department of Agriculture, Department of Industries, Department of Finance, Maharashtra

State Electricity Board (MSEB) and other agency/ agencies would implement the AEZ in a time bound and effective manner.

Table 5. The commodity-wise AEZ established in Maharashtra

Commodity	Districts
Mango (Alphanso)	Ratnagiri, Singhudurg, Thane, Raigad
Mango (Kesar)	Aurangabad, Jalna, Beed, Latur, Ahmednagar and Nashik
Grapes	Nashik, Pune, Ahmednagar, Solapur, Satara, Sangli, Latur, Osmanabad
Onion	Nasik, Pune, Satara, Ahmednagar, Solapur and Jalgaon
Orange	Amravati and Nagpur
Banana	Jalgaon, Dhule, Hingoli, Nanded, Parbhani, Buldhana & Wardha
Pomegranate	Nasik, Solapur, Pune, Ahmednagar, Latur, Beed and Sangli
Floriculture	Pune, Nasik, Sangli

Horticultural Parks

The state also started different commoditywise horticultural parks like floriculture park at Pune, wine park at Nasik (Vinchur) and Sangli (Palus), and food parks at Pune (Chordia-Shirval) and Nagpur (MAIDC and Haldiram).

SWOT ANALYSIS OF HORTICULTURAL REVOLUTION IN MAHARASHTRA

The state achieved remarkable progress in horticulture sector due to some strength in different areas. Further, there is large scope and prospects to strengthen this sector. Inspite of the progress, there are some weaknesses and threats, which hinder the further growth. The opportunities have to be availed by the concerned key players to promote horticultural growth. The SWOT analysis of horticulture revolution in Maharashtra is given below.

Strengths

- Diverse agro-climatic conditions (Nine agro-climatic zones)
- Four State Agricultural Universities (SAUs) with adequate research and extension support
- 33 Krishi Vigyan Kendra's
- National Research Centers (NRCs) for Grape, Citrus, Onion–Garlic, Pomegranate
- Horticulture Training Center, Talegaon, Pune
- Hi-Tech Floriculture Project, Pune
- Single Window System of State Department of Agriculture

- Centers for capacity building and human resource development- VANAMATI, SAMETI and RAMETIs
- Maharashtra State Agricultural Marketing Board (MSAMB)
- Agriculture Produce Marketing Committees at tahsil level
- Agri-Export Zones
- Involvement of private agencies and NGOs in horticultural development.
- Krishi Vigyan Mandals
- Strong co-operative network
- Strong commodity based farmers' organizations
- Proximity to international airport and sea port
- Innovative farmers

Weaknesses

- 84% rainfed area
- 52% drought prone area
- 41.06% light soils
- Small operational land holdings threatening mechanization in horticulture (73.00% small and marginal farmers in 2001)
- Uncertainty in rainfall
- Post-harvest losses of horticultural produce
- Lack of sufficient post harvest infrastructure

Opportunities

- Organic farming in horticultural crops
- IPM, INM, IWM etc.
- Integrated Cropping System
- Access to global market
- Promoting post harvest technology on farmers field
- Processing and value addition
- Mechanization in horticultural sector
- Natural resources management
- Protected cultivation of high value horticultural crops
- Mobilization of women SHGs in horticultural sector (More than 5 lakh SHGs)
- Improving livelihood security of resource poor

Threats

- Market glut

- Global trade competition
- Residual effect on human health
- Incidence of insect-pest and diseases. e.g. Oily spot on pomegranate, Sigatoka of banana, Spongy tissue in mango etc.

CONCLUSION

The present deliberations lead to conclude that India is a leading producer of fruits and vegetables in the world. Concrete efforts from scientists, policy makers and institutional reforms in the country led to substantial increase in area and production of horticultural crops but with low productivity. Maharashtra state contributed significant share in major horticultural crops and created a brand name in international trade. EGS Linked Horticultural Programme and strong research-extension-farmers linkages extended by marketing support have led to horticultural revolution in Maharashtra. Integrated cropping system module, varietal improvement, quality planting material and demand driven production strategies played a vital role in expanding and sustaining horticultural development. Public-Private Partnership based mass dissemination, Information Communication Technology, print and electronic media, farmer-led extension, human resource development, decentralized, participatory and market-led extension approaches are crucial for strengthening and integrating research-extension-farmer and marketing system. Organization and mobilization of client groups are essential to promote farmers participatory research, extension and marketing. Maharashtra State has promoted commodity-based farmer's partnership co-operative organizations and established AEZs for access to international market. Thus, various stakeholders having potential in different areas must work in tandem to form consortium for demand driven supply chain management and export of horticultural produce, which would increase the country's economic growth in general and sustain livelihood security of resource poor in particular.

REFERENCES

National Horticulture Board, 2008. Indian horticultural Data Base at www.nhb.gov.in

Maharashtra State Agricultural Marketing Board, 2009. www.msamb.gov.in

27. FARMERS' PARTICIPATORY OPERATIONAL RESEARCH IN TAMIL NADU

E. Vadivel

Director of Extension Education, Tamil Nadu Agricultural University, Coimbatore – 641 003

ABSTRACT

The turnkey project on Precision Farming in 400 ha (2005-07) and subsequent scale up of the project in 12,800 ha (2007-08), 9400 ha (2008-09) and 20,000 ha (2009-10) in Tamil Nadu not only provided an opportunity to translate the technologies into action but also helped to identify the inaccuracies and inadequacies in the current research management system. There were moments of upbeat when technologies contributed to double or triple the productivity and equally moments of clumsiness when technologies miserably failed to address the common field problems and lacked synergy among themselves. The later phenomenon had created a plat form where field extension scientists and core research scientists repeatedly interacted and debated on 'real time' field issues and the relevance of the technologies. The message from the field extension scientists was clear on one issue: The technology generation system needs to be revamped, revitalised and refocused so as to make the technologies relevant to the real time situations.

INTRODUCTION

The turnkey project on Precision Farming on 400 ha (2005-07) and subsequent scale up of the project to 12,800 ha (2007-08), 9400 ha (2008-09) and 20,000 ha (2009-10) in Tamil Nadu not only provided an opportunity to translate the technologies into action but also helped to identify the inaccuracies and inadequacies in the current research management system. There were moments of upbeat when technologies contributed to double or triple the productivity and equally moments of clumsiness when technologies miserably failed to address the common field problems and lacked synergy among themselves.

The later phenomenon had created a platform where field extension scientists and core research scientists repeatedly interacted and debated on 'real time' field issues and the relevance of the technologies. The message from the field extension scientists was clear on one issue: The technology generation system needs to be revamped, revitalised and refocused so as to make the technologies relevant to the real time situations.

VARIETAL / HYBRID DEVELOPMENT

When the 'culture' is ready for final evaluation before release, Adoptive Research Trails (ART) through line departments in micro plots and Multi Locational Trails (MLT) at different research centres are being conducted. Beyond this, the extension scientists have suggested a third process: Large Scale Evaluation (LSE) of the culture in macro plots (minimum 1 ha) at the farmers holding (Farmer as participant in the evaluation process) organized through KVK system. The 'scientific bias' in ART and MLT process is removed in the Large Scale Evaluation (LSE) by the farmers in 'real time' situation and the farmers observation is unconditional, panoramic and borne out of experiencing the 'culture'.

MANAGEMENT TECHNOLOGY DEVELOPMENT

India has the largest pool of agricultural scientists in the world working in 39 SAUs, 136 research institutions of ICAR and 565 KVKs. Yet there is a strong criticism that there is a mismatch between research output that are currently being made available and what is needed to address the farm problems. There is a strong need to refocus and reorient the technology generation process to remove the mismatch and really make the technologies relevant. Beyond Kharif and Rabi trails, All India Co-ordinated Research trails, the extension scientists have suggested additional processes to make the technology relevant to the farmers need as detailed below:

1. Process of Identification of Researchable Issues

Process of identification of researchable issues shall be through PRA or through discussion within each of the Commodity Group/Association. Participatory Rural Aappraisal mode was very much acceptable by the farmers. In nine districts of Tamil Nadu, Participatory Rural Appraisal (PRA) was conducted separately district wise with 100 precision farmers for each district. One hundred farmers were identified independently by the Joint Director of Agriculture, Deputy Director of Horticulture and Programme Coordinators of KVK who jointly implemented the project in each district with different levels of physical target. The farmers were divided into four groups and each group was assigned with a theme for SWOC analysis. The themes were *viz.,* production under precision system, plant protection, marketing and institutional support. After an hour long discussion and debate facilitated by the scientists in each group, the group convenors made the presentation of the out put of SWOC analysis to the common pool of 100 farmers for second level of refining the identified issues during the discussion. The very involvement of each farmer in the SWOC analysis led to generation of newer concepts, logics and possible solutions from among the farmers. The consolidated report of SWOC analysis of nine districts sufficiently indicated the district specific priorities of researchable issues and the efficiency levels of farmers to handle such technologies. Compared to the segmental analysis by the scientists of different disciplines in institutions, the holistic analysis of the issues related to variety / hybrids, crop production and protection through PRA process has a sound basis and rationale. Strikingly, when the group leaders of one theme made the presentation of the outcome of the SWOC analysis, the members of the other group did not put forth contradictory views but fine tuned, redefined and added one more dimension. The very PRA process helped the farmers to own the issues demanding no TOT efforts from the scientists to inculcate a new idea or concept in the minds of the farmers. The researchable issues emanated out of the PRA became undisputable whenever presented in any other farmers / scientists forums for reference. Eighteen such researchable issues were identified and scouted and FPOR was contemplated during the year 2008-09 (list in annexure I).

2. The Process of Designing the Methodology and Making Critical Observations

The designing of methodology of research when undertaken exclusively by the scientists with in academic / institutional may be as per set norms. But there should be a terminal evaluation by the farmers through large scale testing of technologies in macro plots (minimum one ha) under Farmer Participatory Operational Research. At the institutions, the scientists make critical observation with logical scientific validation while at the Farmers Participatory Operational Research, the farmers make critical observations and the scientists provide the logical scientific explanation either to validate or to invalidate the observations of the farmers. This process helps to remove the 'conditioned' critical observation of the scientists and allow 'open ended' critical observations by the farmers and public alike.

The size of the FPOR experimentation was on minimum one ha and maximum five ha. The experimental 'plot' was the field of a farmer. The farmers role was to conduct the experiment and make critical observation either by himself or by the fellow farmers of the village or in the neighbourhood, one plot 'with technology' and the other plot 'without the technology'. The location map, contact address etc are provided to all the commodity growers associations of the State, sugar factories, line department officials, NGO's so as to ensure that the experimentation was conducted under the 'public eye'.

The scientists were not expected to make any observations and give a lead to the farmers. The scientists observations invariably borne out of preconditioned concepts and ideologies, and discipline biased while the farmers observations were natural, spontaneous but never biased towards any discipline. Farmers weigh the results in the context of technical, economic and social relevance and many a times based on consumer or market preferences. Many of the observations were educative in nature to the scientists as the farmers had taken into account the known and unknown variables also to draw valid observations.

CASE STUDY

Out of 18 FPOR projects implemented, the programme 'on thin wall laterals' was taken for further discussion in this paper as a case study.

i. PRA-Researchable issue identified

The cost of drip system was high (Rs. 80,000 / ha of annual crops) and needs to be reduced to be half (Rs. 40,000) by introducing sub surface thin wall laterals in as much as the cost of lateral is approximately equal to 50% of total cost of the system.

ii. Methodology

Saragapalli village in Hosur district was identified. Mr.Ramareddy was the farmer and the experiment was conducted on two ha of land during 2008-09, Kariff (Tomato) and Summer (Cucumber)

iii. Technology

- One ha sub surface and one ha surface lateral
- Thin wall sub surface laterals (6 mill) (Total system budget : Rs. 40,000)

Vs

- Normal class 2 surface turbo laterals (Total system budget : Rs. 80,000)

OBSERVATIONS BY THE FARMERS

i. System

Normal surface lateral	Sub-surface thin wall laterals
Cost of the system Rs.80,000	• Cost of the system Rs.40,000
1.50 kg/cm^2 pressure is required to run the system	• 1.00 kg / cm^2 pressure is adequate to run the system
Good and efficient pumping system is required	• Ordinary pumping system is sufficient
Clogging is regular problem	• No such clogging problem
Six years of possible life	• Three years of possible life only
Cattle / squirrel / rat damage is prominent	• No such problem
System is closed only at one end. Water flow is only from one end	• Water flow is from both the ends. The laterals are linked to sub main at both the ends.
Release of air valve once in a week	• Release of air valve daily

ii. Soil and water

Normal surface lateral	Sub-surface lateral
• 90 minute run of drip system per day is essential	• 30 minutes run of system per day is essential
• When there is no crop in the field, drip irrigation is not necessary	• Even when there is no crop in the field, daily 10 minutes wetting is necessary to keep the system intact at sub surface.
• Monitoring and evaluating irrigation event is visible	• Soil evaporation and deep percolation are greatly reduced. To that extent no nutrient leaching down
	• Improved in field uniformity can result in better control of water, nutrients and salts
	• Largely un seen, system mismanagement can lead to loss of crop.
	• For direct sown crop, wetting close to seed is not possible in as much as there is reduced upward movement of water

iii. Crop growth and cultivation

Tomato : Kariff crop coupled with rains

• Normal growth and yield was 81.50 MT / ha	• Normal growth and yield was 87.50 MT / ha
• Cucumber : summer crop coupled with dry hot weather, 800 bags @ 100 kg / bag	• 1000 bags @ 100 kg / bag
• Weeding can be done even after hardening of weeds	• Weeding needs to be done when the weeds are comparatively young
• Two to three weedings	• Single weeding only (less weeds than the other) due to less germination
• Hard rooted crops like brinjal, tapioca can be grown	• Hard rooted crops cannot be grown
• Relay sowing is not possible	• Relay sowing possible, ie. Cucumber was sown 20 days in advance of harvest of tomato

Scientific validation of observations

i. System

Sub- surface systems have the potential to operate at lower pressures than surface drip systems (Jerry Neyfeld, 2008). The dripper in thin wall lateral is like a zib and @ 5 drippers / m unlike in surface laterals in which the drippers are placed at 50 or 60 cm intervals. The discharge per unit of time is less and that too in lesser time.

Much work has not been done on the thin wall sub surface laterals in respect of uniformity measurements and methods. Irrigation scheduling and crop water requirement, wetting patterns in different soils, assessment of root intrusion into laterals and control measures and estimates of overall system longevity under Indian conditions. Profitability and economic aspects have not been determined conclusively and will greatly depend on local conditions and constraints, especially availability and cost of water.

The thin wall laterals emit 4-8 to 5.0 litres of water / meter ie @ lph/ dripper in as much as there are 5 strippers / m in thin wall laterals. Thus the sub-surface system qualifies as Low Pressure Irrigation System (LPS), in which water is applied at a slow rate equal to the consumptive needs of plants in a small fraction of a time over which plants metabolise. The percolation loss is prevented as the uptake by plants and supply through sub surface laterals match almost exactly. LPS through sub surface system enables the farmers to make use of same water source and pumping system with low pressure and shall minimise the initial investment and reduce the energy cost and maintenance cost. Additional benefits are longer lateral lengths due to low pressure or flow, longer irrigation duration without creating run off and deep percolation, matching the water and fertilizer application rates to plant uptake rates (Battalwar, 2007) water application at a rate close to plant water uptake using low flow emitter effect the soil water regime and plant response. It is thus expected to improve yield and reduce water from drainage below root zone as stated by Battalwar 2007.

ii. Soil and Water

The sub-surface system required to be run for 30 minutes against 90 minutes in surface drip system. That means sub-surface drip required only 1/3 of water compared to surface drip system. The irrigation water did not drench the soil surface which help to maintain the soil structure. Equally no deep drainage helps leaching of any salts from the soil profile. The sub-surface drip tape depth is critical. Six to eight inches is ideal for red soils. If the tape is too deep, subbing water upwards towards seeds can be problematic. In black soil tape can be deeper i.e. ten to twelve inches (Stewart *et al.,* 1990). High yielding crops produce planting of stubbles which need to be managed. They can be moved towards the centre of the bed and on the top of the sub-surface drip tape. The soil structure is improved and surface is protected.

The most challenging part of sub-surface drip system is wetting the soil prior to or just after planting. This may require up to two mega litres of water / ha depending on how dry the soil is. Pre -watering is time consuming and is usually started three weeks before planting. Seed sorghum uses 2-6 mega litres / ha/ corn requires 9 mega litres, ha and cotton requires 4.5–8 mega litres of water / ha (Stewart Crawford, 2008).

In a study with 15, 20 and 30 cm emitter spacing on onion, the absence of any significant effects of drip tape emitter spacing on onion yield may be due to the fact that sub-surface irrigation was managed to provide roughly similar irrigation amounts and optimal soil moisture conditions in all treatments (Juan *et al.*, 2007).

iii. Crop growth and cultivation

The yield of tomato did not vary during kharif because of intermittent rains but there was yield hike in sub-surface system for cucumber during summer. When compared to surface drip, sub-surface drip had greater yield for sweet corn (Bar-Yosef *et al.*, 1999) tomato (Phene *et al.*, 1987) and asparagus (Sterrett *et al.*, 1990). Yield response for over 30 crops indicated that the sub-surface laterals system enhances productivity by 12-35% compared to surface drip system and the quantum of water use was less in most cases. The depth of laterals ranged from 0.02 to 0.70 m and lateral spacing ranged from 0.25 (Turf grass) to 5 m (Vegetables and trees) (Battalwart 2007).

SUMMARY AND CONCLUSION

Farmers Participatory Operational Research (FPOR) or Participatory Technology Development (PTD) is a creative process of joint experimentation and research by farmers and scientists in discovering ways of improving farmers' livelihoods. The growing number of documented examples in recent years reveal that (PTD / FPOR) is now accepted as a research approach to agriculture and natural resource management (Laurens Van Veldhuizen *et al.*, 2009). The Farmers' Participatory Operational Research or Participatory Technology Development aroused the farmers inherent curiosity and creativity and increased confidence to continue experimentation. As an informal process, Farmers' Participatory Operational Research has resulted in farmer to farmer networking that rapidly and efficiently spread the ideas. This approach greatly reduces the time spent between problem identification and development of applicable solutions. Results from site specific farmer led research and innovation in one locality can rarely be replicated exactly elsewhere but can serve as sources of ideas to farmers in other areas. During the process of implementing 18 FPOR programmes, it was amply confirmed that FPOR or PTD seeks to reinforce the existing creativity and the experimental capacity of the farmers and can be an integral part of community based extension approaches. It really strengthens the links between indigenous and scientific knowledge and it helps build farmers capacity for self reliance.

Annexure I

The process and results of Farmers' Participatory Operational Research (FPOR)

Sl. No.	Name of the Projects	Researchable issues Identification Process	Researchable issues identified	Treatment and Size of the plot	Farmers Observations	Impact
A	**Crop Improvement**					
01.	Sugarcane pre release cultures under precision farming	Brainstorming sessions PRA & SWOC Focus group meeting Seminars	Reality of the yield potential of the newly released varieties and clones is to be assessed under precision system	Pre-released culture under precision protocol Farmers practices 1 ha / location	100 % germination Good tillering Crop yet to harvest	Neighbourhood farmers are visiting the farms and satisfied with the crop growth
02.	Sunflower Hybrid Selection Trials	Interactive meeting Meeting with Oil millers and oilseed merchants	Identification of superior high yielding hybrids in sunflower	Pre release hybrids : CSFH 5060, CSFH 6256 Check hybrids : Sunbred 275 1 ha / treatment	PRH hybrids are high volume weight (good filling) and high seed yield. Both hybrids are 90 days duration which is 5 days earlier than the check hybrid	Farmers observed the improvement of high yield of CSFH 5060 and CSFH 6256
03.	Performance of Maize Pre-Release Hybrids	Farmers group meeting	Need to test the new maize hybrid performance under PF system	New Maize Hybrid under precision system Farmers Practices 1 ha / locations	The experiment farmers obtained the highest yield	Extension workers convinced greatly and became spokesmen of project
B.	**Crop Management**					
04.	Precision farming techniques in Sunflower	Interactive meeting PRA Brainstorming	Efficient use of resources (Water & Nutrients) Increasing the productivity Economizing	Different quantity of water soluble fertilizers through drip system Control– Conventional	Uniform germination, good crop 40-50% water saving More number of	60 % water saving Enhanced nutrient use efficiency Higher productivity

		fertilizer dose	practice with surface irrigation 1 ha / treatment	filled grain Fertilizer use efficiency Weed free situation	Neighbourhood adoption
05. Nutrient Requirement for Higher Productivity in Banana Cv. Ney poovan	Interactive meeting Banana resource farmers meeting	To enhance the nutrient use efficiency of different fertilizers	Fertigation using different Water Soluble Fertilizers Farmers conventional practices as control 1 ha/ locatioin	Banana responds well to fertigation. However, biofertilizer applied unit looks better than the rest.	Trial is going on, neighbour farmers shown interest to adopt the technologies in their field.
06. Mat system of cultivation for continuous harvest of banana cv. Grand Naine	Focus group meeting with banana growers Conducted interactive meeting with supply chain stakeholders at Coimbatore. Banana growers federation feedback	To study the population dynamics for continuous supply of banana through mat system of planting	Allowing suckers at 3 months interval starting from 3rd month after planting (MAP) Allowing suckers at 4 months interval starting from4th month after planting (MAP) 1ha / treatment	Production of suckers after flowering is getting delayed. After flowering, nutrient flow will be more for bunch development than sucker production.	Continuous harvest is expected. Premium price during non-seasons Bunch qualities for export Easiness in cultivation
07. Sub surface drip fertigation in sugarcane at Erode (Water affluent district)	Farmers interactive meeting	Need to cut the cost of drip and introduce mechanization for harvesting cane.	Sub surface lateral In line drip lateral Farmers conventional practices 1 ha / locations	Reduced cost of drip system Water and nutrient use efficiency	Reduced cost of precision system Higher productivity and easy harvesting.
08. Sub surface drip fertigation in sugarcane at Sivaganga District (Water Deficit District)	Farmers interactive meeting	Need to cut the cost of drip and introduce mechanization for harvesting cane.	Sub surface lateral In line drip lateral Farmers conventional practices 1 ha / locations	Reduced cost of drip system Water and nutrient use efficiency	Reduced cost of precision system Higher productivity and easy harvesting.

09. Sub surface drip in precision vegetables	Farmers interactive meeting	Need to cut the cost of cultivation Need to improve the water and nutrient efficiency through precision system	Sub surface lateral In line drip lateral Farmers conventional practices 1 ha / locations	Reduced cost of drip system Improved water and nutrient use efficiency Higher productivity Need to irrigate even no crops in the field	Reduction of cost of cultivation Created big impact among the precision farmers and started adopting this technology
C. Crop Protection					
10. Management of sugarcane red rot disease using P.fluorescens	Focus group meeting PRA Semi-Structured Interview schedule	To assess the efficacy of liquid formulations of Pseudomonas developed by TNAU against rot of sugarcane	Application of liquid formulation of Pseudomonas fluorescens @ 2.0 lit / ha at different stage through drip system Control: Farmers Practices 1 ha / treatment	No. of tillers are found to be more	High yield with reduced diseases incidence in Sugarcane
11. Biological control of Panama wilt and nematode complex disease of banana	Focus group meeting	To assess the efficacy of liquid formulations of Pseudomonas developed by TNAU against wilt nematode complex of banana	Application of liquid formulation of Pseudomonas fluorescens @ 2.0 lit / ha at different stage through drip system Control: Farmers Practices 1 ha / treatment	Size and no. of fingers in banana bunch were bigger and increased	High yield with reduced disease incidence in Banana
12. Weather based forewarning of pest and disease	PRA Farmers Meeting	To create knowledge on the fore warning and forecasting of pest and	Establishment of AWS at block level to develop weather based pest and disease	Right time warning has issued to farming through this system	The precision farmer is able to take decision to go for control measure

		disease	forewarning and forecasting system Fixed plot-1acre\Crop 15 acre		
D. Value Addition					
13. Efficacy of Solar Tunnel Drying of Rosemary	SWOC analysis with Rosemary Growers Association Discussion with Hindustan Lever and ITC, the buyers	Conventional drying Vs solar tunnel drying with evaluation of system efficiency Economization of drying process	Sun drying – open yard Solar tunnel drying – 18.0 m^2 x 3.75 m (67.5 m^2 area) Economics & Quality - colour, oil characteristics of dried rosemary leaves	Ready to use this dryer for drying operation Easy to work with this drying system Less difficulty in solar drying during noon hours due to high temperature above 50°C	Reduced drying time (58%) in rosemary drying process Hygienic drying with good colour in appearance
14. Evaluation, Demon-stration and Popularization of Post Harvest Equipment	Interactive meeting with farmers	Appropriate knowledge and skill needs to be created on Post Har-vest Equip-ment among farmers.	The post harvest equipments developed by TNAU were demonstrated through partic-ipatory mode.	Farmers expressed that they learned the operating process of different pH equipments	It was found that farmers willing to adopt the equipments in the precision system
15. Knowledge dissemination to the farmers on value addition and food safety of agricultural crops	Interactive meeting with the KVK'S and farming community	Value addition of agricultural crops Food safety measures	Training to the farmers at Precision farming focus district	All the project farm-ers expressed that their knowledge level has increased on value addit-ion and food safety measures.	The farmers expressed to adopt the disseminated technologies at home scale and cottage industry level
E1. Social Sciences (Extension)					
16. Participatory Impact assessment of the Precision Farming	PRA Focus group meeting Semi-Structured	Need to identify the impact of the precision farming on techno-	SWOC with Selected precision farmers cluster associations of nine-focus	Precision farming projects has empowered the farmers on technical,	The precision farming project has created great impact not only in Tamil Nadu but

Project	interview Case studies SWOC analysis Farmers feedback	economic and social aspects Need to study the diffusion effects of precision farming	districts Case studies with project and non-project farmers	economical and socially and improved their livelihood system. Increased productivity and improved quality in all the crops with assured markets.	also other States of TN. Policy on Mega Market, WSF and Drip system
17. Technology Assessment and Designing Extension Module for Precision Farming	PRA Focus group meeting Pre-Survey workshop	Need to study the technology assessment and refinement process in the precision farming project.	80 farmers under precision farming in Salem and Namakkal district (representing North Western Zone)	Perceptible impact created by 'fertigation' in Banana crop compared to other crops 'Returns to investment' was attractive in adopting 'Fertigation' for Banana crop.	Banana Farmers empowered technically and economically
E2 Marketing					
18. Market Risks, Coping Strategies	PRA Workshops	Diagnosis of production and marketing risks with farmers	Participatory production and Market analysis Mapping of key risks	The risks in production and marketing could be managed better through collective action	Prioritization of current and future risks in agriculture and allied sectors

REFERENCES

Bar-Yosef. B., Sagiv, B. and Markovitch, T. 1998. Sweet corn response to surface and sub-surface trickle phosphorous fertigation. *Agron. J.*, **81** (3) 443-447.

Battalwar R.V. 2007. Recent trends and development in micro irrigation. *Proc. National Seminar on PPP in accelerated promotion of micro irrigation in India,* September 2007, Jalgaon organized by BIS.

Jerry Neyfield. 1997. Subsurface drip irrigation, Cooperative Extension. Bringing the University to you, Univesity of Nevada, Fact Sheet – 97-13.

Juan E., John, J. and Bob, W. 2007. (Sub-surface drip of onions). Effects of drip tape emitter spacing on yield and quality. *Agric. Water Management,* (**92**) Pp 126-130.

Laurens Van Veldhuizen, Ann Waters – Bayer and Chesha Wettasinha. 2009. Participatory Technology Development where there is no researchers, IDRC, Source Book.

Phene, C.J., Davis, K.R. Hotmacher, R.B. and McCormick, L. 1987. Advantages of sub-surface irrigation for processing tomatoes. *Acta. Hort.*, **200** : 101-114.

Sterrett S.B., Ross, B.B. and Savage Jr., C.P. 1990. Establishment and yield of asparagus as influenced by planting and irrigation methods. *J. A. M. Soc. Hort. Sci.,* **115** (1) 29-33.

Stewart Crawfod 2008, Cotton catchment communication, www.cottoncrc.org.au)

Reddy Rama 2009. FPOR – Precision Farming Farmers, President, Precision Farming Cluster Associations, Sarapally, Hosur, Krishnagiri Dist. Tamil Nadu.

28. EMERGENCE OF FOOD RETAIL CHAINS IN INDIA: IMPLICATIONS ON HORTICULTURE FARMERS

P.G. Chengappa and K.P. Mangala

Former Vice Chancellor, University of Agricultural Sciences, GKVK Campus, Bangalore-560 065

ABSTRACT

One of the greatest challenges facing farmers is adjusting to the changes taking place in the global food and agricultural system due to liberalization and globalization of the economy. These changes warrant farmers to adjust to the changing demands of the market, reduce cost of production and become more competitive, while at the same time meet the increasing demands of consumers for more specific products and ways of producing them, in this context, the emergence of retail chains and their backward linkages with farmers for fresh fruits and vegetables has been analyzed in this paper. The main strategy of this system is to ensure a steady and continuous supply of vegetables to food retail chains and a continuous of flow of income to farmers. This linkage has been able to change the method of farming and the marketing arrangement adopted by the farmers. The study has indicated farmer- food retail chain linkage is beneficial to both parties. The marketing arrangement by food retail chain reduced the market risks and transaction cost to the farmers, helped them to break away from the clutches of traditional brokers/ wholesalers/ commission agents .the farmers realized higher prices as they performed additional marketing functions such as cleaning , grading and packing of produce. Overall, farmers' income increased because of improved agriculture practices, growing of exotic varieties throughout the year and reduction of marketing cost. This format of linkage involved more of large farmers as compared to small and marginal farmers as possessing an irrigation tube well is essential for growing vegetables.

INTRODUCTION

In India, the concept of food retail chains/organized food retailing started in 1990s with the advent of international formats of retailing, especially with the emergence of food retail chains, such as 'Food World', 'Nilgiris', 'Fabmall', 'MTR', 'Apnabazaar', 'Subiksha' and 'Reliance Fresh'. These food retail chains have brought in several changes in the supply chain management and logistics through the use of quasi-formal and formal contracts to ensure timely delivery of products with desired quality attributes. Food retail chains in India, due to several factors like their recent origin, local or regional nature of their operations, existing legislation regarding procurement of agricultural produce, etc., have not been able to change the procurement systems. Most of the organized food retail chains procure their requirements of food grains (cereals and pulses) from the regulated market yards. It is being practiced to comply with the Agricultural Produce Market Committee (APMC) Act, which stipulates that all wholesale marketing activities of agriculture produce should be carried out at designated market yard, by paying the prescribed market fees and commission charges. Food grains that are procured from the wholesalers at the APMC yards are cleaned, sorted, graded and packed at godowns of the retail chains. Most retail chains repacked the commodities under private labels. These food retail chains depend on traditional channels of food grain marketing and that their entry has not led to shrinkage in the supply chain or any significant improvement in marketing infrastructure or marketing practices.

Similar has been the situation for fresh fruits and vegetables (FFV). India's traditional fresh fruits and vegetables marketing is characterized by fragmentation of the supply chain, concentration of market power with the wholesalers, existence of large number of intermediaries, little or no quality control, absence of standards, lack of product innovation, small volume transactions and low inventories. The world-over, despite of food retail chains reaching saturation, the penetration into fruits and vegetables section is limited (Reardon and Berdegue, 2002, Weatherspoon and Reardon, 2003). The situation is more precarious in India, where food retail chains are of recent origin. However, recently few of the food retail chains have established backward linkages with farmers for procuring fresh fruits and vegetables. These linkages have been able to change the method of farming as well as marketing arrangements. Towards this endeavor, the present paper deals with a study conducted to find the impact of the new institutional arrangement through retail food chain on producer's resource-use pattern, income and marketing arrangements.

METHODOLOGY

Farmers operating with fresh fruits and vegetables consolidation centre operated by the leading food retail chain - 'Spencers' in Hoskote near Bangalore, established in 1996 for procuring fresh fruits and vegetables was purposively chosen for the study. This centre collects about 163 locally grown varieties of vegetables (some exotic one also), and to a small extent, fruits also. During 2005, the number of farmers registered with the consolidation centre was small; 19 regular suppliers and 11 seasonal suppliers. To study the impact of new institutional arrangements on producer's resource-use pattern and income, all the nineteen farmers who regularly supply vegetables, were surveyed. To compare this system of marketing with the traditional system of marketing, 30 other farmers from the same area, selected at random, were surveyed for the study.

Logistic regression model was used to identify the factors that have influence on farmers' supply of vegetables to food retail chain consolidation centre.

Logit Regression Function

$$\text{Let } Pi = E\langle Y = 1 \mid Xi \rangle = \frac{1}{1+e^{-2}} = \frac{e^{z}}{1+e^{z}}$$

$$\text{where, } Z = b_o + + b_1 X_1 + b_2 X_2 + b_3 X_3 + b_4 X_4 + b_5 X_5$$

$$\text{Odds Ratio} = \left(\frac{P}{1-P}\right) = e^{b_o + b_i \overline{X}_i + u_i}$$

$$\text{Log} - \text{Odds} = Li = Ln\left(\frac{P}{1-P}\right) = \hat{b}_o + b_i \overline{X} + u_i$$

Pi =	Probability that farmers will supply vegetable to consideration centre
1 – Pi =	Probability that farmer will not be willing to supply vegetables to consolidation centre
Yi =	Farmer willing to supply or not willing to supply vegetables to consolidation centre (willing to supply =1, Not willing to supply = 0)
X1 =	Age (number of years)
X2 =	Education (mumber of years of schooling)
X3 =	Transport vehicle (dummy variable: owning = 1, Not owning = 0)
X4 =	Area under vegetables (acres)

RESULTS AND DISCUSSION

The leading food retail chain (FRC), Spencers through the establishment of a consolidation centre at Bangalore, have introduced a novel agribusiness model for marketing of agricultural commodities. To ensure the quality of produce, consolidation centre provides information on 'Good Agricultural Practices' (GAP) to farmers, who cultivate crops based on these specifications. To ensure good handling of produce, member farmers clean, grade and pack the produce as per retail chain specifications. The packaging materials are provided by food retail chain for specialty products and for general packaging. Every pack is labeled which indicates the weight, date of packing, etc. The farmers selling vegetables to the consolidation centre are responsible for all the post- harvest operations. By shifting responsibilities such as cleaning, sorting, grading and packaging to farmer-vendors, the consolidation centre has been able to reduce the transaction costs of the retail chain. This practice is diametrically opposite to the handling of fruits and vegetables in the traditional markets, wherein they are just dumped in market yards. Thus, a beginning in quality control of fresh fruits and vegetables has been made by the Spencer's through food retail chains.

This linkage has been able to change the method of farming. The small and marginal farmers, through their intensive cultivation have been able to earn higher incomes. In contrast to Kenya, where supermarkets have to deal with a fewer and larger suppliers (Neven and Reardon, 2004), in the present study, food retail chain consolidation centre has emphasized on having supplies from small and marginal farmers, because of their relative high care in managing farm-scale operations due to absence of mechanization in small scale farming. Since food retail chains need a regular supply of small quantities of vegetables, they prefer to establish backward linkages with small and marginal farmers.

Supply Chain Management by Spencer's

The consolidation centre in Hoskote (in the vegetable production belt) collects about 163 locally grown varieties of vegetables (including some exotic ones), and to a small extent, fruits. Farmers from distances of 50-80 km, supply fruits and vegetables to this centre. The concept adopted by Spencer's is 'Ready to Retail', in which agri-products are graded and packed in the required form by the suppliers (farmers). The new model of Spencer's has helped in shrinking the traditional supply chain for fresh fruits and vegetables, as depicted in Fig. 1.

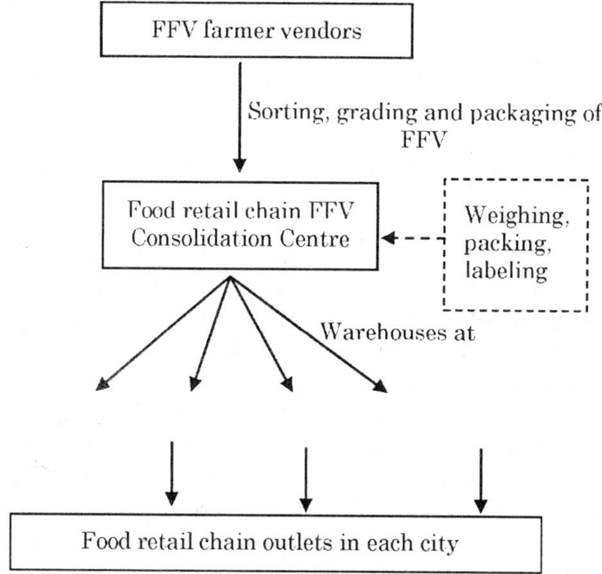

Figure 1. Supply chain adopted by Spencer's for fresh fruits and vegetables (FFV)

The consolidation centre covers a radius of 160 km, and currently handles around 20 tonnes of agri-products per day. At present, it meets only about 70 per cent of its requirement of fresh fruits and vegetables from farmers, and the remaining 30 per cent is procured locally from the Modern Auction System (MAS) market, established by the National Dairy Development Board (NDDB) through a consolidator. The consolidation centre follows the 'Vendor Development' model, which is characterized by the absence of intermediaries in the supply chain, *i.e.,* the farmers themselves are the preferred suppliers. In this model, farmers registered with the consolidation centre, are known as 'vendors', and under each vendor a group (usually 10) of farmer-members (independently) cultivates and supply fruits and vegetables.

The relationships with farmers have been informal, with no written contracts, but are based on oral confirmations of volumes to be delivered. Farmers who wish to register with the consolidation centre should have irrigation facilities. The selection of vendors is also determined by their business management skills. Supply to the Centre also involves more formal transaction methods as well as stringent delivery conditions, frequency of supply and quality standards for the product. The registered farmer-vendors collect the produce from other farmer-members and deliver it to the consolidation centre; quality controls in production and packaging being the responsibility of farmer-vendors. At the centre, packed produce are bar coded and transported to the central warehouse in Bangalore, from where it is further transported to other south Indian cities viz., Chennai, Hyderabad, Thiruvanthapuram. The consolidation centre plans to serve as a captive supply centre to meet the requirements of their own outlets, as well as other retail chains, bulk purchasers and processors, which would ultimately benefit the farmers. With plans to set-up more retail outlets in major cities, the procurement at the consolidation centre is set to increase. The success of this model (backward linkages directly with the farmers and good quality produce) has motivated several agro-processing industries to procure from this consolidation centre.

Quality Control Practices at Consolidation Centre

Quality of produce is maintained at three levels, referred to as QG (Quality Grading), QC_1 (Quality Control) and QC_2 (Quality Care). The QG is the concern of the consolidation centre; the QC_1 is the maintenance of quality of packed products till it reaches the retail outlets, i.e. in loading, transporting and unloading of the produce. The QC_2 refers to the quality to be maintained at the display section of the retail outlets. Fruits and vegetables are graded based on uniformity of size, maturity and colour, physical appearance and freshness. The consolidation centre supplied the materials needed for packaging (for both speciality products and general packaging). Farmers themselves carry out grading and packing; it reduces the number of people handling the produce before it reaches the consumers. At the consolidation centre, each packed product is labelled with details like product name, weight and price; some of them are bar-coded as per the requirement.

Changes in Cultivation Practices Introduced by FRC Consolidation Centre

The FRC consolidation centre has introduced changes in the way crops, particularly vegetables are cultivated. Crops to be cultivated are assigned to each farmer based on farmer's proficiency and history of production, which is documented at the time of enlistment. Investments in irrigation systems are preferred, as it provides the farmers greater control over quality and allows them to produce round-the year. A crop calendar is drawn up, keeping in view the requirements of the FRC retail outlets. Once the crop to be cultivated has been decided, farmers are provided with a package of 'Good Agricultural Practices' (GAP). This package ensures the optimum use of resources with emphasis on minimum use of pesticides. The vendor-leader ensures that the practices are strictly adhered to. Direct supplies by farms also allow the centre to inspect farm and growing practices, first-hand.

There is no formal contract or vertical integration for production or marketing under this arrangement. The centre neither supplies any production inputs nor does it formally agree to procure the produce, which makes the farmers risk-bearers. The centre has no system of providing production credit to the farmers, but helps farmers in procuring inputs from suppliers at reduced rates. Technical guidance on aspects like the time of planting, crop production and management, harvest time, quantity to be harvested per acre, *etc.*, to ensure quality and marketability, are provided by the consolidation centre. Generally, the centre procures the entire quantity of fresh fruits and vegetables supplied by the vendors, except in cases where the specified quality requirement is not met. During the initial stages of establishment, the percentage of rejection in procurement from farmers was high because the farmers were not accustomed to producing good quality produce in a scientific manner. The large-scale rejection of their produce for failing to meet the quality specifications led farmers to change their cultivation practices, following which the rate of rejection reduced and now stands at 8 per cent. The impact of adhering to Good Agricultural Practices (GAP) and production practices such as staggered sowing introduced by the centre has led to increase in the intensity of cultivation as well the production.

Pricing Policy of Consolidation Centre

Prices of fresh fruits and vegetables are determined on the basis of the prices prevailing at different markets in Bangalore. The benchmark price is determined by considering the prices prevailing at Modern Auction System (MAS) market operated by NDDB, Horticultural Produce Cooperative Marketing Society (HOPCOMS) Ltd and Krishna Rajendra market in Bangalore. In this mechanism, consolidation centre ensures a sort of support price even during the glut period in the market, so that farmers do not incur losses. The consolidation centre procures limited quantities from a limited number of farmers. Hence, it has limited liability to each farmer who also cultivates a given crop on a limited area. This produce is bought from food retail chains by consumers, who are quality-conscious than price-conscious. Under this format, the centre ensures input cost plus minimum profit for a limited quantity of produce. During the lean season, farmers are naturally benefited with good prices on par with market with assured market. It was found that farmers preferred to supply their produce to the consolidation centre, as it provided them stable prices and assured market, compared to the highly volatile prices at the wholesale market.

Socio- economic Implications of Linkage of Food Consolidation Centre with Farmers

It was found that younger and educated farmers had entered into tie–ups with food retail chain consolidation centre, which could be due to their enthusiasm and better awareness to take risks and experiment with a new business model. Family size was relatively larger for farmer families associated with the centre compared to the traditional market farmers. Larger family size was advantageous to the consolidation centre, as family labour was totally devoted to post-harvest operating like washing, sorting, grading and packing, labeling and also reduced the cost on hired labour. The land-holding size of FRC farmers was around six acres, while that of non-FRC farmers was two acres. The share of area under well command was also higher for FRC farmers compared to that of traditional market farmers. The gross income from agriculture of FRC farmer was Rs. 1, 72,000 per year while that of traditional market farmer was Rs. 70,000, due to improved agricultural practices and growing of exotic vegetables for FRC round-the-year which provided extra income to these farmers (Table 1).

Table 1. Socio- economic characteristics of farmers

Particulars	Food retail chain farmers	Traditional market farmers
Number of farmers	19	30
Age (years)	39	48
Literate (%)	100	67
Family size (No.)	7	5
Total land holding size (Acres)	6	2
(a) Irrigated land	4.5	1.5
(b) Dry land	1.5	0.5
Bore wells per farmer	2	1
Gross income from agriculture (Rs)	1,72,000	70,000
Category of farmers		
Marginal (< 2.5 acres)	3 (15.8)	28 (93.3)
Small (2.5 -5 acres)	9 (47.4)	2 (6.7)
Large (> 5 acres)	7 (36.8)	0 (0.00)

Note: Figures within the parentheses are percentages to the total number of farmers, in columns 2 and 3.

Cropping Pattern

Along with traditional vegetables, the FRC farmers were also cultivating exotic vegetables, such as broccoli, iceberg, lettuce, parsley, leek, red cabbage, Chinese cabbage, coloured capsicum, green onion, turnip, basil, table radish, etc. Thus, the crop diversity was higher for FRC farmers than traditional farmers. This diversity in crops had increased after their association with FRC consolidation centre, as they had assured market for their produce and their marketing risks were reduced through the new institutional arrangement. Some of these farmers had addition income by growing low value-high volume leafy vegetables like mint, spinach, red amaranthus or coriander. The cropped area of FRC farmers ranged from 500 sq. ft to 10 acres. Exotic vegetables were grown in staggered small multiple plots, to ensure supply round-the-year, as per the requirements of FRC consolidation centre. Seasonal vegetables like cauliflower, carrots, potatoes, tomatoes, etc., are grown on large plots by both FRC and non-FRC farmers.

Comparison of Unit Cost of Production and Net Returns of Vegetable Crops under FRC and Traditional Marketing Channels

In this section, profitability and transaction costs of four major crops, namely, cabbage, cauliflower, carrot, and tomato under the two institutional arrangements have been compared. The differences in profits and transaction costs have been used as indicators of the performance of an institutional arrangement in the marketing of agricultural commodities. Noticeable differences in net return per quintal can be seen for all vegetable crops in Table 2. The increase in net returns was the highest for cabbage growers (48%), followed by cauliflower (40%). Similarly, FRC carrot and tomato farmers realized higher returns by 34 per cent and 18 per cent, respectively. The high net returns for FRC farmers were due to drastic reduction in transaction costs, particularly transportation cost and commission charges. They were also paid higher price by the FRC as they performed additional market functions such as cleaning, grading and packing.

Table 2. Net returns from vegetables production under FRC and traditional marketing channels (2005)

Particulars	Cabbage		Cauliflower		Carrot		Tomato	
	FRC farmers	Traditional market farmers	FRC farmers	Traditional market farmers	FRC farmers	Traditional market farmers	FRC farmers	Traditional market farmers
Yield (tonne/ acre)	33	30	12.5	12	12	13	30	25
Price realized (Rs/tonne)	3490	3000	8430	7000	15500	14000	6540	5500
Input cost/ tonne (Rs)	897 (83)	1039 (60)	1871 (91)	2019 (63)	2589 (77)	2188 (53.5)	1396 (69)	1550 (61)
Transaction cost/ tonne (Rs)	180 (17)	700 (40)	189 (9)	1200 (37)	775 (23)	1905 (46.5)	640 (31)	1000 (39)
Total cost/ tonne (Rs)	1077 (100)	1739 (100)	2060 (100)	3219 (100)	3364 (100)	4093 (100)	2036 (100)	2550 (100)
Net returns/ tonne (Rs)	2413	1261	6370	3781	12136	9908	4504	2950
Net returns / Qtl (Rs)	241	126	637	378	1214	991	450	295
Increase in net returns (%)	48		40		18		34	

Note: Figures in the parentheses are percentages to the total cost

Factors Influencing Farmer's Choice of Different Marketing Channels

The factors influencing the probability of selecting the food retail chain marketing channel as against traditional marketing channel was analyzed using the logistic regression model as it is assumed that the farmers' decision to choose a particular marketing channel binary in nature. The log of odds in favour of selling vegetables, at consolidation centre was positively associated with education, owning transportation facility and area cropped under vegetables but is negatively associated with age. The coefficient of age factor of the farmers was negative, which indicated that with increase in age, the probability of selling vegetables at FRC consolidation centre reduced and probability of selling at traditional market increased. The young farmers preferred FRC marketing channel than the traditional channel. The education of the farmer had a positive impact on selection of FRC marketing channel. With improvement in level of education, probability of selling vegetables at FRC consolidation centre increased. The farmer having own transport vehicle influenced them to sell through FRC marketing channel. With the ownership of transportation vehicle, the chances of selling through FRC marketing channel were 40-times more. The odds ratio was 7:1, indicating that for every one farmer not willing to supply vegetables to FRC consolidation centre, seven farmers were willing to supply (Table 3).

Table 3. Logistic regression coefficients of determinants of supply to consolidation centre

Variables	b	eb	Statistical significance	Elasticity of probability
Age (years)	-0.029	0.971	0.63	-0.15
Education (No. of years of schooling)	2.394	10.958	0.09**	0.29
Transportation (own transportation =1, otherwise=0)	3.681	39.693	0.02*	
Area under vegetables (acres)	1.409	4.093	0.04*	0.44
Constant	-6.44	0.002	0.11	-
Correctly predicted cases			91.8	
Chi- square			48.13	
Odds ratio			7:1	
Probability			0.88	

Note: * Significant at 5 per cent level, ** Significant at 10 per cent level

CONCLUSIONS

The food retail chains are emerging in India through corporate involvement. These food retail chains have not been able to make much impact on the supply chain of fresh vegetables as they continue to depend on the traditional channels of marketing. However, recently few of the food retail chains have established backward linkages involving farmers for procuring fresh fruits and vegetables. Spencer's is one such FRC which has organized a fruit and vegetable consolidation center. This new institutional arrangement has helped farmers to break away from the clutches of traditional brokers/wholesaler/commission agents. The marketing arrangement by FRC, also, reduced the market risks and transaction costs to the farmers. The farmers also realized higher prices as they performed additional marketing functions such as cleaning, grading and packing of produce. Direct supply by farmers allowed the retail chain to simultaneously increase control over quality, supply reliability and price stability. Farmer's income has increased because of improved agricultural practices, growing of exotic varieties throughout the year, reduction of marketing costs.

REFERENCES

Chengappa, P.G. Lalith Achoth. Arapita Mukerjee, Reddy, B.M.R. Ravi, P.C. and Vijayalakshmi Dega 2007. Evolution of Food Retail Food Chains in India. In: *Proceedings: Agricultural Diversification and Smallholders in South Asia*. Ed. Joshi, P., Ashok Gulati and Ralph Cummings Jr., Academic foundation. New Delhi: 385- 405.

Neven, D. and Reardon, T. 2004. The rise of Kenyan supermarkets and the evolution of their horticulture product procurement systems, *Development Policy Review*, **22**(6): 669-699.

Reardon, T. and Berdegue, J.A. 2002. The rapid rise of supermarkets in Latin America-Challenges and opportunities for development, *Development Policy Review*, **20**(4): 371-388.

Weatherspoon, D.D. and Reardon, T. 2003. The rise of supermarkets in Africa: Implications for agrifood systems and the rural poor, *Development Policy Review*, **21**(3): 333-355.

29. THE IMPACT OF EMERGING ORGANISED RETAIL CHAINS ON THE MARKETING OF FRESH FRUIT AND VEGETABLES IN THE DEVELOPING COUNTRIES

Peter J. Batt

Curtin University of Technology, GPO Box U1987, Perth, WA 6845, AUSTRALIA

ABSTRACT

With increasing wealth and greater urbanisation, the share of the food market held by modern retailers is steadily increasing. With centralised procurement systems and dedicated suppliers, modern retailers are able to offer consumers a continuous supply of a diverse range of high quality products at lower prices. Unlike the traditional wet markets, consumers are able to interact with the product at their leisure in the comfort of a clean air conditioned store, to procure value-added goods (processed and semi-processed food) and to not only purchase their goods using credit cards, but to engage in loyalty programs that provide a range of ancillary benefits. For smallholder producers, the emergence of modern retailing provides both opportunities and threats. With a desire to move towards more centralised procurement and/or to source only from preferred suppliers, smallholder farmers risk exclusion from this market. Not only must suppliers meet the stringent food safety and quality standards imposed by the modern retailers, they must maintain a continuous supply and often engage in value-adding activities to present the product for immediate sale. With inadequate technology, infrastructure, market information and limited capital, the challenges of engaging with the modern retailers can be overwhelming. Furthermore, the terms of trade often require deferred terms of payment, which for smallholder farmers used to dealing in a cash economy, can become problematic. Promotional levies, listing fees and volume rebates may also be imposed at the retailers' discretion. While transacting with modern retailers may be desirable in that it offers a guarantee to purchase all that product which meets the buyers specifications, prices are by no means stable, for prices are generally re-negotiated on a regular basis and invariably in a downward direction to enhance the retailers competitiveness in the market. As prices fall, not unexpectedly, there is an increasing propensity for smallholder farmers to renege on contracts and to transact with alternative buyers. In this regard, it is important to recognise and acknowledge the role of the traditional market and the long-term relationships that often exist between smallholder farmers and their downstream market intermediaries. There is evidence to suggest that many consumers prefer to purchase from the traditional market for the product is perceived to be fresher, cheaper and the range is often more comprehensive. Consumers have the opportunity to engage with trusted vendors who can provide the product in the desired quantities and quality and may often extend short-term credit. Most important however, is the convenience and proximity. While it is often more cost effective for modern retailers to transact with the larger growers or to rely on imports, smallholder farmers should not be excluded. Even in the absence of any economies of scale, smallholder farmer's costs of production are often lower because of reduced input costs (including labour) and the adoption of more sustainable production systems. The geographic dispersion of farmers can reduce the risk of crop failure and greatly extend the range and variety of products offered. However, the opportunity for smallholder farmers to engage with modern retailers is largely dependent upon their having access to high quality inputs (seeds, chemicals and fertilisers), credit, agronomic advice and market information.

INTRODUCTION

Worldwide, the market for food is constantly evolving, driven not only by changes in consumer preferences, but also by technology, relationships between actors in the supply chain, prevailing government policies and the business environment (Regmi and Gehlhar 2005). As a consequence, modern retail formats are expanding, driven by shifts in consumers shopping habits, income levels and lifestyles, and the continual erosion of trade barriers which greatly facilitate foreign direct investment.

Throughout the world, food is either sold through retail stores or food service establishments. In 2002, the global value of food and beverage sales exceeded USD 4,096 billion (Regmi and Gehlhar 2005). With over 40% of the total value consumed by the food service sector and with consumers demanding greater convenience, it is highly likely that the value of food service sales will soon overtake retail food sales. At the retail level, food sales can be broken down into fresh (USD 531 billion) and processed (USD 1,762 billion). Processed food products include packaged food (USD 1,148 billion) and beverages (USD 641 billion). Packaged food accounts for about half of total food expenditure in the developed countries, but only a third or less in most developing countries.

In the developed countries, most retail food is sold through supermarkets and hypermarkets. Euromonitor (2004) reports that in the US, 62% of food sales occur through modern retail outlets. In Japan, modern retail outlets account for 58% of food sales and in Western Europe, the figure approaches 56%. The supermarket share of retail food sales in the leading six Latin American countries averages 45-75% (Reardon *et al.,* 2005). Excluding fresh fruit and vegetables, meat and fish, modern retail outlets account for 33% of retail food sales in South East Asia (Indonesia, Malaysia and Thailand), and as much as 63% in East Asia (Korea, Taiwan and the Philippines).

With the rapid expansion of modern retail formats, the structure of the food supply and distribution system is changing. With vastly improved information and logistics technology and the growing importance of food safety, quality and other technical requirements, food producers are being encouraged to participate in fully integrated supply chains. However, many stakeholders are concerned by the emerging power imbalance and the potentially unfavourable terms of trade that may exist between smallholder farmers and the global retailers. While numerous studies have been carried out, initially in Latin America and more recently in central and eastern Europe, Africa and Asia, little progress has been made in identifying the steps that need to be taken to help smallholder farmers adapt to the realities of supplying fresh fruit and vegetables to large-scale retailers.

DRIVERS FOR THE EXPANSION OF MODERN RETAIL FORMATS

Vorley *et al.* (2007) identified how the rapid expansion of modern retail outlets in South East Asia and Eastern Europe has occurred as a result of both strong 'pull' factors; in the form of rising levels of consumer affluence, low levels of competition, low labour and land costs, and favourable legislation; and strong 'push' factors in the form of market saturation and fierce price wars in home markets, combined with strict planning regulations and other policy constraints.

Chen *et al.* (2005) and Reardon *et al.* (2005) identified the primary pull factors as: (i) greater urbanisation, leading to the growth in personal disposable income. This has led to the greater affordability of both refrigerators and microwave ovens which respectively reduced the need to purchase food as frequently and encourage the consumption of more processed and semi-processed food; (ii); more women in the work force leads to a greater demand for one-stop, fast and convenient shopping; (iii) a greater emphasis on food safety and quality. The products purchased from modern retail outlets generally have a superior shelf life, they are more wholesome and as a result of prescribed production systems, contain fewer pesticide residues; (iv) demographic, social and cultural changes which result in the westernisation of lifestyles; a growing number of nuclear families and one person households; (v) the more widespread use of credit cards; and (vi) increased travel, which exposes consumers to a wider range and variety of fresh fruit and vegetables. Economic growth has also led to significant investments in infrastructure and logistics capabilities. Modern retailers invariably implement their own procurement and logistics systems which not only ensure more consistent supply, but greatly improve efficiency through reducing post-harvest losses. The use of electronic data exchange interfaces is becoming increasingly common between large retail chains and their preferred suppliers, enabling retailers to send out orders, verify the receipt of orders, to schedule delivery and provide data on sales to enable suppliers to manage inventory.

The recent financial crises in much of Asia has made local retailers cheap to acquire (Vorley *et al.,* 2005) and the in-flow of foreign direct investment, induced by market liberalization, has been a key determinant in the introduction and diffusion of modern retail formats in Asia (Reardon *et al.,* 2005). Increased capitalisation and the buying power of the global chains have provided huge incentives for global retailers to leverage scale, distribution logistics and information technology systems. With extensive experience in modern retailing, procurement, stock optimisation, the control of quality standards, cool storage and the preservation of shelf life, the expansion of international retail chains in Asia has forced local retailers to enhance their services and efficiency. The global expansion of their operations has also provided modern retailers with an opportunity to capture consumer spending from other sectors by using a greater variety of retail formats.

According to Deloitte (2008), retailers with operations in ten or more countries posted an average net profit margin of 4.7% compared with 3.1% for those operating in only one or two countries. In 2006, retailers in the emerging markets enjoyed above average growth. In Latin America and South Africa, retail sales increased by 12.9% and 12.8% respectively. While sales growth in the Asia/Pacific was a more modest 8.1%, excluding Japan, retail sales increased by an average of 15.2%. With more than one billion people, India is expected to become the next retail battlefield. Even although India remains relatively closed to foreign direct investment and its business environment is riddled with obstacles, as the leaders of India's huge business conglomerates turn their attention towards retailing, a modern retail infrastructure is rapidly emerging.

IMPACT OF MODERN RETAIL EXPANSION ON SMALLHOLDER FARMERS

Traditionally, most of the fresh produce in Asia has been distributed through a complex network of wholesale and retail markets. In the wholesale market, prices are determined primarily by supply and demand. However, prices are highly variable day-by-day and even hour-by-hour, and both the quantity and quality of the produce offered for sale is highly variable, occasioned by the weather and the seasonality of supply (Folley, 1973). For modern retailers, these variations in quality and price are unacceptable, for it makes it impossible to adequately price the product or to engage in any promotion or merchandising of the product (Batt, 2006). Furthermore, the anonymity which is often associated with the wholesale marketing of generic food products makes it impossible for the buyer to be able to assure downstream customers that the food is safe, or that it has met prescribed environmental and ethical standards. Consequently, in order to maintain product quality and to provide a more consistent supply, modern retailers are purchasing greater quantities of fresh produce from preferred suppliers.

Preferred suppliers are those who can maintain a reliable supply of good quality product that meets the buyers prescribed specifications at a competitive price (Chen *et al.,* 2005). Furthermore, preferred suppliers should react favourably to unforeseen needs such as suddenly accelerated or decelerated volumes of business, changes in product or delivery specifications, service problems or any other legitimate request (Leenders and Fearon, 1993). Preferred suppliers will warn ahead of time of material shortages, logistical impediments and anything else that may affect the buyers operations. They will provide technical support and other expertise when requested by the customer or whenever the supplier believes it can assist the customer to become more competitive. Preferred suppliers may also take responsibility for developing a product category, launching new product variants, new and improved methods of packaging, or taking greater responsibility for unsold produce (Vorley *et al.,* 2007).

Not unexpectedly, these arrangements are often problematic for smallholder farmers acting independently. Product quality cannot be assured because of the lack of appropriate inputs (seeds, chemicals and fertilisers), the application of inappropriate technology, inappropriate agronomic and post-harvest advice, inadequate access to packing sheds and cold storage facilities, the lack of refrigerated transport, poor roads and prevailing methods of packaging and post-harvest handling (Chen *et al.,* 2005; Reardon *et al.,* 2005; Murray-Prior *et al.,* 2006).

However, in their transactions with modern retailers, the most difficult constraint for smallholder farmers to overcome is their inability to implement appropriate quality assurance systems. Invariably, retailers use quality assurance systems as a means to standardise product requirements from many suppliers, who often come from many different regions and many different countries (Reardon *et al.,* 2005). Standards specify and harmonise the product and delivery attributes, thereby enhancing supply chain efficiency and lowering transaction costs. Where public standards are deficient, particularly with regards to food safety, these quality assurance systems ensure that appropriate steps have been taken to protect the retailer's integrity, their brand and market share. As competition intensifies, these quality assurance systems are used as a means to differentiate the retailer's offer quality by promoting superior product quality attributes.

The increasing use of standards and preferred supplier arrangements will present challenges for smallholder farmers. The high standards demanded often require investments that smallholder farmers are unable to make with limited or no access to credit (Vorley *et al.,* 2007). In meeting the prescriptive standards often laid down by the retailers, smallholder farmers are often unable to guarantee a safe water supply, provide toilets and hand washing facilities, construct packing sheds with concrete floors, or to maintain the standard of book keeping that traceability demands (Chen *et al.,* 2005). Furthermore, there are serious doubts about the ability of smallholder farmers acting independently to be able to afford the costs associated with the implementation of quality assurance systems and the annual costs of third-party certification (Reardon *et al.,* 2005).

Irrespective of whatever systems of production smallholder farmers may employ, they must realise that a proportion of their produce will fail to meet the buyer's specifications. This will require them to make alternative arrangements to dispose of it through other channels or even to throw it away (Chen *et al.,* 2005). Batt and Parining (2000) demonstrate how many smallholder farmers prefer to transact with local traders and market intermediaries even although they offer a lower price because they purchase the entire crop, irrespective of quality. Murray-Prior *et al.* (2006) advocate a dualistic marketing strategy whereby smallholder farmers supply both the traditional market (a price driven chain) and the value driven chains (modern retail). Such not only overcomes the problem of disposing of product which fails to meet specifications, but it provides an alternative market outlet should the buyer choose not to renew the contract or, in a worst case scenario, where the buyer either defaults or withdraws completely from the market.

In other instances, poor roads and the lack of communication results in smallholder farmers being unaware of the demand for their products and alternative market requirements because they have limited opportunities to leave their communities (Murray-Prior *et al.,* 2006). In most cases, smallholder farmers rely on trusted market intermediaries for their information. However, few traders provide farmers with the desired level of market information and there are limited opportunities to obtain information from alternative sources because of isolation. A lack of price information and poor negotiation and bargaining skills make it difficult for smallholder farmers to negotiate a mutually acceptable price (Chen *et al.,* 2005).

Isolation and poor roads can accentuate the problems associated with maintaining a continuous and reliable supply of fresh produce and greatly accentuate post-harvest losses (Chen *et al.,* 2005; Murray-Prior *et al.,* 2006). Transport costs often account for the majority of marketing costs due to poor roads and inadequate transport facilities. The problem can be further aggravated by the lack of appropriate post-harvest storage facilities and the need for farmers to participate in traditional religious holidays and community obligations.

While the lack of inputs, appropriate production and post-harvest technology, and poor infrastructure puts pressure on price, of far greater consequence is the price competition that often emerges between retail chains and indeed, between the traditional market and the modern retail market. With centralised procurement strategies and preferred supplier programs, modern retailers have the capacity to import fresh produce when domestic farmers are unable to supply. While this is seldom a desirable long-term position, both economically and socially, the availability of imported produce invariably establishes a price that local suppliers must match. As a consequence, the prices offered by modern retail buyers are often insufficient to enable smallholder farmers to recover their costs (Chen *et al.,* 2005).

Because of the need to ensure prices are consistent across all retail stores, modern retail chains will often set the purchase price for a fixed period (Chen *et al.,* 2005). While mechanisms exist to renegotiate the price if market prices decline during the period, no such provision exists to renegotiate prices upwards. In addition, suppliers are often required to pay promotional fees when a product is featured in an in-store promotion and discounts are payable when new stores are opened. Invariably, penalties are levied when suppliers fail to meet predetermined volume and quality expectations.

However, the major problem for smallholder farmers is the inordinate delay often associated in making payment: it is not usual for modern retailers to pay suppliers 30 to 90 days after delivery (Chen *et al.,* 2005). Where the sale of fresh produce is the major source of income to meet anticipated household expenses, such delays are unacceptable, thus forcing smallholder farmers to accept lower prices in exchange for cash. Not unexpectedly, when traders in the traditional market offer higher prices, smallholder farmers will often forgo their commitment to supply modern retailers. While such short-term opportunistic trading is entirely rational for smallholder farmers, there is the potential, in the long term, for such practices to undermine their status as preferred suppliers. While fresh produce buyers in Asia have traditionally operated under informal verbal contracts, which are renewed upon continuing good performance, this system is being replaced by formal written contracts which demand on-time delivery and the delivery of products with the desired quality attributes (Reardon *et al.,* 2005). While these contracts also provide an incentive for suppliers to stay with the buyer and, over time, to make investments which are often specific to the retailer, for a number of reasons, most smallholder farmers tend to be risk averse (Chen *et al.,* 2005).

Herein lies the major constraint and perhaps the root cause for many of the impediments smallholder farmers face in trying to meet the needs of both modern retailers and the traditional market: the lack of finance. This will not only restrict the capacity of smallholder farmers to adopt practices and to make investments that facilitate the movement of fresh produce along the supply chain, but it will also restrict their capacity to provide many of the value-added services that modern retailers demand in processing, packing and bar coding the product, or investing in the infrastructure and electronic data exchange interfaces that are becoming increasingly common between the large retail chains and their preferred suppliers (Reardon *et al.,* 2005).

Almost invariably, the ability of smallholder farmers to transact either directly or indirectly with modern retailers will depend upon their ability to establish collaborative marketing groups. Irrespective of the organisational form that any collaborative marketing group may take, Murray-Prior (2007) suggests that the two critical ingredients for success are: (1) a comparative advantage; and (2) a reasonable level of trust among the members of the community who seek to establish the collaborative marketing group. Vorley *et al.* (2007) suggest that the key success factor is the ability of the group to impose discipline on its members in order to deliver consistently against the promises that have been made to downstream customers.

However, collaborative marketing groups often suffer from bureaucracy and the inability to coordinate activities (Vorley *et al.,* 2007). All too often, they are created purely to provide countervailing bargaining power, with little consideration towards enhancing customer value or customer service. Inappropriate organisational structures are often imposed that seriously constrain the capacity of the organisation to respond to market signals. Lele (1981) suggests that the reasons for the failure of collaborative marketing groups often results from the inability to understand the actual constraints and solutions facing smallholder farmers. Insufficient emphasis is often placed on the internal prerequisites of leadership, management, experience and technical know-how. In other instances, external factors such as poor infrastructure, poor market information or inappropriate government policies can constrain the operation of collaborative marketing groups (Murray-Prior 2007). The lack of incentives, both for the smallholders who patronise the collaborative marketing group and those who it employs, can rapidly lead to under investment.

IMPACT OF MODERN RETAIL EXPANSION ON TRADITIONAL MARKETS

Despite the rapid expansion of modern retail formats, supermarkets are not the dominant retailers in all markets and nor are they likely to become the dominant actors for some time (Vorley *et al.,* 2007). In those countries where growth rates in urbanisation and household income are slow, modern retail formats will expand at a more modest rate (Chen *et al.,* 2005). Especially with regard to fresh fruit and vegetables, Reardon *et al.* (2005) report that the modern retailers attempt to displace the traditional markets is much slower in the fresh produce sector. There is evidence to suggest that modern retailers generally commence their operations in-country with staples, packaged and processed food and non food products, and then, over time, they expand into meat and dairy and eventually into fresh produce (Vorley *et al.,* 2007). No doubt this is largely the result of needing to develop the production base, to introduce and to implement quality assurance systems on farm, and to develop entirely new supply chains and logistics systems.

For some modern retailers, this means establishing centralised procurement and distribution centres. However, Chen *et al.* (2005) suggest that this is only viable when the chain services a minimum of 20 hypermarkets. Hence, at least initially, modern retailers will need to identify and to develop a number of preferred suppliers which may include specialised wholesalers and concessionaires.

Given the difficulties in transacting with smallholder farmers, either individually or collectively, and the various logistical problems associated with poor roads and the lack of infrastructure, it is more desirable to wash, sort, grade and pack closer to the retail store than at the farm level (Chen *et al.,* 2005). For the traditional wholesale markets, the expansion of modern retailers and the emerging institutional market provides an opportunity to create value-added logistics platforms. However, in order to better service their downstream customers needs, a significant investment is often required to improve facilities such as cool storage, clean water, electricity, office space and parking space.

At the retail level, it is evident that the traditional wet markets offer a number of competitive strengths that mean that they are able to maintain their market share despite the competition from modern retail outlets. Chen *et al.* (2005) report that in most of Asia some 80-90% of consumers prefer to purchase fresh produce from traditional markets. Consumers perceive that the produce is not only fresher and cheaper, but the vendors generally offer a more comprehensive range. Chamhuri and Batt (2009) infer that participants who shop in the traditional retail outlets are more satisfied with their purchase as they are able to negotiate a price with vendors, whereas prices in the modern retail outlets are fixed. For some consumers, transacting with preferred vendors and developing long-term relationships will not only provide access to credit, but also provide an assurance that they are getting good quality service. Not only can they buy the desired quantity, but vendors will often go out of their way to source particular products or to fulfil specific requests. Furthermore, increasing numbers of consumers are dissatisfied in their transactions with modern retailers as a result of poor customer service, long check-out queues, ignorant and impolite staff, and the non availability of advertised goods.

According to Chamhuri and Batt (2009), convenience is a factor that attracts consumers to shop from both the traditional and modern retail outlets. For most consumers, the traditional market is not only more convenient, but closer to their home. On the other hand, while those who purchase from modern retail outlets do not always live near the supermarkets, convenience for them means one stop shopping, and anything that saves or simplifies work and brings comfort to the consumer. For these shoppers, convenience means that the store is well organised, it is easy to find a parking space, trolleys and baskets are provided, products are well presented with appropriate signage, there is a wide selection of products, and with longer trading hours, consumers can shop anytime.

IMPACT OF MODERN RETAIL EXPANSION ON GOVERNMENT

Vorley *et al.* (2007) suggest that the expansion of modern retailers presents three major issues for government; (1) the need to invest in traditional markets; (2), the need to control supermarket power; and (3) the need to support producer organisations.

Despite the dramatic expansion in modern retail formats, there is increasing evidence to suggest that traditional markets are quite resilient (Vorley *et al.,* 2007). Throughout much of Asia, smallholder farmers continue to sell to traditional market outlets for a number of reasons including the services that traders and market intermediaries provide in transporting, sorting and grading the produce, the provision of market information and cash advances, immediate payment (in cash) and the long-term trading relationships which substantially reduce the risk and uncertainty associated in transacting with unfamiliar exchange partners.

Improving the traditional wholesale and retail markets will not only help the many independent traders and retailers compete with the modern supermarket chains, but also provide the many smallholder farmers with more marketing options (Vorley *et al.,* 2007). At the consumer level, with appropriate government support, the traditional market can ensure that basic levels of food safety and quality are not only maintained but improved. Throughout the world, public safety standards have lagged behind the private standards established by retailers and food manufacturers (Chen *et al.*, 2005). Incentives to modernise can be created through the enforcement of public regulations regarding food safety, weights and measures, taxes, competition and occupational health and safety, thus making traditional markets better places to shop.

One of the biggest problems that results from abandoning traditional markets in favour of the modern retailers is that one source of market power is replaced by another (Vorley *et al.,* 2007). It is the role of government to balance consumer interests with those of other stakeholders through a range of regulatory instruments which may include: foreign investment policy; competition law; limits on store size, location or business hours; or minimum local content requirements.

The first step in regulating the abuse of buyer power in the developing countries is to reform as many as possible of the trading practices that adversely impact on smallholder farmers (Vorley *et al.,* 2007). This includes deferred payment terms, unfair refusals to receive goods and the unfair return of goods, reducing prices after purchase, and coercive demands for slotting fees and promotional allowances.

If smallholder farmers are to participate more equitably in the transformation of modern retail markets, the fundamental issue is to build the capacity of smallholder farmers to provide a sufficient quantity of produce that meets the customer's specifications, reliably and consistently. While it is possible to readily improve both the quality and the safety of the fresh produce cultivated through the provision of appropriate inputs, technology and agronomic advice (Chen *et al.,* 2005), it is important to recognise that modern retailers and institutional buyers will transact with smallholder farmers only through market intermediaries, wholesalers or producer organisations (Vorley *et al.,* 2007).

Traditionally, development interventions have focused on building producer organisations that deal directly with retail buyers, effectively bypassing traditional market intermediaries. However, in most instances, more can be gained by strengthening the relationships between farmers and market intermediaries (Vorley *et al.,* 2007). Market intermediaries absorb much of the risk and perform many of the value-added services that modern retailers demand, but if smallholder farmers are isolated from the market, deprived of market information and unaware of their marketing options, resentment and distrust will invariably weaken these linkages. Hence there is a need for government to upgrade and improve the dissemination of market information (Chen *et al.,* 2005).

While collective action can enable smallholder farmers to transact either directly or indirectly with modern retailers, such is unlikely to happen without a substantial investment in training and management systems, which should be provided collectively by government agencies, NGO's and the private sector (Vorley *et al.,* 2007). In addition, if smallholder farmers are to comply with the standards expected by modern retailers, they must invest and thus the need for finance must be addressed (Chen *et al.,* 2005). However, nothing will happen without an appropriate investment in infrastructure and marketing systems. In this respect, government has a critical role to play in providing the appropriate enabling environment.

REFERENCES

Batt, P.J. 2006. Expanding the quality concept to satisfy consumer demand. In Batt, P.J. (ed) Proceedings of the First International Symposium on Fresh Produce Supply Chain Management. *Acta Hortic.* **699**: 83-90.

Batt, P.J. and Parining, N. 2000. Price-quality relationships in the fresh produce industry in Bali. *International Food* and *Agribusiness Management Review,* **3**: 177–187.

Chamhuri, N. and Batt, P.J. 2009. Consumer Choice of Retail Outlet: Focus Group Interviews in Malaysia. **In** Oppenheim, P.P. (ed) Proceedings of XVI International Symposium on Horticultural Economics and Management, *Acta Hortic.* **831**: 237-246.

Chen, K., Shepherd, A.W. and da Silva, C. 2005. Changes in Food Retailing in Asia: Implications of Supermarket Procurement Practices for Farmers and Traditional Marketing Systems. AGSF Occasional Paper 8. FAO Rome.

Deloitte. 2008. Global Powers of Retailing; Standing out from the crowd. http://www.deloitte.com.

Euromonitor International. 2004. http://www.euromonitor.com.

Folley, R.R.W. 1973. *Intensive Crop Economics,* Heinemann, London

Gehlhar, M. and Regmi, A. 2005. Factors Shaping Global Food Markets. **In** Regmi, A and Gehlhar, M. (ed) *New Directions in Global Food Markets,* ERS/USDA: 5-17.

Leenders, M.R. and Fearon, H.E. 1993. *Purchasing and Materials Management,* Tenth Ed. Irwin.

Lele, U. 1981. Cooperatives and the poor: a comparative perspective, *World Development* **9**: 55-72.

Murray-Prior, R. 2007. The role of grower collaborative marketing groups in developing countries, *Stewart Postharvest Review,* **6**: 9.

Murray-Prior, R., Batt, P.J., Rola-Rubzen, M.F., McGregor, M.J., Concepcion, S., Rasco, E.T., Digal, L.N., Montiflor, M.O., Hualda, L.T., Migalbin, L.R. and Manalili, N.M. 2006. Global Value Chains: a Place for Mindanao Farmers? **In** Batt, P.J. (ed) Proceedings of the First International Symposium on Fresh Produce Supply Chain Management, *Acta Hortic.* **699**: 307-315.

Reardon, T., Timmer, C.P and Berdegué, J.A. 2005. Supermarket Expansion in Latin America and Asia. Implications for Food Marketing Systems. **In** Regmi, A. and Gehlhar, M. (ed) *New Directions in Global Food Markets*, ERS/USDA: 47-61.

Vorley, W., Fearne, A. and Ray, D. (ed), 2007. *Regoverning Markets; A Place for Small-Scale Farmers in Modern Agrifood Chains?,* Gower Publishing Limited, Hampshire.

30. VALUE CHAIN MANAGEMENT IN INDIA: AN OVERVIEW & PERSPECTIVES

Dr. A.P. Paul

Managing Partner, Business Resources International, International Projects, Financial & IT Services, Evertsenstraat 22 and 23, 2665 AE Bleiswijk, The Netherlands
[Former Director General, All India Management Association, New Delhi]

ABSTRACT

As explained by a well known VCM Centre in Canada, "Value Chain Management" [VCM] is a business philosophy based on taking advantage of opportunities to create and capture value that traditional supply systems cannot. It is about all businesses, wherever they operate in relation to the endmarket, leveraging their combined competencies and strengths to more effectively and efficiently, supply consumers in a rapidly evolving business environment.

To dispel any doubts, VCM concepts and its applications take into account supply chain management concepts and systems, as are known and applied the industrial sector, especially engineering industries. However, unlike in the industrial sector, when SCM is applied to agriculture, VCM concepts and its applications, perforce has to take into account, totally different factors, dimensions, business processes, role and role relations amongst supply chain partners etc. These have been explained in the paper.

VCM takes into account, the operational efficiency of all related business processes, encourages and facilitates equitable, voluntary sharing of business risks, costs, profits, chain digitalization, fair trade practices, etc amongst different partners of the entire supply chain – from seed to shelf – to serve the potential consumer.

Increasingly, "VCM concepts and practices" are widely accepted, applied to agribusiness in countries like USA, Canada, Australia, the UK, various countries in Europe etc and being, constantly improved. At policy levels in the retail food chains supply of the corporate sector, it is often seen as part of "Corporate Responsibilities" and adopts and implements, various strategies for its effective application and implementation, supported by In-Company Training and Development Programmes. There are a wide range of full-time Graduate and Post Graduate Programmes for students, various Executive Development Programmes for the corporate sector on VCM, have in operation for several years, already in Agricultural Universities abroad. These are fully supported by massive Research & Development Programmes undertaken by highly qualified and experienced faculty, field work etc., including many international projects and programmes.

Consequently, more recently, several developing countries in Africa, South America etc, where an increasing number of small farmers' organizations, small producers' cooperatives etc dealing with items like coffee, rice, cocoa, etc – once at the total mercy of vested interests – are finding meaningful use of VCM. It has helped them to secure better economic share, improve agricultural practices, implement fair trade practices and facilitate, overall economic and social development, by relevant applications of VCM.

In our Paper, we have referred to a list of our tentative findings and conclusions of our informal investigative efforts in India. It was noted that various concepts and applications of "Value Chain Management" are relatively new to the agricultural sector as well as to the organized retail food sector in India. Compared to other management subjects, qualified and experienced faculty on VCM are not readily available in India. There was a dearth of R&D studies, local case studies, teaching materials, audio-visual aids on Value Chains of different crops in agriculture in India. VCM would help to improve productivity at the farm level, upgrade crop and product quality, save costs, lower consumer price levels etc.

However, the situation is fast changing for the better in India – from policy level recognition to follow-up action at ground level, on a wide variety of crops in different dimensions, across the country. These are in keeping with the vast changes that are taking place in the agriculture sector, the swift developments in the food supply chains, changing patterns of consumer demand and expectations especially in urban areas, emergence of supermarket chains, infrastructure improvements, logistics and transportation systems, storage and warehousing, etc.

In this context, it is noted that, under its "National Agriculture Innovation Programme" [NAIP], the Indian Council of Agricultural Research, New Delhi has approved and initiated 68 Projects under Component-2 of NAIP with numerous Agricultural Universities, Research Agencies & NGO's in India. Many other concurrent developmental efforts are being planned or being implemented.

The paper concludes with specific Recommendations, to develop and apply relevant concepts and applications of VCM for further consideration, discussion and appropriate action, inter alia, about the setting up "Centers of Excellence for Value Chain Management in India".

INTRODUCTION

It should be noted that in recent years, due to rapid changes in food supply chains and its different types of networks in our different States, logistics systems, consumer expectations etc, there have been several strategic changes in agriculture and related agribusinesses in India. These require the rapid development of new competences, skills and abilities, starting from our agricultural research scientists, the rural and seasoned grower and farmer for the supply of their various perishable products with despatch and efficiency to our local mandis and domestic marketplace for their potential consumers - whether in rural areas or urban retailers.

Further, in export marketing of horticulture products from India, due to rapid globalisation there are urgent needs for compliance to international market regulations *e.g.,* HACCP, GlobalGAP, *etc.*, amidst intense, often ruthless, competition from powerful well-established players. These call for greater efficiencies and consistent high quality products as imperatives along with greater amongst all partners of the value chain. Therefore, developing nations like India are faced with serious challenges. Food safety, health factors and integrity have increasingly become critical issues and choices for survival are faced by all stakeholders of food supply chain networks.

As Wageningen International states, "such developments set the stage for new strategic choices for food supply chain networks to forward and backward integration, redesigning product offerings and business processes. Therefore international firms respond by creating supply chain networks with different brands, quality, information, services, etc. as well as systems for better governance, coordination and tracking and tracing. In turn, this requires implementation of new organizations forms and personnel competences."

Simultaneously, in the domestic market, it was by highest levels of government, that "the organised retail trade in India has the potential to add over Rs. 2,00,000 Crores [US$ 45.0 billion] business by the year 2010, generating employment for some 2.5 million people in various retail operations and over 10.0 million additional workforce in retail supplier activities – contract production and processing, supply chain and logistics, retail real estate development and management, *etc.* [Ref: 'Inclusive Value Chains in India by Malcolm Harper"; 2009].

Hence, in our search for improvements in agriculture and increasing globalisation, it is strongly suggested that, inter alia, the VCM approach sounds promising to facilitate upgrades and appears worthy of consideration, as a key to all round improvements for our agribusiness in the future.

WHAT IS VALUE CHAIN MANAGEMENT [VCM]?

As explained by a well known VCM centre in Canada, "Value Chain Management" is a business philosophy based on taking advantage of opportunities to create and capture value that traditional supply systems cannot. It is about all businesses, wherever they operate in relation to the end market, leveraging their combined competencies and strengths to more effectively and efficiently, supply consumers in a rapidly evolving business environment"

At the outset, to dispel doubts, it should be understood that VCM concepts and its applications take into account Supply Chain Management [SCM] concepts and systems, as are known and applied the industrial sector, especially engineering industries.

However, unlike in the industrial sector, when SCM is applied to agriculture, VCM concepts and its applications, perforce has to take into account, totally different factors, dimensions, business processes, role and role relations amongst supply chain partners etc.

Salient Features of VCM

Conceptually, VCM represents the creative confluence and growing interface, interaction and integration between the disciplines of agricultural sciences and agribusiness management which facilitates improved and efficient distribution and provision of consistent quality food products, at equitable price levels to the common consumer in the marketplace.

In agriculture, with commodity crops such as rice, cotton, tea, coffee, rubber *etc*, in horticulture with fruits such as mangoes, apples, grapes etc or vegetables such as brinjals, cauliflowers, tomatoes etc or with flowers such as roses, carnations, anthurium *etc.,* - each of the aforementioned, have varying soil, climatic requirements, distinct agricultural cultivation practices, grower groups and its own widespread supply chains and business economics. To this list, one can also add many other agribusiness items such as fish, meat items, dairy products *etc.*

In the event, it becomes part of food supply chains, which further resonates with distinct features, as each of these items relate to vastly differing demand and supply patterns, catering to varying locations, consumer incomes, preferences, interests, tastes *etc.,* in India or elsewhere.

VCM concepts and its applications encounter significant and distinctive quantitative and qualitative factors and core features of agriculture and its various crops/products and goes beyond it. These include such factors as rural and local cultivation practices, relevant agricultural technology, weather dependent, affecting environment and ecology, land use, irrigation facilities, pesticide controls. growth of organic farming, storage and warehousing issues, logistics and movement, food security, health factors, marketing techniques related to, time to ripen, better quality, in acceptable colour, taste, aroma, softness, packaging preferences of potential consumers, competitive price levels *etc.*

Often stated, synonymously with Food Supply Chains Management and starting with Value Chain Mapping, VCM takes into account, the operational efficiency of related business processes, within well-defined parameters of global security and transparency.

VCM encourages voluntary sharing of data and information, facilitates trust in transactions, equitable, business risks, costs, profits, chain digitalisation, fair trade practices, etc. amongst different partners of the food supply chain - from seed to shelf – so as to serve the consumer.

As consumers, unlike manufactured consumer goods, all of us have an immediate stake in VCM, as food security issues and lower price levels in day to day consumer food items, are a source of constant concern, for one and all.

GLOBAL OVERVIEW ON VCM

Increasingly, "VCM concepts and practices" are becoming widely accepted and applied to agribusiness in countries like USA, Canada, Australia, the UK, various countries in Europe *etc.* At corporate policy levels, it is often seen as part of "Corporate Responsibilities" and adopts various strategies for its implementation, supported by training and development programmes.

It may be noted that the "Global Commerce Initiative" [GCI], based in Germany is, sponsored by leading international Supermarket Chain companies in the food retail sector which increasingly support, appropriate VCM initiatives. I was informed by FICCI, that they have been involved with GCI efforts, since its inception. Further, when GCI held an International Conference in Mumbai, last year, FICCI was involved. It was attended by senior executives from leading FMCG's and supermarket chains in India.

The development and applications of VCM are increasingly supported and promoted by a massive network of teaching/training/research programmes which conduct full-time teaching programmes on Food Supply Chains and VCM at various Agricultural Universities and VCM centres during the past decade in the US, Canada, the UK, Australia, various countries in Europe, especially The Netherlands as well Executive Development & In-Company Training programmes for all managerial levels in the corporate food sector, supermarket chains *etc*.

More recently, several developing countries in Africa, South America, China, Egypt *etc,* where an increasing number of small farmers' organisations, small producers' cooperatives etc dealing with items like coffee, rice, cocoa, fish, cashew nuts *etc*., - once at the total mercy of vested interests - are finding meaningful use of VCM. It has helped them to secure better economic shares, improve agricultural practices, implement fair trade practices and facilitate, overall economic & social development, by relevant applications of VCM.

Such recorded experiences are fully documented in the references and reading lists below. These are of special significance to individual/joint family farmers, Growers' organisations, NGO's, SHG's, banking and micro finance agencies etc in India and cannot be overlooked.

TENTATIVE, INITIAL FINDINGS & RESULTS ABOUT VCM IN INDIA

To continue from our background notes, under para 1, as mentioned earlier, during his periodic visits to India from NL, between January and June 2008, Dr. Paul contacted and had discussions with key, senior personnel at the Ministry of Agriculture, New Delhi, related nodal government agricultural institutions such as ICAR, selected Agricultural Universities and the Retail Food Sector in India. He was also in dialogue with agriculture science experts and senior faculty at the Wageningen University [WUR] in The Netherlands as well as with senior officials at the Royal Dutch Embassy in New Delhi, so as to examine and proceed with likely project possibilities of VCM in India. Their advice, guidance, cooperation and encouragement is gratefully acknowledged by the author.

The initial findings of informal, investigative efforts in India led to encouraging, positive and supportive responses and serious interest from a target group of concerned persons in various institutions.

Tentative and Initial Findings were

- It was found that most persons in the target group were familiar with Supply Chain Management [SCM] concepts and practices in manufacturing industry, especially in engineering companies. IIM, Bangalore has a "Centre of Excellence in SCM", perhaps rightly, was not enthused about VCM.
- All concerned expressed serious concerns about low productivity in our agricultural sector, especially about price rise issues *etc*; hence, they were anxious to see improvements made in the management systems relevant to agriculture and agribusiness practices in India, if possible.
- Few were aware about "Value Chain Management", as it applies to agriculture and/or to the retail food supply chains in India
- Compared to other functional management subjects, qualified and experienced faculty on VCM are not readily available in India. Hence, one did not come across teaching programmes on VCM and its applications.
- There was a dearth of R&D studies, local case studies, teaching materials, audiovisual aids etc on value chains of different crops of agriculture in India
- There was urgent need for more information about concepts of VCM and its relevant applications in horticulture and retail food industry in India
- Consequently, there were expressions of serious interest to learn more about "VCM and its applications in India", provided such information or suitable training programmes were made available; and if relevant.

Tentative Conclusions were

- There is urgent need for local research studies and information to provide a data base in the agriculture sector to promote VCM and its relevant applications.
- VCM will have to be practical, relevant & beneficial so as to improve local agricultural business practices, needs, requirements and serve the needs of growers and consumer demand and preferences.
- VCM should help to improve productivity at the farm level, upgrade crop and product quality, save costs, lower consumer price levels etc.

VCM IN AGRICULTURE [ESP. HORTICULTURE] AND FOOD SUPPLY CHAINS MANAGEMENT IN INDIA

Even though various concepts and applications of "Value Chain Management" are relatively new to the agricultural sector as well as to the organised retail food sector in India, it should be noted that the situation is slowly, changing for the better in our country.

As we are serious to proceed any further with VCM, our first priority will be the need to develop & have access to a reliable, up-to-date database about the "Value Chains" for all crops in the agriculture sector of India, inter alia, to facilitate relevant applications of VCM.

In this context, it is noted that, under its "National Agriculture Innovation Programme" [NAIP], the Indian Council of Agricultural Research, New Delhi has approved and initiated 58 Projects under Component-2 of NAIP with numerous Agricultural Universities, Research Agencies & NGO's covering the entire country.

It covers extensive studies on "Value Chains" of a wide variety of crops, on a countrywide basis. Hence, in the coming years, India will possess a solid data base of authentic data and up to do date information for agricultural extension services including value chains, etc. It is hoped that in the next two to three years, the results of such work will be available for use in India.

Prof. Malcolm Harper in his pioneering publication on "Inclusive Value Chains in India"– Linking the smallest producers to modern markets - [2009] includes 14 carefully researched and data based case studies about inclusive value chains in India. It includes selected fresh fruit and vegetables, commodity crops, fisheries, honey, coffee and poultry and unusually, non-food artisan products like Mojari footwear.

The above publication is truly, a treasure house of information and data based of intense research studies from committed experts and researchers covering different industries. It covers work related to VCM concepts, its challenges, relevant applications and summary evaluation in India.

To mention, a few references: Case studies on Namdhari Fresh, ITC Choupal Fresh, Spencer's Retail, a BASIX Intervention, Kohinoor Foods, Small broiler farming in Kerla, Operation Mojari *etc.*

Apart from the above mentioned, to mention a few illustrations: the growing network of companies of the National Dairy Development Board, [NDDB] especially its subsidiary, SAFAL, which deals with fresh vegetables and fruits, Reliance Fresh, Field Fresh, Tata Star Bazaar, Mahindra Shubh Labh outlets, Metro Cash and Carry, Haryali Kisan, Big Bazaar, Nilgiri's and many more, are making steady growth and progress, as they continue to modernise the food supply chains operations. They would be well-advised to see, where VCM could complement and strengthen, their respective operations.

Recent developments and changes in the mandates of Agricultural Produce Marketing Committees [APMC] in Karnataka and other states, with the special emphasis and additional budgetary allocations in agriculture in our next Five Year Plan, the plea for setting up a national spot exchange for electronic trading of farm produce etc – all these and many related developments, are indicative of major policy level efforts & new strategies for implementation aimed, to improve the management of agriculture sector.

At the same time, contract farming is slowly, gaining importance in India, VCM and its applications can enhance value addition for all chain partners. Hence, it is increasingly important and gaining recognition.

Currently, from limited published literature available, it is noted that MNC's and large supermarket chains in the food retail sector in India, are increasingly taking into account the realities of VCM, despite numerous difficulties and constraints imposed, often by political interference by vested interests, misinformed ideologues etc.

As is well-known, numerous Self Help Groups [SHG's], NGO's and Micro Finance agencies, NABARD, well–established agribusiness consulting companies [*e.g.*, Technopak] *etc.*, are spreading rapidly into interior rural farming communities and areas, in all our states. In all such efforts, VCM applications can be of immense value to help small farmers' organisations and often, save them from the vicious clutches of traditional, vested interests, money lenders *etc.* Often, they work in tandem with FMCG companies in their quest for rural market penetration at the bottom of the consumer pyramid, a la Prof. C.K. Prahlad.

VCM can be of considerable relevance to upgrade and enhance the operations, performance and contributions of numerous consumer Cooperative Organisations in India. However, when the suggestion was brought to the attention of the concerned official in the Ministry of Agriculture, it was sadly, dismissed!

The reference and reading list provided at the close of this paper provides specific additional documentation about VCM and its applications from a number of developing countries in South America, Africa, Kazakhstan etc, for further study and consideration.

LEARNING FROM INTERNATIONAL VCM PROJECT EXPERIENCES

Hence, learning from data based international experiences and evaluation about VCM and its applications especially in several developing countries seems prudent and worthwhile.

It should be noted that with VCM, suitably modified and made relevant similar applications can be readily considered in India.

To summarise such possibilities:

- VCM provides more equitable rewards to rural growers
- VCM promotes better field cultivation practices and improvements of productivity in agriculture
- VCM encourages improved logistics management and facilitates better storage, warehousing, transportation, on-time delivery systems *etc.*
- VCM encourages cost savings at different stages in the supply chain by chain digitalization
- VCM supports fair trade practices, in our vast network of informal marketing of farm products in our different states and geographies
- VCM strengthens better quality control on products with greater shelf-life for consumers and better food security in a society which

suffers from chronic food adulteration practices especially in its PDS networks.

- VCM facilitates sharing and spreading of data and information about value created greater devolution of risk & reward factors, greater transparency *etc.*, in its agricultural practices, as well as to various partners in the food supply chains.

Hence, we submit, there should not be undue fears about positive, medium and long-term results and benefits to agriculture and agribusiness in India. If other developing countries can gain from it, surely, with suitable checks and balances, VCM can be suitably modified, where necessary and made relevant to suit for our own needs and requirements for its applications in India. It requires political will as various improvements to value added aspects and factors will reveal itself in the food supply chain, to the advantage of all partners in the food supply chain, including the consumer.

PROS AND CONS : SUMMARY EVALUATION

Like most new ideas, [relatively new to India; but in existence and with proven track records in numerous countries abroad], surely one must anticipate and will confront numerous challenges in the possible adaptation and applications of VCM in India.

Our biggest stumbling blocks are - as always - our traditional mindsets, conservative attitudes and resistance to change. We are also aware that these are mostly due to ignorance, superstition, lack of exposure to new concepts and practices, learning from success experiences and political will for rapid implementation.

In the process, there are huge lost "opportunity costs" at the same time, it should be recognised that these are also, in reality, business opportunities for generating employment, earnings, growth and development.

Allow me to draw your attention to the dramatic changes that are sweeping India - from the rapid growth in communications, IT applications, automobile, air and other transport sectors, e-Governance etc, in recent years.

In this context, may I submit that in the long run, VCM concepts and its applications can extend well-beyond horticulture and floriculture? With suitable customisation and appropriate IT software, VCM can become a valuable tool and applied to other types of agribusiness, such as animal husbandry, aquaculture, poultry/pig farming etc.

Further, it can be noted in passing, based on intense IT software development efforts, an "agriMORE Platform", as an IT tool to facilitate implementation of VCM, has been developed and already, in commercial operation for horticulture in The Netherlands.

Admittedly, VCM is not going to be an overnight transformation in agricultural and related agribusiness practices in India. It is likely to be painfully slow, with foreseeable problems of apathy, wilful obstruction, *etc.* All such new concepts and applications mean implantation of change processes, which will be slow in a country with such complex diversity. Yet, move we must. India can ignore such major international experience and improvements in performance in the agricultural sector, at our own costs, delayed action and imperil to provide opportunities to millions below the poverty line.

UAS EXPERIMENT IN VCM

In this context, a "Workshop on VCM"- a pioneering effort was conducted at UAS, Bangalore.

The Programme was organised under the pioneering guidance and support of Vice Chancellor of University of Agricultural Sciences, GKVK, Bangalore and conducted under the supervision of Professor and Head of the Department of April. Marketing, Cooperation and Business Management and senior faculty of UAS for their students with joint expert collaboration of Chair, Food Supply Chains Group at Wageningen University and Research and the author, who served as Programme Consultant, on 2nd and 3rd April 2009.

An excellent Training Manual with practical guidelines was provided by WUR, the Netherlands.

RECOMMENDATIONS FOR FOLLOW-UP

Few recommendations for further consideration, discussion, decision-making and appropriate action, are as follows:

a. Institutional Support and Long-term Planning

Continued support, guidance and encouragement from concerned officials at the Ministry of Agriculture, National Mission on Horticulture, ICAR, IIHR and other national and state level government agencies, are required, to promote relevant concepts and applications of VCM in India. It is a key starting point in our efforts for better understanding and further improvements. Much work has been initiated and are at different stages of implementation

b. Institutional Networking

Support services of various networking agencies, to mention a few: National Institute of Agricultural Extension Management [MANAGE], Indian Institute of Rural Management, Anand, Indian Institute of Plantation Management, Bangalore, ICRISAT, Hyderabad, various Commodity Boards [*e.g.,* Tea, Coffee, spices *etc.*], UPASI, NABARD and such nodal agricultural institutions and agencies could network and complement such VCM related efforts.

Our Public Distribution Systems and networks have a lot to learn from VCM applications.

The co-operatives movement in India, especially various Consumer Cooperatives Organisations will be well-advised to examine and utilise VCM, as it considers best.

c. VCM in Agribusiness Management Education

To develop a core of faculty trained in VCM, trainers, consultants and students, through a series of two / three day, residential seminars on VCM with the support of and at selected Agricultural Universities and/or Training, Research and Development Institutions in various parts of India; International Technical Assistance on VCM can be considered and made available.

Urgent need to plan and develop local case studies, reference and reading materials, data and documentation, audio visual aids etc for proposed training and educational programmes on VCM. Such teaching programmes could be augmented by a series of follow-up by "Pilot Farm Extension Projects" using VCM concepts and its relevant applications.

Most of the Business Management Departments in Agricultural Universities have various teaching programmes, which could give greater attention, than hitherto, to the role of food supply chain management to include VCM concepts and its applications

It is high time to seriously consider the setting up "Centres of Excellence for Value Chain Management in India". Depending on resources,[and with suitable international assistance, where required], it could be independent or developed, within the administrative framework of selected Agricultural Universities/Departments of Agribusiness Management or related institutions of agribusiness, *etc.*

d. Co-operation with Corporate Sector, *etc.*

In keeping with PPP models of growth, it is imperative for VCM is to progress, the proactive support, advice and guidance from FICCI/CII/ ASSOCHAM, SCOPE and other Employers' Associations and perhaps, progressive Trade Unions [*i.e.,* not stubborn, ideological and narcissists caught in a time warp!] in selected cities in India.

AIMA with its Local Management Associations [LMAs] could develop and conduct, a series of Introductory Seminars, aimed to sensitize top management on VCM and its relevant applications in the corporate sector.

Urgent need for the corporate retail food sector in India, to fully understand the significance of VCM and its relevant applications in their sector.

Planning, development and implementation of In-company training programmes, for middle management, supervisory and operator levels, conducted by the respective Training and Development Departments; with assistance from LMAs, consulting and training organizations, NGOs *etc.*

e. Multi-media Support

It is a matter of profound regret that in India, (which is primarily a developing nation dependent on agriculture and the monsoons), there is a glaring imbalance of media support and exposure on critical issues, analysis, reporting and commentary between reporting on the corporate sector and agriculture. Reporters of the rural and agricultural scene are rare. Mr. P. Sainath a news reported of the 'Hindu' daily stated in public; there will be hordes of reporters covering a fashion show in Mumbai or New Delhi [if I might add, mostly for the goodies!] than having a single reporter in a rural village area, to report on myriad injustices, suicides or superstitions and more importantly, small success stories and minor breakthroughs in rural agriculture, especially achieved by innovative young farmers!

The silence is deafening and such continued neglect needs urgent attention and correction. It seems a sad blind spot in the reporting of our society. Where are all our regional language news media experts? This is a wakening call to the media masters of India.

It is my considered view, that there is urgent need for many more Barka Dutts and Karan Thapars, to continue to provide a balanced picture of developments, contributions and failures of our vast hinterland that is rural India.

CONCLUSION

Lest India be left behind in its continued improvements in agriculture and agribusiness, it is hoped that our agricultural scientists and management experts blend their respective domain expertise in an appropriate manner and thereby, contribute their creative team efforts for achieving further improvements in our food supply chains, through VCM applications.

REFERENCES

List of "Full Proposals Approved under Component-2 of National Agriculture Innovation Programme [NAIP] of ICAR, New Delhi"; 2008

Summary Notes on "Workshop on VCM" on 2nd & 3rd April 2009 at UAS, Bangalore.

Inclusive Value Chains in India - Linking the Smallest Producers to Modern Markets; Malcolm Harper; 2009; www.worldscientific.com

Social & Environmental responsibility in progress - Making Quality Systems Work for Poverty Alleviation, Biodiversity Conservation & Company Performance - Conference Report; Organizers: Hivos-Oxfam Novib Diversity Fund, IUCN NL and Dutch Ministry of Foreign Affairs; 2008; Contact: Gigi Manicad at www.oxfamnovib.nl

Cooperative Development: Peace & Security in South Asia – Papers presented at the International Conference organized by Centre for Rural & Industrial Development from 27-29 November 2008; Edited by Dr. Neetu Gaur & Dr. Vijay Laxmi; March 2009; CRRID, Sector 19-A, Madhya Marg, Chandigarh - 160 019, India (All Rights Reserved).

Producer Organisations & Market Chains - Facilitating trajectories of change in developing countries; Edited by Giel Ton, Jos Bijman & Joost Oorthuizen; 2007; www.WageningenAcademic.com

Tropical Food Chains – Governance regimes for quality management – Edited by: Ruerd Ruben, Martinus van Boekel, Aad van Tilburg & Jacques Trienekens; 2007; www.WageningenAcademic.com

Quantifying the Agri-Food Chain; Edited by Christien J.M.Onderstijn. Jo H.M. Wijnands, Ruud B.M. Huirne and Olaf van Kooten; 2006; www.springer.com

Source: Indian Council of Agricultural Research, New Delhi: 2008.

31. LINKING SMALLHOLDER PRODUCERS TO MARKET: THE NEED FOR EFFICIENT SUPPLY CHAIN MANAGEMENT

Peter J. Batt and Bich Nhu Le

Curtin University of Technology, GPO Box U1987, Perth, WA 6845, AUSTRALIA

ABSTRACT

With increasing affluence and the expansion of modern retailing and food processing, fresh produce buyers are forging enduring long-term relationships with their preferred suppliers. Such relationships provide customers with assurances that the product will be delivered on time, in the quantities required, to predetermined specifications which govern not only the product quality but the manner in which the product has been produced. For the buyers, such relationships reduce transaction costs, there is less wastage and risk and uncertainty are greatly reduced. For the suppliers, such relationships provide an assured market but they do not always guarantee a higher price. Such relationships will only develop where there is an equitable sharing of value in the exchange, an element of trust and a long-term commitment. Invariably, for smallholder farmers, their participation in such value-added chains is contingent upon the formation of formal or informal groups, thereby providing some economies of scale and a greater capacity to meet the needs of their downstream customers. Trust and social capital is instrumental in the formation of these groups. Trust and satisfaction in the exchange will be derived primarily from the knowledge that smallholder farmers are not being taken advantage of by downstream market intermediaries. This will require market intermediaries to provide timely information on market prices and where appropriate, technical advice and micro-finance to facilitate production. However, there is always the risk that smallholder farmers will forgo long-term relationships to secure immediate short-term benefits by transacting with alternative buyers who offer a higher price. Furthermore, on occasions, it may be necessary for smallholder farmers to transact with alternative buyers to dispose of that product which fails to meet specifications and/or where the collapse of failure of the focal customer leaves smallholder farmers without a buyer. Various mechanisms are available to provide a degree of flexibility and at the same time, to ensure that contractual obligations are fulfilled. Thus, it is not only possible but indeed desirable for these value-added chains to co-exist in parallel with traditional supply chains. Ultimately, for smallholder farmers to participate in these modern value-added chains there must be a willing buyer who is able to pay a price that is sufficient to adequately compensate the producer. With such an assurance, collaborative farmer groups can confidently embark upon a production plan to maintain sufficient supply. However, smallholder farmers must have access to the desired technologies, agronomic advice and micro-finance to secure good quality seed, chemicals and □andanna□rs and downstream buyers must refrain from the use of coercive market power, for this will have a significant negative impact on trust and the smallholder farmers' commitment to the relationship.

INTRODUCTION

Throughout much of Asia, the marketing and distribution of fresh fruit and vegetables is experiencing a period of unprecedented change. Globalisation has not only resulted in a dramatic expansion in the market share held by the modern supermarket chains, but with increasing personal disposable income, there is a significant shift in consumption from fresh food to processed food (Gehlhar and Regmi, 2005). Furthermore, with rising income comes a greater desire for more variety, more expensive and higher value food (OECD–FAO, 2005). Product diversity is increasing as producers, food manufacturers and retailers strive to meet the demands of increasingly sophisticated consumers (Martech, 2005). Products are being differentiated on the basis of what they contain and or what they do not contain (Centrec 1999). Some attributes are based on the methods of production, the place of production, the way in which the product has been processed or handled, and the impact that either the production or processing of the product has had on the environment and the community.

Traditionally, most of the fresh produce in Asia is distributed and sold through a complex network of wholesale and retail markets. In the wholesale market, prices are determined primarily by supply and demand. However, prices are highly variable day-by-day and even hour-by-hour, and both the quantity and quality of the produce offered for sale is highly variable, occasioned by the weather and seasonality of supply (Folley 1973). Given the inherent variability in quality, produce must be made available on the market floor for inspection by potential buyers. However, this method of sale not only results in a substantial reduction in product quality, but as the product itself is handled a multiple number of times, further reductions in product quality are inevitable.

For modern retailers and the food processing sector, these variations in quality and price and the inconsistent supply are unacceptable, for it makes it impossible to adequately price the product or to engage in any promotion or merchandising of the product (Batt, 2006). Furthermore, the anonymity which is so often associated with the wholesale marketing of generic food products makes it impossible for the buyer to be able to assure downstream customers that the food is safe, or that it has met prescribed environmental and ethical standards. Consequently, in order to maintain product quality and provide a more consistent supply, institutional buyers are purchasing greater quantities of fresh produce from preferred suppliers.

THE EMERGENCE OF LONG-TERM RELATIONSHIPS

Quality, a competitive price and reliability of delivery are generally regarded as the most important criteria by which buyers evaluate potential suppliers (Cunningham and White 1973; Lehmann and O'Shaughnessy 1974; Dempsey 1978; Wilson 1994). However, institutional buyers are influenced by both rational and emotive variables in choosing between alternative offers (Hutt and Speh 1995). Rational motives are usually economic, whereas emotional criteria are generally concerned with organisational status, security, risk avoidance, social, political or environmental concerns.

Where there is no difficulty in accurately specifying the exact nature of the product and there are several reliable suppliers in the market, a buyer can simply choose that supplier who offers the lowest price from among all those suppliers who fulfil the functional requirements (Hakansson *et al.* 1977). However, where a number of alternative suppliers have equally one another in terms of quality, delivery and price, various attributes such as the supplier's reputation, financial position, communication and attitude towards the buyer may become decisive (Dempsey, 1978). Cardozo and Cagley (1971) and Puto, Patten and King (1985) demonstrate that institutional buyers are strongly attracted to well known or existing suppliers, for current suppliers are perceived as being less risky.

Uncertainty is the key dimension affecting organisations, inter-organisational relationships and the costs of transacting (Achrol *et al.* 1983). Uncertainty prompts firms to establish and to manage relationships in order to achieve stability, predictability and dependability in their relationships with others (Oliver 1990). Performance is thought to improve when more relational marketing structures are introduced in response to high levels of uncertainty. For buyers, engaging in long-term relationships with suppliers has resulted in various economies of scale and improvements in both supply and demand. There are obvious economies of information arising from the reduced amount of time spent in price negotiation, less time searching for new suppliers and tendering, and less time monitoring the quality of the products consigned to the buyer (Wilson 1996; Fearne *et al.* 2001). Buyers can anticipate improved access to a more reliable supply of inputs, improved product quality and performance, and a higher level of technical interaction in the form of developing and launching new products. With increased mutual dependence, buyers can expect greater stability and loyalty which may result in reduced lead times, greater attention to common problems and priority in times of scarcity, reduced inventory levels and more stable prices.

For the growers, while these arrangements may not necessarily result in a better price, the relationship provides access to shelf space, a guaranteed market for produce of an acceptable quality and security through the knowledge that the partners have a financially secure trading arrangement (Hughes and Merton 1996). In what is arguably a mature and even declining industry, preferred suppliers are rewarded with volume growth, which makes it worthwhile to invest in the economies of scale and improved technology. Various benefits may also accrue from the exchange of sensitive market information. Preferred suppliers get feedback from customers on the consumer acceptability of new varieties and new products and get greater product support and assistance in new product development (Fearne *et al.* 2001).

However, as competition increases and prices generally decline, in order to remain competitive, buyers are giving much greater consideration towards improving the performance of their supply chains (King and Venturini 2005).

EFFICIENT SUPPLY CHAIN MANAGEMENT

A supply chain is a network of organisations involved through upstream and downstream linkages, in the different processes and activities to produce superior value in the form of products and services for the ultimate consumer (Fearne *et al.* 2001; Reardon and Berdegué 2002; Cadilhon *et al.* 2006).

Using the network approach (Hakansson and Snehota 1995), the organisations (actors) involved in the supply chain include growers, market intermediaries, wholesalers and retailers, their input and service suppliers, and ultimately, the consumer (Wheatley and Peters 2003). The main processes (activities) are the handling, conditioning, storing, packing, transport and trading of fresh and/or processed products.

Linkages refer to the interdependency that exists between the actors and the activities that they perform and how these activities are linked to each of the other actors (Ford *et al.* 1998). As the output of one firm becomes an input for another, in fresh produce supply chains, all participants must understand how the quality of the product will deteriorate irrevocably at each stage of the supply chain with inappropriate handling (van der Vorst *et al.* 2007; White 2000).

Value is the worth in monetary terms of the economic, technical, service and social benefits a customer receives in exchange for the price that it pays (Anderson and Narus 1999). Value is achieved when the proper function is secured for the proper cost (Hutt and Speh 1995). Buyers typically consider product and supplier attributes in a manner that seeks to minimise the various problems associated with purchase and use (Wilson 1994). Simpson, Siguaw and White (2002) describe how suppliers have the potential to positively (or negatively) affect customer inventories, product quality, cost and delivery times. As prudent supplier selection leads to improved buyer performance, price is becoming less important in the decision to purchase (Batt 2006). Nevertheless, a supplier must be cost competitive even if it offers a superior quality product or more reliable delivery.

In simple terms, supply chain management is concerned with the sharing of information, in order to save time, reduce costs (manufacturing, inventory, distribution and waste), increase effectiveness (accurate targeting of consumer needs and wants), and to add value (Fearne *et al.* 2001). The efficiency of a supply chain ultimately depends upon the efficiency of each individual actor and the linkages that are established between them (Herlambang *et al.* 2006; Power 2005; Humphrey 2005). Effective supply chain management requires cutting costs, assuring consistency from day to day, and raising product quality and diversity. Cutting costs, in turn, requires an improvement in all aspects of procurement, including product and transaction costs. This is done by means of improvements in coordination and logistics systems, contracts with wholesalers and producers, and private standards which specify the quality, safety, volume and packaging of the products (Reardon and Berdegué 2002).

Coordination

Supply chains are complex systems with numerous activities spread over multiple organisations and sometimes over lengthy time horizons. To improve the performance of the supply chain it is necessary to overlay a coordination system that brings these multiple functions and organisations together as a part of a unified system (Arshinder *et al.* 2008).

The whole concept of supply chain management relies heavily on the notion that there is sequential interdependence between what are essentially independent firms (Thompson 1967). This kind of interdependence involves direct relationships between actors ordered in a serial fashion: one actor's input is another actor's output. Inventory management, logistics, purchasing and procurement, production planning and performance measures attempt to optimise sequential production processes and operations, while efficient governance mechanisms attempt to reduce transaction costs and appropriate property rights in downstream or upstream stages in the chain (Lazzarini *et al.* 2001).

Interdependence is a necessary condition for successfully capturing and leveraging inter-functional and inter-organisational integration (Heide 1994). Integration of supply chains has been described as attempting to elevate the linkages within each component of the chain, to facilitate better decision making and to get all the pieces of the chain to interact in a more efficient way (KIT *et al.* 2006). The most effective of these networks will be those that are able to get the mix of information requirements, physical logistics and collaboration right, and to provide shared benefits to the partner organisations (Power 2005). However, if the benefits of successful coordination are to be achieved, four variables are considered crucial: trust, decision-making, information sharing and goal congruence (Lejeune and Yakova 2005).

Trust

Trust is an important factor in shaping effective and efficient supply chains (Biénabe *et al.* 2007; Batt 2003a; Batt 2006b; Handfield and Bechtel 2004). Anderson and Narus (1990) view trust as the belief that an exchange partner will perform actions that will result in positive outcomes for the firm and will not take unexpected actions that may result in negative outcomes. Moorman, Deshpande and Zaltman (1993) define trust as the willingness to rely on an exchange partner in whom one has confidence.

While both of these definitions view trust as a behavioural intention that reflects reliance on the other partner, both definitions, in part, capture quite different aspects of the construct. Moorman, Deshpande and Zaltman's (1993) definition of trust as a belief, a sentiment or an expectation about an exchange partner, results from the partner's expertise, reliability and intentionality. This component of trust, which Ganesan (1994) describes as credibility, is based on the extent to which the buyer believes that the supplier has the necessary expertise to perform the activity effectively and reliably. However, trust also relates to the focal firm's intention to rely on their exchange partner. Ganesan (1994) describes this component as benevolence, because it is based on the extent to which the focal firm believes that its partner has intentions and motives beneficial to it. A benevolent partner will subordinate immediate self-interest for the long-term benefit of both parties and will not take actions that may have a negative impact on the firm (Geyskens *et al.* 1998).

Heide (1994) considers inter-organisational trust to be a governance mechanism that mitigates opportunism in exchange transactions characterised by uncertainty and dependence. Trust reduces the need for structural mechanisms of control (Achrol 1997) and, over time, both firms learn to become more interdependent. In building trust in buyer-seller relationships, Sako (1992) finds it necessary to differentiate between trust at three levels. Contractual trust is an expectation that the exchange partner will abide by its written or oral contractual obligations and act according to generally accepted business practice. Competence trust is derived from the assumption that the entrusted firm will carry out the activities competently and reliably. Goodwill trust arises where both parties have developed mutual expectations that the other will do more than what it is formally committed to perform. While history, cumulative interaction and transference may build the lower forms of trust, the development of goodwill trust will require not only the absence of exploitation and coercion, but also a history of demonstrated good intentions. Not only must the buying firm demonstrate its trustworthiness, but it must also test the supplier's trustworthiness. Swan, Trawick and Silva (1985) indicate how competence, customer orientation, honesty, dependability and likeability facilitate the development of trust between exchange partners. Moorman, Deshpande and Zaltman (1993) argue that the interpersonal factors that most affect trust include perceived expertise, sincerity, integrity, tactfulness, timeliness and confidentiality. Crosby, Evans and Cowles (1990) contend that mutual disclosure, a cooperative rather than a competitive intention and the style and intensity of the communication between individuals is critical in establishing and maintaining interpersonal relationships.

Decision making

The type of decision-making process conducted in a supply chain will impact on the nature of the long-term relationships. Buyer driven chains are more regulated and characterized by higher levels of governance and vertical coordination between producers, supplier-integrators, processors and retailers (Vorley *et al.* 2007). In these supply chains, the power is concentrated in the hands of a small number of retailers and food processors (KIT *et al.* 2006).

Power resides in the ability of one firm to make another do what he/she would not have otherwise done (Gaski 1984). According to French and Raven (1959), power is derived from the more dependent firm's perception of the dominant firm's ability to mediate rewards, mediate punishment, its legitimate right to prescribe behaviour, some specific knowledge or expertise and the extent to which the more dependent firm identifies with the dominant firm. More recently, Johnson *et al.* (1993) have classified power as either mediated power (reward, coercion and legal legitimate power) or non-mediated power (expertise, referent, information and traditional legitimate power).

Kaplinski and Morris (undated) describe how activities, actors, roles and functions are coordinated through power asymmetry. More powerful actors in the supply chain assume responsibility for the inter-firm division of labour, monitoring outcomes, linking the discrete activities between actors, establishing and managing relationships between the various actors and organising logistics to maintain networks. However, coordination does not require a single firm to engage in these roles: indeed, there may be a multitude of firms coordinating the flow of product and activities along the chain.

Ogbonna and Wilkinson (1998) argue that the power to control is not only dependent on the possession of power, but also the extent to which other exchange partners have countervailing market power. The presence of countervailing power forces exchange partners to differentiate between the possession of power and its use. Noordewier *et al.* (1990) describe how, over time, relational norms may emerge between the parties that limit the exercise of power. Granovetter (1985) describes how even in the presence of substantial power imbalances, exchange partners may prefer to operate in ways that are informed by habit, custom and practice. Relationships may be so embedded that the more powerful firm risks the loss of its reputation should it be found to have engaged in opportunistic trading. Conversely, the frequent use of mediated power is likely to damage relational norms, cooperation and accommodation between channel partners (Brown *et al.* 1995). Overt attempts to directly influence weaker party's through the use of mediated power is generally viewed with considerable disfavour, for it will lead not only to conflict, but the relative attractiveness of alternative exchange partners will increase.

Information sharing

Communication in marketing channels serves as the process by which persuasive information is transmitted (Frazier and Summers 1984), participative decision-making is fostered, programs are coordinated (Anderson and Narus 1990), power is exercised (Gaski 1984) and commitment and loyalty are encouraged (Anderson and Weitz 1992). Communication enables information to be exchanged that may reduce certain types of risk perceived by any one of the party's to the transaction. Any uncertainty about a customers or suppliers organisational structure, viability, method of operation, technical expertise or competence can be resolved by communication between the parties. Communication not only improves the supplier's credibility, but may also provide a convenient and simple means of gaining knowledge of the supply market (Cunningham and Turnbull 1982). Communication facilitates other elements of the interaction such as adaptations by suppliers and customers to the design or application of a product, or the modification of production, distribution and administrative systems by either party.

As channel members in relational channel structures are more interdependent, a higher level of communication is necessary because exchange partners need to share more information to coordinate activities (Mohr and Nevin 1990). The establishment of structural mechanisms that provide real-time information and accurate feedback regarding each partner's actions will not only enhance behavioural transparency and strengthen cooperation, but greatly improve day-to-day performance. Communication clarifies customer's needs and expectations, assists in problem solving and conflict resolution, improves performance measures and creates competitive advantage (Tummala *et al.* 2006).

Information and communication technologies provide the backbone to support the sharing of information in real-time for effective decision making by supply chain partners (Tummala et al. 2006). Normally, small producers get their information from neighbours, friends and relatives; interest groups within the community; local traders, or input suppliers and buyers; the private sector or agribusinesses; government organisations and non-government organisations (Hitchcock 2006). The most effective of these networks are those that are able to get the mix of information requirements, physical logistics and collaboration right (Power 2005). The knowledge and information that is shared can replace much of the guesswork and lead to greater efficiencies.

However, it is not only market information that must be exchanged. If the chain is to be efficient, actors need to be aware of new innovations and to incorporate these into their business (KIT *et al.* 2006). The key agents for knowledge transfer vary from chain to chain. The lead firm is not always responsible for enhancing technical competence along the supply chain as producers sometimes receive targeted technical support from buyers or wholesalers who may provide various incentives to meet downstream customers requirements such as implicit contracts, lower risk, a price premium or similar inducements (Humphrey 2005). Support for upgrading or new technologies may come from a variety of sources outside the value chain including extension services, international NGOs, development agencies and multilateral bodies. Buyers are more likely to offer technical support to smallholder producers when there is: (i) scarcity of supply; (ii) alternative sources of supply are restricted by the non availability of land; (iii) alternative sources of supply are restricted by transport costs; (iv) particular localities have significant advantages over competitors; and (v) firms trade on their image as being socially responsible or supporters of smallholder producers (Humphrey and Memedovic 2006; Humphrey 2005).

Knowledge transfer can also be driven by the farmer's themselves, for producers have a detailed knowledge of what works best in their fields (KIT *et al.* 2006). They can share these experiences among themselves, identify best practices, establish experiments and they can make study trips to large-scale producers, research institutes and experimental centres. In this way, formal scientific knowledge will be combined with practical knowledge, which will not only boost efficiency in the chain, but make the producers more attractive business partners.

Goal congruence

Sustainable long-term relationships often require substantial changes in behaviour and attitude based on common aims, trust, cooperation, mutual dependency, a joint problem solving approach, team decision making and increased interdependence (Fearne *et al.* 2001). Some of the greatest costs in developing long-term relationships are in modifying habits, the knowledge of the people involved and their willingness to modify existing business practices, policies, values and processes (Tummala *et al.* 2006). Goal congruence is defined as the extent to which actors in the network perceive that their own goals will be satisfied by the accomplishment of network goals (Lejuene and Yakova 2005).

Cooperation in a working relationship implies a joint effort, team spirit and collaboration to achieving both intra-firm and inter-firm goals. For the exchange partners, there is an expectation of a balanced transaction, reciprocity and mutuality over time (Anderson and Narus 1990; Morgan and Hunt 1994). The extent to which firms will cooperate is dependent on: (i) the degree to which the parties believe that they can simultaneously achieve their goals; (ii) the existence of mutual agreement between the parties concerning their actions in achieving individual goals; (iii) the clarity of the information processed by the exchange partners; (iv) the establishment of mutually accepted norms upon which the achievement of individual goals are approved/disapproved; and (v) the acceptance of norms of exchange which protect the exchange parties from opportunistic behaviour (Childers and Ruekert 1982).

According to Macneil (1979), norms of exchange are shared expectations as to how trading partners will and should behave. Cannon and Perreault (1999) suggest that cooperative norms such as flexibility, solidarity and mutuality reflect how trading partners expect to work together. These norms do not mean that one partner must acquiesce to the other needs, but rather, it is an acknowledgement by both partners that they need to work together to be successful. Spekman *et al.* (1988) identify how cooperative relationships establish open lines of communication, nurture and sustain longer relationships and develop mechanisms to solve differences by building trust. Cooperative compatibility is one of the most important factors in sustaining long-term relationships because a cooperative culture is necessary to enable supply chain actors to solve problems and improve operations (Tummala *et al.* 2006).

Contracts

Contract farming is an alternative system for the production and supply of land based and allied produce by producers under advance contracts (Singh 2008). The essence of such arrangements is the commitment to provide an agricultural commodity of a specified type, in a specified quantity, at a specified time, price and place to a known buyer.

Contracts can be of three types: (i) procurement contracts under which only sale and purchase conditions are specified; (ii) partial contracts wherein only some of the inputs are supplied by the contracting firm and produce is bought at a predetermined price; and (iii) full contracts under which the contracting firm supplies all the inputs and the farmer becomes a supplier of land and labour (Singh, 2008).

There are various reasons for both producers and farm product processors/marketers to prefer contracts over vertical integration or arms-length transactions. For both parties, a contract can be terminated at reasonably short notice. For producers, contracts give access to additional sources of capital and a more certain price by shifting part of the risk of adverse price movement to the buyer. Producers also gain access to new technology and inputs which might otherwise be beyond their reach. For the processor or marketer, contracts make smaller demands on scarce capital resources and impose less of a burden on labour relations, ownership of land and production activities. Food processors can minimise their overhead costs of production by operating their plants at or near full capacity as contracting gives an assured and stable supply of raw materials. Contracts also help improve product quality by introducing incentives and penalties (Singh, 2008).

However, it is not the contract *per se*, but the relationship that is crucial in determining the success of contract farming. Many contract farming initiatives fail due to either poor design or poor implementation. Adverse selection and moral hazard problems pose challenges that need to be managed in order to ensure the farmer delivers under the contract terms and conditions. Major conditions for the successful interaction between producers and the buyer include a guaranteed market for the farmers produce; an effective repayment mechanism; the provision of market information for producers to effectively bargain; large volumes of transactions through collaborative marketing groups to lower transaction costs; cooperation; and no alternative source of raw material for the buyer. Building trust with the producers demands mutual respect, a fair and transparent negotiation process, a realistic assessment of benefits, a long-term commitment, equitable sharing of risk, and sound business plans. Innovative pricing mechanisms like bonus payments at the end of the season, shares in company equity, dividends, producer fixed pricing and quality based pricing which rewards performance, also help contract performance (Singh 2008).

Private standards

The development of quality standards greatly improves supply chain efficiency by providing an assurance of quality (King and Venturini 2005) which potentially lowers transaction costs (Reardon *et al.* 2005). Quality management assures the customer that both the product and the system under which it has been produced will meet their expectations. Quality assurance increases the attractiveness of smallholder producers as business partners through the activities that they undertake to: (i) sort the product into more homogeneous grades; (ii) implement quality control systems at critical points in the production system to assure food safety; and (iii) implement quality assurance systems that protect the environment and the rights of workers (Humphrey, 2005; Humphrey and Memedovic, 2006; Reardon *et al.,* 2005).

Standards in agribusiness supply chains operate at multiple points. They are created, adopted, applied and verified by different actors at different points along the chain (Humphrey and Memedovic 2006). However, an important distinction must be made between those mandatory standards that are developed and/or adopted by governments and those collective or private standards that are developed by individual firms, groups of firms or business associations. Of greatest concern, in a more liberal trading environment, are the increasing complexity of mandatory standards and the increasing importance of private standards (Humphrey 2005; King and Venturini 2005). Buyers use private standards to standardize product requirements across multiple suppliers, who often come from different countries. Private standards are designed to ensure that public standards are met in all the markets in which the buyer operates. Private standards may cover detailed quality specifications as well as food safety requirements and environmental considerations which normally go well beyond public food safety regulations (Reardon *et al.* 2005).

As the certification process involves a combination of baseline and periodic follow-up inspections, strict record-keeping procedures and audits, quality assurance can be very expensive to implement. Humphrey and Memedovic (2006) note two other problems for smallholder producers: (i) the need for technical training and support to achieve certification; and (ii) the high costs of adjusting production systems to meet the new requirements. These costs include making changes to production systems and purchasing new capital equipment (Biénabe *et al.* 2007).

LINKING SMALLHOLDER PRODUCERS TO INSTITUTIONAL MARKETS

There are numerous impediments in linking small independent producers to market. Smallholder producers often live in remote areas, far from good roads and markets. The physical environment can be difficult: rainfall may be erratic, the soils poor and crops attacked by pests and diseases (KIT *et al.* 2006). In the production and marketing of fresh fruit and vegetables, Irianto and Harwanto (2009) identify the key problems as the lack of infrastructure, technical information, the lack of finance and social problems. Kelly *et al.* (2003) add that the lack of a basic education, poor market information systems, grades and standards also present barriers to the participation of smallholder producers in modern institutional markets. Murray-Prior (2007) identify how poverty, high levels of illiteracy, ill health, low social and political status and power make it difficult for smallholder producers to embrace the changes required for them to compete in markets that require high levels of management skills, sophisticated production technologies and economies of scale. Limited access to physical inputs, high transaction costs, problems with land tenure and law and order present additional impediments.

In particular, smallholder producers often experience considerable difficulty in meeting the standards demanded for food safety and quality due to a lack of; (i) knowledge regarding quality and food safety requirements, post-harvest handling and packaging; (ii) training to cultivate and handle the produce correctly; and (3) capital to invest in technology such as production systems, cooling, transport and communication equipment (Riisgaard *et al.* 2009). Where the product is highly perishable and requires specialised production, handling and transport, both the capital investment and the investment in research, development and marketing is likely to preclude smallholder producers (Kirsten and Sartorius 2002). The uncertainty in quality and reliability of supply can put price pressure on smallholder producers who are unable to store the product (Hobbs and Young 2001).

Smallholder producers generally lack business skills, market knowledge and political empowerment, which might otherwise enable them to overcome many of the institutional impediments (Shepherd 2007). Poor access to external resources such as information about prices and other market-related information is a particular issue for smallholder producers (KIT *et al.* 2006; Hitchcock 2006). Smallholder producers typically have few assets to withstand the effects of price risks, opportunism (Biénabe *et al.* 2007), price cuts and the cancellation of orders (Riisgaard 2008). Their asset base is generally insufficient to meet the volume, consistency and year-round availability of downstream customers and to accommodate the long delays often associated with payment (Biénabe *et al.* 2007).

While smallholder producers can occupy many positions in the supply chain as producers, partners, integrators and co-owners, it is difficult for smallholder producers to move from one position to another in the supply chain (KIT *et al.* 2006). Smallholder producers are typically linked with other supply chain actors, notably the immediate buyers of their products. Without first improving their capacity to add value to their market offer, the performance of the supply chain as a whole will decline. Almost invariably, in order to establish enduring long-term vertical linkages with other supply chain actors, smallholder producers must first form some enduring horizontal linkages.

Collaborative marketing groups

Murray-Prior (2007) define a collaborative marketing group as a group of farmers who have organised to collectively market their produce. This broad definition includes structures such as cooperatives, grower associations and partnerships.

Collaborative marketing groups play a key role in organising economic activity for smallholder producers. They not only facilitate relationships with other supply chain actors (KIT *et al.*, 2006), but by working together, it becomes possible for smallholder producers to standardise and aggregate production, ensure compliance with the prerequisite quality standards and deliver products on time and at a competitive price (Biénabe *et al.*, 2007).

However, collaborative marketing groups may also form in an attempt to increase bargaining power in a situation of market failure arising from *ex ante* market power, *ex post* market power and asymmetric information (Milford, 2004; Nilsson, 2001; Cook and Plunkett, 2006).

Oktaviani (2004) identify how collaborative marketing groups are often formed to take advantage of government programs, either because the government has provided legal and taxation advantages for the establishment of collaborative marketing groups, or because the government has promoted, supported and often established collaborative marketing groups as part of their development programs.

Historically, most collaborative marketing groups struggle to sell their products, as they lack coordination mechanisms and operate with little information about market conditions and consumer preferences. At times, these groups are faced with an oversupply, which often depresses market prices for their products (Bolling and Gehlhar, 2005). Furthermore, their chances of success are often decreased by a lack of external prerequisites including: a legal and institutional environment to support collaborative marketing; physical infrastructure such as roads and telecommunication facilities; access to inputs such as seeds, fertilisers and chemicals; social services such as education, health and housing; access to credit; and support from extension services to provide technical and management advice (Murray-Prior, 2007).

Lele (1981) suggests that one of the main reasons for the failure of collaborative marketing groups is insufficient emphasis on the internal prerequisites of leadership, management, experience and technical know-how. A key factor is trust (Murray-Prior, 2007). Trust in a community relates to the social homogeneity of the community, with greater levels of homogeneity and trust having a positive effect on collective effort. Members of a collaborative marketing group need to trust their organisation and its management and leaders. This leads to lower transaction costs and provides the collaborative group with an advantage over other forms of organisation.

The enabling environment

Responsible firms need to develop stable, long-term relationships if they are to deliver quality products to their customers cost effectively. They therefore have a major interest in supporting the development of smallholder producer groups, helping them to organise, to educate and train them to produce the desired product, and to ensure that they approach the market in a businesslike way (KIT *et al.,* 2006).

To improve their position in the supply chain, smallholder producers need to improve their management of information. Improved record keeping on the use of labour and farm inputs is necessary to give a proper understanding of the costs involved and to enhance the producer's capacity to negotiate price. Traceability means keeping records to guarantee the buyer on the source of the product and the inputs that were used. Market information involves knowing about prices and trends in the market so that the producers can bargain with potential buyers (KIT *et al.,* 2006). Producers need to develop a chain vision – an understanding of how supply chains work. They need to view the chain as a network of specialised firms that need to interact with each other to make money. They need to acknowledge the position of other chain actors and respect that their interests are also legitimate. They should understand the need for cooperation, even although the interests of buyers and sellers will often be opposed.

Producers and their organisations may improve their position in the chain in several different ways through: (i) process upgrading by using new technologies or management methods (Humphrey, 2005; KIT *et al.,* 2006); (ii) product upgrading (Humphrey, 2005) through planting new varieties, discontinuing the use of agrochemicals or using standardised packaging and labelling, product diversification and quality control (KIT *et al.,* 2006); (iii) functional chain upgrading by taking on new activities in the chain; and (iv) chain upgrading by establishing a new value chain (KIT *et al.,* 2006). Humphrey (2005) also notes the need for greater agility within the chain which requires improved systems and greater emphasis on non-contractible parts of the relationship, such as reliable delivery, trust, flexibility in supply and the ability to innovate.

However, before smallholder producers can address a market efficiency gap, they must first overcome a market access gap. There must be some basic infrastructure such as roads and communication facilities and the producer must have access to land, water, seeds, capital etc. (KIT *et al.,* 2006). Infrastructure includes transport and telecommunication systems, multipurpose technologies for packaging and product preservation, third-party providers of logistics and information services, and organisations that create and transfer knowledge. Government needs to provide the laws and regulations that govern commercial practices, food safety and product quality, trade, labour practices and intellectual property.

To become a crop specialist, the producer should not only have technical production skills – irrigation, integrated crop and pest management, land preparation, but also the ability to plan activities, maintain accounts, manage cash flow, lead and motivate workers (KIT *et al.,* 2006). Producers should be willing to learn and innovate constantly. In this way, producers can grow and evolve as the market changes.

All of these skills and assets imply that smallholder producers are organised. Most private companies prefer working with organised producers because it reduces their risks and transaction costs. Organisation is therefore a basic asset that needs to be developed – it is the backbone of the whole process of supply chain development (KIT *et al.,* 2006).

REFERENCES

Achrol, R.S., Reve, T. and Stern, L.W. 1983. The environment of marketing dyads: a framework for comparative analysis. *J. Marketing,* **47**(Fall): 55-67

Achrol, R.S. 1997. Changes in the theory of inter-organisational relations in marketing: toward a network paradigm. *J. Acad. of Marketing Sci.,* **25**(1): 56-71.

Anderson, J.C. and Narus, J.A. 1990. A model of distributor firm and manufacturing firm working relationships. *J. Marketing,* **54**(1): 42-58.

Anderson, J.C. and Narus, J.A. 1999. Business market management. Understanding, creating and delivering value. Prentice Hall.

Anderson, E. and Weitz, B.A. 1992. The use of pledges to build and sustain commitment in distribution channels. *J. Marketing Res.,* **29**: 18-34.

Arshinder, Arun Kanda and Deshmukh, S.G. 2008. Supply chain coordination: Perspectives, empirical studies and research directions. *Int. J. Production Econ.,* **115**: 316-335.

Batt, P.J. 2006a. Expanding the quality concept to satisfy consumer demand. In: Batt, P.J. (ed) Proceedings of the First International Symposium on Fresh Produce Supply Chain Management. *Acta Hortic.* **699**: 83-90.

Batt, P.J. 2006b. Principles of supply chain management and their adaptation to the Asian horticulture sector. In Batt, P.J. and Cadilhon, J-J. (ed) Proceedings of the international symposium on fresh produce supply chain management. RAP Publication 2007/21. FAO: 25-38.

Batt, P.J. 2003. Incorporating measures of satisfaction, trust, and power-dependence into an analysis of agribusiness supply chains. In: Johnson, G.I. and Hoffman, P. (ed) *Agriproduct supply-chain management in developing countries.* ACIAR Proceedings 119: 27-43.

Biénabe, E., Boselie, D., Collion, M-H, Fox, T., Rondot, P., van de Kop, P. and Vorley, W. 2007. The internationalization of food retailing: opportunities and threats for small-scale producers. In: Vorley, W., Fearne, A. and Ray, D. (ed). *Regoverning Markets*: A place for small-scale producers in modern agrifood chains. Gower Publishing Limited. Hampshire.

Bolling, C. and Gehlhar, M. 2005. Global food manufacturing reorients to meet new demands. In Regmi, A. and Gehlhar, M. (ed). *New directions in global food markets.* Economic Research Service. USDA: 62-73.

Brown, J.R., Lusch, R.F. and Nicholson, C.Y. 1995. Power and relationship commitment: their impact on marketing channel member performance. *J. Retailing,* **71**(4): 363-392.

Cadilhon, J-J., Moustier, P., Poole, N.D., Tam, P.T.G and Fearne, A.P. 2006. Traditional vs. Modern Food Systems? Insights from Vegetable Supply Chains to Ho Chi Minh City (Vietnam). *Development Policy Rev.,* **24**(1): 31-49.

Cannon, J.P. and Perreault, W.D. 1999. Buyer-seller relationships in business markets. *J. Marketing Res.,* **36**: 439-460.

Cardozo, R.N. and Cagley, J.W. 1971. Experimental study of industrial buyer behaviour, *J. Marketing Res.*, **8**: 329-334.

Centrec. 1999. Forces influencing the evolution of agricultural value chains. (available at www.centrec.com)

Childers, T.L. and Ruekert, R.W. 1982. The meaning and determinants of cooperation within an interorganisational marketing network. In: Bush, R.F and Hunt, S.D. (ed) *Marketing theory: philosophy of science perspectives*. AMA Chicago: 116-119.

Cook, M.L. and Plunkett, B. 2006. Collective entrepreneurship: An emerging phenomenon in producer-owned organizations. *J. Agric. Appl. Econ.,* **38**(2): 421-28.

Crosby, L.A., Evans, K.R. and Cowles, D. 1990. Relationship quality in services selling: an interpersonal influence perspective. *J. Marketing,* **54**(July): 68-81.

Cunningham, M.T. and White, J.G. 1973. The determinants of choice of supplier, *European J. Marketing*, **7** (Winter): 189-202.

Cunningham, M.T. and Turnbull. P.W. 1982. Inter-organisational personal contact patterns. In Hakansson, H. (ed), *International marketing and purchasing of industrial goods. An interaction approach,* IMP Project Group, Wiley: 304-315.

Dempsey, W.A. 1978. Vendor selection and the buying process, *Industrial Marketing Management*, **7**: 257-267.

Fearne, A., Hughes, D. and Duffy, R. 2001. Concepts of collaboration: supply chain management in a global food industry. In Sharples, L., Ball, S.D. and Eastham, J.F. (ed) *Food and drink supply chain management - Issues for the hospitality and retail sectors*: Butterworth-Heinemann.

Folley, R.R.W. 1973. *Intensive crop economics*, Heinemann. London.

Ford, D., Gadde, L-E., Hakansson, H., Lundgren, A., Snehota, I., Turnbull, P. and Wilson, D. 1998. *Managing business relationships*. Wiley. Chichester.

Frazier, G.L and Summers, J.O. 1984. Interfirm influence strategies and their application within distribution channels. *J. Marketing*, **48** (Summer): 43-55.

French, R.P. and Raven, B. 1959. The bases of social power. In: Cartwright, D. (ed), *Studies in social power*. University of Michigan Press, Ann Arbor, MI.

Ganesan, S. 1994. Determinants of long-term orientation in buyer-seller relationships. *J. Marketing*, **58**(April): 1-19.

Gaski, J.F. 1984. The Theory of power and conflict in channels of distribution, *J. Marketing*, **48** (Summer): 9-29.

Gehlhar, M. and Regmi, A. 2005. Factors shaping global food markets. In: Regmi, A. and Gehlbar, M. (eds). *New directions in global food markets*. Economic Research Service. USDA: 5-17.

Geyskens, I., Steenkamp, J.B. and Kumar, N. 1998. Generalisations about trust in marketing channel relationships using meta-analysis. *Int. J. Res. Marketing,* **15**(3): 223-248.

Granovetter, M. 1985. Economic action and social structure: the problem of embeddedness. *American J. Sociol.,* **91**(November): 481-501.

Hakansson, H., Johanson, J. and Wootz, B. 1977. Influence tactics in buyer-seller processes. Industrial *Marketing Management*, **5**: 319-332.

Hakansson, H. and Snehota, I. 1995. *Developing relationships in business networks*. Thompson. London.

Handfield, R.B. and Bechtel, C. 2004. Trust, power, dependence and economics: can SCM research borrow paradigms? *Int. J. Integrated Supply Management,* **1**(1): 3-32.

Heide, J.B. 1994. Interorganisational governance in marketing channels. *J. Marketing*, **58** (January): 71-85.

Herlambang, T., Batt, P.J. and McGregor, M.J. 2006. Developing an effective food chain management in a developing country: a case study on Manalagi mango fruit supply chain in Indonesia. *16th Annual IAMA Food & Agribusiness Forum, Symposium & Case Conference October* 12-15. Buenos Aires.

Hitchcock, D. 2006. From market to producers or linking farmers to markets: what comes first - the hen or the egg? *Southeast Asia Sub-Regional Seminar on Enhancing Capacities of NGOs and Farmer Groups to Link Producers to Market*, 9-12 May 2006. Bali - Indonesia.

Hobbs, J.E. and Young, L.M. 2001. Vertical Linkages in Agri-Food Supply Chains in Canada and the United States. Ontario, Canada: Agriculture and Agri-Food Canada.

Hughes, D. and Merton, I. 1996. Partnership in produce: the J. Sainsbury approach to managing the fresh produce supply chain. *Supply Chain Management,* **1**(2): 4-6.

Humphrey, J. 2005. Shaping Value Chains for Development: Global Value Chains in Agribusiness. GTZ Trade Program. Available from http://www2.gtz.de/ dokumente/ bib/05-0280.pdf.

Humphrey, J. and Memedovic, O. 2006. *Global value chains in the agrifood sector*. Vienna: United Nations Industrial Development Organization

Hutt, M.D. and Speh, T.W. 1995. Business marketing management. A strategic view of industrial and organisational markets. Fifth Ed. Dryden Press.

Irianto, B. and Harwanto. 2009. Linking small banana producers in Lumajang District to better markets. In: Oppenheim, P. (ed) Proceedings of the Sixteenth International Symposium on Horticultural Economics and Management. *Acta Hortic.* **831**: 115-122.

Johnson, J.L., Sakano, T. Cote, J.A. and Onzo, N. 1993. The exercise of interfirm power and its repercussions in US-Japanese channel relationships. *J. Marketing,* **57**(April): 1-10.

Kaplinsky, R. and M. Morris (not dated), *A handbook for value chain research*, IDS, University of Sussex.

Kelly, V., Adesina, A.A. and Gordon, A. 2003. Expanding access to agricultural inputs in Africa: a review of recent market development experience. *Food Policy,* **28**(4): 379-404.

King, R.P. and Venturini, L. 2005. Demand for quality drives changes in food supply chains. In: Regmi, A. and Gehlhar, M. (ed). *New directions in global food markets*. Economic Research Service/USDA: 18-31.

Kirsten, J. and Sartorius, K. 2002. Linking agribusiness and small-scale producers in developing countries: is there a new role for contract farming? *Development Southern Africa*, **19** (4): 503-529.

KIT, Faida and IIRR. 2006. *Chain empowerment: Supporting african producers to develop markets*: Royal Tropical Institute, Faida Market Link and International Institute of Rural Reconstruction.

Lazzarini, S.G., Chaddad, F.R. and Cook, M.L. 2001. Integrating supply chain and network analyses: the study of netchains. *J. Chain and Network Sci.* **1**(1): 7-22.

Lehmann, D.R. and O'Shaughnessy, J. 1974. Difference in attribute importance for different industrial products. *J. Marketing*, **38**(April): 36-42.

Lele. U. 1981. Co-operatives and the poor: a comparative perspective. *World Development,* **9**: 55-72.

Macneil, I.R. 1980. The new social contract: An inquiry into modern contractual relations. Yale University Press, New Haven, CT.

Martech Consulting. 2005. Trends that impact New Zealand's horticultural food exports. Growing futures case study series. (available at www.martech.co.nz)

Milford, A. 2004. Coffee, co-operatives and competition: The impact of Fair Trade. Bergen, Norway.

Mohr, J. and Nevin, J.R. 1990. Communication strategies in marketing channels: a theoretical perspective. *J. Marketing*, **54**(October): 36-51.

Moorman, C., Deshpande, R. and Zaltman, G. 1993. Factors affecting trust in market research relationships. *J. Marketing*, **57**(January): 81-101.

Morgan, R.M. and S.D. Hunt (1994), The commitment-trust theory of relationship marketing, *J. Marketing,* **58** (July): 20-38.

Murray-Prior, R. 2007, The role of grower collaborative marketing groups in developing countries. *Stewart Postharvest Revi.* 6: 9.

Nilsson J. 2001. Organisational principles for co-operative firms. *Scandinavian J. Management*, **17**(3): 329-356.

Noordewier, T.G., John, G. and Nevin, J.R. 1990. Performance outcomes of purchasing arrangements in industrial buyer-vendor relationships. *J. Marketing*, **54** (October): 80-93.

OECD–FAO. 2005. OECD–FAO Agricultural outlook 2005 – 2014: Highlights. Paris and Rome.

Ogbonna, E. and Wilkinson, B. 1998. Power relations in the UK grocery supply chain. Developments in the 1990's. *J. Retailing and Consumer Services*, **5**(2): 77-86.

Oktaviani, R. 2004. Economic rationale, challenges for and future development of cooperatives in Indonesia. In: Trewin, R, (ed) *Cooperatives: Issues and trends in developing countries*. ACIAR Technical Report No. 53. Canberra: 78-84.

Oliver, C. 1990. Determinants of inter-organisational relationships: integration and future directions. *Acad. Management Rev.* **15** (2): 241-265.

Power, D. 2005. Supply chain management integration and implementation: a literature review. *Supply Chain Management: An Int. J.* **10**(4): 252-263.

Puto, C.P., Patton III, W.E. and King, R.H. 1985. Risk handling strategies in industrial vendor selection decisions. *J. Marketing*, **49** (Winter): 89-98.

Reardon, T. and Berdegué, J.A. 2002. The rapid rise of supermarkets in Latin America: challenges and opportunities for development. *Development Policy Rev*. **20**(4): 371-388.

Reardon, T., Timmer, C.P. and Berdegué, J.A. 2005. Supermarket expansion in Latin America and Asia implications for food marketing systems. In: Regmi, A. and Gehlhar, M. (ed) *New directions in global food markets*. Economic Research Service/USDA.

Riisgaard, L. 2009. Global value chains, labour organization and private social standards: lessons from East African cut flower industries. *World Development*, **37**(2): 326-340.

Riisgaard, L., Bolwig, S., Matose, F., Ponte, S., du Toit, A. and Halberg, N. 2008. A Strategic Framework and Toolbox for Action Research with Small Producers in Value Chains. *DIIS Working Paper 2008/17*. Copenhagen.

Sako, M. 1992. Prices, quality and trust: Interfirm relations in Britain and Japan, Cambridge University Press.

Shepherd, A.W. 2007. Approaches to Linking Producers to Markets. A Review of Experiences to Date. *Occasional Paper*. FAO. Rome

Simpson, P.M., Siguaw, J.A. and White, S.C. 2002. Measuring the performance of suppliers: an analysis of evaluation processes, *J. Supply Chain Management* (Winter): 29-41.

Singh, S. 2008. Leveraging contract farming for improving supply chain efficiency in India: some innovative and successful models. In Batt, P.J. (ed) Proceedings of the Second International Symposium. Improving the performance of supply chains in the transitional economies. *Acta Hortic.*, **794**: 317-324.

Spekman, R.E., Kamauff, J.W. and Myhr, N. 1988. An empirical investigation into supply chain management: a perspective on partnerships. *Supply Chain Management*, **3**(2): 53-67.

Swan, J.E., Trawick, I.F. and Silva, D.W. 1985. How industrial salespeople gain customer trust, *Industrial Marketing Management*, **14** (August): 203-211.

Thompson, J.D. 1967. Organizations in action: Social science bases of administrative theory. McGraw-Hill. New York,

Tummala, V.M.R., Phillips, C.L.M. and Johnson, M. 2006. Assessing supply chain management success factors: a case study. *Supply Chain Management: An Int. J.* **11**(2):179-192.

Van der Vorst, J.G.A.J., da Silva, C.A. and Trienekens, J.H. 2007. *Agro-industrial supply chain management: Concepts and applications*. FAO.

Vorley, W., Fearne, A. and Ray, D. 2007. Restructuring of agri-food systems and prospects for small producers. Gower Publishing Limited.

Wheatley, C. and Peters, D. 2003. Who benefits from enhanced management of agri-food supply chains? In Johnson, G.I. and Hoffman, P. (ed) *Agriproduct supply-chain management in developing countries*. ACIAR Proceedings 119: 113-123.

White, H.M.F. 2000. Buyer-supplier relationships in the UK fresh produce industry. *British Food J.*, **102**(1): 6-17.

Wilson, E.J. 1994. The relative importance of supplier selection criteria: a review and update, *Int. J. Purchasing and Materials Management* (Summer) : 35-41.

Wilson, N. 1996. Supply chain management: a case study of a dedicated supply chain for bananas in the UK grocery market. *Supply Chain Management*, **1**(2): 28-35.

32. SUCCESS STORY: TECHNOLOGY ADOPTION AND THE IMPACT ANALYSES OF PUMMELO AGRIBUSINESS DEVELOPMENT IN THE PRODUCTION CENTRE OF EAST JAVA, INDONESIA

Yuniarti, Pudji Santoso Suhardi and R. Endah

Senior Researchers, Assessment Institute for Agricultural Technology (AIAT) East Java, Malang, Indonesia.

ABSTRACT

This success story is a result of literature study on some assessment reports of pomelo done by Assessment Institute for Agricultural Technology (AIAT), East Java, other report on pomelo farming and field observation in the centre production of pomelo in Magetan regency and East Java. Pomelo assessments done by AIAT East Java were (1) Assessment and Development of Farming System of pomelo in Magetan regency in 1997/98-2000 to find a pattern of pomelo agri business development in this area, (2) Analyses on Technology Diffusion and Impact of the Assessment of farming system of pomelo towards the farming performance in Magetan regency in 2002 to know the diffusion level and impact of the assessment of farming system of pomelo and (3) the Impact Analysis of an Assessment result of village agro-industry (cassava flour, sugar-coated pomelo skin and corn chips) in East Java in 2006 to know the impact of the assessment result of three village agro-industries.

The success of the first year harvesting could motivate the farmers to extend their farming and cultivate their farms well as the recommendation. Their was increasing of pomelo plants as much as 100,000 trees on the area of 150-240 ha and increasing of farmers group as 19 groups came from 13 new villages. The success of pomelo agri business development is the result of hardworking synergies between the involved institutions, based on their mission. Impact value of the improved technology adoption in 2001 reached more than Rp. 1.6 billion. This pomelo agribusiness development pattern can be used as an example and guideline for the other commodities agribusiness development.

INTRODUCTION

Pummelo (*Citrus grandis*, L.) is a location-specific commodity in East Java with the production centre in Magetan regency. This regency is the national centre of pomelo, supplying the need of this commodity in Indonesia, including Jakarta, Bandung, Surabaya, Semarang and Bali (Astuti, 2003). Pomelo farming is the expected support of the farmers in the production centre to increase their family income. Around 85% of their total income come from pomelo farming.

The pomelo plants in the production centre were people plantation with individual management. The exist plantation in 1997 was in the form of small areas of production in spreading location so that it was difficult both in managing the farm and marketing the produce. Maintaining was varied and less optimum, it was because the technology which was applied without any guidance. The total trees were 141,472 with lots of varieties. Besides, the management of post harvest handling as well as the processing were not optimum, the large marketing opportunity was not efficiently used and also there were difficulties in seeking tho working capital.

The pomelo production centre in Magetan regency could be found in district Sukomoro and Bendo where 85% of the farmers' income came from this farming. The width of the farm owned by the farmers was between 0.35 – 0.40 ha. Pomelo had been grown by the farmers for 12 years.

District Kawedanan and Takeran were the development areas of pomelo. The owned land in both places were smaller, they were 0.32 ha consisted of 0.25 ha of irrigated land and 0.07 ha of dry land. Pomelo in irrigated land was intercropped with food crops using the pattern of rice-rice/corn-peanuts, while in the dry land, the pattern was corn-peanuts.

In 1997, most of the plants were attacked by gummosis diseases (*Diplodia natalensis*) with the incidence of the disease (39–93%). At the beginning of harvesting time, damage caused by fruit fly (*Dacus* sp.) and fruit borer (*Citripestis sagitiferella* Morr.) resulted in the decreasing of marketable fruits as much as 70%.

THE DEVELOPMENT AND ADOPTION OF THE TECHNOLOGY PACKAGE

1. Application of the Technology Package

Based on the problems faced on pomelo farming, during three years (1997/1998–2000) AIAT East Java together with Local Government of Magetan regency doing an assessment to find a pattern of pomelo agribusiness development in this area. Operational steps taken were (1) Problems identification using PRA (Participatory Rural Appraisal) method, (2) Colonization of production pockets to become economic scale of carpet plantation, (3) Consolidation of farmer's farm management to become carpet plantation own by the farmer's group, (4) Planting only 2-3 superior varieties, (5) Application technology package of integrated farming management of health pomelo, (6)

Guarding application of farming management technology through intensively extension and guidance, and (7) Optimization of the activities on sub system of in-put production preparation, production process, post harvest handling, processing and marketing, so that the use of all the resources become more efficient. Total village involved in this activities were 11 villages with 19 farmer's group cooperator, where the numbers of members were between 25 to 30 farmers per group, using total area of 300 ha. The recommended technology package included the result of some researches on Farming System of Citrus (excluded pomelo) and based on the local farmers experience.

The components of the recommended technology were: (1) Controlling gummosis disease by trimming, maintaining and painting of the trunk using generic fungicide of Californian porridge with sulfur as the active material, (2) Controlling fruit fly and fruit borer by cutting the life cycle of those pests, by throwing and burring the infected fruits into the soil, followed by fruits wrapping before they were two months old using polypropylene plastic.

In fact, improving the technology component as one of the agribusiness node influenced other nodes. The success of the first year harvesting could motivate the farmers to extend their farming, so that in 2000 the total number of plants doubled compared to those in 1997 (Table 1).

Table 1. The numbers of pomelo plants in the planting season of 97/98-99/00 in Magetan Regency

Source of the seed	Number of plants (trees)	The increasing of new plants (trees)			
		97/98	98/99	99/00	Total
Extension office	-	20,000	20,020	28,500	68,520
Farmer-self	-	19,000	40,000	13,500	72,500
Total	141,472	39,000	60,020	42,000	141,020

Source: Santoso, *et. al.*, 2002.

Understanding of the technology and farmers ability in individually applying the recommended technology was not effective without coordination, so that the guidance of the technology application that previously based on the farmer individually, then in 1998 changed into based on the farmer's group.

Ariyanto, 2003 reported that total number of pomelo plants in 2001 reached 448,800 trees in 1,222 ha area of the production centre. Pomelo production in 2002 reached 25,032 ton, increased around 12% compared to those in 2001 as much as 22,335 ton (Astuti, 2003).

2. The Technology Application Guidance

The technology application guidance was done as the assistance by researchers, extension staffs, key farmers and related institutions to the farmers and agribusiness agents. The assistance included training of the field workers, program socialization, demonstration plot, built an agro-clinic, farmer's guidance and technology exhibition.

The detailed of those activities were;

1. Training was done to increase the understanding of the field workers. The knowledge on pomelo farming of the field workers from the local extension office was limited, so it was important to put the technical staff from the assessment institute in the field to accelerate adoption of the recommended technology by the farmers.
2. Program socialization was done in order to increase the self-belonging of the involved institution staffs to participate fully in the assessment activities and improving the pomelo farming. The socialization was done cooperatively with the walking studio of "Special Radio of Agriculture Information" and radio of "Local Government Broadcast" in Magetan regency. Besides, the activity was done through guiding and media printing such as leaflets and brochures.
3. Demo-plot of "Improving the Maintaining of Productive Pomelo Farm" was built in district Sukomoro and Bendo, each was 2-5 ha, while the Demo-plot of young pomelo farm was built in the new area, i. e. in district Kawedanan and Takeran, each was 2-3 ha. Those demo-plots were built with the hope that it would be an example of the application of the recommended technology, in order to accelerate the diffusion process to the other farms.
4. Pomelo Agro-clinic was a clinic for pomelo plants which was established in Sukomoro, functioned as the place for consulting the farmers and field workers, extension staffs and researchers. Agro-clinic also opened for consulting at the same time with live-broadcast of "Special Radio of Agriculture Information" in a week in the assessment location.
5. "The Pomelo Technology Exhibition Week" had a purpose to disseminate the result of the assessment in the first year, increase the self-belonging of the program, and also inform the potential and prospect of the development of pomelo agribusiness in Magetan regency. This activity was ended by the discussion program.

During "The Pomelo Technology Exhibition Week" in July 1998, there was a training about the free-diseases pomelo seedling joined by 20 participants included the field workers and the seed growers who will produced 40,000 of pomelo seedlings. Besides that, there was a contest of the best maintaining farm to motivate the farmers to cultivate their farms well as the recommendation. The winner was announced in The discussion program.

Before the discussion program, there were other supporting program such as (1) Pomelo fruit contest, (2) Contest on creative serving of fresh pomelo fruit and their products, (3) Contest on skill and knowledge of pomelo agribusiness, and (4) Contest on `Princess of Pomelo 1998'. At the exhibition session, the women farmers interested to be trained in the technology of jam and sugar-coated pomelo skin making.

"The Pomelo Technology Exhibition Week" was evaluated to be successful in combining the process of technology transfer with the feeling of self-belonging to the assessment program, knowing the potential and the prospect of pomelo agribusiness development, and also increasing their pride toward their produce in the effort to create Magetan regency as the pomelo centre in East Java. This program was hoped to be the annual agenda of the local government of Magetan regency to maintain and develop the motivation of the farmers.

THE IMPACT OF THE TECHNOLOGY PACKAGE ADOPTION

1. Increasing the Farmers' Response

The farmers' response to the new technology was high enough, especially in district Sukomoro and Bendo, it was higher than that in district Kawedanan and Takeran (Table 2). In 1998/1999 increasing of numbers of guided villages and farmers more than 100% comparing in 1997. In 2001, the numbers of adopter farmers were 417 persons with the impact area of 172 ha. The most interesting thing from this situation was the farmer groups asked the guidance by themselves. Most of the plants were new plants which were bought by the farmers, it was done after they saw the success of the harvesting in 1998 and know the potential and prospect of pomelo agribusiness as the source of economy growth in the area.

Table 2. Increasing the number of village and farmer groups who got the guidance in pummelo agribusiness in 1997-1999 in Magetan Regency

Districts	Number of villages			Number of farmer groups		
	97/98	98/99	Addition	97/98	98/99	Addition
Sukomoro	3	9	6	7	19	12
Bendo	3	8	5	7	12	5
Kawedanan	2	4	1	2	4	1
Takeran	3	4	1	3	4	1
Total	11	25	13	19	39	19

Source: Santoso, *et. al.*, 2002.

Moreover, there was an increasing in planting pomelo by the farmers who did not grow pomelo before. During 1997-1999 the numbers of grown pomelo noted as many as 100,000 trees or equaled to 150-240 ha. Those plants were 60% funded by the farmers with the coordination of Local Agriculture Extension Office in the efforts to dense the production area to create the carpet plantation as commercially scale business.

2. Increasing the Health of Pummelo Farm

At the beginning of the assessment, about 89-93% pomelo plants were infected by gummosis diseases (*Diplodia natalensis*) with the scores of 24-35%. As the price of the pesticide was very expensive, the alternative use of the Californian porridge was a package that was recommended by the researchers; since it could be made by the farmers themselves using the raw material of sulfur and lime stone (Supriyanto, *et al.*, 1999). The use of the Californian porridge could decrease the incidence and severe disease up to 55% and 18% for each. In 2002, total farmers used this Californian porridge reached 90% (Pudji *et. al.*, 2002).

The attack of the fruit borer (*Citripestis sagitiferella* Morr.) achieved 34.1%, could be decreased little by little to 0.9% in 2000. The fruit borer was controlled mechanically by picking the infectious fruits and then buried into the soil, the other alternative was the use of the insecticide rationally. The similar pattern was done to the attack of fruit flies. The controlling of the fruit flies was done by the trap of methyl eugenol which was applied at once and continuously. The success of the pest and disease controlling could increase the productivity and raise the farmers' motivation to develop their farming.

3. Increasing the Productivity and Income of the Farmers

The impact assessment of the pomelo farming system development in Magetan regency was done in 2002. The cooperator farmers' productivity on pomelo could achieve 19,000 kg /year/ha, while the non-cooperator farmers only achieved 14,250 kg/year/ha in average; in other words, it increased 4,750 kg/year/ha (46%). The pomelo farming income of the cooperated farmers was higher than that of the non-cooperated; it was each Rp 34,200,000.- and Rp 18,988,500.-/year/ha, in other words it increased Rp 15,211,500.-/year/ha (80%). The impact of the physical production of the assessment was 817 ton valued as much as Rp 1,470,600,000.-. The assessment of the pomelo agribusiness in three years spent the cost of Rp 156,497,000.- so the net value of the impact of the assessment in 2001 was Rp 1,314,103,000.- (Table 3). The three varieties of the pomelo was officially released with the name of Sri Nyonya, Nambangan and Magetan.

Table 3. The impact of the assessment result of the pummelo agribusiness to the productivity, farming income, number of the adopter farmers and width of the impact area in 2001 in Magetan Regency, East Java

Description	Value
Productivity (kg/year/ha)	
Cooperator farmers	19000
Non-cooperator farmers	14250
The difference between a and b	4750
Income of the Pamelo Farming (Rp./year/ha)	
Cooperator farmers	34,200,000
Non-cooperator farmers	18,988,500
The difference between a and b	15,211,500
Number of the adopter farmers (person)	417
The width of the impact area (ha)	172
The production impact (ton)	817

The value of the impact (Rp.)	1,470,600,000
The cost of the assessment (Rp.)	156,497,000
The net impact value (Rp.)	1,314,103,000
The added value of the sugar-coated pomelo skin (Rp.)	320,000,000

Source: Santoso *et al.*, 2002.

The activity of the pomelo agribusiness assessment also affected the existence of the farmer groups and the establishment of the Magetan Pomelo Association. The number of the farmer groups in 1997/1998 was 11 farmer groups and increased into 39 in 2001. In 2000 the association covered the providing of the production facilities especially the seedling and the place of the members' harvesting fruits which will be marketed together.

In 2000, the socialization and training in making sugar-coated pomelo skin were done for the woman farmers. In 2001 the production of the sugar-coated pomelo skin achieved 500 kg/year with the production value of Rp 6,000,000.- and in 2002 the production achieved 2000 kg/year valued Rp 320,000,000.-, where the price was Rp 160,000.-/100 kg. In 2008, 70% of the product was sold in Magetan and the rest (30%) was marketed to other city such as Madiun and Surabaya. In Surabaya the producer also dropped this product to super markets (Person. Comm., 2008).

4. The Empowering of the Agribusiness Nodes

The success of the rationality of the variety number was greatly depended on the seed growers in producing the seedlings with the determined variety. The supply of the seedling in the planting season of 2000/2001 was more enough compared to the previous year. That was because of the increasing of the seed growers' ability and the availability of the scion of the three varieties of the released pomelo. The high market potential of the seedling resulted increasing the motivation of the seed grower to produce qualified seedling. The easiness of getting the other production in-put was also important in order to fasten the process of recommended technology adoption, for example the easiness and the availability of sulfur, methyl eugenol and polypropylene plastic.

The skin of the infectious pummelo fruit and fruit resulted by thinning treatment were also can be used as the sugar-coated pomelo skin, and it had a high selling value. It needed the support of the processing technology. The Magetan Pummelo Association with the support of the local government had built the home industry of the sugar-coated pomelo skin in district Sukomoro. The machine was designed by the AIAT East Java researchers. In 2006, marketing area of this product were Magetan, Madiun, Ngawi and Surabaya (Santoso *et al.,* 2006).

The marketing of the pummelo fruit had three channels: (1) farmers - collecting traders - consumers around Magetan, (2) farmers - collecting traders–intercity traders - retailer outside Magetan - consumers outside Magetan, and (3) farmers - collecting traders - intercity traders – wholesalers - retailers outside Magetan - consumers outside Magetan. Interested packing, both from bamboo or carton - hard paper could increase the added value. So that it was hoped that the high quality of pomelo fruit could go to the new market segment. The Magetan Pomelo Association was hoped to use this opportunity, such as open the pomelo kiosks in the train station in Surabaya, Juanda Airport and supermarkets.

The problems faced by the farmers in applying the recommended technology were the production facilities, consolidation of application in the field and the fund limitation. The Magetan Pomelo Association which was established in 24 December 1998 was hoped to answer the problems and also to improve the farmers' role in the marketing system. At the beginning of 1999, 8 villages in district Sukomoro and Bendo covered 17 farmers groups had received farm's credit as much as Rp. 5,220,000.-/ha which consisted of 55% for production facilities and the rest was 45% for the farm maintenance (Table 4).

Table 4. The number of the farmer groups who recei ed farm's credit 1999 in district Sukomoro and Bendo, Magetan Regency

District	Number of villages	Number of farmer groups	Number of members	Width (ha)	Total of farm's credit (Rp. 000.-)
Sukomoro	7	14	427	310,5	1,620,810
Bendo	1	3	120	114,5	597,690
Total	8	17	547	425,0	2,218,500

Source: Santoso, *et al.*, 2002.

Impact value of the improved technology adoption in 2001 reached more than Rp. 1.6 billion. This pomelo agribusiness development pattern can be used as a guidance for the other commodities agribusiness development.

CONCLUSION

The successful of pomelo agribusiness development in Magetan regency was the result of synergic hard working between the involved institution based on their mission. The assessment of AIAT East Java lasted for three years, included the improving of the production technology, post harvest handling and the institution (planting only released variety, colonization of pomelo planting area, consolidation of farm management and institution development) with some methods of technology transfer. The impact value of the improved technology adoption in 2001 reached more than Rp. 1.6 billion. This pomelo agribusiness development pattern could be used as a guidance for the other commodities agribusiness development.

To support the achieved success in the field, the several things should be followed up such as:

1. Always refer to the determined steps; they are seedling optimization, number of variety rationalization, colonization of production area, consolidation of farm management, optimizing harvest and

post harvest handling, marketing efficiency and also improving the institution and the nodes of the pomelo agribusiness.

2. The Magetan Pomelo Association should be able to be the controller of the nodes of the agribusiness which started to develop to make them more efficient in using the natural resources to create the centre of pomelo agribusiness.
3. Increasing the promotion in order to open the new market segment especially for the high quality fruits.
4. Establishing the packing house of pomelo fruits was necessary.

REFERENCES

Astuti, Palupi P. 2003. Otonomi. Kabupaten Magetan. http://www2.kompas.com/kompas-cetak/0311/04/otonomi/664358.htm

Ariyanto, Gesit., 2003. Otonomi. Menjadikan Sektor Peternakan dan Pertanian Tumpuan Pendapatan.

http://www2.kompas.com/kompas-cetak/0311/04/otonomi/669234.htm

Santoso, P., Agus Suryadi, Nugraha Pangarsa and Yuniarti, 2002. Analisis difusi teknologi dan dampak SUP Pamelo (*Citrus grandis* L.) terhadap kinerja usahatani di Kabupaten Magetan. Jurnal P2TP vol. 5 no. 1, January 2002. p 65 – 80.

Santoso, P., Ali Yusran, Purwanto, Yuniarti, Suhardi, Agus Suryadi dan Rika Asnita, 2006. Laporan Akhir: Analisis Dampak Hasil Pengkajian Agroindustri Pedesaan (Tepung Kasava, Manisan Kulit Jeruk Pamelo dan Marning Gepeng). di Jawa Timur. BPTP Jatim, Malang.

Supriyanto, A., E. Legowo, P. Santoso, M. Sugiarto, Djoema'ijah, Hardiyanto, Suhardi, M.E. Dwiastuti, A. Triwiranto, O. Endarto, Sutopo, D.P. Saraswati, B. Victor, Setiono dan S. Nurbanah, 1999. Pengkajian dan Pengembangan Sistem Usaha Pertanian Pamelo di Kabupaten Magetan. Prosiding Seminar Hasil Penelitian/Pengkajian BPTP Karangploso. P. 92 – 104.

V
Government Policy and Development Programmes

33. HORTICULTURAL DEVELOPMENT AND SUSTAINABLE GROWTH OF AGRICULTURE SECTOR IN INDIA: A CRITICAL ANALYSIS

Mahendra Singh and V.C. Mathur

Division of Agricultural Economics, Indian Agricultural Research Institute, New Delhi-110012

ABSTRACT

The study analyses the growth and variability in area, production and yield of various sub-sectors of horticulture in India during 1991-92 to 2005-06 and diversification status of horticultural sub-sectors during 2005-06.The changes in consumption pattern of various food items between 43rd and 63rd rounds of the NSSO among rural and urban groups was also analysed. Apart from this, the changes in exports of various horticultural products were also examined during 1996-97 to 2006-07 in India. The study suggests that the region-wise causes in variability in area and yield of sub-sectors of horticultural crops should be identified and state-wise diversification status should be used to frame suitable policy to strengthen backward and forward linkages in horticulture sector and ensure overall sustainable growth of the agriculture sector in the country.

INTRODUCTION

Horticulture sector is playing seminal role in national income, employment and food security in the country. Vegetables, fruits, plantations crops and spices contributed around 60, 31, 7 and 2 per cent, respectively, of the value of total horticultural production in 2006-07 (Singh and Mathur, 2008). The share of horticulture sector in value of agricultural exports during triennium 2003-05 was about 18 percent, which increased to 37 percent in 2006-07 (GoI, 2008). This sector has achieved 3.54 percent per annum growth in value of gross output compared to 1.39 percent in cereals during Xth Five Year Plan in the country (Table 1). The working group on horticulture, plantation crops and organic farming for the XI Five Year Plan (2007-12) redefined horticulture as "science of growing and management of fruits, vegetables including tubers, ornamental, medicinal and aromatic crops, spices, plantation crops their processing, value addition and marketing".

Table 1. Growth rate in output of various sub-sectors of Indian agriculture, 1951-52 to 2006-07.

(Gross value of output at 1999-2000 prices) (Percent)

Period	Cereals	Pulses and oilseeds	Fruits and vegetables	Other crops	All crops	Live stock	Fishery
1951-52 to 1967-68	4.19	2.98	2.67	2.42	3	1.02	4.68
1968-69 to 1980-81	3.43	0.97	4.82	2.98	3	3.26	3.08
1981-82 to 1990-91	3.52	5.41	2.84	1.71	2.97	4.78	5.74
1991-92 to 1996-97	2.36	2.92	6.07	2.18	3.09	4.00	7.05
1997-98 to 2006-07	1.39	1.43	3.54	3.70	2.36	3.61	2.93

Source: Eleventh Five Year Plan (2007-12) Document, Volume-III, Planning Commission, Government of India.

The proposed contribution of various agricultural sub-sectors to achieve the targeted growth of over four percent in agriculture during the present 11th Five Year Plan is presented in Table 2. It shows that horticulture and livestock sub-sector would play seminal role to achieve targeted growth of 4.1 percent at the end of 11th Five Year Plan. The proposed growth in horticulture and livestock sectors would two times of growth in crop sector during the present XIth Five Year Plan.

Table 2. Projected growth scenario in various agriculture sub-sectors for 11th Five Year Plan, India.

(percent)

Agriculture sub-sector	AGDP share	Proposed growth rate per annum
Crops	46	2.7
Foodgrains	26	2.3
Oilseeds	6	4.0
Other crops	14	3.0
Horticulture	21	5.0
Livestock	25	6.0
Fisheries	4	6.0
Forestry/logging	4	0.0
Total		4.1

Source: Report of the Working Group on crop Husbandry, Agricultural Inputs, Demand and Supply Projections and agricultural Statistics for the Eleventh Five Year Plan (2007-12), Planning Commission, New Delhi.

The elasticity of gross domestic product from agriculture (GDP Ag) with respect to various determinants during 1981-2004 is presented in Table 3. It shows that one unit expansion in area under fruits and vegetables contributes half percent of growth in GDP from agriculture. The elasticity of area under fruits and vegetables was highest among all determinants considered in analysis. The growth in area under fruits and vegetables was highest during 1981-91 but declined during 1992-2004.

Table 3. Elasticity of GDP agriculture with respect to selected variables and their growth rates in India, 1981-2004.

Factor	Growth rate in each factor			
	Elasticity	1981-2004	1981-1991	1992-2004
Technology	0.308	2.69	2.93	1.81
Public investments	0.174	-1.40	-4.07	1.76
Private investments	0.128	3.93	4.05	2.83
Area under fruits and vegetables	0.458	2.75	3.06	2.30
Fertilizer	0.122	5.02	8.18	3.20
GDP agriculture		3.04	3.29	2.55

Source: Report of the Steering Committee on Agriculture and Allied Sectors for Formulation of the Eleventh Five Year Plan (2007-12), Planning Commission, New Delhi.

Keeping in the mind the importance of horticulture sector in achieving the desired growth rate (4.1 percent) during XIth Five Year Plan, the present study was undertaken specifically to analyse the growth and variability in area, production and yield of various sub-sectors of horticulture sector and diversification status; examine the changes in consumption pattern of various food items between 43rd and 63rd rounds of NSSO (National Sample Survey Organization) in rural and urban groups; and the changes in exports of various horticultural products during last decade in country.

METHODOLOGY

The study is based on secondary data published by various government organizations. The study covers a period of 15 years from 1991-92 to 2005-06. Assuming that the impact of the WTO agreement on multilateral trade would be evident two years after adoption of the provisions by member states in January 1995, the total period was sub-divided into two sub-periods, a pre-WTO period from 1991-92 to 1996-97 and a post-WTO period from 1997-98 to 2005-06.

Estimation of the degree of diversification

The Herfindahl index was used to measure the degree of diversification The index was computed as follows:

$$HI = \sum_{i=1}^{n} P_i^2 \qquad i = 1, 2, \ldots n$$

Where, P_i is the proportion of area under ith crop in the total system; and n is the number of crops in the system. The value of the Herfindahl index is ranges between 0 to 1, where 0 indicates perfect diversification and 1 refers to perfect specialization; thus the index bears inverse relationship with diversification.

RESULTS AND DISCUSSION

The growth and variability of area, production and yield of major horticultural sub-sectors are presented in Table 4. Substantial growth has occurred in area of all the sub sectors during entire period (1991-92 to 2005-06) but maximum growth was observed case of spices followed by fruits, plantation crops and vegetables, respectively. However, variability also moved in the same direction, indicating the unstable nature of area expan-sion on account of price volatility. Area growth was highest for plantation crops and lowest for vegetable crops during pre-WTO period (1991-92 to 1996-97), while variability also followed the same trends. During post-WTO period (1997-98 to 2005-06) highest growth in area was observed for fruits crop followed by spices. The trend in variability also followed the same direction, indicating that area expansion encourages variability. A negative yield growth in fruits crops coupled with positive production growth during post-WTO period implies that total gain in production has come from area expansion.

Table 4. Growth and variability of major horticultural sub-sectors in India, 1991-92 to 2005-06.

(Percent)

Crop/sector	Area			Production			Yield		
	Entire period	Pre-WTO period	Post-WTO period	Entire period	Pre-WTO period	Post-WTO period	Entire period	Pre-WTO period	Post-WTO period
				Compound growth					
Fruits	3.83	3.65	4.95	3.48	7.27	2.89	-0.34	3.49	-1.96
Vegetables	2.27	0.36	2.49	4.04	4.71	3.40	1.73	4.34	0.89
Plantation	2.30	4.50	1.79	2.07	5.36	1.81	-0.23	0.83	0.01
Spices	3.90	1.73	4.87	6.28	6.14	7.70	2.29	4.33	2.69
Total	3.16	2.24	3.74	3.82	5.66	3.27	0.65	3.34	-0.45
				Variability					
Fruits	18.98	7.14	15.84	16.54	13.41	11.05	7.60	8.40	7.48
Vegetables	11.09	5.61	7.48	18.32	8.69	11.18	9.18	9.46	5.05
Plantation	10.39	8.39	5.39	13.35	10.23	12.40	8.42	5.37	10.23
Spices	27.76	7.10	27.42	30.18	12.43	23.44	16.96	8.30	17.87
Total	14.73	4.59	10.98	17.16	10.28	10.47	6.82	6.96	6.15

Source: Author's calculations with data from GoI (2007) Indian Horticulture Database, National Horticulture Board, Ministry of Agriculture, New Delhi.

The status of diversification in major horticultural crops growing states in the year 2005-06 is presented in Table 5. Tripura, West Bengal, Assam and Madhya Pradesh had lowest indices in the individual horticultural sub-groups namely fruits, vegetable, plantation crops and spices respectively. This implies that these states are ranked first with respect to diversification among the all states as the value of Herfindahl index and diversifications have an inverse relationship.

Table 5. Status of the horticultural diversification in Indian States, 2005-06.

State/UT	Fruits	Vegetables	Plantation	Spices
Andhra Pradesh	0.412	0.197	0.487	0.455
Arunachal Pradesh	0.286	0.673		0.511
Assam	0.223	0.576	0.518	0.222
Bihar	0.287	0.225		0.228
Chattishgarh	0.314	0.211		0.340
Goa	0.336	1.000	0.547	
Gujarat	0.218	0.162	0.685	0.472
Haryana	0.253	0.262		
Himachal Pradesh	0.312	0.230		0.281
Jammu & Kashmir	0.453	0.242		0.488
Jharkhand	0.228	0.000		
Karnataka	0.286	0.245	0.471	0.278
Kerala	0.313	0.684	0.685	0.565
Madhya Pradesh	0.264	0.147		0.385
Maharashtra	0.224	0.229	0.800	0.765
Manipur	0.363	0.215		0.587
Meghalaya	0.260	0.331	0.541	0.415
Mizoram	0.310	0.144	1.000	0.570
Nagaland	0.265	0.253	0.702	0.299
Orissa	0.334	0.169	0.582	0.331
Punjab	0.323	0.305		0.657
Rajasthan	0.353	0.241		0.302
Sikkim	0.528	0.319		0.654
Tamil Nadu	0.375	0.344	0.613	0.245
Tripura	0.183	0.669	0.500	0.352
Uttar Pradesh	0.801	0.342		0.416
Uttrakhand	0.263	0.203		0.251
West Bengal	0.222	0.217	0.412	0.520

Source: Author's calculations with data from GoI (2008). Agricultural & Processed Food Products Export Development Authority (India), Ministry of Food processing Industries, New Delhi.

The trends in value of consumption of various food items is presented in Table 6, it shows that the value of vegetables and fruits shared 12 and 3 percent in rural groups and 11 and 5 percent in urban groups respectively during 2006-07. The highest growth was recorded in consumption of vegetables and fruits among all food items during 1987-88 to 2006-07 in rural and urban India. However, variability in consumption was relatively higher in rural followed by urban groups in these items, which implied that, the consumption of fruits and vegetables still relatively income elastic than cereals in rural group.

Table 6. Trends in value of consumption of various food items in India, 1987-88 to 2006-07.

(Percent)

Item group	Share in total food expenditure 2006-07		Growth in expenditure between 1987-88 to 2006-07		Variability in expenditure during 1987-88 to 2006-07	
	Rural	Urban	Rural	Urban	Rural	Urban
Cereals and cereal substitutes	32	23	20	24	32	36
Pulses & their products	7	6	26	26	39	39
Milk & milk products	15	19	30	30	41	41
Edible oil	7	7	28	23	41	35
Egg fish and meat	7	7	36	29	48	41
Vegetables	12	11	36	31	47	42
Fruits	3	5	36	32	47	43
Sugar, salt & spices	8	7	25	24	37	36
Beverages refreshments & Processed food	9	14	36	32	47	43
Food total	100	100	27	28	39	39

Source: Household Consumer Expenditure in India, 2006-07, NSS 63rd Round: National Sample Survey Organization, Ministry of Statistics and Programme Implementation, Government of India, January, 2008.

Note: 43rd Round: July 1987 to June 1988; and 63rd Round: July 2006 to June 2007.

Table 7. Changes in India's exports of major horticultural commodities, 2001-02 to 2005-06.

Commodity / Period	Value (US$ million)		Share in agricultural exports (%) TE 2005-06	Share in total merchandise exports (%) TE 2005-06	Percent change in value in TE 2005-06 over TE 2001-02
	TE 2001-02	TE 2005-06			
Floriculture products	27	59	0.6	0.06	119
Spices	359	419	4.8	0.51	17
Cashew	464	500	5.7	0.6	8
Fruits / vegetables seeds	15	16	0.2	0.02	7

Fresh fruits	81	204	2.3	0.25	152
Fresh vegetables	100	202	2.3	0.25	102
Processed vegetables	45	84	0.9	0.1	87
Processed fruits and juices	105	98	1.1	0.12	-7
Horticultural Commodities	1196	1582	17.9	1.91	32
Agricultural & Allied products	5839	8736	100	10.62	50

Source: Author's calculations with data from Foreign Trade & Balance of Payments, Various issues, Centre for Monitoring Economy, Mumbai.

Table 8. Trend and variability in value of exports of horticultural products of India, 1996-97 to 2006-07.

(Percent)

Item	Growth	Variability
Floriculture	20.4	89
Fruits & Vegetables Seeds	6.7	33
Total for Floriculture & Seeds	16.2	71
Fresh onions	16.5	69
Other Fresh Vegetables	16.3	53
Dried Nuts(Walnuts)	5.4	25
Fresh Mangoes	10.1	36
Fresh Grapes	16.2	74
Other Fresh Fruits	14.7	49
Total for Fruits and Vegetables	14.5	55
Dried & Preserved Vegetables	0.3	41
Mango Pulp	15.5	51
Pickle & Chutney	15.1	58
Other Processed Fruits & Vegetables	17.2	62
Total for Processed Fruits & Vegetables	15.2	55

Source: Author's calculations with data from GoI (2008). Agricultural & processed Food Products Export Development Authority (India), Ministry of Food processing Industries, New Delhi.

Among the horticultural commodities, cashew ranked first followed by spices, fresh fruits and fresh vegetables, respectively, in export values during the period TE 2001-02 to TE 2005-06 (Table 7). In terms of temporal changes, the maximum increase in triennium ending (TE) 2005-06 over TE 2001-02 was observed in value of exports of fresh fruits followed by floriculture products.

Table 8, shows that the maximum growth in export was observed in floriculture with highest variability, implying there is ample scope for export of floriculture products the suitable solution to minimize the variability.

CONCLUSIONS

This study examined the growth and variability in area, production and yield of various sub-sectors of horticulture in India during 1991-92 to 2005-06 and diversification status of sub-sectors during 2005-06.The changes in consumption pattern of various food items between 43rd and 63rd rounds in rural and urban groups was also analysed. Apart from this the changes in exports of various horticultural products was also examined during 1996-97 to 2006-07 in India. The study suggests that the causes in variability in area and yield of sub-sectors of horticultural crops should be identified and state-wise diversification status should used to frame suitable policy to strengthen backward and forward linkages in horticulture sector and overall sustainable growth of the agriculture sector in the country.

REFERENCES

APEDA. 2008. *Agricultural & Processed Food Products Export Development Authority (India)*. Ministry of Food Processing Industries, Government of India Website: http:// apeda.com

CMIE, Foreign Trade & Balance of Payments (various issues). Centre for Monitoring Economy Mumbai

CSO, 2006, 2008. Statewise estimates of value of output for agriculture and livestock (1990-91 to 2002-03).

Central Statistical Organization, Ministry of Statistics and Programme Implementation, Government of India.

GoI, 2008. *North Eastern Region Vision 2020*. Ministry of Development of North Eastern Region.

NHB, 2007. *Indian Horticulture Database*. National Horticulture Board, Ministry of Agriculture, Government of India, New Delhi.

Planning Commission, 2007. Report of the Working Group on Horticulture, Plantation Crops and Organic Farming for the XI Five Year Plan (2007-12). Government of India, New Delhi.

Planning Commission, 2007. Report of the Working Group on crop Husbandry, Agricultural Inputs, Demand and Supply Projections and Agricultural Statistics for the Eleventh Five Year Plan (2007-12). Planning Commission, New Delhi.

Planning Commission, 2007. Report of the Steering Committee on Agriculture and Allied Sectors for Formulation of the Eleventh Five Year Plan (2007-12). Planning Commission, New Delhi.

Planning Commission, 2008. *Eleventh Five Year Plan (2007-12) Document, Volume-I &III*. Government ofn India, New Delhi.

RBI, 2007. *Handbook of statistics on the Indian economy*, Reserve bank of India, 2006-07.

Singh, M. and Mathur, V.C., 2008. Structural changes in horticulture sector in India: Retrospect and Prospect for XIth Five Year Plan. *Indian J. Agric. Econ.*, **63**(3):332-348.

34. INTERNATIONAL HORTICULTURAL DEVELOPMENT SCENARIO – IMPLICATIONS FOR INDIA

C. Prasad[1] and P.C. Bhatia[2]

1. Former Deputy Director General (Agril. Extn.), Indian Council of Agricultural Reserch, EB-106, Maya Enclave, New Delhi-110064, India
2. Former Asstt. Director General (Agro.),
Indian Council of Agricultural Research, A1/83 Janak Puri, New Delhi

ABSTRACT

The recent developments and achievements in horticulture in different parts of the world call for greater optimism for suitable support and encouragement for strengthening different dimensions of horticulture at national and international level. The horticultural crops, fresh or processed, have become foreign exchange earners everywhere. India with the tropical climate in south, temperate in North and sub-tropical in the hilly regions can grow virtually all types of horticultural crops and plants, which are commonly grown all over the world. India is looking forward for doubling the horticultural productivity in near future in order to help achieve the targetted agricultural growth rate of four percent vis-à-vis around 2.5 percent at present. In the past, horticulture could not get the importance it deserved mainly owning to the thrust on the cereal crops, which were urgently needed for meeting the national food security. Ultimately the efforts resulted into the Green Revolution in 1960s. India plans to increase food production to 300 million tons by 2012 including fruits and vegetables production of 58.74 and 109.05 million tons respectively from the current level (2006).

Horticulture has been given priority attention in the country since ninth Five Year Plan of the Government of India (1997-2002). The launching of Technology Mission for Integrated Development of Horticulture in the North East and Himalayan States by the Ministry of Agriculture in 2001-02 was a land mark for the promotion of horticulture in these states. The National Horticulture Mission launched in 2005-06 shows the aggressive leadership of the government to harness the potentiality of horticulture to its fullest for the food and nutritional security of the country. The prominent countries in horticultural development are China, India, Brazil and the USA; they occupy an area of 10456, 5510, 2386 and 1215 thousand hectares under fruits with the production of 89355, 58740, 36952 and 28144 thousand million tons respectively. Sustained development of horticulture everywhere is a sine qua non for nutritional security. This paper is devoted to transfer of technology (TOT) in horticulture in all its dimensions–research, manpower development, effective field programmes, training & management for accelerating productivity, production, processing, storage, management & marketing including exports.

INTRODUCTION

Historically, agricultural development in India or elsewhere was oriented towards the production of cereal, food and fodder crops. Horticultural crops remained the domain of private sector grown in small holdings and were traditionally poorly managed. This picture has now changed with growing economy and increasing health as well as nutritional consciousness and subsequent monetary returns stemming from increased consumption of fruits and vegetables.

In India, after attaining independence in 1947, major emphasis was laid on achieving self-sufficiency in food production. Development of high yielding wheat cultivars and high production technologies and their adoption in areas of assured irrigation paved the way towards food security ushering in the Green Revolution in the 1960s. However, it gradually became clearer that horticultural production, for which Indian topography and agro-climates are well suited, is an ideal method of achieving sustainability of small holdings, increasing employment, improving the environment, providing an enormous export opportunity and above all achieving nutritional security. As a result, due emphasis on diversification to horticultural crops was given only during the most recent decade. It is concluded that India has a good natural resource base and an adequate R&D infrastructure. The demand for horticultural produce is rising due to increasing population, enhanced economy, changing food habits, awareness of high nutritional need and greater emphasis on adding values and exporting.

Fruits, nuts, and vegetables play a significant role in human nutrition especially as sources of vitamins, minerals, and dietary fiber. Their contribution as a group is estimated at 91% of vitamin C, 48% of vitamin A, 27% of vitamin B6, 17% of thiamine, and 15% of niacin in the U.S. diet. Fruits and vegetables also supply 16% of magnesium, 19% of iron and 9% of the calories. Legume vegetables, potatoes, and tree nuts (such as almond, filbert, pecan, pistachio, and walnut) contribute to proteins in the diet, and their proteins are of high quality as to their content of essential amino acids. Nuts are a good source of essential fatty acids, fiber, vitamin E, and minerals. Other important nutrients supplied by fruits and vegetables include folacin, riboflavin, zinc, calcium, potassium, and phosphorus. They are also used in preventing and curing many serious diseases like cancer, stroke, hypertension, diabetes and cardiological disorders.

WORLD AND INDIAN SCENARIO

Area, Production and Productivity

Major fruits producing countries percentage share (2006): As far as the major fruit producing countries are concerned, China ranks 1st, India

2nd and Brazil comes at the 3rd place. China is at the top for apple and vegetable production while India ranks first for producing mango, banana and papaya. For vegetables production China ranks first. The data for 2006 percentage share shows that India produces 40.00% and 24.50% mango and banana respectively, while China's share for production of Apples is 38.70%. The next in order are USA (5.2%), Italy (3.4%), Spain (2.9%), Indonesia (2.9%) and Mexico (2.8%).

Area and productivity: India occupies 5.51 million and 7.16 million hectares under fruit and vegetable crops as compared to world figure of 53.17 million hectares under fruits and 52.62 million hectares under vegetables. As per the data reported for the year 2005-06, the production of fruits and vegetables in the world stands at 536.59 and 917.56 million tones respectively, whereas India produces 58.74 (10.94%) and 109.05 (11.88%) million tones of fruits and vegetables respectively. The areas under fruits and vegetables are 10.36% and 13.61% of the world. In case of total productivity or production of fruits and vegetables at the world level, banana, grapes, oranges and potato are very high, some producing 76.4, 67.69, 64.28 and 327.3 million tones respectively in the year 2003-06. The production and ranking in the world for important fruits and vegetables is presented in Fig. 1.

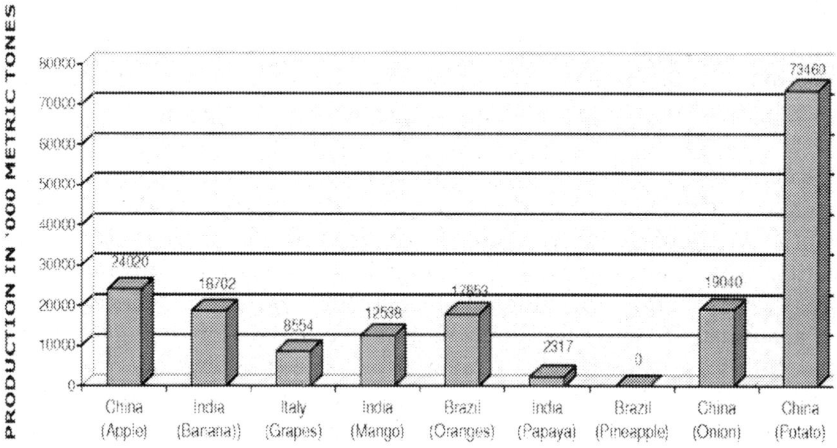

Fig. 1. Highest (1st Rank) of different countries for important fruits and vegetables.

Productivity of fruits and vegetables mt/ha.: The perusal of the data indicates that for mango, Brazil has the highest productivity of 16.0 Metric tones/ha followed by China and India producing 8.5 and 6.2 mt/ha respectively. Though India contributes 40.1% production share to world produ-ction, productivity needs to be augmented. Similarly for grapes though the productivity in India per hectare is 25.4 mt/ha being the highest in the world, the acreage and production is quite low as compared to Italy, USA, France and China. With regard to apple, France has a productivity of 39.1 mt/ha, the highest in the world followed by Italy, China, USA and Argentina producing 38.4, 37.0, 28.8 and 28.3 mt/ha respectively. India has productivity of 7.5 mt/ ha for apple which is a matter of concern.

Indian Scenario

The horticulture contributes 29.5 per cent to GDP of agriculture. The productivity has gone up by three times in banana and by 2.5 times in potato. In India the per capita consumption of fruits and vegetables is only around 46 gram and 130 gram against a minimum of 92 gram and 300 gram respectively recommended by Indian Council of Medical Research, New Delhi and National Institute of Nutrition, Hyderabad. Diversification and value addition will be the key words in the Indian horticulture in the 21st century. Today horticultural crops cover about 25 per cent of total agricultural exports of the country. The corporate sector is also showing greater interest in horticulture.

Institutes and Projects

The research efforts in the past were mainly concentrated on crop improvement, propagation of seed/planting materials, agro-techniques, crop protection and post-harvest management. Horticulture research in India is about four decades old. Systematic research on fruit, vegetable and ornamental crops began in 1954 with the initiation of independent institutions and programmes. The following institutes/organizations are important in this regard.

Indian Council of Agricultural Research (ICAR): The ICAR acts as a repository of information and provides consultancy on agriculture, horticulture, resource management, animal sciences, agricultural engineering, fisheries, agricultural extension, agricultural education, home science and agricultural communication. It has the mandates to co-ordinate agricultural research and development programmes and develop linkages at national and international levels. ICAR has 5 Deemed Universities, 48 Research Institutes, 5 National Bureaus, 11 Project Directorates, 30 National Research Centres, 77 All India Coordinated Research Projects, one Central Agricultural University, 53 Zonal Research Stations, and 567 Krishi Vigyan Kendras, besides 41 State Agricultural Universities. The distribution of 567 KVKs, SAUs/CAU (382), ICAR Institutions (44), NGO's (90), Public Sector (3), State Government (33), Deemed Universities/Central Universities/Other Educational Institutions (15). Horticulture Division of ICAR spearheads the horticulture research in India, which is carried out through 10 National Institutions, 12 National Research Centres and 13 All India Coordinated Research Projects.

National Horticulture Board (NHB): National Horticulture Board was set up as an autonomous society by the Government of India in 1984 with a mandate to promote integrated development in horticulture to help in coordinating, stimulating and sustaining the production and processing of fruits and vegetables and to establish a sound infrastructure in the field of production, processing and marketing with a focus on post-harvest management to reduce losses.

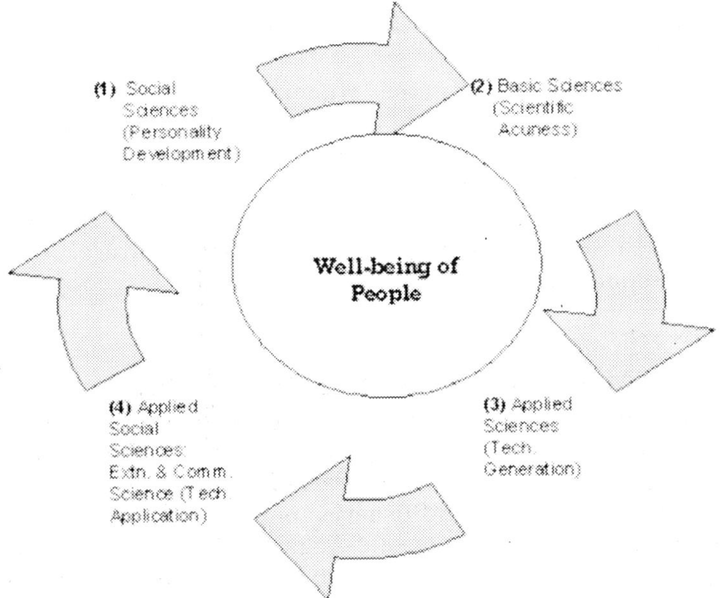

Fig. 2. Theorizing social sciences – as first

Technology Mission for NER & Himalayan States: In order to improve livelihood opportunities and to bring prosperity to the North Eastern Region (NER) including Sikkim, Government of India launched a Technology Mission (TM) in 2001-02 for Integrated Development of Horticulture. Considering the potential of horticulture for socio-economic development of Jammu & Kashmir, Himachal Pradesh and Uttaranchal, Horticulture Technology Mission was extended to these States from 2003-04. The Mission is based on the "end-to-end approach" taking into account the entire gamut of horticultural development with all backward and forward linkages in a holistic manner.

National Horticulture Mission (NHM): National Horticulture Mission is a centrally sponsored scheme launched by Government of India during the year 2005-06 (Tenth Five Year Plan). The major objective of this scheme is to provide holistic growth of horticultural sector in India and to enhance horticultural production.

Agricultural and Processed Food Products Export Development Authority (APPEDA): This authority (APPEDA) was established by the Government of India under the Agricultural and Processed Food Products Export Development Authority Act passed by the Parliament in December, 1985. In accordance with the Agricultural and Processed Food Products Export Development Authority Act, 1985, (2 of 1986) the functions assigned to the APPEDA cover: development of industries, registration of persons as exporters, fixing of standards and specifications for the scheduled products, carrying out inspection of meat and meat products, improving of packaging of the scheduled products, improving of marketing, promotion of export-oriented production and development, collection of statistics, and training in various aspects of the industries.

Other organizations: The other organizations addressing to the issues of fruits and vegetables production include: National Agricultural Cooperative Marketing Federation, Central Food Technological Research Institute, Indian Institute of Packaging, Spices Export Promotion Council, Cashew Export Promotion Council, The Agricultural Horticultural Society of India, The Horticultural Society of India, and National Horticultural Research and Development Foundation (NHRDF).

Global Network

The Food and Agriculture Organization (FAO): The Food and Agriculture Organization of the United Nations leads international efforts to defeat hunger and serving both developed and developing countries. FAO helps developing countries and countries in transition, modernize and improve agriculture, forestry and fisheries practices and ensure good nutrition for all. Established in 1945 as of 8 August 2008, FAO has 191 members states along with the European Community and the Faroe Islands, which are associate members.

International Society for Horticultural Science (ISHS): The aim of the ISHS is "...to promote and encourage research and education in all branches of horticultural science and to facilitate cooperation and knowledge transfer on a global scale through its symposia and congresses, publications and scientific structure."

World Trade Organization (WTO): The WTO is the only body making global trade rules with binding effects on its Members. It is not only an institution, but also a set of agreements. The history of the organization dates back to 1947, when the General Agreement on Tariffs and Trade (GATT), was set up to reduce tariffs, remove trade barriers and facilitate trade in goods. Over the years, GATT evolved through eight rounds of multilateral trade negotiations, the last and most extensive being the Uruguay Round (1986-1994). The WTO came into being on 1st January 1995, following the conclusion of the Uruguay Round. GATT then ceased to exist, and its legal texts were incorporated into the WTO as GATT 1994.

ORGANIC HORTICULTURE

In the past two decades, the organic market while remaining a niche sector, has grown to a share of 1 to 1.5% of total spending on food in some countries, most notably in the EU. Among these products, organic fruit and vegetables have been expanding the fastest. Consumer demand is growing and in some countries it is expected that organic food sales will reach 5-10% of the market share in near future. The market for organically produced crops and commodities has become more structured as more and more countries have adopted uniform standards for organic food production and for organic certification and labelling, and as the mainstream supermarket trade has committed itself to developing organic food ranges. FAO has been giving due emphasis for strengthening the aspects of organic farming of horticultural crops through: supporting, workshops, group meetings, symposiums, preparing guidelines for organic cultivation, inter-organizational consultations to develop viable strategies for production of organically produced seeds and plant materials, publication of proceedings and recommendations of various expert groups highlighting futuristic approaches in this regard.

PRODUCTION CONSTRAINTS

The research agenda in horticulture is by design relevant to national plans and priorities and research programmes are normally formulated keeping in view the thrust areas in development. As per the Planning Commission 10th Five Year Plan document, GOI, New Delhi, the major technology related constraints contributing to low productivity of horticultural crops and inferior quality of produce are: vast majority of holdings are small and un-irrigated, large tracts of low and unproductive plantations needing replacement/ rejuvenation, low productivity of crops due to inferior genetic stocks and poor management, heavy post-harvest losses and low utilization in processing sector, lack of advanced technologies for post-harvest handling, processing and marketing of produce, poor quality of seeds and planting materials and their assessment mechanism, preponderance of old and senile orchards and their poor management practices, small and uneconomic average farm size of the orchards, high order of perishability of horticulture produce, leading to high degree of losses, lack of modern and efficient infrastructure facilities, poor technological support and poor post-harvest management practices, under developed and exploitative marketing structures, absence of adequate standards for quality produce, inadequate research and extension support to address specific problems of horticulture crops and their linkages with farming community and industry, large scale variations in credit support and tax structures for diverse commodities, instability of prices, poor risk management; and lack of authentic up-to-date database and poor data collection and information system. Losses caused by biotic stresses are very high and due to pesticide residue problems development of eco-friendly IPM strategy is more relevant in horticulture.

EXPORT AND IMPORT STATUS OF FRUITS AND VEGETABLES

India is the second largest producer of fruits after China and contributes 11% of all vegetables and 15% of all fruit production in the world. Majority 65-70% of fruits and vegetables produced in India are consumed domestically.

India has been perhaps justly renamed as the `fruit and vegetable basket of the world`, a factor that weighs fascinatingly upon the cultivation of fruits in the country. India serving as home to a supreme variety of fruits and vegetables, holds an exclusive position in production figures amongst other countries. More than 90 percent of India`s exports in fresh products travels to west Asia and East European markets. India`s exports of fresh fruits and vegetables has amplified by 2437.12 crore rupees in 2007-08 financial year. This amplification list merrily incorporates the products like walnut, fresh mangos, fresh grapes and umpteen other fresh fruits and vegetables.

India produces 41 percent of world's mangoes, 23 percent bananas, 24 percent cashew nuts and 36 percent green peas. Exports of mangoes, grapes and mushrooms have started going to the United Kingdom, Middle East, Singapore and Hong Kong.

India had exported 54350.80 metric tones of fresh mangoes with the value of 127.42 crore rupees in 2007-08. The major varieties of mangoes exported include - Alphonso, Kesar, Amrapali, Banganpalli, Kesar, Dusheri, Langra, Chausa, Mallika and. Swarnrekha. The major markets for Indian mangoes comprise U.A.E, Bangladesh, U.K, Saudi Arabia and Nepal. India is variedly approximated to account for around 60 percent (9.5 million tones) of the world`s mango production of 15.7 million tones.

FOOD PROCESSING

The fruit cultivation and production in India has recorded a growth rate of 3.9 percent, whereas the fruit processing sector has grown at approximately 20 percent per annum. However, the growth rates have been expansively higher for frozen fruits and vegetables (121%) and dehydrated fruits and vegetables (24 percent). There exist more than 4000 fruit processing units in India, with an combined capacity of more than 12 lakh metric tone (less than 4 percent of total fruits produced). It is reckoned that approximately 20 percent of the production of processed fruits is intended for exports, the rest supplies to the defence, institutional sectors and household consumption; fruit cultivation in India, however, depends a lot upon mango and mango-based products, constituting 50 percent of exports. 'In India less than 2 per cent of the fruits and vegetables produced are processed as against 65 per cent in the US, 70 per cent in Brazil, 78 per cent in the Philippines, 80 per cent in South Africa and 83 per cent in Malaysia.

QUALITY CONTROL ISSUES

In the light of recent changes in the international trade, to exploit the potential for export of horticulture produce to its full capacity, efforts are required to evolve and develop a full-fledged Quality Assurance Mechanism. The quality control mechanism and procedures should ensure: to comply with food safety and consumer protection requirements at all stages of food production and its processing, to observe all the relevant food safety requirements and to take suitable action taken to prevent risk to the final consumer, no product be exported which does not comply with the general requirements of the food preservation and quality by-laws, and complying with food safety requirements.

EXTENSION APPROACH AND METHODOLOGY

Basic Issues: India is looking for the second phase of Green Revolution, which very much depends on the growth of rainfed agriculture. Lately the pace of agricultural growth has declined from about 3.0 percent to around 2.5 percent. When the rain God is favorable, we harvest good crops, but in one or two droughts, our buffer stock dwindles. We have imported wheat lately from Australia. We have not achieved stability and sustainability in our green revolution and the major stumbling block is the weak Extension Education System in the country.

Three pillars of agriculture India are research, education and extension. While the investment in research and education is appreciable, the investment in extension education, considering the budgets of the Extension Division of the ICAR as well as the Directorate of Extension of the Ministry of Agriculture, Govt. of India have not matched the needs of the extension system for effectively serving the farming communities. After T&V extension system of the World Bank, virtually there is a vacuum. Therefore the bridge between S&T and the farming communities – the extension system has been staked and weakened. In view of this, the major reform required is to merge the Directorate of Extension (DOE) of the Ministry of Agriculture, and the Division of Extension of the ICAR into an autonomous Council for transfer of technology for a balanced and aggressive extension all subject-matter areas with emphasis on horticulture. This alone can provide the most needed "Voice & Teeth" to the national extension networks. An aggressive leadership is needed for extension system at the national level, which is unfortunately badly lacking. Down below the said Council, it would have two arms: **(i)** ICAR First-line extension system including KVK system; and **(ii)** the State Departments of Agriculture with ATMA network. The states should develop a cadre of extension education personnel, which they do not have today – their extension staff are doing the extension services, not extension education. There is another alternative if the government is bold enough to adopt, the USA System of Agricultural extension – education, research and extension totally integrated. As per initial policy of the Govt. of India, to have only one SAU per state, they devised the present mechanism, but in view of more than one SAU in each of the major states – reducing the area jurisdiction and increasing the scientific staff for each university, the USA integrated model should work and will be economical (Prasad C., 2009).

Demand & Market Driven Extension: For connecting marketing with the TOT, the value addition should be done by post-harvest technologies including food processing – partially or fully. The AMUL/NDDB model is an excellent example of holistic extension in dairying – organizing the dairy farmers to marketing of the products – all in one Chain, an innovation – A to Z, approach. Such efforts have been made in Bulgaria in case of vegetables and fruits, what they call as promoting agro-industrial complex; in 1970s they had over 119 such complexes in that small country. Using para – extension workers by training educated rural youth is another devise to strengthen bottom-up extension; they will replace the village level extension workers. The promotion of Public-Private Partnership (PPP), though potential, still remains to be taken full advantage of.

To make demand driven extension, a basic change in extension approach has to be brought about: we need to develop 'model farms, model farmers and model villages and they should become the hub of all extension activities: bottom-up (Prasad C., 2009).

E-Agricultural Extension: This is a generation of information communication technology (ICT) for fastest possible technology transfer. Information and Communication Technology (ICT) is proved to be powerful tool to reach the down-trodden, neglected farming communities. The ICT tools could be used for dissemination of the agricultural and allied technologies in time with minimum budget. ICT and e-connectivity tools could be used to deliver the information to all the farming communities effectively without losing or distortion of the information and opportune time. It has an added advantage which resolves the limited staff crisis in the department. Hence there is need to strengthen the extension system and delivering mechanism by e-linking and interfacing with SAUs, KVKs, Research Institutes/Stations, NGOs, Farmers Associations, private companies and relevant institutions.

SKEWED DEVELOPMENT

After over 60 years of independence, India is still struggling on poverty and hunger for at least over 25 percent of the rural population; they are deprived, vulnerable and handicapped. The salvation has to come from agriculture and cottage industries including agro-industries. So far there has been skewed development in the country focusing on economic gains only, consequently the social side of development has not only been ignored, but in fact forgotten. This has led to "mental poverty" among well-to-do people – greed for wealth and power leading to corruptions and chaos in the society. Leadership, which should drive people to prosperity, peace and salvation is, by and large, self-centered. Such trend has to be arrested by bringing the positive social values in leadership in general and the development professionals and social workers in particular. A balancing act has to be done between economic and social development very seriously.

Nobel laureate Amartya Sen has so definitively demonstrated that empowerment leads to entitlements and entitlements lead to enrichment. The skewed distribution of the fruits of the impressive progress, the country has made, requires immediate course correction because economic progress in itself cannot be our national goal.

Social sciences develop the personality of people with good mind-set, positive values and socialistic approach. Such professionals then can lead the basic and applied sciences without undermining the focus, goals and the prime targets – the poor. This concept has to be introduced in the educational systems both informal and formal. Mind it, education has a long gestation period (Prasad C., 2009).

PROFESSIONAL MANAGEMENT – A MISSING LINK IN AGRICULTURE

'The issue of governance has in the recent times emerged at the forefront of the development agenda. Good governance is one of the most crucial factors required if the targets of the Tenth Five Plan are to be achieved. The universally accepted features of good governance are the exercise of legitimate political powers; and formulation and implementation of policies and programmes that are equitable, transparent, non-discriminatory, socially sensitive, parti-cipatory, and above all accountable to the people at large. Experience shows that while good governance can help secure human well-being and sustained development, it is equally important to recognize that poor governance could well erode the individual capabilities as well as institutional and community capacities to meet the needs of sustenance' (Planning Commission, 2002-07). Management is a social science; it is defined as working with people of an organization/institution devoted to certain mandates, specific objectives and goals. Management would mean working and managing men, money and allied resources of an organization/ institution for achieving its goal most effectively and efficiently.

HRD CRUCIAL

The Human Resources Development (HRD), has become very crucial and critical. On this basic issue, the Planning Commission, GOI states: 'Today, there is a realization borne out of the lessons of experience that investment in human capital is probably as much, if not more, important than "hard" capital investment. Human development is now seen not only as an end in itself, but also as a vital means to achieve the end of economic growth and prosperity for all. The Planning Commission has brought out in the first year of the Tenth Plan a National Human Development Report to signal the recognition of this.' In advanced countries, 10-15 percent budget of a department/projects is spent on the training to the incumbents/staff. Training, entrepreneurial and capacity building facilities in terms of R&D Institutions, Centres as well as manpower are available, but the urgency of quality manpower, self-employment and creating commercial agriculture through agri-business, have to be realized; training infrastructure hence have to be re-organized and strengthened.

In fact, differentiating education and training has become essential in view of the scope and importance of training of varieties of kind. As a matter of fact, time has come to recognize training as the fourth main functions of the SAUs and the ICAR Institutions. The slogan should be 'teaching, research, extension and training.'

HORTICULTURAL INNOVATIONS

Protected farming - Israel model, holistic farming (AMUL & Bulgarian Models), contract farming, and precision farming have shown the promise to change the horticultural domain drastically. They have the potentiality as shown in Andhra Pradesh and Tamil Nadu to bring the second phase of Green Revolution.

REFERENCES

Agricultural Statistics at a glance, 2007. Directorate of Economics and Statistics, Department of Agriculture and Cooperation, Ministry of Agriculture, Government of India.

Chadha. K.L., 1998. Horticultural Research in India – Infrastructure, Achievements, Impact, Needs and Expectations, Division of Fruits and Horticultural Technology, Indian Agricultural Research Institute, New Delhi, India.

Department for Environment, Food & Rural Affairs (DEFRA), 2002. Quinquennial Review of Horticulture (International).

FAO Expert group workshop on the preparation of technical guidelines on the organic cultivation of tropical and sub-tropical fruits, Kuala Lumpur –Malaysia – 22nd - 26th July, 2002.

Ghosh, S.P., 1999. Research preparedness for accelerated growth of horticulture in India, *J. Appl. Hortic.*, 64-69.

Gosselin Andre *et. al.* (1988). Policy issues in horticulture, World conference on horticultural research, 17-20 June, 1998, Rome, Italy.

Hassan. P.M.S., Saeed P.M.B., Segeiron, H.E., Sehgeiron Eizin and George, R.A.T., *ISHS Acta Horticulture* 143, – VIII African Symposium on horticultural crops.

Horticultural Research International, 2001-02. Annual Report and Accounts.

Indian Horticulture Database, 2006. National Horticulture Board, Ministry of Agriculture, Government of India, 85, Institutional Area, Sector-18, Gurgaon-122015

Kadar Adel, 2001. Importance of Fruits Nuts & Vegetables in Human Nutrition & Health, Dept. of Pomology, University of California, Davis, *Perishables Handling Quarterly*. Issue No. 106, May, 2001.

Kulecho I.K. & Weatherhead K., 2006. Issues of irrigation of horticultural crops by small holder farmers in Kenya, *Irrigation Drainage*. **20**: 256-266.

Lee, S.K. and Kader, A.A., 2000. Preharvest and post-harvest factors influencing vitamin C content of horticultural crop, *Postharv. Biol. Technol.*, 20:207-220.

Leis M. and Martinell. **A.** Needs and Expectations of Horticultural Research by the necessary operations in Italy. ISHS –World Conference on Horticulture Research.

Petit M. PolicyIissues in International Funding of Horticultural Research.

Perkins-Veazie, P. and Collins, J.K., 2001. Contribution of nonvolatile phytochemicals to nutrition and flavor, *Hort. Technology* **11**:539-546.

Prasad C., 2009, Fifth National Extension Education Congress-2009 on Extension Perspective in Changing Agri-Rural Environment in March 5th-7th, 2009 at Kanpur, Uttar Pradesh.

Produce for Better Health Foundation, 1999. Dietary guidelines: the case for fruits and vegetables first, Produce for Better Health Foundation, Wilmington DE.

Singh H.P., 2008. Triggering agricultural development through horticultural crops, *Indian J. Agric. Econ.*

USDA.2000.Nutrition and your health: dietary guidelines for Americans. *Home and Garden Bull.* **232**, USDA, Washington D.C. (website: www.usda.gov/cnpp).

35. VEGETABLE SYSTEMS IN CENTRAL ASIA AND THE CAUCASUS: RESEARCH AND DEVELOPMENT TO IMPROVE LIVELIHOOD SECURITY

Ravza Mavlyanova

Regional Coordinator for Central Asia and the Caucasus, AVRDC – The World Vegetable Center, 6, Osiyo Str., Tashkent, 100000, Uzbekistan

ABSTRACT

The region known as Central Asia and the Caucasus is comprised of eight independent countries: Armenia, Azerbaijan, Georgia, Kazakhstan, Kyrgyzstan, Tajikistan, Turkmenistan and Uzbekistan. The land area of the region covers 4.2 million km^2, of which about 70 percent is classified as agricultural land. Approximately 74.4 million people live in the region, with 16-33 percent living below the regional poverty line of US$ 2 per day. From 30 to 50 percent of poor people live in rural areas. The region has an annual economic growth rate of 0.9 percent. Agriculture contributes up to 7–34 percent of GDP, and employs up to 32 percent of the population. After the collapse of the Soviet Union in 1991, all eight countries have struggled with weak agricultural sectors hampered by inadequate management, a lack of effective irrigation technology and other modern equipment, few production receiving points and processing facilities, poor seed production, and a lack of investment. From 1993 to 2004 cultivated areas of vegetable crops increased from 491,000 to 638,000 ha, and gross production increased from 7.4 million t to 10 million t. Average productivity of vegetable crops (19.7 t/ha) is below potential. There is little off-season production, with only 15 percent of total vegetable production available from November to March. Vegetable production in the region has good prospects, provided the required infrastructure is created, production is intensified and diversified, improved varieties and appropriate technologies are introduced and adopted, postharvest operations and marketing are strengthened, and capacity building is carried out. Donor and private sector financing for vegetable research is limited. All eight countries need support for vegetable crop genetic resources collection and conservation, germplasm enhancement, crop diversification, mountain agriculture, seed production, low-cost technology development, socioeconomic and policy research, and training. To assist in the development of market-oriented vegetable production systems and promote vegetable research strategies for NARS in the region, AVRDC – The World Vegetable Center established the Central Asia and

Caucasus Regional Network for Vegetable Systems Research and Development in 2006. Network partners collaborate through periodic consultation and meetings, coordinated regional variety trials, releasing new varieties, training, farmer field days, capacity building, information dissemination, and knowledge sharing. In a future CACVEG will enlarge the activity on various directions and involve more partners.

INTRODUCTION

Political instability and economic crises constrain sustainable agriculture in Central Asia and the Caucasus. The poorest countries in the region still risk not achieving the UN Millennium Development Goal of halving poverty by 2015. In some low-income Commonwealth Independent States (CIS) countries, such as Georgia, Moldova, and Tajikistan, poverty rates hover between 15 and 20 per cent. None of the low-income CIS countries, except Azerbaijan, is expected to halve poverty by 2015. Armenia, Georgia, and Uzbekistan have made some progress, but the Kyrgyz Republic and Tajikistan need to accelerate their efforts to achieve the target (UN 2005).

Approximately 74.4 million people live in the region, with 16-33 per cent living below the regional poverty line of US$ 2 per day (PFU 2008). From 30 to 50 per cent of poor people live in rural areas. The region has an annual economic growth rate of 0.9 percent (World Fact Book 2008).

After the collapse of the Soviet Union in 1991, the newly independent states in the region faced serious economic and social problems. Economies collapsed, and GNI (Gross National Income) per capita declined by an average of almost 50 per cent between 1991 and 2000. Poverty and food insecurity increased. Approximately one-third of the region's people depend on agriculture for their livelihood, and agriculture contributes up to 7–34 percent of GDP (FAO 2006). Major crops are wheat and cotton. Horticulture also plays an important role in a food security.

All countries in the region face constraints to vegetable production, including weak agricultural management infrastructure, lack of effective irrigation technology and other modern equipment, few production receiving points and processing facilities, poor seed production, and lack of investment.

Due to wear and the removal of the main production parts, most agricultural machineries are no longer serviceable: one-third of the equipments have been in use for more than two amortization periods. Labour productivity for processing agricultural raw materials has diminished 2-3 times compared with similar enterprises in developed countries, and more than 50 per cent of food processing operations are carried out manually; fruit and berries, for instance, are dried with traditional labour methods (FAO/GFAR 2002).

The governments of the eight countries in the region seek to stabilize agriculture, reduce poverty, and ease the transition to a market-oriented economy. A number of important laws and decrees regarding agricultural development have been adopted. Land laws and land cadastre (registration) have been adopted to reflect ownership and use, leases, users' liabilities for land conservation and soil fertility improvement, and efficient water resource use. Collective land was redistributed and thousands of private farms have been established in each country. Legislative processes are developing biodiversity conservation and use guidelines, enhancing quarantine inspection, establishing intellectual ownership rules, etc. Governments also are setting priorities for development and research.

After independence, many countries achieved grain safety. The objective now is to reduce poverty, improve nutrition, and increase farmer incomes in rural areas through agricultural diversification and sustainable development of the vegetable production sector. Extension services and credit facilities for farmers are increasing, but farmers need training in new technologies, management skills, and market-oriented entrepreneurship. Manpower training in agricultural science, management, and agribusiness has to be intensified. Support is required to disseminate international public goods including research and food safety technologies. By fostering synergy and cooperation among local, regional, and national governments and local and international organizations, the countries in Central Asia and the Caucasus can begin harmonizing agricultural policy, prioritizing research, diversifying their economies, and entering regional and international markets.

REGION PROFILE

Central Asia and the Caucasus is located geographically at 39–48° N and 43–75° E. There are five countries in Central Asia (Kazakhstan, Kyrgyzstan, Tajikistan, Turkmenistan, and Uzbekistan) and three countries in the Caucasus (Armenia, Azerbaijan, Georgia). Land area is 4.2 million km^2, of which about 70 per cent is classified as agricultural land. About 88 per cent is permanent pasture, 7 per cent arable rainfed, and 5 per cent arable irrigated (FAO 2006).

The region's agroecological zones encompass highlands, deserts, steppes, and plains: cold semi-arid lowland rainfed, lowland irrigated, lowland rainfed, lowland semi-arid rangeland, and mountains. Precipitation falls mostly in the autumn-winter period, when temperatures drop to below zero and vegetation dies back, and in spring through the middle of May. Agricultural crops can be grown under rainfed or irrigated conditions. Different types of soils are found throughout the region: brown, chestnut, chernozem (black earth), sierozem (grey desert soil), mountain-meadow, and others. There are also saline soils, where several different types of melons are cultivated.

VEGETABLE SYSTEM IN CENTRAL ASIA AND THE CAUCASUS

Vegetable Production

Vegetable production can help insure food security, employ the population, diversify crops and diets, and increase incomes. From 1993 to 2004, cultivated areas of vegetable crops and gross production generally increased in the region, though at different rates in each country (Table 1).

Table 1. Trends in area, production, and yield of vegetables in Central Asia and the Caucasus by country (1993-2004)

Country	Area (000 ha)		Production (000 t)		Yield (kg/ha)		Growth rate during 1993-2004 (% per annum)		
	1993	2004	1993	2004	1993	2004	Area	Production	Yield
Armenia	22.1	26.6	449.8	672.0	20353	25263	1.0	3.1	2.0
Azerbaijan	35.2	106.0	534.6	1392.5	15188	13137	12.8	11.4	-1.4
Georgia	31.1	47.1	411.3	447.0	13240	9490	5.0	1.3	-3.7
Kazakhstan	97.0	157.7	990.4	2442.7	10210	15490	5.1	11.3	6.1
Kyrgyzstan	16.5	42.2	278.3	706.4	16867	16759	5.4	9.5	4.1
Tajikistan	32.2	38.2	592.2	568.1	18368	14888	1.9	-0.1	-2.0
Turkmenistan	45.5	53.2	534.3	526.1	11737	9889	1.1	-0.2	-1.3
Uzbekistan	212.2	167.1	3667.6	3303.8	17283	19771	-1.8	0.3	2.2
Total	491.8	638.0	7458.5	10058.6	15164	15766	2.8	4.1	1.3

Source: FAOSTAT, 2006.

Vegetable productivity and quality depend on agroecological conditions, cultivation zones, and other factors. At the moment, productivity of vegetable crops, with an average yield of 19.7 t/ha in 2007, continues to remain below potential. Post harvest losses are significant due to a lack of infrastructure for harvesting, handling, transportation, storage, and processing.

Per capita levels of fruit and vegetable production for countries in the region is shown in Table 2, with the lowest levels in Georgia and Tajikistan.

Table 2. Production (kg/person/year) in Central Asia and the Caucasus countries in 2003-2005

Country	Food items		
	Tomato	Onion	Vegetables, other
Armenia	76	14	150
Azerbaijan	52	20	104
Georgia	29	6	85
Kazakhstan	32	22	128
Kyrgyzstan	31	21	104
Tajikistan	30	27	69
Turkmenistan	54	19	84
Uzbekistan	51	20	80

Source: FAOSTAT, 2006.

Ecological conditions enable farmers in the region to raise both cold-resistant crops (cabbage, carrots, beet), and more thermophilic vegetable crops (pepper, melon, cucumber). The majority of countries have 240 no-frost days.

The long vegetative period and high summer temperatures allow farmers to raise a variety of vegetables, melons, and gourds, from ultra-early to late-maturing varieties. Vegetables are grown in open ground during spring, summer, autumn, and early winter. In southern zones, it is possible to grow cold-resistant crops such as cabbage, onions, and root crops on open ground throughout the winter. Tomato and cucumber are raised in glass and plastic sheet greenhouses during the winter-spring and autumn-winter periods. By constructing protective shelters (greenhouses, tunnels, hotbeds) farmers recently have begun to diversify their crops, growing cauliflower, peppers, eggplant, and leafy vegetables.

The region is the center of origin of several important vegetables (cabbage, carrots, onions, melon). More than 30 species of vegetable crops are grown, but tomato, watermelon, cabbage, and onions make up 60 per cent of all production. Farmers in the region tend to raise only certain kinds of vegetables, despite the generally good growing conditions that could support a diversity of vegetable crops. Therefore, it is important to select varieties appropriate for a range of seasonal and ecological conditions, and to employ technologies that can extend cultivation into the off-season.

Only 15 per cent of the region's total vegetable production is available from November to March. The lack of vegetables in the off-season leads to unbalanced diets and poor nutrition. A joint FAO/WHO Expert Consultation on Diet, Nutrition, and the Prevention of Chronic Diseases recommends the intake of a minimum of 400 g of fruit and vegetables per day (WHO/FAO 2002). Few people in the region attain this level of vegetable consumption in winter; most must rely on vegetables that have been pickled, salted, or processed in other ways for storage.

Although vegetable production is becoming more efficient as countries in Central Asia and the Caucasus make the transition to market economies, none of the countries produce enough vegetables to export. Statistical data on exports are not published.

With the break-up of the Soviet Union, the region's uniform seed production system was disrupted. Today, the lack of good seed is reflected in low productivity levels and the poor quality of produce. There are not enough seed farms in the region and the material base is out-of-date. Many countries have no data on seed production of for vegetable crops.

Peri-urban and Urban Agriculture

Vegetable production at the household level makes a significant contribution to the region's rural and urban vegetable supply chain. Majority of the population lives in rural areas, where most people have home gardens to produce vegetables for the family; the surplus is sold in local markets and in nearby cities. Men, women, and children are involved in vegetable cultivation. The participation of all family members in vegetable production increases the family's income, and raises the social status of women.

Small private farms contribute to the region's vegetable supply. These small-scale enterprises do not require large capital investments, but they can provide farm families with a good income through local sales. Private farms that produce vegetables are not officially organized, although all countries in the region have established farmer associations to work with local authorities. The potential for export will increase as more farms begin producing vegetables and fruit; raising productivity and diversifying production will thus become more important. However, small-scale farmers lack opportunities for value-added vegetable processing.

Agro-firms gradually are being created to exploit seasonal opportunities for income generation. Wealthier farmers and traders now cultivate vegetables in the off-season in gas-heated greenhouses.

The regional population is growing, and vegetable production must be increased to supply people with fresh, safe produce throughout the year. Vegetable production is developing more intensively around large cities, which have more roads and modes of transport, markets, and the greater purchasing capacity of the urban population. There is some very limited direct delivery, in which farmers supply vegetables to schools, hospitals, restaurants, or other large establishments.

Challenges

The market reforms of the first ten years of independence largely have been insufficient to build a thriving agricultural sector in the region. However, there is untapped potential for increasing vegetable production in Central Asia and the Caucasus, which can help decrease poverty, improve diets and health, promote food security, and balance trade. A reorganized market system promoting the intensification and diversification of vegetable production and processing will require all producers, large or small, to increase efficiency and productivity within the framework of a joint common policy.

Agriculture must be modernized as the region makes the transition to a market-oriented economy. This will require a different, market-focused approach to research and the introduction and development of technology. The range of vegetables available in the region is limited and supplies decline drastically in the off-season. Appropriate low-cost technologies for cool-season production and improved varieties of nutritious species for year-round, safe production are urgently needed. Access to a diversity of germplasm must be increased for the creation of high-quality varieties and sustainable seed production. The need for postharvest facilities and operations will grow in importance as vegetable production increases.

Marketing research can help farmers better understand the potential and the risks of vegetable production. Some form of financial support is needed now through donor-funded projects or microcredit facilities until the vegetable sector is strong enough to rely solely on commercial production. Creating a market infrastructure for private enterprise, farmers, and pre- and post harvest agribusinesses will promote vegetable production for domestic consumption and for export, which will in turn encourage the development of rural infrastructure. This will provide employment for the rural population, raise incomes, and slow down the rural migration to cities.

Training and financial support for training is needed to disseminate technical knowledge and improve farm and agribusiness management skills. Effective training for small producers is just beginning in the region; it will require cooperation from all regional NARS and partnerships with other actors in the global economy.

RESEARCH

After independence, all countries in the region reoriented agricultural research to serve a market-oriented economy. Scientific institutes now work under the ministries of agriculture. Academies of agricultural sciences were kept in Georgia and Tajikistan. Each country has its scientific research institutes working on vegetable production, with experimental stations in various ecological zones (Kalb and Mavlyanova, 2005). Professional training is carried out through agricultural institutes and universities.

All countries in the region have agricultural research development strategies linked to national priorities for science. However, only 4.3 per cent of the total resources allocated for agricultural research go to vegetable research, and the vegetable subsector receives a lesser share of human resources, only 3.4 per cent (Mubarik *et al.,* 2006).

The most important constraint to regional vegetable production is limited financing for agricultural research, followed by lack of equipments such as computers, tractors, and up-to-date laboratory devices, and a dearth of skilled and unskilled labour. Laboratory methodologies and seed cleaning techniques need to be improved. Protected structures and varieties developed specifically for protected production are needed.

State agricultural commissions run variety trials and process the release of new varieties, but in some countries research institutes must pay for state variety testing, which further constrains the release of new varieties.

The economic difficulties experienced in the early years of independence caused a sharp reduction in support for agricultural research in the region. As a result, few research institutes have adequately trained staff, especially young specialists. There is little coordination between regional research institutions, although the informal links among scientists are growing.

To overcome these constraints, countries in Central Asia and the Caucasus need to develop laws and improved means of financing vegetable research, including grants. Cooperation will be required from governments, donors, and the private sector to promote progress. As noted in the Issik-Kul Declaration of 2001, stronger links must be built with international organizations for scientific research and sustainable agriculture.

Priorities for research in all eight countries include vegetable crops genetic resources collection and conservation, germplasm enhancement, crop diversification, mountain agriculture, seed production, low-cost technology development, socioeconomic and policy research, and capacity building (PFU 2008).

REGIONAL NETWORK FOR VEGETABLE SYSTEMS RESEARCH AND DEVELOPMENT (CACVEG)

AVRDC – The World Vegetable Center has organized international and interdisciplinary coalitions of public and private actors in Southeast Asia, South Asia, sub-Saharan Africa, and Central Asia and the Caucasus to generate and distribute vegetable germplasm, seed, and technology.

The center opened a regional office in Tashkent, Uzbekistan in 2005 within the framework of the Consultative Group on International Agricultural Research (CGIAR) Program for Central Asia and the Caucasus. In line with the Asian Development Bank's Central Asia Regional Economic Cooperation Program, regional NARS representatives organized a workshop on "Improved Income and Nutrition in Central Asia and the Caucasus through Enhanced Market- and Trade-oriented Vegetable Systems Research and Development" from 25-27 April, 2005. The following key actions were identified:

1. Form a vegetable systems R&D network in Central Asia and the Caucasus to promote regional and international collaboration.
2. Examine current needs of the vegetable sector to set R&D strategies and policies for participatory development.
3. Collect, characterize, conserve and use vegetable germplasm of both exotic and indigenous species.
4. Enhance vegetable variety development (both hybrids and open-pollinated varieties), release and dissemination by both the public and private sectors.
5. Develop the indigenous seed and seedling business of new and improved varieties to promote vegetable productivity and marketing.
6. Develop and disseminate appropriate cost-efficient production technologies to generate income.
7. Identify and introduce small-scale postharvest technologies.
8. Establish regional information system on prices, quality standards, and regulations for domestic, regional and international markets.
9. Strengthen the vegetable R&D capacity of NARS through appropriate training (internships, field days, courses, etc.) and information dissemination.

Eight national coordinators were appointed to participate in the network by the governments of Armenia, Azerbaijan, Georgia, Kazakhstan, Kyrgyzstan, Tajikistan, Turkmenistan, and Uzbekistan. The network held its first meeting in August 2005 in Tashkent, where all signed an agreement to establish the Central Asia and Caucasus Regional Network for Vegetable Systems Research and Development (CACVEG) to assist in the development of vegetable production systems oriented to market and trade, and to promote vegetable research strategies for all national agricultural research systems (AVRDC 2006).

Specific Objectives of the Central Asia and Caucasus Regional Network for Vegetable Systems Research and Development

1. To promote sustainable and improved production of quality and safe vegetables for consumption, processing, trade and to develop indigenous vegetable-related industries within Central Asia and the Caucasus.
2. To foster collaborative research partnerships among the national agricultural research systems in areas of common interest, to attain better and more efficient use of the expertise, technologies, germplasm and scarce resources available within the region.
3. To facilitate the generation and adoption of improved technologies for vegetable systems R&D through elaboration of joint projects, collaborative research, information exchange, and scientific consultation.
4. To strengthen the technical proficiency of vegetable professionals with scientific capability to respond to the national and regional needs for efficient and sustained vegetable production.
5. To strengthen linkages for acquisition and exchange of relevant information and technologies among CAC countries and outside the region.

The Network's Primary Functions

1. Coordinate and foster collaboration and complementary research and technical exchanges, taking into account the institutional strengths, weaknesses and needs of each partner country.
2. Through consultation and periodic network meetings, identify major research needs of common concern, develop collaborative plans, and monitor and evaluate collaborative R&D.
3. Improve the mechanisms for strengthening communication and rapid exchange level of technology and information derived from collaborative R&D.
4. Identify training needs and conduct group and individual training programs.
5. Facilitate financial and technical support to enable NARS to play lead roles to solve specific regional research problems.
6. The Steering Committee will: set policy and priorities, and approve the strategic plan of the network; review and approve the annual work plans and budgets of the network; monitor and assess progress of the network and report to members; and mobilize resources within and outside the region for the network.

7. National representatives serve on the CACVEG steering committee; AVRDC is the network's executive agency and serves as liaison among network member countries, the steering committee, the CGIAR Program Facilitation Unit (PFU-CAC) Program, NGOs, and the private sector.
8. The Executive Agency has the following responsibilities: seek additional financial support jointly for CACVEG from the donor community, and draft project proposals for possible funding support to CACVEG; implement agreed network activities jointly with NARS; provide scientific and administrative support to CACVEG; prepare agendas and reports of working sessions of the Steering Committee; and formulate the network workplan and budget in consultation with the Steering Committee.
9. Subject to relevant national policies and availability of resources, the network members are to: facilitate exchange of germplasm; facilitate participation of vegetable professionals in R&D and training activities; participate in the exchange of information, technologies and expertise within CAC; and execute agreed programs and provide venues and support for agreed activities.
10. In addition to other activities which may be agreed upon later, CACVEG is to engage in the following core activities: germplasm collection, conservation, characterization and exchange, in close collaboration with appropriate international agencies; testing of improved varieties, segregating lines and other relevant materials; collaborative research on agreed topics, with individual countries taking lead roles in areas where they have interest and comparative advantage; seminars, conferences, workshops and training courses on topics of common interest; information exchange, including publication of newsletters, bibliographies, bulletins, and manuals; collection and establishment of baseline data on vegetables; and capacity building.

The network helps to unite eight less developed countries in shaping and managing a new regional economy through scientific cooperation and the development of a sustainable system of vegetable production.

The Activity Within CACVEG

AVRDC – The World Vegetable Center possesses a rich diversity of vegetable germplasm that can be used and adapted to local conditions. Collaboration within the framework allows for effective germplasm introduction and faster variety development and testing, bringing new varieties into production in a shorter time. Annual regional varietal trials in different agroecosystems were initiated in 2005. AVRDC collaborates with the main research institutes in the region, including: Scientific Center of Vegetable and Industrial Crop (Armenia), Azerbaijan Research Institute of Vegetable Growing (Azerbaijan), Research Institute of Crop Husbandry (Georgia), Research Institute of Potato and Vegetable and Melon Crops

(Kazakhstan), Kyrgyz Research Institute of Crop Husbandry (Kyrgyzstan), Institute of Horticulture and Vegetable Growing (Tajikistan), Research Institute of Crop Husbandry (Turkmenistan), Uzbek Research Institute of Vegetable, Melon Crops and Potato and Uzbek Research Institute of Plant Industry (Uzbekistan) and universities, including Tashkent State Agrarian University (Uzbekistan) and Kyrgyz Agrarian University (Kyrgyzstan).

The collaborative efforts have produced positive results. More than 1000 lines of 15 vegetable crops from AVRDC are now used in research; promising accessions (early maturing, higher yielding, resistant to diseases, fruit quality, etc.) have been selected in each country and seeds of promising varieties multiplied to conduct competitive trials.

Fifty promising lines of ten vegetable crop species were trialed in competitive variety trials in 2008. Industrial variety testing is ongoing in some countries. A total of 28 varieties of six crop species including tomato, sweet and hot pepper, eggplant, vegetable soybean, mungbean, and bean varieties passed state variety trials in Armenia, Kazakhstan, Tajikistan and Uzbekistan (Table 3).

Table 3. AVRDC regional varietal trials in Central Asia and the Caucasus (2005-2008)

Country	Crop species	Number of accessions				
		Evaluated	Revealed	Competitive trials	State trials	Released
Armenia	8	143	41	25	13	-
Azerbaijan	7	78	5	2	-	-
Georgia	3	14	7	-	-	-
Kazakhstan	6	324	3	2	2	-
Kyrgyzstan	4	45	5	3	-	-
Tajikistan	12	62	13	6	4	-
Turkmenistan	5	37	8	1	-	-
Uzbekistan	13	321	41	11	9	4
Total	15	1024	123	50	28	4

Information is shared through exhibitions, presentations, interviews, articles, field guides, newsletters, posters, leaflets, and a database. Recipes are evaluated and promoted to enhance micronutrient availability among the regional population.

In 2008, at the Second Vegetable R&D Network Steering Committee Meeting, country representatives noted that the network has improved collaboration among regional agricultural institutions and increased the exchange of experience and knowledge among vegetable specialists in the region.

Early maturing, high-yielding mungbean varieties 'Zilola' and 'Marjon', vegetable soybean varieties 'Ilkhom' and 'Universal', and high yielding hot pepper 'Uchkun' have been included in Uzbekistan's State Register for 2007-2009. Seed multiplication of promising and release varieties was conducted to supply farmers through agreements with partner institutes and business firms. The new varieties will help farmers increase yield and raise their incomes. Field demonstrations of AVRDC's promising varieties have been organized with the Farmers' Association of Uzbekistan.

Uzbekistan provided grants for joint projects of AVRDC and the Uzbek Research Institute of Plant Industry on "Complex evaluation of valuable-marketable traits of vegetable soybean accessions and selection of promising varieties for transferring to the state variety trial" (2006-2008) and "Complex evaluation of vegetable germplasm with unique traits and promising lines for submission to the state varietal trial" (2009-2011).

One postgraduate student working on tomato in Armenia has been defended his thesis for PhD. Two postgraduate students have been conducting research on vegetable soybean and hot pepper since 2008, and one more postgraduate student began research work on yard-long bean in Uzbekistan in 2009.

Twenty-three specialists working in vegetable research from eight countries were trained in international and regional training courses and have received certificates. Regional review and planning meetings in vegetable variety selection and adoption are conducted annually. Farmers' Field Days are conducted annually in all eight countries to demonstrate promising and released vegetable varieties and to strengthen collaboration with farmers for seed multiplication. Farmers' training was conducted in Uzbekistan on improved cultural practices for vegetables. In a future CACVEG will enlarge the activity on various directions and involve more partners.

CONCLUSION

To improve research capacity, regional research institutes should participate in international projects and grants, and seek support and financing for education and training at the state, regional, and international level. Closer collaboration with farmers' associations, production processing organizations, scientific institutions, and nongovernmental organizations will help disseminate research. Promoting new vegetable varieties to commercial outlets such as markets, supermarkets, and shops will increase demand, and require faster multiplication of seed.

Increasing the supply of high quality, high yielding vegetable varieties adapted to biotic and abiotic conditions of Central Asia and Caucasus will add diversity to diets, increase farmers' incomes, and improve the overall health and economic standing of people in the region. Supported by national and regional policies to promote intensification and diversification, vegetable production and processing can further the region's export activities and participation in international trade.

REFERENCES

AVRDC – The World Vegetable Center. 2006. *Proceedings of the CAC Vegetable R&D Network First Steering Committee Meeting*, Tashkent, Uzbekistan, 9 August 2006. AVRDC – The World Vegetable Center, Shanhua, Taiwan, 120 p.

FAO. 2002. Joint WHO/FAO Expert Consultation on Diet, Nutrition and the Prevention of Chronic Diseases, 2002: Geneva, Switzerland.

FAO. 2006. Compendium of Food and Agriculture Indicators 2006.

FAO/GFAR. 2002. Summary Report. *Regional Workshop on Post-Harvest in Central Asia and the Caucasus*. Almaty, Kazakhstan, 20-22 February 2002.

FAOSTAT. 2006.

FAOSTAT. 2007.

Kalb TJ, Mavlyanova RF (eds). 2005. Vegetable production in Central Asia: Status and Perspectives. *Proceedings of the workshop*, 13–14 June 2003, Almaty, Kazakhstan. AVRDC – The World Vegetable Center. Shanhua, Taiwan, 134 p.

Mubarik A, Mavlyanova R, Wu M, Farooq U, Lin L, Kuo CG. 2006. Setting Research and Development Priorities for Market-oriented Vegetable Production Systems in Central Asia and the Caucasus. In: *"Increasing Market-oriented vegetable production in Central Asia and the Caucasus through collaborative research and development,"* 25-27 April 2005, Tashkent, Uzbekistan. AVRDC – The World Vegetable Center, Shanhua, Taiwan, p. 105-137.

Program Facilitation Unit (PFU), 2008. A Decade of Partnership for Sustainable Agricultural Development in Central Asia and the Caucasus. CGIAR PFU-CAC. Tashkent. Uzbekistan. January 2008. Fourth Edition. 32 p.

UN., 2005. The United Nations Millennium Development Goals Report, 2005. http://MDG 083105 revised.indd 18.

World Fact Book. 2008.

36. CONTRIBUTION OF MYANMAR HORTICULTURE IN SUPPORTING THE LIVELIHOOD SECURITY AND ECONOMIC GROWTH

San San Yi

Researcher Manager, Horticulture Division, Myanmar Agriculture Service, Ministry of Agriculture and Irrigation, Union of Myanmar

ABSTRACT

Myanmar is a predominantly agrarian economy based on crops production. According to the statistical data, Myanmar is self-sufficient in food and rice production at the national level (FAO Statistical Year Book 2005/06 Vol.2). According to the production statistics of Myanmar, the horticultural crop production is fifth rank in the agricultural sector. In Myanmar more than a hundred kinds of vegetables and over sixty kinds of fruits are grown in different agro-ecological regions of Myanmar (MAS, 2005). Horticulture crops are widely grown in the whole country but mostly all of the quantities are domestic consumes due to rather need of quality for exports in the past. The sown area of horticulture crops has increased from 4.78 million hectare in 1988-1989 to 8.03 million hectare in 2006-2007. The local demand of horticulture product has staidly improved due to the increase of population, living standard and implementation of an open market economy. Myanmar horticulture crops occupy about 5% of the total cropping area of the country and it has more than 1.2 million hectares (0.58 million hectares for fruits and 0.64 million hectares for vegetables) in 2007-2008 (CSO, 2007). Through 1990s and up to now the implementation of an open-trade policy substantially increased the volume of agriculture commodities including horticulture products to external trades particularly to China. The average annual growth rate of vegetables production in Myanmar is 5.4 percent (AVRDC, 2004).

Indigenous vegetables also need effective implementation program that is how many provide to daily food. Myanmar household survey pattern of expenditure for fruits and vegetables for urban and rural families are to be (8.89%) and (9.45%) respectively. Mandal (2006) estimated that consumption of vegetable in Myanmar may roughly be 130g pre day per head which was based on capable productivity of the present vegetable and fruits. Myanmar horticulture is indeed in an increasing good shape at present than in the past decades, however it would need further and constant reforms, adjustments and manipulations in order to move horticulture constantly upwards and sustained. Myanmar will be a country of food basket in the region in future thus research and development of major food crops including horticulture need more assistance and cooperation by international institutions. In this respect, private, public and international organizations may need to strengthen the cooperation and coordination for the sake of the development of horticulture production, processing and marketing for the local, regional and global food security and economic sustain.

Myanmar is located in South East Asia, lying between 10° and 28° N latitudes and 92° and 101° E longitudes. The country is favored in land resources with a total land area of 67.66 million hectares. At present, the net sown area of the country is recorded as 12.61 million hectares (31.17million acres). Total crop sown area is about 20.41 million hectares (50.42 million acres) having the Myanamr cropping production with cereals occupying over 50% of land area under cultivation due to open market economy for rice, followed by pulses, oilseed crops covering each about 20% of land area. The remaining food crops, industrial crops, medicinal and horticulture crops combined are about 15% of total cultivated land area. The horticulture crops occupy about 5% of the total cropping area with more than 1.1 million hectares (0.58 million hectares). It is also estimated that there still exists 0.30 million hectares (0.73 million acres) of fallow land and 5.97 million hectares (14.76 million acres) of cultivable waste land making a total of 6.27 million hectares (15.49 million acres) land area available for crop production.

In Myanmar agriculture plays an important role in contributing to the maintenance of employment in rural areas and in the whole food and non-food production chain. Current contribution of this sector to the national (Gross Domestic Product) GDP is 40.2 percent and 12.14 percent of total export earnings, remaining 49.8 percent is contributed by other sectors (MOAI, 2008). Average per capita income in the country is US$ 56, where 61 percent of labor force employed in agriculture sector especially in rural and urban areas (CSO, 2008). Most of agriculture lands are currently cultivated by small scale farmers. The average size of holding is 2.27 hectares. 61 percent of total net sown area is held by the farm size less than 4.05 hectares.

MYANMAR AGRICULTURE AND HORTICULTURE CROPS

Myanmar is located in South East Asia, lying between 10° and 28° N latitudes and 92° and 101° E longitudes. The country is favored in land resources with a total land area of 67.66 million hectares. At present, the net sown area of the country is recorded as 12.61 million hectares (31.17 million acres). Total crop sown area is about 20.41 million hectares (50.42 million acres) having the Myanamr cropping production with cereals occupying over 50% of land area under cultivation due to open market economy for rice. Then follow pulses, oilseed crops covering each about 20% of land area. The remaining food crops, industrial crops, medicinal and horticulture crops combined are about 15% of total cultivated land area. The horticulture crops occupy about 5% of the total cropping area of the country and it has more than 1.1 million hectares (0.58 million hectares for fruits and 0.74 million hectares for vegetables) in 2007-2008 (CSO, 2007).

It has typically tropical and subtropical climate and there are three seasons, (1) the rainy season (mid-May to mid-October), (2) the dry cold season (mid-October to mid-February) and (3) summer season (mid-February to mid-May). The temperature variation is not so much in the south part but in the central dry zone, the temperature in the hot season may become 42°C and in the cold season it is about 10-15°C. In the hilly regions it has more cool temperature (about 30°C in maximum and 5-7°C in minimum) than other parts of country.

Myanmar is agriculture base country and will continue to be the same in future because of possessing diverse agro-climatic condition and abundant land, water and plant natural resources. Therefore Myanmar agriculture plays an important role in contributing to the improvement of employment in rural areas in the whole food and non-food production chain. Current contribution of it to the national Gross Domestic Product (GDP) is 40.2 percent and 12.14 percent of total export earnings, remaining 49.8 percent is contributed by other sectors (MOAI, 2008). Average per capita income in the country is US$ 273, where 57 percent of labour force is employed in agriculture sector especially in rural and urban areas and Myanmar population employment structure can be seen in Figure 1 (CSO, 2007). Most of agriculture lands are currently cultivated by small scale farmers. The average farm size is 2.27 hectares. 61 percent of total net sown area is held by the farm size less than 4.05 hectares (CSO, 2007).

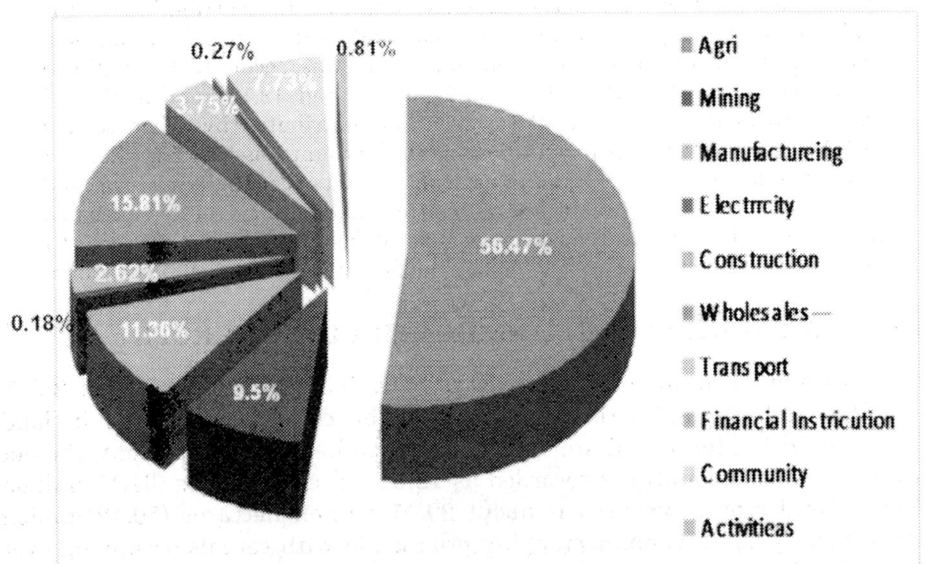

Fig. 1. Distribution of employed population in different occupations in Myanmar (1999)

PRODUCTION SITUATION OF FRUITS AND VEGETABLES IN MYANMAR

Horticulture crops are widely grown in the whole country but mostly all of the quantities are domestic consumes due to rather need of quality for exports in the past. In Myanmar floriculture and horticulture are separate departments but they are under the same umbrella of the Myanmar Agriculture Service (MAS), Ministry of Agriculture and Irrigation (MOAI). Through 1990s and up to now the implementation of an open-trade policy substantially increased the volume of agriculture commodities including horticulture products to external trades (Figure 2). The significant amount of horticulture products (mango, watermelon, dried taro, onion, chilli etc) were exported through border trades to neighboring countries particularly to China (Figure 3). The local demand of horticulture products has improved in recent years with the increase of population and living standard (Figure 3). MAS has started introduce Good Agriculture Practice (GAP) under the assistance of ASEAN corporation to intend to produce safe and high quality products. Therefore in the near future horticulture products will be more exported to external markets which may improve livelihood income and agricultural competitiveness of rural households.

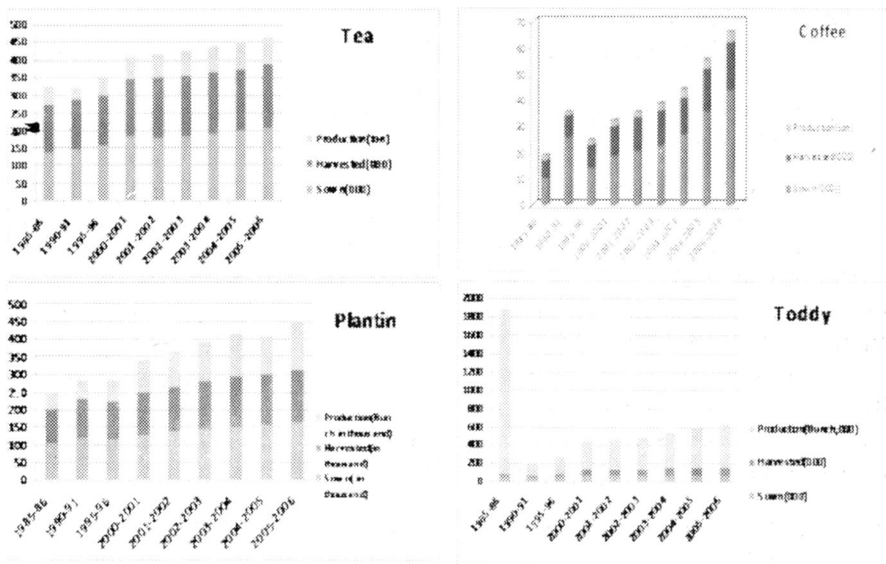

Fig. 2. Sown acreages, harvested acreages, production of selected crops

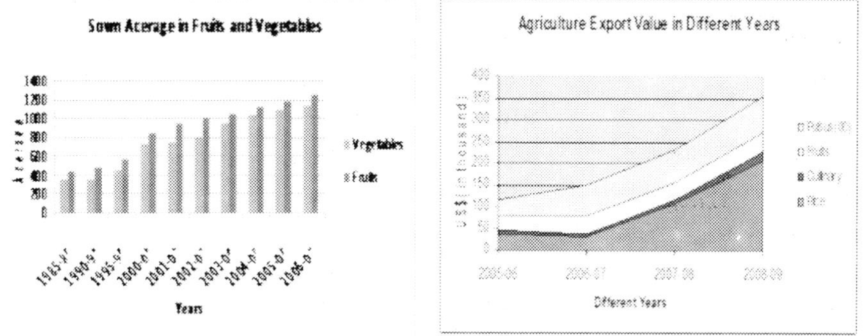

Fig. 3. Production and export trend on fruit and vegetable

The MOAI especially in MAS and Department of Agriculture Research (DAR) have established several plant tissue culture laboratories since 1980s for conducting clonal propagation, promoting plant breeding in cereal crops and conservation for orchid and some horticulture crops. Comparing with the neighboring countries, however, the success of technology is lower significantly. Therefore, in order to attain the quality and well harmonized seedling production efforts are being made by MOAI to transfer the technology to the private sector. On the other hand institutional and private are needed capacity buildings to be promoted and conducted in this area.

In Myanmar more than a hundred kinds of vegetables and over sixty kinds of fruits are growing in different agro-ecological regions of Myanmar (MAS, 2005). Moreover, a large number of indigenous vegetables and fruits are well adapted in it. According to the production statistics of Myanmar, the horticultural crop production is fifth rank in the agricultural sector, after the rice, pulses, oil seed, and industrial crops. According to favorable growing seasons, vegetable crops can be separated into two main groups; rainy season vegetables and dry season vegetables. Because of differences in agro-ecological climates in different regions, growing season differs from one region to another. Consequently some vegetable crops can be seen in the market throughout the year.

Vegetable crops consumed by Myanmar people can be classified into four groups: kitchen crops, seasonal vegetables, vegetables produced year-round basis, and vegetables grown naturally as wild species. The average annual growth rate of vegetables production is 5.4 percent and it is the highest among neighboring countries such as Thailand (1.0%), Banglades (2.8%) and Pakistan (4.6%) (AVRDC, 2004). The Table 1 and 2 showed the production statistics of major vegetable and fruits crops grown in Myanmar in the year 2007-08. The mentioned areas in table do not include naturally grown, wild crops and perennial type of vegetables and fruits such as drum stick, acacia and several vegetables in the regions.

Table 1. Production statistics of some major fruits crops in Myanmar (2007-08)

Sr.	Name of Fruits	Sown Area (ha)	Harvested Area (ha)	Total Production (ton)
1.	Mango	62969	62900	413215
2.	Banana	57824	57820	883282
3.	Orange	9162	9160	11461
4.	Pummelo & Citrus	4856	4855	12611
5.	Durian	6833	6830	24691
6.	Pineapple	16501	16500	159917
7.	Custard apple	3412	3410	11452
8.	Litchi	3597	3590	25813
9.	Grape	892	890	9655

10. Apple	1729	1720	1703
11. Pear	974	970	17173
12. Damson	1315	1315	10022
13. Walnut	221	220	2807
14. Cashew nut	61095	61095	13486
15 Plum	12773	12770	230320
16. Chestnut	50	48	100
17. Tamarind	16766	16755	97961
18. Strawberry	89	78	920
19. Coconut	46643	46350	364689
20. Lime	14256	14254	33461
21. Others	190529	190500	
Total	512486	512030	

Source (CSO, 2007)

Table 2. Production statistics of some major vegetables crops in Myanmar (2007-08)

Sr.	Name of vegetables	Sown Area (ha)	Harvested area (ha)	Yield (ton/ha)	Total production (ton)
1.	Chilli	129027	129022	0.89	115467
2.	Tomato	105934	105934	11.66	1235571
3.	Onion	70934	70932	14.52	1029841
4.	Potato	36676	36658	14.43	528977
5.	Garlic	28667	28667	7.00	200565
6.	Cabbage	28463	28463	15.77	448752
7.	Mustard	33069	33069	6.98	230898
8.	Cauliflower	23432	23432	13.24	310252
9.	Radish	19808	19808	12.58	249121
10.	Bottle gourd	19210	19210	11.52	221344
11.	Lettuce	7758	7758	6.71	52065
12.	Carrot	2313	2313	10.11	23385
13.	Asparagus	372	372	4.11	1529
14.	Other vegetables	225849	225839		
	Total	746585	746544		4851562

Source (CSO, 2007)

SOCIO-ECONOMIC ASPECTS IN MYANMAR HORTICULTURE

The horticulture production is essentially a farm business enterprise that benefits thousands of families in urban, peri-urban and rural communities in Myanmar. Growing vegetables provide self-employment to families who are engaged in all aspects of the business; raising nurseries beds, transplanting, crop management, harvesting, preparation for the market, and even selling. Most vegetable growers used to employ their family labour so as to reduce production cost and remain competitive in local markets. In peri-urban and available water resources areas, the main source of income is from vegetable production that has to some extant sustained their economy and livelihood. Small scale vegetable cultivation in Myanmar has made a contribution of income and poverty reduction, but yet research has to be documented in this process.

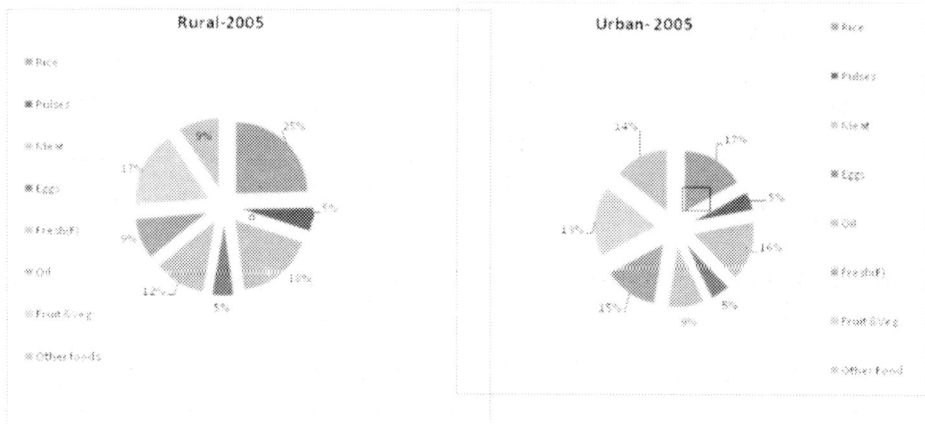

Fig. 4. Percentage of distribution of average monthly expenditure by group

In rural areas farmers used to collect wild and indiginerous vegetables from their farms and nearby jungles so that they may be able to use for home consumption and get extra income by selling out in near markets. It is also need effective implementation program that provides to daily food. Vegetables and fruits grown in home sted and school gardens could be a valuable source of food for the family and the school, and could make an important nutritional contribution particularly to micronutrient intake. Home garden can be raised with spare family labour with the participation of women and children. It is therefore, important for most households and virtually every school to devote more time to growing vegetables. A community garden near source of water at the village is also a useful adjunct to the villagers' own backyard gardens. In cities, even the small piece of available land behind the house could, with the assistance of waste water, yields a valuable supply of vegetables all year round. Even people living in flat can grow certain varieties in pots kept on their verandas. Those types of producing vegetables support not only for daily family food consumption, saving daily household expenditure but also surpluses for additional income. According to the Figure 4 from household survey pattern of expenditure for fruits and vegetables for urban and rural families are to be (8.89%) and (9.45%) respectively. It indicates that rural families spend more for vegetables and fruits than urban areas.

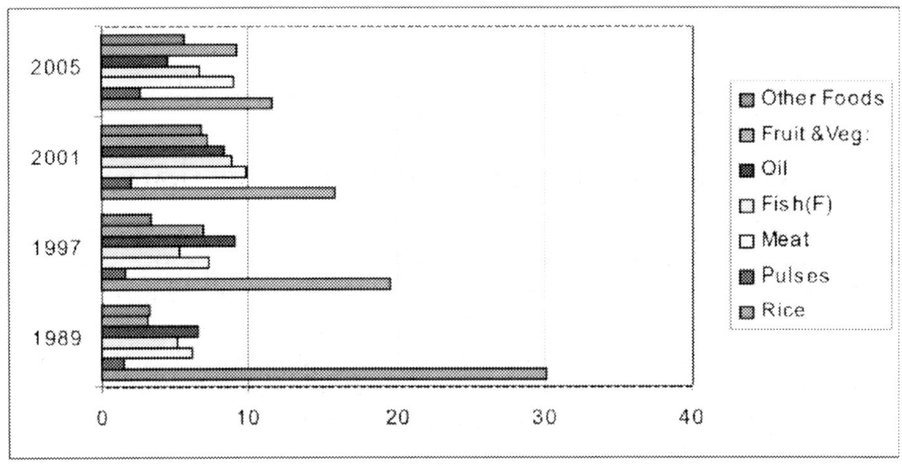

Fig. 5. Percentage of distribution of average household expenditure in different years.

COST-BENEFIT RATIO OF MAJOR VEGETABLES PRODUCTION

Fruit and vegetable production is usually profitable and perishable compared to stable crops. Their product has high values and has income generation potential and their production is attractive especially for small scale farmers. The cost benefit of major vegetables and rice for the year 2007 are abstracted in the Table 3. It is also observed that 50-60% of the total production cost has been invested for vegetable inputs like seeds, manure, fertilizers, insecticides, farm manure, plastic, net, bamboo sticks and other material costs. But some vegetable production has lower advantages than stable rice when labour and input are the limiting factors. Mostly small-scale farmers are said to be suffering from heavy costs input such as seeds, fertilizers and pesticides and uncertainty of market for their products. These farmers are forced to use open pollinated varieties and traditional land races.

Table 3. Cost of production and profit of some major vegetables (2007)

Sr.	Crop	Cost/acre	Income/acre	Profit	Cost-benefit ratio
1.	Lettuce	632,000	900,000	268,000	1: 1.42
2.	Cabbage	818,000	1125,000	306,900	1: 1.38
3.	Cauliflower	869,100	1225,000	355,900	1: 1.41
4.	Tomato	161,570	1,800,000	184,300	1: 1.11
5.	Onion	664,200	1,500,000	835,800	1; 2.26
6.	Garlic	688,300	1,000,000	311,700	1: 1.45
7.	Chilli (dry)	275,800	690,000	414,200	1: 2.50
8.	Potato (rain)	607,300	810,000	202,700	1: 1.33
9.	Potato (winter)	640,100	879,120	239,020	1: 1.37
10.	Yard long bean	584,800	744,000	159,200	1: 1.27
11.	Lady's finger	372,300	464,000	91,700	1:1.25
12	Rice (up land)	130,000	180,000	50,000	1:1.4

13	Rice(low land)	132,000	210,000	78,000	1:1.6
14	Wheat	134,600	200,000	56,400	1:1.5
15	Maize	126,650	162,500	35,850	1:0.77

Source: Horticulture Division of MAS, Unit- Kyats

FRUITS AND VEGETABLES CONSUMPTION IN MYANMAR

Vegetables and fruits are essential daily Myanmar diet that comprises of the staple rice and an accompaniment of meat, fish paste, and several vegetables as garnish that provides bulk nutrients and fiber which could help prevent major diseases, such as cardiovascular diseases and certain cancers. They are the most important sources of vitamins, macro- and micro-nutrients for the country people. In recent years, vegetables and fruits have gained great popularity among local people for their health bebefit. Also rapid changes in broad range of vegetables preference have brought about, rising per capita income and the improvement of nutrient knowledge. On the other hand, some indigenous vegetables provide medicinal effects to human health. Besides, chilli, onion, potato, garlic and ginger are commonly known as major culinary crops which are consumed on daily basis by Myanmar families. Indigenous vegetables play significant role in food security in both rural and urban regions. They can service as main food or secondary condiment to dishes prepared from indigenous varieties which have valuable source of energy and nutrients. These indigenous vegetables and fruits in Myanmar appear to resolve widespread under nutrition and drastically reduce vulnerability. However, information horticulture diversity in production and consumption across the local region are very scarce in each crop.

Mandal (2006) estimated that consumption of vegetable in Myanmar may roughly be 130g pre day per head which was based on capable productivity of the present vegetable and fruits. It was rather less than WHO/FAO recommended rate. Consumption rate of individuals may depends on food availability and their purchasing power. Normally seasonal vegeta-bles are cheaper in price at their peak harvest season so that consumers used to have more vegetables than other periods. Therefore all out efforts need to be strengthened to boost the production and productivity of vegetables and fruits in Myanmar.

CURRENT SITUATION OF HORTICULTURE

Keeping in view of importance of horticulture in both livelihood security and economic support it is necessary to ensure sustainable horticulture production since there are many small scale farmers engaged in this sector. But there are many limitations in production of Myanmar horticulture and they are;

1. Insufficient Horticulture Production, Poverty and Malnutrition

One of the central problems to be addressed is enhancing farm family income (poverty alleviation) and nutrition through vegetable production and/

or seed production of indigenous vegetables. Increasing vegetable and indigenous vegetable seed production will help enhance community food security by diversifying diets and sources of income.

There have been no sustained efforts with vegetable crops which have focused on farm-level participatory research, technology transfer, farmer to farmer approaches in pesticide safety, integrated pest management or sustainable production systems- this in spite of Myanmar's strong tradition of vegetable production and consumption. There have also been very weak parallel efforts at lack of post harvest technology and wholesale market infrastructure development for horticulture crops.

2. Underutilization of Indigenous Vegetables

Local indigenous vegetables undoubtedly have the greatest potential to improve nutrition with the least cost. Myanmar's strong tradition of consumption of a great variety of green leafy indigenous vegetables should be encouraged and augmented through making seed easily available while reinforcing the value and health benefits from these species.

There are some in-country programs to date by few international and INGOs and NGOs that have used or distributed vegetable seed (indigenous or otherwise) as a part of nutrition or community health programs in rural villages and even in urban areas, but, still require wider scale for further development.

There are also very less efforts to promote or multiply seed of open-pollinated (non-hybrid) indigenous vegetables in spite of their major contributions to diet and nutrition. The few existing government agencies and private sector efforts are focused on the production of more expensive F1 hybrid vegetable seed for export or use by local producers with better access to capital and market near urban areas.

Awareness of health and nutritional benefits of indigenous vegetables should be raised through seed distribution and promotional efforts associated with normal activities of village level health and /or Mother and Child Welfare Associations and other village level associations. This will be one of the possible ways to reduce malnutrition in the country. In Myanmar, rural communities are currently characterized by environmental deterioration. Many are among the poorest locations in the nation. Sustainable horticulture presents an opportunity to rethink the importance of family farms and rural communities. In combination with other economic development policies, sustainable horticulture practices and policies can help community that meet employment, education, health, cultural and spiritual needs of people.

3. Unsafe and Ineffective Pest Management

While pesticides are still used in the production of major agronomic crops like rice, they are commonly used in commercial vegetable production. Pesticide abuse and misuse are arising problems in Myanmar as they have been in most other Southeast Asian countries. There is an urgent need to promote using safer and more sustainable pest management practices in commercial vegetable production areas. The misuse of often dangerous pesticides is a major problem as most of the vegetables are consumed as raw and fresh. Some may remain certain level of pesticide residues and chemicals harmful to the users may remains in the due to improper use; lack of technical know-how on type of pesticides and its efficacy; and improper treatment with chemicals before marketing for better price of the products. There is also urgent need to enforce the existing pesticide law, regulation and procedures, otherwise illegal importation; trading and use of dangerous pesticides will be growing day by day.

Myanmar Agriculture Service and association like Myanmar Fruits and Vegetables Production and Marketing Association are now encouraging vegetable farmers to follow Good Agricultural Practices (GAP), and are working closely with ASEAN countries for safe and healthy production systems that are both economically and environmentally sustainable. But the activities under GAP program further needs to be accelerated for larger scale educational awareness of GAP to producers, traders and sellers in order to supply safe and healthy products to the consumers through existing extension services and media.

4. Shortage of Trained Extension Staff and Researchers

Human resources development is an important foundation for building any program. To improve and maintain trained personnel both in public and private sectors compartmentalization and regional isolation of research and extension are very serious constraints in horticulture production and other agriculture sectors in Myanmar. The country now faces generation gap as well-trained officials entering supernavations and some have already retired. In addition, in-country, in-service training programs and advanced degrees for extension workers or researchers in vegetable or other horticultural crops have been few. Thus appropriate human recourses development strategy and plan need to be developed.

5. Current Status of Horticulture Research and Development

Horticulture research and development programs are being conducted mainly by MAS and Department of Agricultural Research (DAR) of MOAI with its own budget allocation. Vegetable research in Myanmar has not developed well yet but several researches have been initiated and implemented with the assistance of FAO, JICA, KOICA and other international organizations like ARC (AVRDC). Horticulture researches are prioritized on germplasm collection, varietal improvement programs, agro-techniques, pest and disease management and post-harvest technology. Records of MAS reveales that over 987 varieties of germplasm and 165 varieties of indigenous vegetable crops have already been collected. Some promising vegetable varieties were also released by MAS and DAR.

There are also private vegetable seed production and improvement projects in Myanmar. Private companies such as Known-you-seed (Thailand), Chia Tai, Myanmar Uni Seed, etc. are producing mostly hybrid vegetable seeds to be used locally as well as for export. Off-season production and year-round production of some vegetable varieties are making progress. Use of poly-mulch, rain-cut (plastic tunnel) and foliar spray are increasing yearly to attain quality products. In addition, special vegetable production zone of about 620 hectares has been established near Yangon city ensuring regular supply to city population.

The application of plant biotechnology in vegetable crops development is in infant stage. Virus free planting material production and mini tuber production for potato through tissue culture is initially introduced at the laboratories of MAS. Maintenance of some varieties with tissue culture is underway.

CHALLENGES AND CONSTRAINTS

Horticulture production in Myanmar has been increasing due to high local demand and export potential to neighboring countries. The challenging problem is to produce safe and healthy vegetable products because most vegetable crops are consumed as raw. Indiscriminate use of pesticides in vegetable production tends to be dangerous to the lives of human beings and animals. Even though chemical analysis on residuals of pesticides are being normally made by the authorities concerned and recommendations for the use of hazardous pesticides are duly made for safety production of vegetables but still it calls for urgent need to enhance better understanding of the safer production and marketing systems to those who are involved in the whole chain from the field to the table.

The other challenging factor is low capability of vegetable seed production in Myanmar. The available of quality vegetable seeds and fruit seedlings also remain a major bottle neck. Most vegetable seeds are imported from neighboring countries and all are hybrids and expensive. The commercial vegetable producers are using hybrid seeds for better quality and quantity. Authorities concerned have few capacities to test, certify and recommend the legal and illegal incoming vegetable seeds to be in line with the existing procedures and practices for the certification and registration of imported vegetable seeds.

The production of horticulture crops using good agricultural practices in Myanmar has just introduced and practiced under the framework of ASEAN cooperation. Increased practice of good agricultural production on vegetable crops demands wider education and transfer of the techniques to be taken care by both private and public sectors.

The organic vegetables production has also potential since Myanmar has still virgin and waste land across the country. But there is a need to certify by the internationally recognized institutions.

Major constraints in vegetable production include shortage of well-trained personnel in vegetable crops, lack of pre and post harvest handling technology and related facilities and marketing. To achieve success and competing goal, farmers, researchers, input suppliers, farm workers, processors, agriculture technicians, consumers, policy makers should actively participate in each every step. A wide diversity of strategies and approaches are necessary to create a more sustainable horticulture system.

FUTURE PERSPECTIVE

Horticulture production in Myanmar needs to be improved in terms of quantity and quality for domestic and international markets. In particular, processing of surpluses which are wasted at all the times and the development of pre and post-harvest handling, processing and marketing will be major field of interest in future by the private sector once there is high demand for processed items and competitive in international markets. There exists many areas favorable for vegetable seed production for tropical and temperate vegetable varieties and thus vegetable seed production business seems to emerge strong in the very near future.

The cooperation and collaboration with AVRDC-ARC in the areas of collection, evaluation and utilization of indigenous vegetables will also be accelerated in future. Sustainable horticulture research and education (SHRE) programs, and demonstration projects are required to be strengthen through the technical assistance of international organizations, INGOs and bilateral programs to improve the existing horticulture production and marketing systems.

CONCLUSION

Myanmar horticulture is indeed in an increasing good shape at present than in the past decades, however, it would need further and constant reforms, adjustments and manipulations in order to move horticulture constantly upwards and sustained. The horticulture production in Myanmar will be as important as major food crops for domestic use and increase export. It is serving for the necessities of livelihoods security and good shape of economic growth as well. Myanmar will be a country of food basket in the region in future thus research and development of major food crops including horticulture need more assistance by international institutions. Myanmar has huge potential for horticulture production in terms of quantity as well as for quality. In this respect, private, public and international organizations may need to strengthen the cooperation and coordination for the sake of the development of horticulture production, processing and marketing for the local, regional and global food security.

REFERENCES

Asian Vegetable Research and Development Center (AVRDC), 2004. ARC Training Report 2003. A complication of research reports. p. 44

CSO 2007. Central Statistical Organization, Myanmar

Mandal, B.C, 2006. Manual on seed production of major vegetables of Myanmar, Food and Agriculture Organization of United Nations, Office of Representation in Myanmar

MAS, 2005. Myanmar Agriculture Services. Annual Report, Myanmar.

37. GOVERNMENT POLICY, BASIC ISSUES AND FORMIDABLE CHALLENGES IN PROMOTION OF HORTICULTURE AND PLANTATION CROPS

A.K. Mehta

Assistant Director General (Agricultural Extension), Indian Council Agricultural Research, New Delhi

ABSTRACT

Horticulture accounts for 30% of India's GDP from 8.5% of the cropped area. Globally, India occupies 2nd position after China in fruit and vegetable production but is at first place in production of mango, banana, litchi, papaya, pomegranate, sapota, aonla & okra and largest producer, consumer and exporter of spices. Among the plantation crops, India is the largest producer of cashew nuts. In horticulture, research has to prioritized in genetics biotechnology, integrated and sustainable production systems, post harvest handling, storage marketing and consumer education. These technological up-gradations will help India to emerge as a major producer of horticultural products and thus secure market access for its agro exports. Producer's strategy need to be designed based on financial resources, managerial skills and entrepreneurial capacity. To increase the production and productivity of fruit crops, there is need to use vegetables in intercropping which will help in increasing the income of farmers during the period when the fruit orchard has not become commercially viable. Development of horticulture sector has to be accompanied by the growth of agro-processing industry. Based on studies of international market for export, India can lead in export of non-traditional fruits like cassava, sapota, litchi and guava. To achieve the desired results, thrust is to be provided in research for development of improved varieties/hybrids in fruit, vegetables, plantation crops, medicinal plants and aromatic plants; need to be develop suitable varieties for growing under protected cultivation; production technology for matching export requirements; horticultural based cropping system; efficient water management; post harvest technology; integrated management of nutrients diseases and pests; mechanization for different garden operations; research on market intelligence and use information technology. In vegetables, research is to be prioritized on saturation of area with F1, hybrids, protected cultivation of high quality vegetable for export and promotion of technologies for round the year. In spices, there is need to develop high yielding disease free varieties. In floriculture, need of development of varieties suitable for export. In medicinal and aromatic plants, there is need for commercializing of plant required in large quantity for domestic and international markets. In coconut, plantation in large area is old and senile, so need of re-plantation on large scale. Certain policy initiatives are required like abolishing of VAT, octroi on plastic raw material, amending of forest law of cutting of trees as it is effecting rejuvenation of horticultural plantation and need to create seprate corpus fund for horticulture. Horticulture is being handed by multiple Ministries which often results overlapping with each other. So it should be under one Ministry i.e. Ministry of Agriculture. Numbers of policy initiatives have been taken by Government to promote horticulture. The most important being launching of National Horticulture Mission. Similarly, while realizing the potential of horticulture in the North-Eastern region and other special category states, a centrally sponsored scheme called Technology Mission for Integrated Development Horticulture in Noith Eastern States, Sikkim, Jammu & Kashnir, Himachal Pradesh and Uttarakhand has been launched. The technology mission address all the aspects of horticulture development through its four missions namely, Mini-Mission I, Mini-Mission II, Mini-Mission Ill and Mini-Mission IV. In addition National Horticulture Board is involved in the development of high quality horticulture farms in identified belts. Govt. of India has identified some prominent crops as high value crops of great economic importance and is taking all possible steps and initiative to commercialize the sector.

INTRODUCTION

Horticulture accounts for 30% of India's agricultural GDP from 8.5% of cropped area. The horticulture sector has over the years, emerged as main stay of growth in agriculture and made a significant contribution to agricultural GDP. Diversified and accelerated agricultural growth is critically dependent upon the development of horticulture sector. It plays an important role in improving the productivity of land, generating employment, ameliorating the economic conditions of farmers, entrepreneurs and enhancing exports. India's horticulture sector is fascinatingly diversified and covers a wide range of fruits, vegetables, flowers, spices, medicinal and aromatic plants and a variety of plantation crops like coconut, arecanut, cashew nut and cocoa. Presently, horticultural crops occupy around 20.08 million ha area of the country producing 207.01 million mt of crops (2007-08).

Fruits

The global fruit production during 2004 was 511 million mt. During the past 15 years, production of fruits in China has gone up by 246 per cent, which helped China to become the largest producer of fruits (83.24 million mt) in the world from its third largest producer position and contributes 16 per cent share in total production. India is the second largest producer of fruits with production of 63.5 million mt and contributes 10 per cent share in global fruit production. Brazil occupies the third position, with 36 million mt and contributes seven per cent in global production.

India occupies first place in production of mango, banana, litchi, papaya, pomegranate, sapota and aonla; second place in limes and lemons and fifth place in pineapple production of the world. The productivity of grape is highest in the world. About 41 per cent of world's mango, 29 percent of world's banana and 30 percent of the world's papaya is produced in the country. India has higher national average productivity in banana and sapota compared to world average productivity, while in citrus, mango, apple, guava, pineapple, papaya, India has substantially low productivity in comparison to the world average.

India accounts for an area of 5.70 million ha under fruit crops with a production of 63.5 million mt. Among various states, Maharashtra ranks first and contributes 27 per cent in area and 21.5 per cent production. Andhra Pradesh ranks second in area and production contributing 13 and 16 of fruits.

Vegetables

India is the second largest vegetable producer after China with 11% production share in the world and Brazil is the third. It is also the largest producer of okra and second largest producer of most of the other important vegetable crops at global level, namely brinjal, cabbage, cauliflower, pea, onion and tomato and third largest producer of potato in the world. However, it falls behind in productivity in most of the crops except for tomato where India is ranked at number one. In okra, it is at par with world productivity and in cauliflower it is quite close to the average world productivity.

The area under vegetables increased from 5.59 million ha in 1991-92 to 7.80 million ha during 2007-08. The production in this period increased from 58.53 to 125.88 million mt. During the same period, productivity of vegetables increased from 10.5mt/ha to 15.0 mt/ha.

While West Bengal continues to be the leading state in area and production, the productivity is higher in Tamil Nadu followed by Uttar Pradesh and Bihar. Potato has the highest share in total vegetable production of the country (28.8%) followed by brinjal (8.6%), tomato (8.5%), tapioca (7.8%), onion (7.4%) and cabbage (6.1%).

Floriculture

India has made noticeable advancement in production of flowers and is second largest producer of flowers after China. Floriculture during 2007-08 is estimated to cover an area of 1.61 lakh hectares with a production of 8.70 lakh mt of loose and 43.41 million (number) cut flowers.

Tamil Nadu is the leading producer of loose flowers closely followed by Karnataka both in terms of area and production. The productivity per hectare was highest in Bihar (17.05 mt) followed by Haryana (11.55 mt). On the other hand the productivity of flowers was the least in Rajasthan (0.59 mt). 77% of area under floriculture crop is concentrated in seven states comprising of Tamil Nadu, Karnataka, Andhra Pradesh, West Bengal, Maharashtra, Haryana, Uttar Pradesh and Delhi.

In cut flower production, West Bengal leads with a production of 896 million cut flowers followed by Karnatàka with 413 million cut flowers. Other major cut flower producing states include, Uttar Pradesh, Maharashtra, Gujarat, Himachal, Uttarakhand, *etc.*

Medicinal and aromatic Plants

The medicinal plants based industry is growing at the rate of 7-15% annually. According to a conservative estimate, the value of medicinal plants related trade in India is to the tune of about Rs. 5,000 crores per annum while the world trade is about 62 billion US dollars and is expected to grow to the tune of 5 trillion US dollars by the year 2050. The present international situation in the production and trade of essential oils and aroma chemicals is very complex and vibrant. The Govt. of India has identified and documented over 9500 plant species considering their importance in the pharmaceutical industry.

The Indian Systems of Medicine have identified 1500 medicinal plants, of which 500 species are mostly used in the preparation of drugs. The medicinal plants contribute to cater 80% of the raw materials used in the preparation of drugs.

Spices

India is the largest producer, consumer and exporter of spices and spice products in the world and produces more than 50 spices. The spices production in India is of the order of 4.10 million mt from an area of about 2.66 million ha during 2007-08. Area and production of spices in the country have registered substantial increase over the last fifteen years with average annual growth rate of 2.7% and 6.2% in area and production respectively. Chilli is the major spice crop occupying about 29 per cent of area under cultivation and contributing about 34 per cent of total spices production in the country. Turmeric accounts for 14% of production and 6% of area, while garlic accounts for 19% of production and 5% of area. Seed spices contribute 17% of production and occupy 41% of area while pepper contributes 2% of production and occupies 9% of area of the total spices in the country.

Rajasthan occupies major area under spices owing to seed spices cultivation in the State followed by Andhra Pradesh, Kerala, Karnataka, Madhya Pradesh and Gujarat. Andhra Pradesh, which is the foremost state producing chilli and turmeric in the country, ranks first in terms of production followed by Rajasthan, Madhya Pradesh, Tamil Nadu, Orissa, Karnataka and Kerala.

Plantations Crops

There are two groups of plantation crops namely coconut, arecanut, oil palm, cocoa, cashewnut and tea, coffee and rubber. India is the largest producer and consumer of cashew nuts. It is estimated that total production of plantation crops during 2007-08 is around 12.04 million mt from an area of 3.22 million hectare.

POLICY AND STRATEGIES

Development of agriculture in India needs some critical management inputs particularly that of supply chain management - collaboration among various stakeholders along with efficient vertical and horizontal integration. The horticulture sector in particular has to prioritize development of research in the issues of genetics, biotechnology, integrated and sustainable production systems, post-harvest handling, storage, marketing and consumer education. Diversification offers an attractive option and a major source of pushing up growth of agricultural sector. While technological up-gradation and associated institutional changes are identified as thrust areas for future development of the horticulture sector, exports are considered to be most important for the growth of the sector. India can look forward to emerge as a major producer of horticultural products and thus secure reasonable market access for its agro exports, which are largely dependent on the competitive technologies that will help in enhancing export potential. This development will also help in overall growth of the economy through generation of extra foreign exchange, creating employment opportunities and also upliftment of the small and marginal farmers, with definite positive implications on income and employment. The government should create a positive environment that will ensure a mutually beneficial relationship between farmers and organized sector. Along with investment in infrastructure, development of extension activities and linkages with farmers is also an important area where the government can play an influential role. After the successful trials of SAFAL Market in Bangalore, many state governments have expressed their desire to establish similar markets after they have amended their state APMC Acts (NDDB, 2004-05).

The two golden rules for successful development of the horticulture sector are to ensure consistency in supply and provide recorded and demonstrated traceability of products. Thus, production strategies are the most crucial in strategy development. The basic production factors required for high value horticulture production are non-restrictive policies in favour of horticultural development; suitable and controllable climate conditions, availability of labour and horticultural growing skills, basic local general infrastructure for transportation and marketing, *i.e.,* access to road, train, sea, air transport, telecommunications, power and water; and basic local horticultural infrastructure, *i.e.,* access to horticultural inputs and services. Producer strategy should be designed based on financial resources, managerial skills and entrepreneurial capacity. The development strategy should be based on innovation. Production innovations initially focused on efficiency and effectiveness in order to increase yields and lower costs. Now it is important that the production innovations should focus on developing sustainable production techniques, and also focus on adding value in terms of packaging and processing.

India being the land of small and marginal farmers and studies have been advocating the fact that small farmers are going to feed India, it is important to mobilize and help them to diversify to meet the increasing domestic demand of horticulture products. As identified from African nations the small farmers are the key to initiate the horticultural revolution and with technical change and increase in international competitiveness large scale operations and vertical integration takes place. Thus to sustain the growth and development of the small land holder, farmers should be monitored to identify, select and support horticultural and entrepreneurial smallholders. Linking small farmers with high value urban and export markets would lead to the development of the rural sector. An uncontrolled expansion of horticultural production should not be encouraged. Horticultural crop diversification should be encouraged by intercropping horticultural with non-horticultural crops. This will yield more food, more income and better soil health. To increase the production and productivity of fruits and vegetables, introduction of vegetables in the crop rotation and adoption of recommended practices is very important. The use of vegetables in intercropping also helps in increasing the incomes of farmers during the period when the fruit orchard has not become commercially viable. The diversification plan of the horticulture sector needs to identify potential crops area wise and the area under low yielding vegetables and fruits should be shifted to more productive and profitable one.

There is a strong need to strengthen the research on horticultural crops to develop demand-driven technology by improved variety, pest management, etc., in both public and private sectors. These technologies should be quickly disseminated through government institutions, NGOs and even private participants by encouraging farmers' participation and upgrading their technical capabilities. A small targeted group of talented stakeholders and small entrepreneurs should be identified and assisted in obtaining access to funding, technical assistance and supply chain partners. This small group should be actively encouraged to develop business plans and feasibility studies. The horticultural development requires a minimum set of basic production factors, an optimal crop management infrastructure, post-harvest infrastructure, entrepreneurial management and horticultural expertise, logistical infrastructure and supporting financial infrastructure.

Development of horticultural sector should be accompanied by the growth of the agroprocessing industry. The opportunity exists to promote the industry by intensifying production of a required, appropriate variety of tomato, cucumber, mango, pineapple, lemon, *etc.*, for the products like ketchup, sauce, juice and pickles. Thus the production strategy should target not only meeting the domestic and export demand of fresh products but also of the processed products. There is the need to improve postharvest operations related to handling, storage and marketing of fresh and processed produce. Volumes saved in post-harvest losses are actually the surpluses generated, without additional cost. The horticulture sector has an immense potential of generating employment. Additional employment can be generated by developing the horticulture based agro processing units. This sector needs to be developed as an organized industry and has to be run collectively by all the stakeholders with farmers as entrepreneurs.

The sale of fruits is generally through pre-harvest contractors, so that the farmer gets an advanced payment and covers his risk. Vegetables are usually sold through commission agents and very little of pre-harvest contracting is done. Due to this the net returns are generally low. Farmers devote more time to their field crops rather than to the orchards. If the farmer does the marketing of his produce himself then the net returns to him would double. The marketing cost of fruits and vegetables is almost 50 per cent of the total cost of production, thus, there is a need to set up institutional agencies that can advance credit to farmer and motivate them to market the produce themselves. Post-harvest losses in horticultural crops range from 15-50 per cent. At micro level these losses increase the marketing cost of the product and at macro level they also reduce the per capita availability. Thus there is need to develop technologies, methods and mechanics to reduce these losses. There is a need to remove the distortions in the present supply chain and create better integration between the different links of the supply chain and reduce the losses. This will result in a net gain to producers, consumers and to the nation.

Farmers usually procure inputs from the retail market and end up selling their produce in the wholesale market. Buying at retail price and selling at wholesale price is the most uneconomic way doing the business. Thus the involvement of an institutional structure in coordinating the demand of individual farmers can reduce the total cost of inputs to them. The market needs to be demand driven rather than supply driven. The price of the produce should not be based on the prevailing wholesale price but on the basis of cost of cultivation of that produce. Farmers should be their own price setters rather than price followers. There is also an immediate need to integrate the production, marketing and processing processes of the produce to get maximum benefits from fruits cultivation.

The infrastructural problems pertaining to the cold storage facilities are dual as some places don't have the cold storage while other places have the problem of underutilization of the existing cold storage. The utilization is even lower than 30 per cent of the total capacity in many cases. There are problems with price structure in the processing units; the price offered by them does not justify the prevailing wholesale price or even the cost of production of the produce. Development of competitive international transportation, linked to domestic air, road and rail transport would help in the reduction of post-harvest losses. High air freights are also a hindrance for cost-effective exports.

For the exports, a large fluctuations in the production of fruits and vegetables causes problems in being a regular trade partner. The window of international demand for the horticultural products is very small. Thus a planned strategy is to be made to target the markets during that period. India can lead in the export of non-traditional fruits like cassava, sapota, litchi and guava. Further, successful implementing of the core marketing strategies will help in future expansion of the domestic and international markets. In this regard the potential states have been identified in the study. But the exports face certain barriers too. Although the Uruguay Round of

1986-1994 and the subsequent WTO agreements along with talks under the Doha Round have brought some significant changes to global agricultural trade, but with respect to horticultural development there are major protection barriers both tariff and non-tariff. To enhance exports there is a need to develop air transport cargo system specialized for fresh fruits and vegetables, along with the airports, roads and rail connectivity with the area of procurements. The country's capability to generate surpluses for exports depends on its ability to tab the potential of small farmers. Thus there is a need to develop the credit structure and farmer export cooperatives. Awareness should be created among the farmers to understand the requirements of the importing countries. These non-tariff barriers need to be understood to get a breakthrough in the market access. For this, assistance from APEDA and exporters' association as well as training to the farmers is necessary. Quality control and longer shelf life are crucial for exports. Organic production of fresh fruits and vegetable is important to capture markets in Europe.

THRUST AREAS OF RESEARCH

Keeping the above status in view, following thrust areas in research can be taken up;

- Development of improved varieties/hybrids of fruits, vegetables, plantation crops, medicinal and aromatic plants with high production potential, biotic and abiotic resistance and suitable for introduction processing export with high nutrient value.
- Emphasis on indigenous minor fruits and seed spices. Need to develop varieties suitable for growing under protected cultivation in vegetables and flowers.
- Development of appropriate horticultural based cropping systems for different agro-climatic areas. Production technology, for matching export requirements and diversified use including processing and value addition.
- Protected cultivation for export-oriented vegetables and flowers.
- Research on efficient water management including micro-irrigation and fertigation.
- Post-harvest technology including value-addition and product diversification of important horticultural crops.
- Developing micro propagation of important horticultural crops for mass multiplication of quality planting materials. Devising technology for genetic manipulations in desired crops for introducing desirable traits of yield, quality and stress tolerance
- Advanced research on national disease problem such as mango malformation, guava wilt, citrus decline, spongy tissue in mango, root wilt in coconut, *Phytophthora* diseases of different horticultural crops.
- Floriculture research for quality production of cut-flowers, post-production handling and transportation of floricultural products.

- Integrated management of nutrients, diseases and pests of important horticultural crops to reduce input costs, environmental pollution and to avoid pesticide-residue problems.
- Hi-tech horticulture should receive attention for improving production, productivity and quality in different horticultural commodity.
- Concerted efforts be made to conduct experiments in organic farming systems to make the findings more scientific.
- Research on market intelligence, use of information technology, e-commerce be taken up on different commodity.
- Mechanization for different operations in gardens should be made to make the operations more efficient and reduce the drudgery on farm labourers.

The development programmes are running in the country under the aegis of Ministry of Agriculture, Ministry of Commerce and Ministry of Food Processing Industries. The programmes under these ministries can be categorized into (i) programmes to be continued as such, (ii) programmes to continue with minor modifications and (iii) programmes to be restructured.

a. National Mission on Organic Farming
b. Establishment of Cashew Board
c. National Bee Board
d. Establishment of Planting Material Authority for Horticulture
e. National Vegetable Development Programme
f. National programme on development of commercial floriculture, medicinal and aromatic plants.

There are several components, which are being undertaken by both National Horticulture Mission, Technology Mission for North East Region and National Horticulture Board. There is a need to reorient such programmes so as to avoid overlapping.

Fruits Crops

Indigenous and unexploited fruits possessing nutritional, processing value for *e.g.*, jamun, bael, custard apple, fig, phalsa, date palm *etc.*, need to be given focused attention.

Emphasis needs to be given to promote cultivation of suitable crops in dry-land, arid regions and wasteland.

Introduction of high yielding strains/varieties of temperate fruit crops on a large scale for increasing fruit production and productivity through new orchards as well as replacement of old and uneconomical orchards.

Promoting farming system approach and diversification by intercropping with medicinal and aromatic crops.

Vegetable Crops

Vegetables contribute a major share in ensuring food and nutritional security of population. However, no meaningful programmes have been launched on vegetables in recent times. The following aspects of vegetable crops need to be taken on priority.:

- Saturation of area with F1 hybrids
- Promoting leafy vegetables for nutrition particularly in villages and tribal areas
- Protected cultivation for high quality vegetable for export
- Promoting kitchen gardening
- Focus on crops having short supply
- Promotion of technologies for round the year production of crops like onion

Spices

- Need to develop high yielding disease free varieties

Floriculture

Suitable programme needs to be launched for development of elite new varieties suitable for export on PP mode so as to slowly replace the patented varieties.

India need to reorient its exports based on pot plants and cut follage. Incentives for identification, domestication and commercialization of indigenously available follage and pot plants need to be given.

Medicinal and Aromatic Plants

Herbal gardens are being promoted to lay emphasis on sensitization of MAP cultivation. Any meaningful programme would call for commercializing those plants that are required in large quantities in domestic and international markets. Some species like Noni (*Morinda citrifolia*) which is gaining popularity needs to be given priority to exploit its potential.

Priority needs to be given in areas of new drug development from plants, development of pharmacopeia standards for quality, safety, efficacy and scientific validation of therapeutic claims.

Coconut

Coconut plantations in large area have become old and senile. Focus priority therefore should be accorded to large-scale replantation programmes.

There is need for focused attention on farm level processing, product diversification, byproduct utilization and value addition.

Tree life insurance and welfare schemes for coconut climbers.

Still conventional varieties are being planted that have lot of disadvantages. There is a need to take up large scale plantation of Dwarf hybrids under high density planting to improve productivity and farm income.

INFRASTRUCTURAL ISSUES

Government should create common infrastructure for post harvest management, vapour heat treatment (VHT), irradiation centers with equal participation from Central and State Governments for promotion of horticulture exports. The operational part and maintenance may be entrusted preferable to some private agencies to ensure their uninterrupted functioning and proper maintenance.

Create special seed production zones on the lines of SEZ's for hybrid seed production of vegetable/flower seeds for domestic and export markets.

Provide physical infrastructure and financial support for the newly established commodity associations for strengthening of their activities.

FINANCIAL ISSUES

Special financial support for specially selected crops of different states like litchi in Bihar; Banganpally mango in A.P., Dasheri mango in UP, Alphanso mango in Maharashtra; pineapple of NER; Kinnow in Punjab; passion fruit in North East and grapes in Southern States to develop core competence to promote exports.

POLICY INITIATIVES

In order to reduce cost of cultivation VAT, Octroi *etc.,* and import duties on plastic raw material need to be abolished. Excise duty on micro irrigation system needs to be reduced considerably.

There is urgent need to amend forest laws regarding cutting of trees as these are hampering the task of rejuvenation of horticultural plantation.

Economy of North East Region is largely based on salaried population and therefore very few entrepreneurs have come up to take horticulture activity. Besides, cost of transportation is extremely high both within and outside the region. Government should therefore provide incentives to promote entrepreneurship besides, transport subsidy should also be provided to the farmers in the region for moving goods within and outside the State. For goods originating from NER and underdeveloped states like Bihar, Chattisgarh and Jharkhand to provide level playing field to the farmers to market their produce.

There is need to create a separate corpus fund for horticulture to meet any emergency requirements on project base so as to protect the farmers in distress.

ORGANIZATIONAL REFORMS

The programmes related to horticulture are at present handled by multiple Ministries including Ministry of Agriculture, Ministry of Commerce, Ministry of Food Processing Industries and Ministry of Rural Development which often overlap with each other. In order to bring synergy among the various programmes and meaning to the enormous expenditure being made by these ministries on similar programmes, there is an urgent need to bring all the programmes under an umbrella *i.e.,* Ministry of Agriculture.

SCHEME AND INCENTIVES

Various policy initiatives and measures have been taken up by both the Central and State Governments, in order to develop conducive business environment for apt promotion of entrepreneurship in horticulture sector. The emphasis for horticulture development is mainly focused through the developmental programmes, of the Department of Agriculture and Cooperation.

The Department of Agriculture and Cooperation of the Ministry of Agriculture is the nodal department for over-viewing horticulture development in the country. The Division of Horticulture was carved out of the Crops Division in 1981 and a position of Horticulture Commissioner was created in 1985. The Division is vested with the responsibility of over-seeing the overall development of horticulture at national level.

The Plan investment for horticulture development increased significantly from the VIII Five Year Plan onwards and resulted in considerable strengthening of the horticultural development programmes in the country. The allocation for horticulture increased many folds during the X Plan with the launching of Technology Mission for Integrated Development of Horticulture in North East Region, National Horticulture Mission, Micro irrigation Programme. At present the horticulture crops programmes form around 30 percent of the total outlay for agriculture development of the Department of Agriculture and Cooperation.

The most important being, the launching of the 'National Horticulture Mission' as a Centrally Sponsored Scheme with a view to promote holistic growth of the horticulture sector through an area based regionally differentiated strategies, which include research, technology promotion, extension, post harvest management, processing and marketing, in consonance with comparative advantage of each state/region and its diverse agro-climatic feature. Its underlying approach is a technology driven cluster approach with focused attention on competitive horticulture crops. The other main objectives of the Mission are to:

- Enhance production and productivity of horticulture crops;
- Increase coverage of crops under improved/high yielding cultivars;
- Improve nutritional security and income support to farm household;
- Enhance production of high value and low volume horticultural products for exports
- Strengthen infrastructure facilities for marketing and export;
- Establish convergence and synergy among multiple on-going and planned programmes for horticulture development;
- Promote, develop and disseminate technologies, through a seamless blend of traditional wisdom and modern scientific knowledge;
- Create opportunities for employment generation for skilled and unskilled persons, especially unemployed youth; *etc.*

To achieve the above objectives, the Mission seeks to adopt the following strategies:

- Ensuring an end-to-end holistic approach, having backward and forward linkages, covering production, post harvest management, processing and marketing to assure appropriate returns to growers/producers.
- Promoting R&D technologies for production, post-harvest management and processing
- Enhancing acreage, coverage, and productivity through diversification from traditional crops to plantations, orchards, vineyards, flower and vegetable gardens.
- Extension of appropriate technology to the farmers for high-tech horticulture cultivation and precision farming.
- Assisting in setting up post harvest facilities such as pack house, ripening chamber, cold storages, controlled atmosphere (CA) storages *etc.* processing units for value addition and marketing infrastructure.
- Adopting a coordinated approach and promotion of partnership, convergence and synergy among R&D, processing and marketing agencies in public as well as private sectors, at the national, regional, state and sub-state levels.
- Where appropriate and feasible, promoting National Dairy Development Board (NDDB) model of cooperatives to ensure support and adequate returns to farmers.
- Promoting capacity-building and Human Resource Development at all levels.
- Strengthening infrastructure facilities such as soil and leaf analysis labs, survey and surveillance of pest and diseases, green house, poly houses, micro irrigation, plant health clinics, vermicompost, *etc.*

Realizing the potential of horticulture in the North-Eastern Region and other special category states, a Centrally sponsored scheme called 'Technology Mission for Integrated Development of Horticulture in North Eastern States, Sikkim, Jammu and Kashmir, Himachal Pradesh and Uttarakhand' was launched. The Mission/scheme is based on the 'end-to-end approach' taking into account the entire gamut of horticulture development in a holistic manner, with all the backward and forward linkages. It aims at establishing convergence and synergy among numerous ongoing governmental programmes through horizontal and vertical integration of these programmes, in order to ensure adequate, appropriate, timely and concurrent attention to all links in the production, post harvest and consumption chain. The 'Small Farmers' Agri-Business Consortium (SFAC)' is involved in coordinating this scheme. State level SFACs have also been constituted in most of the implementing states for monitoring and implementing the programme at the grass-root level.

The objectives of the Mission are to:-

- Improve the production and productivity of horticulture crops by harnessing the potential of the region;
- Give special emphasis on'Low Volume, High Value, Less Perishable Horticulture Crops';
- Develop a horticulture based farming system, thereby providing viable and ample opportunities for employment, especially for women, besides improving the productivity of land;
- Address the issues of production of quality planting material, of organic farming, of efficient water management as well as of plant health, along with proper development and demonstration of technologies; *etc.*

The technology Mission addresses all the aspects of horticulture development through its four Mini Missions, namely:

Mini-Mission-I

This involves research activities and aims to supply nucleus/basic seed and planting material of horticultural crops, with a view to provide technology support. This involves standardization of production and protection technologies, and technology refinement and training through on-farm trials. The 'Indian Council of Agricultural Research (ICAR)' is the nodal agency for coordinating and implementing the programmes under this mini mission.

Mini Mission-II

This mission primarily aims at increasing the production and productivity of the horticulture produce in the region through integrated implementation of various governmental programmes based on compact area development approach. The major components of this mini mission are: area expansion; creation of water sources; on-farm water management; production of planting materials; transfer of technology through training; demonstration, publicity and training of trainers; popularization of organic farming and use of bio-fertilizers; promotion and popularization of agricultural equipments; promotion of integrated pest management; establishment of Plant Health Clinics; establishment of Tissue/Leaf Analysis Laboratories; entrepreneurial development of women farmers; *etc.* Department of Agriculture and Cooperation is the nodal department for coordinating this mini mission. The mission is implemented by agriculture/horticulture departments of the states/UTs.

Mini Mission-III

This Mission aims at creating infrastructural facilities for post harvest management, marketing and export. Its activities are mainly coordinated by National Horticulture Board (NHB). The Department of Agriculture and Cooperation is the nodal department for implementation of this mini mission. The existing schemes of the NHB, DMI, NCDC, APEDA, NAFED *etc.*, either with the existing outlays or with enhanced outlays, are to be implemented under this mini mission. Considering the gap, new components have been proposed in marketing. The major components are:

- Post harvest management
- The existing approved schemes of National Horticulture Board (NHB) with additional outlays to be to be implemented in a more focused way in the mission mode
- Strengthening of nutritional status in rural areas
- Strengthening of marketing infrastructure
- Improvement of services in wholesale markets
- Development of rural primary markets
- Development of *Apni* Mandis
- Promotion of Agmark in domestic trade by strengthening the state grading laboratories and agricultural marketing information network
- Forward and Backward Linkages to enhance productivity and quality of produce
- Pilot Project of Alternate Marketing System in order to improve farmers' income
- Schemes of NCDC to assist cooperative societies for marketing of horticulture products
- Schemes of APEDA for fright subsidy and for packaging development
- Market intervention through National Agriculture Cooperative Marketing Federation (NAFED) and TRIFED.

Mini Mission-IV

This mini mission aims at addressing all the issues of processing and marketing of processed products. The Ministry of Food Processing Industries is responsible for coordinating the activities and its schemes are being integrated into this mission. The mini mission envisages promotion of new units; upgradation and modernization of the existing units; as well as market promotion, research and development.

NATIONAL HORTICULTURE BOARD

National Horticulture Board is involved in the development of high quality horticulture farms in identified belts and make such areas vibrant with horticulture activity, which in turn, will act as hubs for development of commercial horticulture through production and post-harvest management; development of post harvest infrastructure; strengthening market information system and horticulture database. The Board is also assisting research and development programmes to develop products suited for specific varieties with improved methods and horticulture technology. The main schemes that are being pursued by NHB are:

1. Development of commercial horticulture through production and post harvest management
2. Capital investment subsidy for construction/modernization expansion of cold storage and storages for horticulture produce
3. Technology development and transfer for promotion of horticulture
4. Market information services for horticulture crops
5. Horticulture promotion services

Plantation Crops

Plantation crops in India are considered to be the main segment of the horticulture crops. They are the mainstay of agrarian economics in many states and union territories (UTs) of the country. They play an important role in the agricultural and industrial development of the country as a whole. They contribute a significant amount to the national exchequer and country's exports by way of excise and export earnings. They also provide direct and indirect employment to large number of people in the country, and thus tries to supplement the poverty alleviation programmes, especially in rural sector.

Plantation crops constitute a large group of crops. The major plantation crops include coconut, arecanut, oil palm, cashew, tea, coffee and rubber; and the minor plantation crops include cocoa. India is the largest producer and consumer of cashew nuts. The total production of cashew is around 0.57 million tones from an area of 0.24 million hectares. India also occupies number one position in arecanut production.

India is the third largest producer of coconut and leads 90 coconut-producing countries in the world. The area under coconut plantation in India has been distributed over 18 States and 3 UTs, under different agro-climatic conditions. India is a premier coir manufacturing country in the world. Wide range of coconut products, edible and non-edible, are available for both domestic and export market. Tender coconut water concentrate is another product, apart from soft drinks, which is manufactured and marketed successfully.

But, in India, plantation crops have been continuously facing the problem of lack of investment and depressed yields, and are in great need of modernization. Their total coverage is comparatively less and they are mostly confined to small holdings. Thus, the Government of India has identified some prominent crops as high-value crops of great economic importance. It is taking all possible steps and initiatives to commercialize the sector.

References

Mattoo, A; Mishra, D. and Narain, A., 2007. From competition at home to competing Abroad – A case study of India's Horticulture: Oxford University Press

Surbhi Mittal, 2007. *Can Horticulture be a Success Story for India.* Indian Council of Research on International Economic Relations (CRIER), New Delhi.

VI
National and International Technical Cooperation

38. FAO EXPERIENCE OF TECHNICAL COOPERATION AMONG DEVELOPING COUNTRIES (TCDC)

Gavin Wall

FAO Representative in India and Bhutan, New Delhi, India

ABSTRACT

The experience of FAO in technical cooperation among developing countries (TCDC) is recounted and the role of TCDC in the FAO Special Programme for Food Security (SPFS) and the model of National Programme for Food Security (NPFS) is explored. The operational modalities of the FAO approach are described in detail in order to clarify the role of the SSC partners and to underscore the level of institutional support that is considered important in SSC programmes.

The relevance of SSC in horticulture is examined; this highlights that the delivery of fresh fruit and vegetables to consumers is a technology intensive undertaking. The opportunity to learn from the experiences of other countries that have faced similar constraints in developing horticultural value chains is considered to be self-evident. The potential for the provision of technical expertise within a SSC programme is then examined by considering some of the changes that have taken place in the architecture of international aid. The potential for new bi-lateral donors to draw on their national expertise and for the new private philanthropic organisations to utilise TCDC approaches, which are consistent with their preference for market-led approaches, is highlighted.

INTRODUCTION

EVOLUTION OF THE SPECIAL PROGRAMME FOR FOOD SECURITY

The Special Programme for Food Security (SPFS) was launched after its unanimous approval by the 106th Session of the FAO Council in May/June 1994. The programme was endorsed by the Heads of State and Governments during the World Food Summit held in Rome in November 1996. The objective of the SPFS is to assist Low-Income Food-Deficit Countries (LIFDCs) to improve their national food security through rapid increases in productivity and food production, as well as through reduction in year-to-year variability in production, and to improve access to food through a multi-disciplinary and participatory approach on an economically and environmentally sustainable basis.

The SPFS was initiated with an emphasis on four components, *viz;* (i) improve water control; (ii) intensify crop production sustainably; (iii) diversify production; and (iv) carry out participatory constraints analysis. Although the SPFS began in a few countries (15 in 1995) and with a restricted budget (USD 3.5 million), there are now more than 100 countries where SPFS activities have been implemented and more than USD 775 million have been mobilized for these activities.

Since the World Food Summit, many interacting forces took place worldwide to raise the level of international commitment to end hunger. In line with the Millennium Summit held in 2000, the recommendations of an independent evaluation made in 2001 and the "World Food Summit: *five years later*" in 2002, the SPFS began to move in a new direction in support of the National and Regional Programmes for Food Security (NPFS and RPFS) to achieve food security on a large scale. To date, NPFS activities are already under implementation in 15 countries, and programmes are approved in four additional countries. Eleven further NPFS have been formulated and are currently under review, while in 20 countries they are under active formulation. The RPFS, on the other hand, is under implementation in four sub-regions and is at different stages of formulation/review in 18 other countries.

THE SOUTH-SOUTH COOPERATION INITIATIVE

The initiative was launched in 1996, within the framework of the SPFS. Its objective is to enhance an effective solidarity among developing countries and allow the recipient countries to benefit from the relevant experience and expertise in food production of more advanced developing countries. The unique value of south-south knowledge sharing is that it provides access to know-how generated in the south, and is therefore of particular relevance to the specific development needs and potentials of the south. The SSC does not replace bilateral efforts or north-south cooperation, but aims at reinforcing and complementing these channels.

Since its inception, South-South Cooperation (SSC) has proved to be a successful initiative with 39 Tripartite Agreements signed; 1 473 experts and technicians working in 62 countries have participated in these agreements. The success of SSC has led the United Nations to advocate for a greater implementation among its various agencies[1].

The SSC has been an integral part of the SPFS support to member countries, and today, as FAO moves towards support for large-scale and comprehensive NPFS and RPFS, the demand for SSC is increasing, as

1 An overview of recent SSC initiatives promoted by the UN is available at: http://tcdc.undp.org/news.asp#Public-private

countries seek to build on specific strengths and experiences from successful food security programmes. Moreover, the context in which SSC is taking place is changing rapidly and presenting new development challenges as well as new opportunities for the SSC.

With the beginning of the NPFS and RPFS, the demand for technical assistance from the recipient countries within the framework of SSC has also increased. It is in this context that FAO has entered into a strategic alliance with the People's Republic of China on SSC in support of the NPFS and RPFS, and discussions are proceeding with a number of other advanced developing countries to conclude similar agreements.

The terms and conditions for the use of experts and technicians under the SSC initiative are different from those governing the Partnership Agreement concerning the use of experts for Technical Cooperation among Developing Countries (TCDC)[2]. While TCDC experts are generally assigned individually for short-duration consultancies, the SSC initiative is intended to field a significant number of experts and technicians for a longer duration, in order to: (i) create a critical mass and thereby maximize its impact; and (ii) ensure an adequate transfer of know-how and practical training to the host country counterparts. The conditions for remuneration between the two programmes are also different.

THE SSC APPROACH USED BY FAO

Overview

The SSC initiative enables the delivery of technical assistance from more advanced developing countries to specific recipient countries participating in the SPFS. This technical assistance consists of policy and programme development experts positioned at the national and decentralised levels, joined by technicians at the local level with strong practical field experience.

The broad range of technical areas to be covered corresponds to the comprehensive nature of NPFS and RPFS as well as new emerging issues of common interest to the south, such as trade facilitation. The duration of the cooperants' assignment is flexible to accommodate the need for short-, medium- and long-term assistance, depending on the need of the host country. Whenever possible, cultural and linguistic similarities between the provider of expertise and the recipient country should be taken into account to facilitate programme implementation. Such flexibility is also expected to result in the participation in the programme of some donor countries or funding institutions.

It is expected that the teams will not only introduce improved ways of bringing about sustainable and replicable agricultural development, but also,

---

2 For further reading, please refer to "Use of experts for Technical Cooperation among Developing Countries (TCDC) and Countries in transition TCCT) – Guidelines on Policy, Conditions and Procedures" (Rev. I – June 1997).

through their commitment and example, serve as an important stimulus for change within the communities to which they are assigned. The number of experts and technicians to be provided under this initiative should be determined on a case-by-case basis, but it should be sufficient to achieve a critical mass in a given country, with adequate site coverage to represent the main agro-ecological zones of the country.

In addition to the posting of cooperants, the SSC cooperating countries also have the opportunity to provide training to the technicians and experts from the host countries on NPFS and RPFS issues through universities, higher education institutes and other mechanisms, so as to enable them to learn about emerging issues, methods and practices in the food and agricultural sector.

Ideally, the SSC arrangements should be envisaged during the initial formulation of the NPFS and RPFS in a given country or region, and be fully integrated with the programme from the outset. However, considering the time required to establish the necessary contacts and to arrive at a mutually acceptable agreement, there may be cases in which the details of SSC can only be worked out after the launching of NPFS/RPFS activities.

Operational Modalities

As mentioned above, the increased volume and changing nature of the SSC have implications for the modalities through which the SSC is implemented. Both the cooperating and host governments are expected to play a prominent role, with greater responsibilities, in initiating and strengthening the SSC, while FAO's role will be limited to facilitating the development of the SSC agreements, providing technical assistance as required, and only in certain cases providing limited financial support.

The following steps are taken to initiate the SSC:

i. Country request

A recipient country (host country) or an advanced developing country (cooperating country) presents an official request to FAO, expressing its intention to participate in the SSC initiative under the NPFS or RPFS. The request should generally also specify the name(s) of the country (ies) with which the country concerned wishes to undertake the SSC.

Upon receipt of this request, FAO informs the countries, seeks their formal agreement and arranges for a joint SSC formulation mission. FAO also informs potential donors which might financially support the programme.

The requests for SSC within the framework of a strategic alliance should be addressed directly to the cooperating country with copy to FAO, to enable the cooperating country to include this assistance among the priorities of its bilateral assistance to the country concerned. The modality for negotiating and receiving the SSC within the framework of a strategic alliance may differ depending on the institutional mechanism in the recipient and cooperating countries concerned. In the case of the People's Republic of China, for example, it is necessary that: (i) the authorities dealing with bilateral issues in the recipient countries send a letter of request to the embassy of China; and that (ii) the request for receiving SSC assistance from China be among the host country's priorities, agreed through its bilateral discussions with the Chinese authorities.

ii. Formulation

Subsequent to the agreement of both sides, a joint FAO/cooperating country formulation team is fielded to work with the host country's experts to draw up the SSC Agreement. The tasks of this tripartite[3] mission will vary, depending on the extent to which the host country has already progressed with the NPFS and RPFS implementation, but they will typically cover:

- identification of the regions concerned and selected sites (the number depends on local circumstances but should be sufficient to allow the geographical representation of the main food-producing regions and agro-ecological conditions);
- preliminary identification of activities in accordance with the NPFS and RPFS related technical areas, in the appropriate sites;
- agreement on technical areas of expertise, composition and number of expert and technician teams, and schedule of the fielding;
- preparation of draft standard terms of reference for policy and programme development experts at the national and decentralized levels and of field level technicians;
- finalization of the necessary institutional, logistical and other arrangements, covering in particular the provision of counterpart staff, appropriate lodging, transport and office facilities as well as arrangements for programme management, monitoring and evaluation;
- preparation of detailed cost estimates and agreements on cost-sharing (see point iii) below);
- preparation of a joint report containing adequate information to enable the formulation of a draft SSC Agreement;
- identification of areas in which the cooperating country could offer training opportunities, and definition of training modalities, including the number of potential applicants;
- review of national medical facilities and identification of administrative requirements to implement the Medical Coverage for Non-Staff (MCNS) scheme for medical claims.

iii. Preparation of the South-South Cooperation Agreement

FAO prepares the draft Tripartite Cooperating Government/Host Government/FAO SSC Agreement, based on the information provided in the joint

3 This mission may also include the representative of a donor.

formulation report. The agreement is then submitted for endorsement to the authorities in the cooperating and host countries, which are also requested to authorize their representation to FAO to sign it on behalf of the government concerned.

The final agreement is signed in FAO headquarters by the representatives of the cooperating and host countries, and FAO. Arrangements could be made for the signature of the agreement in the host or cooperating country, if needed. Samples of the different types of agreement are attached (Annex. 1, 2 and 3). They are provided as a basis for the formulation mission and may be amended and modified according to any specific case.

iv. Implementation

The duration of the SSC Agreement is two to three years with possibility of further extension if needed. Its implementation includes the following steps:

- selection and clearance of SSC experts and technicians;
- designation of counterpart staff: to ensure capacity building and dissemination of knowledge, the host country should try to assign at least three counterparts to each SSC expert and technician;
- identification of housing and office facilities by the host country;
- preparation of the teams to be stationed in the sites, following language and inception training and briefing on both the cooperating and the host countries, as specified in the Tripartite Agreement;
- participative preparation of detailed plans for technical areas that fall within the comprehensive NPFS or RPFS, as appropriate, for the various sites;
- supervision of programme implementation, drawing primarily on national human resources;
- practical training of farmers and others involved in the programme;
- reporting on progress achieved.

v. Expertise

The composition of the teams under an SSC Agreement will depend on local needs and circumstances, such as:

(a) An SSC Coordination Team, composed of four/five experts, one in each of the main technical areas relevant to the project implementation to be positioned in the capital city, or in the SPFS/NPFS/RPFS sites, to coordinate and provide technical backstopping for the technicians' activities. The most senior expert will also act as Team Leader with overall responsibility for the SSC team. These experts, in addition to holding an advanced university degree at M.Sc or Ph.D. level in their respective fields, should also have several years of practical work experience. They would, in consultation with the national authorities, draw up the overall work plans for the SSC implementation and provide technical supervision to the technical teams. They will also be responsible for the preparation of a periodic progress report on the SSC team's work in the country.

(b) The bulk of the expertise will consist of technicians assigned to the rural communities participating in the SPFS or NPFS/RPFS, who will work directly with the farmers. The technicians should hold a vocational training certificate with substantive practical experience.

vi. Monitoring and Evaluation

The SSC will be monitored at three levels. The overall coordination, monitoring and evaluation of the SSC initiative are the responsibility of FAO/TCOS. TCOS keeps a database on the programme that captures information, *inter alia,* on: number of agreements signed, lists of cooperating and recipient countries, number of experts and technicians envisaged under each agreement, number of experts and technicians in the field, and SSC experts' and technicians' technical areas.

At the country level, the implementation of the SSC initiative will be the responsibility of the SPFS/NPFS/RPFS team, with the FAO country office playing a monitoring role. Subject to the availability of funds, a National Professional Officer (NPO) may be assigned to the FAO Representation in countries having 50 SSC cooperants or more to monitor and report on SSC activities, in collaboration with the NPFS National Coordinator. It has to be noted that the overall programme planning, detailed annual work plan with each expert and technician, individual staff appraisal as well as the compilation of QPIR (Quarterly Programme Implementation Report) and six-monthly PPR (Programme Progress Report) will be responsibility of the National Programme Coordinator, in collaboration with the SSC team leader.

At the local level, the beneficiaries will be community groups, small commercial farms, government staff, and training and research institutions. The SSC experts and technicians will deliver their services directly to the groups involved in the planning, implementation, monitoring and evaluation of the programme, and in the appraisal of the SSC cooperants.

Each SSC programme will have clearly defined objectives at the various levels regarding the technical assistance provided by the SSC experts and technicians. In this context, key working areas will be: transfer, demonstration and adaptation of technologies; practical training of farmers and national personnel such as extension agents, and community mobilization. Specific indicators regarding these objectives will be developed *i.e.,* technology adoption rate, technology adoption quality and quantity, attendance, participation, satisfaction, change of knowledge, additional training requests, technical performance, benefits of the new techniques, implementation quality and timeliness, service delivery efficiency and effectiveness, national personnel's technical and managerial capacities, *etc.* The experts and technicians will be responsible for implementing their specific work plans, which are established involving the beneficiaries at the local level as well as the programme coordinator at the national level, to whom they will also report quarterly.

vii. Training

As mentioned above, the SSC cooperating countries are encouraged to provide training on the NPFS and RPFS related fields to the experts and technicians from the host countries through universities and higher education institutes. The training will concern emerging areas of common interest to the south, including application and certification of good agricultural practices, setting and implementation of quality and safety standards, value chain management, application of new technologies, etc. If applicable, the programme and modalities for such training will be defined during the SSC formulation mission and will be described in the Tripartite Agreement as appropriate.

TERMS AND CONDITIONS

The SSC is subject to a Tripartite Agreement jointly prepared and signed by the Cooperating Government, the Host Government and FAO. This Agreement shall specify the responsibilities of each party as well as the conditions of assignment of the experts and technicians (precise terms of reference, required fields of competence, location of the assignment, etc.). In cases where a donor wishes to contribute to the SSC directly, by meeting the costs involved fully or partially, it may become a quadripartite agreement. The SSC expert/technician's contractual employment relationship is the responsibility of the Cooperating Government. FAO intervenes only as a facilitator in the context of the SSC programme and in the preparation of the Tripartite Agreement.

The agreements signed outside a strategic alliance are called standard agreements. Their modalities are the results of the negotiations between the Cooperating Government, the Host Government and FAO (and exceptionally the donor) on the basis of a draft standard text. In the case of an SSC agreement falling under a strategic alliance, different terms and conditions will apply, and the modalities will be subject to the host country depending on its economic wealth and its cooperation ties with the cooperating country.

The standard conditions of the SSC experts' and technicians' assignment are;

- While the experts and technicians will continue to receive all their normal salary and other emoluments in their home country, during the entire duration of their assignment in the host country, they will receive a monthly subsistence allowance of USD 900 for each technician and USD 1 500 for each expert (including during time-off accruing at 2.5 days per month) to be paid in a convertible currency on a bank account opened in the host country in the name of each expert and technician. Any derogation to this procedure will need to be negotiated during the formulation mission and reflected in the Tripartite Agreement.
- A one-time grant of USD 450 is provided to each expert and technician to cover the costs of their installation in the host country.

- Each expert and technician benefit from a medical insurance coverage under FAO's MCNS scheme, without prejudice to the status of the experts and technicians as employees of the cooperating country.

THE RELEVANCE OF SSC IN HORTICULTURE

Advances and innovations in science and technology have been essential catalysts of growth for developing societies; most readily apparent in the growth surge in Asia in recent decades. This situation applies equally in horticulture where production technologies combine with post-harvest technologies and new agribusiness models to enable small-scale farmers to be competitively meet the expectations of markets for fresh fruit and vegetables. In a recent lecture1, Dr. Calestous Juma, Professor of the Practice of International Development at Harvard's Kennedy School of Government, explored the role that emerging technologies can play in fostering economic growth and improving human welfare in Africa and re-examined opportunities for development cooperation in the light of the current global food and financial crises. Professor Juma drew on experiences on the linkages between economic growth and technological innovation in fields such as agriculture, health, energy, education and environmental management. He suggested that the world could be entering a new age of science and technology diplomacy that can underpin the convergent interests of the countries of the south.

Horticultural crops are an attractive option for small-scale farmers because they can fetch higher prices in the market than, say, grains and are labour intensive. However, in horticultural farming, where prices are rarely regulated, financial viability depends as much on business and marketing skills as on the farmer's technical expertise in crop production. Horticultural products are mainly sold fresh; some are eaten raw while others are cooked. Some horticultural products have traditionally been processed when no other form of storage was available, *e.g.*, dried fruit and jams. As society develops and becomes more affluent the market for processed and prepared horticultural products develops. A market also develops for horticultural products such as flowers and house and garden plants which are sold for purely aesthetic reasons. Increased wealth also brings with it an increased demand for product diversity in the form of new crops, off-season supplies and different flavours.

Important characteristics of horticultural crops are:

- that they are mainly eaten for their contribution to the flavour and interest of food and for the supply of minor but essential nutrients, especially vitamins;
- that they are not basic food commodities; people will put off buying if the price is too high;

1"Engineering and Economic Growth in Africa" a lecture at the National Academy of Sciences, Washington DC January 9 2009.

- that consumption levels vary, depending on the selling price and the income of the buyer,
- that many of the crops are not traded in large volumes and there is a limited market;
- that the products are perishable, which means there is always a reduction in quality if they are not sold immediately, usually leading to a fall in value;
- that there is a wide range and variety of horticultural products and so if one product is too highly priced the consumer will generally buy another;
- that the products are normally traded in a very free market where price is determined by supply and demand.

The foregoing emphasises the complexity of horticulture value chains and serves to underscore the need for technical inputs throughout the value chain if the risks faced by small-scale farmers are to be managed by them. Because many of the factors that affect the risk of price fluctuations and food losses are influenced by the business models in the post farm part of the value chains, an understanding of how agribusiness works in developing countries is critical in delivering technical assistance. This is where TCDC experts can play an important role.

HOW NEW PLAYERS CAN FILL THE GAP

The gap between poor countries' pressing development needs and official development aid has created financial and institutional space for new aid givers. The new players can be classified in two groups:

New bilateral donors from the South: Providing assistance to poor countries is no longer the sole province of rich countries. Transition economies and middle income countries now give to poor countries. At last count, 29 such countries have established or are building aid programs. The new bilateral donors include small donors like Thailand, Brazil and some of the new members of the EU, medium size donors like Korea and Turkey and large donors like China, India and Saudi Arabia, which have annual aid programs of $1 billion or more.

Private organizations: Private organizations include a vast array of actors including tens of thousands of philanthropic foundations, tens of thousands of NGOs and hundreds of thousands of religious groups and community based organizations. These organizations mediate resources directly from rich individuals in rich countries to development activities in poor countries.

Unlike traditional donors, the new bilateral donors give little in the form of debt relief (as they have small outstanding liabilities), and little in the form of technical cooperation (as the domestic consulting industry is small). Correspondingly, a greater fraction of their aid is in the form of projects and programs. One of the largest new donors, China, provides its assistance in the form of turnkey projects, providing planning, finance, manpower and training to implement projects completely. China has built on its own strong history of infrastructure project management to focus its aid, and prides itself on short project preparation and implementation periods compared to traditional donors.

Although the new bilateral donors may have small domestic consulting industries they do have government employees and some employees of the larger NGOs that have credible experience of direct relevance to developing countries. This suggests that the use of experts and technicians under SSC agreements is both technically feasible and a cost efficient means of providing assistance.

39. VEGETABLE SCIENCE INTERNATIONAL NETWORK (VEGINET): THE WAY FORWARD

Prem Nath* and Anupam Varma**

** Dr. Prem Nath Agricultural Science Foundation, Bangalore, India*
***Indian Agricultural Research Institute, New Delhi, India*

ABSTRACT

About 700 participants from 37 countries across the globe, of the International Conference on Vegetables (ICV) held on 11-14 November, 2002, at Bangalore, India, discussed over 600 papers and posters presented on its 13 theme areas of scientific, technological and economic importance. It was recognised that this phenomenal knowledge and much more of it that is globally available and will be available in the future should be collected, consolidated and easily retrieved and shared for appropriate use by the stakeholders. Therefore, the General Assembly of the Conference ICV-2002 decided to establish the Vegetable Science International Network (VEGINET).

The goal of VEGINET will be to strengthen partnership and inter-institutional cooperation among the member organizations of the vegetable sector towards improved production and utilization of vegetables. The main objectives of the VEGINET will be to strengthen / promote vegetable research and development by facilitating inter-and intra-regional and inter-institutional cooperation; promote partnership between public and private sector for improving vegetable production; facilitate development of human resources; promote improved and sustainable production of vegetables for food; promote improved and sustainable production of vegetables for food and facilitate building of sustainable infrastructure from farm-to-the-consumer and processor.

The rising challenge, structure and membership. Partnership, human resource development, mandate and strategy, activities, and resource generation of the VEGINET will be explained during the presentation.

INTRODUCTION

The dual challenge of poverty and hunger, accentuated by reduction in per capita land and water availability, is much greater today than ever before. It is estimated that by 2030 nearly 80% of the global population, half of which is already suffering from malnutrition and under nutrition, will be living in developing countries. Further, it is expected that by 2030 nearly half of the population of these countries will be migrating to urban areas for livelihood. Modernization of vegetable production, marketing and utilization system can make a big difference in economic empowerment and living conditions of the resource poor farmer's world over. This will also help in reducing migration to urban areas and minimize socio-economic stress, besides improvement in the quality of life of the farming communities. Therefore, there is a need to invest in 'Vegetable Revolution' by combining modern technology for improved more nutritious vegetables resistant to pests and diseases and suitable for progressive agronomic practices. This requires concerted efforts of the scientists and policy makers, supported by a strong political will. To take the lead, a concept paper for the establishment the Vegetable-Science International Network (VEGINET) – a network of vegetable scientists and research institutions for active cooperation in research and development, human resource development and dissemination of knowledge with the goal to promote vegetables to fight global hunger and malnutrition, was presented in the International Conference on Vegetables (ICV-2002), Bangalore, India in November, 2002 (Varma, *et al.*, 2003; Nath, 2005. a & b; Nath 2009). This initiative aimed at the objectives of the Technical Cooperation among Developing Countries as illustrated by Gustafson (2002) and Wall (2009).

Over six hundred participants of the ICV-2002, representing 37 countries, resolved to establish VEGINET so that the phenomenal global knowledge for vegetable improvement, production and utilization is shared equally by all the countries for sustainable nutrition security. VEGINET has been a topic of cutting-edge research based on real life problems and situations for curricular development. An important aspect of these studies is to develop models for knowledge sharing for use in conserving natural resources, biodiversity, developing solutions to the problems related to international trade, and connect various agencies interested in international trade for developing strategies to compete well in international markets (Palaniswamy, 2009).

The Goal

The goal of the VEGINET is to strengthen partnership and inter-institutional cooperation among the member organizations of the vegetable sector towards improved production and utilization of vegetables.

Objectives

The main objectives of the VEGINET are to:

- Facilitate inter-and intra-regional and inter-institutional cooperation through exchange of scientific, technological and related economic information.
- Identifying priority areas of cooperative research and development.
- Facilitate development of human resource through better education and training.

- Promote improved and sustainable production and utilization of vegetables to achieve:
- Food security
- Nutritional security, and
- Economic empowerment of the farming community.

THE APPROACH

It is time to recognize the magnitude of problems and the challenges they present to find economically sound and sustainable solutions. This can be achieved by improved scientific, technological, economic and managerial capabilities. Under the existing situation, the most rapid means of increasing and improving food supplies is to harness the appropriate multi-disciplinary knowledge and human competence not only to produce but also to protect and add value to ensure full realization from the efforts of the vegetable growers. Stronger, more sustainable and viable linkages between production, post-harvest conservation, transport, processing and marketing are most urgently needed to maximize returns from all inputs to the producers and consumers.

Actual estimates of the post harvest losses in vegetable production are not known, but according to conservative estimates nearly 50% of the produce is lost due to diseases and pests, lack of transportation facilities and poor storage. Even if 25% of these losses could be prevented, nutrition security can be achieved with the same level of production, both quantitatively and qualitatively without additional stress on limited natural resources. In this process, much greater value can be added to the produce and also large employment generated in the total system. This will increase availability of nutritious vegetables, generate better income and reduce poverty along the whole chain, especially in the rural areas and make better quality food available to the consumer at lower prices. The impact of this on other sub-sectors of the vegetable production and utilization system would be enormous. A real socio-economic transformation can be brought about by a sound programme of integrated development of the vegetable system (along with that of other perishables) and lead the agricultural economy to dramatic improvement. This approach calls for a clear new policy and strategy of action, with provision for course correction in sound implementation programme. It also calls not only for building but maintaining and continuously improving the infrastructure, in most of the developing countries. Inter-institutional and inter-country cooperation facilitated by the VEGINET can be a powerful approach in this total effort.

The overall approach of the VEGINET will be to:

- Help in identifying priority areas for research and development.
- Help in the development of competent human resource for better production, protection, processing, marketing, distribution and utilization of vegetables.
- Help in formulating collaborative programmes of action.

- Harness knowledge and competence to achieve the objectives.
- Help in exchange of information through interactive internet portal and organization of international workshops, symposia and conferences.

THE STRATEGY TO PROMOTE VEGETABLE PRODUCTION AND UTILIZATION

The strategy to promote vegetable production and utilization will be to help the developing countries in formulating policy, designing strategy and developing programmes of action nationally, inter-regionally and globally to increase availability of vegetables in a sustainable manner. The success of this approach will depend on the active participation of existing Centres of Excellence in building inter-institutional and inter-country networks to solve specific problems of mutual interest. The problem solving projects should include professional bodies of scientists and technologists and associations of farmers, traders and industrialists. This will accelerate use of research results. The immediate need is to identify the Centres of Excellence for vegetable research interested in fruitful collaboration to solve the problems of mutual interest. Cooperation between the Centres of Excellence can have a multiplier effect in improving the vegetable production and utilization systems.

HUMAN RESOURCE DEVELOPMENT (HRD)

A major emphasis of the VEGINET should be to help build institutional capabilities for enhancing HRD competence, through short and medium term training as well as through long term education leading to doctoral and postdoctoral fellowships for visiting fellows and entrepreneurs, from within and outside the network, at the Centres of Excellence. Special efforts will have to be made for the Least Developed Countries (LDCs) to provide international fellowships for Master's degrees in areas of their priority and to build capacity for vegetable science.

THE MANDATE AND STRATEGY

1. Help identify priority areas for Development Research and formulate joint programmes of action to improve human competence to production, protection, processing marketing, distribution and better use of vegetables and other perishable produce as dietary components.
2. Help developing countries to formulate policy, design strategy and programmes of action nationally, inter-regionally and globally to increase availability of vegetables in a sustainable manner.
3. Using the existing Centres of Excellence build inter-institutional and inter-country networks to solve specific problems of mutual interest. The joint projects for solution of these problems should include professional bodies of scientists and technologists and

associations of farmers, traders and industrialists. This will accelerate use of research results.

4. Facilitate creation of National and Regional Technology Exchanges for collection, retrieval, and dissemination of knowledge to the users.
5. Encourage, promote and facilitate international scientific and technological cooperation through exchange of personnel.
6. Publish newsletters, journals, reports, pamphlets and books in specific areas of vegetable science & technology and related areas for participating institutions and members to help in maintaining closer links.
7. Develop and maintain an interactive internet portal for quick exchange of information and feed-back from the stakeholders.
8. Provide assistance in building of new institutions and strengthening the existing institutions to address technological problems.
9. Promote and strengthen center(s) or special programmes for human resource development for improving vegetable production and their utilization.
10. Create and maintain a global database of professionals and/or establish a Human Resource Bank, from which expert consultants can be drawn by institutions and countries.
11. Create and maintain database connected manufacturers and suppliers.
12. To act as a consultative body on vegetable science, technology and related economic matters that have an international or inter-regional dimension.
13. Establish commissions and/or other functional bodies of VEGINET, to deal with specific scientific, technological and economic problems of the vegetable system.
14. Take on any other activity to deal with the problem of vegetable science & technology that would contribute to raising production, ensure better crop and product protection and help to raise farm incomes for improving the quality of life.

THE PROJECTED ACTIVITIES

The VEGINET should organize Consultations, Workshops and Conferences in priority areas of importance to its members. This will strengthen linkages and build new concepts to help formulate more realistic action plans and demonstration of results.

Some of the areas identified for this purpose are:

- Development of policies, strategies and integrated programmes of action to increase availability of vegetables.
- Production technology and productivity

- Development and use of value added germplasm
- Hybrid seed development and production
- Genetic improvement of vegetables
- Biotechnological applications
- Biosafety of GM vegetables
- Crop protection technology in the field and at post-harvest level
- Post-harvest management of fresh produce and processed products
- Organic vegetable production and means of its natural protection
- Development of vegetables for better nutrition
- Under-utilized vegetables and by-products for food and feed
- Marketing and trade infrastructure
- Technical co-operation among the countries having similar climate and ecological protection.

Publications

Apart from the publications emerging out of Consultations, Workshops and Conferences, the VEGINET should bring out its periodic newsletters and other publications based on success achieved by its members. It will help create databases on exchange of plant materials, advanced technologies, traditional knowledge and capacity building.

The Management System

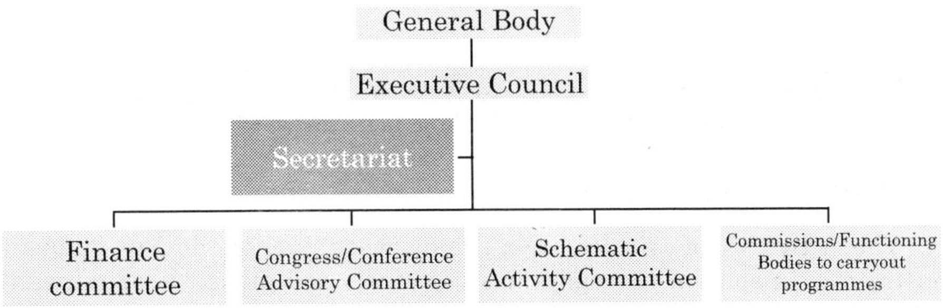

Fig. 1. The organizational structure of VEGINET

The organizational structure of VEGINET (Fig. 1) consists of General Assembly, Executive Council, Finance Committee, Congress/Conference Advisory Committee, Scientific Activity Committee, Commissions/ Functioning Bodies to execute VEGINET programmes, and a supporting Secretariat. The General Assembly of VEGINET is the policy making and programme planning body. The General Assembly consists of the founder members, individual members and one representative of each member institution. The Executive Council, elected by the General Assembly implements the decisions taken by the General Assembly and reports to it.

The primary duty of the Executive Council is to conduct and manage official affairs of VEGINET between the meetings of the General Assembly. This includes such specific responsibilities as preparation of budget, collection and disbursement of funds and supervision of its programme of work and of the VEGINET Secretariat. The Executive Council may also be authorized to take specific policy decisions which would be endorsed / ratified by the General Assembly.

MEMBERSHIP

The membership of the VEGINET is open to individual scientists and organizations actively engaged in vegetable research, technology development, biotic and abiotic stress management, processing, marketing, planning and policy formulation and governance of vegetable production and utilization systems. The earlier fee structure (Varma *et al.,* 2003; Nath, 2007) did not attract sufficient membership, Hence, a simplified revised membership structure was suggested as mentioned below (Table 1) and was approved by the General Assembly on November 11, 2009.

Table 1. Membership fee structure of VEGINET in US $

Member	Developed Countries		Developing Countries	
	Earlier fee structure	Present fee structure	Earlier fee structure	Present fee structure
Institutional Member	3000 to 10,000*	1000	3000 to 10,000*	250
Individual Member	1000 to 5000	100	50 to 500	25
Founder Member	250	50	100	12.50 (IRs 600)

*Depending on the number of scientists working in an institute

RESOURCE GENERATION

The activities of the VEGINET have been limited due to budgetary constraints. In-spite of the ambitious objectives and plans the activities the activities of VEGINET could not be taken up in full swing because of limited resources, mostly limited to the membership fee. We appreciate and acknowledge the generous support provided by the PNASF by which some of the activities of VEGINET could be kept alive. The resource mobilization seems to be the priority in implementing the activities of VEGINET. Mobilization of adequate financial resources is essential for the success of VEGINET. This could be achieved from contribution of countries, bilateral and multilateral agencies, including the public and private sectors.

EXECUTIVE COUNCIL

The 2nd General Assembly of November 11, 2009 which met during the International Conference on Horticulture (ICH-2009), nominated and elected the first Executive Council of VEGINET and the Nomination Committee which was formed during the First General Assembly of November 9, 2002 and acted during 2002-2009 was dissolved to allow the first Executive Council to function. The members elected were as follows;

Chairperson -Dr. Prem Nath, Chairman, Dr. P.N. Agricultural Science Foundation (PNASF), Bangalore INDIA; **Advisor FAO**-Ms. Alison Hodder, Senior Officer, Horticultural Crops Group, AGPC, FAO, Rome, Italy; **Regional Vice Chairpersons**-Dr. Ahmed T. Moustafa, Head and Regional Coordinator (APRP), ICARDA, United Arab Emirates; Prof. Michael Boehme, Humboldt University of Berlin, Germany; Prof. J.S.Dias, Technical University of Lisbon, Portugal; Dr. Ravza Mavlyanova, Coordinator for Central Asia and the Caucasus, AVRDC-The World Vegetable Center, Uzbekistan; Dr. Usha R. Palaniswami, Chair, Division of Natural Sciences and Mathematics, Excelsior College, Albany, New York; **Secretary General** - Dr. Anupam Varma, INSA Senior Scientist, Indian Agricultural Research Institute, New Delhi, India; **Assistant Secretary General**-Dr. Aminuzzaman Talukder, Country Director for Cambodia and Regional Food Security Advisor, Asia-Pacific, Helen Keller International, Cambodia; **Treasurer**-Dr. O.P.Dutta, Director, R&D, Namadhari Seeds Pvt. Ltd., Bangalore, India; **Members representing different regions**-Professor Acram Taji, Queensland University of Technology, Queensland, Australia; Dr. O.O. Fadina, Department of Crop Protection and Environmental Biology, University of Ibadan, Ibadan, Nigeria; Dr. Grisana Linwattana, Senior Research Scientist, Horticulture Research Institute, Bangkok, Thailand; Prof. Zetian Fu, China Agricultural University, Beijing, China; Ms.Yuniarti, Assessment Institute for Agricultural Technology (AIAT), Malang, Indonesia; Dr. San San Yi, Assistant Research Manager, Vegetable Research and Development Center, Yangon, Union of Myanmar.

CONSTITUTION OF THE VEGINET

The constitution of VEGINET was developed by the Nomination Committee and was presented and ratified by the second General Assembly held during the International Conference on Horticulture (ICH-2009) scheduled November 11, 2009 (Huria, 2009).

Activities undertaken during 2002-09

Guided by the Nomination Committee and depending on the budget and resources available, the VEGINET Secretariat undertook the following activities during the period 2002-2009.

1. VEGINET on Website

The VEGINET remained active on the PNASF website: www.pnasf.org beginning the year 2003 and through these years. The Secretariat, hosted by PNASF responded to the queries received across the globe.

2. VEGINET in Newsletters

The news and activities of the VEGINET appeared in ten issues of the PNASF Newsletters published between 2003 and 2009. These were distributed among the members of VEGINET and PNASF.

3. VEGINET Brochure

The brochure was prepared, published and distributed among the interested participants and VEGINET members and during the seminars and conferences.

4. Participation in International Conferences

In order to disseminate the information on its existence and activities of the VEGINET, the Convener participated and made presentations during the following conferences;

ASHS Annual Conference held during July, 2005, in Las Vegas, USA.

ISHS 3rd International Cucurbit Symposium held during September, 2005 in Townsville, Australia.

ISHS IV Balkan Symposium on Vegetables held during September, 2008 in Plovdiv, Bulgaria. The VEGINET brochure was distributed during conferences which aroused interest among the participants of the conference from different countries.

5. Proposed Second International Conference on Vegetables in Fortaleza, Brazil

As agreed by the General Assembly during the first International Conference on Vegetables (ICV-2002) held during November, 2002 in Bangalore, the VEGINET took initiative along with the FAO to hold the second International Conference on Vegetables (ICV) during 2007 in Fortaleza, Brazil with EMBRAPA as host, which could not happen.

6. VEGINET in USA

On the recommendation of the Nomination Committee of the VEGINET, a new chapter of the VEGINET was opened in the United States of America. In order to increase membership and to promote linkage and activities among the scientists and growers in the USA, Dr. Usha R. Palaniswamy, a life-member of VEGINET, was appointed as focal point. The VEGINET was presented during the Annual Conference of the American Society of Horticultural Science (ASHS).

A poster presentation was made by Dr. Usha R. Palaniswamy during Annual Conference of ASHS held during July 2007 at Scottsdale, Arizona. The title of the paper was "USDA/CSREES International Science and Education (ISE) Competitive Grant Programme-2006 funding to establish strong network for vegetable improvement, Vegetable Science International Network (VEGINET)-USA Chapter. (Palaniswamy and Nath, 2007)

A poster presentation was made during the Annual Conference of the ASHS during July, 2009 held at Illinois, Missouri. The title of the paper was "USDA-CSREES-ISE Program Funded VEGINET-USA in Participation of International Conference on Horticulture (ICH-2009) with the theme *Horticulture for Livelihood Security and Economic Growth*". (Palaniswamy and Nath, 2009)

7. The International Conference on Horticulture (ICH-2009)

The VEGINET actively coordinated with the PNASF, Bangalore, India in originally conceiving the idea of holding and in preparing for the International Conference on Horticulture (ICH-2009). The VEGINET is the Co-sponsor with the PNASF in holding the International Conference on Horticulture (ICH-2009).

8. Linkage with other Institutions / Organizations

The VEGINET received support from the FAO in organizing the second International Conference on Vegetables (ICV-2007) scheduled during 2007 in Fortaleza, Brazil, which could not happen.

The VEGINET was invited to participate in the first General Assembly of the Global Hort, held on 12th November, 2006 in New Delhi. The VEGINET received approval of grant from the Global Hort for some activities of the ICV-2007.

The VEGINET has received support from the Indian Society of Vegetable Science and encouragement from the International Society of Horticultural Science.

9. Membership

The VEGINET has received voluntary support from the individuals and institutions from the developed and developing countries who supported the idea of Technical Cooperation among Developing Countries (TCDC). At the present, the members of the VEGINET represent India, USA, Thailand, Nepal, Cameroon, Nigeria, South Africa, Zambia, Senegal and some other countries.

10. Resource Generation

The activities of the VEGINET have been limited because of the budget constraints. In-spite of the ambitious objective and plan it could not be implemented because of limited resource mostly limited to the membership fee. The considerable support received from the PNASF in maintaining the VEGINET Secretariat has survived the activities of the VEGINET. The resource mobilization seems to be the priority in implementing the activities of VEGINET.

11. VEGINET Nomination Committee and Secretariat

As approved by the General Assembly of the ICV-2002 held during November, 2002, the Nomination Committee consisted of the following members;

- Dr. Prem Nath, Convener (India);
- Mr. Wilfried O. Baudoin, International Adviser (Italy);
- Dr. M. L. Anothai Choomsai, Member (Thailand);
- Dr. O. A. Denton, Member (Nigeria);
- Dr. H. A. B. Parpia, Member (India);
- Dr. R. K. N. Singh, Member (Canada);
- Dr. G. Anais, Member (Caribbean);
- Representative (Latin America).

The Secretariat hosted by the PNASF continued to respond to the VEGINET Nomination Committee established by the General Assembly of the First International Conference on Vegetables (ICV-2002) held during November, 2002. It continues to respond to the queries of the interested participants from different countries. The Secretariat assisted in the preparation of the draft constitution of the VEGINET which was cleared by the Nomination Committee and printed for members to comments for improvement.

THE WAY FORWARD

Now that the First Executive Council of VEGINET is in place programmes related to its objectives will get greater attention and move forward to have a greater impact of this initiative. The proposed activities of VEGINET have been highlighted earlier in this paper. There is a need to prioritize the proposed activities. The emphasis may be given to (a) identify the priority areas, which require international cooperation for capacity building for research and development, in vegetable production and utilization system, (b) identify the centres of excellence to help in capacity building and conducting research in areas of mutual interest, (c) bring out periodic newsletters and other publications based on success stories of its members, (d) develop and maintain an interactive internet portal for quick exchange of information and feed-back from the stakeholders, (e) create databases on exchange of plant materials, advanced technologies, traditional knowledge, and a global database of professionals to establish a Human Resource Bank, (f) prepare base papers covering specific scientific, technological and economic problems of the vegetable system, and (g) organize the Second International Conference on Vegetables (ICV) and develop policy to organize ICV at regular intervals.

A strong and active VEGINET will not only revolutionize vegetable production across the world, but also play an important role in achieving nutrition security and economic empowerment of the farmers, particularly in the developing world.

REFERENCES

Gustafson, Daniel. 2002. Trends and Issues in Technical Cooperation among Developing Countries; *Proceedings of international conference on vegetables* (ICV-2002) November 11-14, 2002, Bangalore; Pages: 861-864.

Huria, V.K. 2009. VEGINET Constitution presented to TCDC session, ICH-2009, November 9-12, 2009, Bangalore.

Nath, Prem. 2005. Vegetable Science International Network (VEGINET), Program and Abstracts-ASHS Annual Conference-2005, Las Vegas, USA; July 2005 Hort Science, **40** (4); Page 1111.

Nath, Prem.2005.Vegetable Science International Network (VEGINET), *Proceedings of ISHS 3rd international symposium on cucurbits*, September 11-17, 2005, Townsville, Australia; Pages: 55-58.

Nath, Prem. 2009. Vegetable Science International Network (VEGINET), *Souvenir, ICH-2009* November 9-12, 2009; Pages 12-16.

Palaniswamy, U.R. and Nath, Prem. 2007. USDA/CSREES International Science and Education (ISE) Competitive Grants Program 2006 Funding to Establish Strong Network for Vegetable Improvement Vegetable Science International Network (VEGINET)-USA Chapter, Annual Conference ASHS 2007, July 16-29, Scottsdale, Arizona, USA.

Palaniswamy, U.R. and Nath, Prem. 2009. USDA/CSREES-ISE Programme Funded. VEGINET-USA in participation of International Conference on Horticulture (ICH-2009) with the theme-Horticulture for Livelihood Security and Economic Growth. Annual Conference ASHS 2009, July 2009, Illinois, Missouri, USA.

Varma, A., Parpia, H.A.B. and Nath, Prem. 2004. Vegetable Science International Network. In: P. Nath, Gaddagimath, P.B. and Dutta, O.P. (Eds), *Food Security and Vegetables – A Global Perspective*, Dr. Prem Nath Agricultural Science Foundation, Bangalore, India. Pages: 389-398.

Wall, Gavin. 2009. FAO Experience of Technical Cooperation among Developing Countries; *Abstract Book, ICH-2009* November 9-12, 2009 Bangalore; Page: 469.

AUTHOR INDEX

PNASF INTERNATIONAL AND NATIONAL AWARD

The Dr. Prem Nath Agriculture Science Foundation (PNASF), Bangalore announced the second International and National Gold and Silver Medal Awards for known national and international personalities who had contributed significantly to agriculture/horticulture, food and nutrition security. These awards were presented by the Chief Guest, H.E. the Governor of Karnataka, Shri H.R. Bhardwaj, during the inaugural session of the International Conference on Horticulture (ICH-2009) on November 9, 2009, Bangalore, India. The first such award was announced and presented on November 11, 2002 during International Conference on Vegetables (ICV-2002) in Bangalore, India.

International Gold Medal Awards

- **Prof. Jules Janick, James Troop Distinguished Professor of Horticulture, Purdue University, USA - Promotion of Horticultural Science**

The Dr. Prem Nath Agricultural Science Foundation International Gold Medal award for the year 2009 was conferred on Prof. Jules Janick, for his meritorious contribution to World Horticulture benefiting the wellbeing of the humanity globally. Prof. Jules Janick has and continues to have a long and distinguished career devoted to horticulture in all its facets.

- **Dr. Mahmoud Solh, Director General ICARDA, Syria - Promotion of Dryland Horticulture**

The award for the year 2009 was conferred on Dr. Mahmoud El Solh, for his relentless services to the cause of Arid Horticulture for promoting social, economic and nutritional wellbeing of developing countries.

- **Mr. Wilfried O. Baudoin, Former Leader Horticulture Group, FAO, Rome - Promotion of Vegetable Science**

The award for the year 2009 was conferred on Mr. Wilfried O. Baudoin for his life long outstanding contribution to world horticulture by guiding and implementing FAO projects on horticulture across the globe. The 'Hortivar' programme initiated by him continues to be a torch bearer for amateur and experienced gardeners worldwide.

- **Dr. R.S. Paroda, Former Director General ICAR, New Delhi - Promotion of Agricultural Research**

The award for the year 2009 was conferred on Dr. R.S. Paroda, for his outstanding contribution in the field of Agricultural Research and Management for the betterment of humanity.

National Gold Medal Award

- **Dr. Anupam Varma, Former National Professor , ICAR , New Delhi - Promotion of Plant Protection Research**

The award for the year 2009 was conferred on Prof. Anupam Varma, for his outstanding contribution to fundamental and applied aspects of plant virology and horticulture.

- **Dr. S. Bisaliah, Former Vice Chancellor , University of Agricultural Sciences, Bangalore - Promotion of Economic Agriculture**

The award for the year 2009 was conferred on Dr. S. Bisaliah for his outstanding contribution in the field of Agricultural Economics and Agricultural education.

- **Prof. R.S. Deshpande, Director, Institute of Social and Economic Change, Bangalore- Research on Agriculture Policy**

The award for the year 2009 was conferred on Dr. R S Deshpande, for his relentless services and leadership in agricultural development policy and planning at national and international level.

- **Dr. K.V. Peter, Former Vice Chancellor, Kerala Agriculture University, Trissur - Promotion of Horticulture Science**

The award for the year 2009 was conferred on Prof. K.V. Peter for his outstanding contribution in horticulture science in the country.

National Silver Medal Award

- **Dr. K.R.M. Swamy, Former Head, Division of Vegetable Crops, Indian Institute of Horticultural Research, Bangalore- Promotion of Vegetable Research**

The award for the year 2009 was conferred on Dr. K.R.M.Swamy for his outstanding contribution to the cause of Vegetable and Plantation crops research in India.

- **Dr. B.S. Prabhakar, Former Senior Scientist, Indian Institute of Horticultural Research, Bangalore - Promotion of Vegetable Agronomy**

The award for the year 2009 was conferred on Dr. B.S. Prabhakar for his outstanding contributions towards the promotion of vegetable production technologies in the country.

- **Dr. D.P. Kumar, Professor of Horticulture, University of Agriculture Sciences, Bangalore - Promotion of Horticulture Science**

The award for the year 2009 was conferred on Dr. D.P. Kumar for his outstanding contribution towards the promotion of horticultural crop research and development activities in Karnataka.

ICH 2009 Silver Plaques

- **Dr. P. G. Chengappa, former Vice Chancellor, University of Agricultural Sciences, Bangalore- Outstanding Academic Leader in Agriculture**

The International Conference on Horticulture (ICH-2009) Plaque of Honor Award was conferred on Dr. PG. Chengappa for his outstanding contribution to Agricultural Economics and Agricultural Education in India.

- **Dr. O.P. Dutta, Secretary, Scientific Programme Committee, International Conference on Horticulture 2009-Outstanding Scientific Programme Leader in Horticulture**

The award was conferred on Dr. O.P. Dutta for his relentless services and outstanding leadership in developing the Technical Programme of ICH 2009 at Bangalore, India.

- **Dr. B. Raju, Chairman Logistic Committee, International Conference on Horticulture 2009- Outstanding Contribution to Conference Logistics**

The award was conferred on Dr. B. Raju for his relentless services and outstanding leadership in organizing the logistics of ICH 2009 at Bangalore, India.

Dr. Prem Nath Agricultural Science Foundation (PNASF), Bangalore, India

To promote agricultural education, research and sustainable development with a focus on food and nutrition security

PNASF is a Charitable Trust registered with the Government of Karnataka, India. It is established to assist and encourage Indian agricultural scientists to address India's agricultural problems and at the same time use their skills for the benefit of developing countries. It facilitates interaction of knowledge and ideas among leading professionals in agriculture, development and planning, on issues of national and international importance related to agriculture and food.

PNASF Supports

1. Assistance, organization and/or participation in national, regional or international conferences, workshops, expert consultations, seminars etc;
2. Assistance to small field projects through grant received;
3. Assistance in the dissemination of technical messages through technical publications;
4. Assistance to young agricultural scientists in research through modest grants;
5. Assistance to post-graduate students in agriculture through awards, scholarships etc and
6. Development of linkages or partnership with national, regional and international agriculture and related institutions and societies

For further enquires please contact

Dr. Prem Nath Agricultural Science Foundation
9, 1st Cross, 1st Main, 1st Block,
Rajmahal Vilas Extension 2nd Stage,
Bangalore – 560 094, Karnataka State, INDIA
Tel: + 91 – 80 – 2341 5188, Fax: + 91 – 80 – 2351 1555
E-mail: drpremnath@vsnl.net ; Website: www.pnasf.org